Regional Environmental Changes in Siberia and Their Global Consequences

D1796286

Springer Environmental Science and Engineering

For further volumes:
http://www.springer.com/series/10177

Pavel Ya. Groisman • Garik Gutman
Editors

Regional Environmental Changes in Siberia and Their Global Consequences

 Springer

Editors
Pavel Ya. Groisman
National Oceanic and Atmospheric
Administration (NOAA)
National Climatic Data Center
Asheville, NC, USA

Garik Gutman
The NASA Land-Cover/
Land-Use Change Program
NASA Headquarters
Washington, DC, USA

ISSN 2194-3214   ISSN 2194-3222 (electronic)
ISBN 978-94-007-9645-4   ISBN 978-94-007-4569-8 (eBook)
DOI 10.1007/978-94-007-4569-8
Springer Dordrecht Heidelberg New York London

Printed on acid-free paper

Springer is part of Springer Science+Business Media (www.springer.com)

Preface

The vast region of Siberia is expected to experience potentially rapid land-cover changes, which are among the earliest indicators of the Earth's response to climate warming. Moreover, climate change affects both the boreal ecosystems and socio-economic infrastructure. Projections of the future climate for the end of the twenty-first century indicate further temperature increases as well as increase in frequencies of various extreme events. Future trajectories of Siberian ecosystems will strongly depend upon a number of large-scale social and economic decisions. On the other hand, changes in Siberia are predicted to affect the climate and people on a global scale. The contributions in this volume demonstrate the utility of satellite data over Siberia for monitoring changes and trends over time, including the associations among vegetation productivity, surface temperature, biogeochemical and water cycles, and air pollution under changing climate conditions and their relation to human population dynamics.

This is a compilation of results from studies on Siberia by the institutions of the Russian Federal Service for Hydrometeorology and Environmental Monitoring (RosHydroMet) and of the Russian Academy of Sciences (in particular, of its Siberian Branch); the US universities and agencies (NASA, NOAA, and NSF); the *International Institute for Applied Systems Analysis* (IIASA, Austria); Friedrich-Schiller University, Germany; Danish Meteorological Institute; and other European institutions.

The volume was written by an international team consisting of scientists from the USA, Europe, and Russia under the auspices of the Northern Eurasia Earth Science Partnership Initiative (NEESPI). It should be of interest to those involved in studying recent and ongoing changes in Siberia, be they senior scientists, early career scientists, or students.

The book is dedicated to the memory of Dr. Don Deering – the founder of the NEESPI and its first project manager. Dr. Don Deering (NASA Goddard Space Flight Center) passed away on February 15, 2010. Below is a brief retrospective of Don's role in establishing and promoting the NEESPI program.

In the summer of 1999, Don and Dr. Garik Gutman, the program manager of the NASA Land-Cover and Land-Use Change Program (LCLUC), agreed that it would be nice if there were an interdisciplinary program in Siberia and that the ongoing NASA projects in Siberia could form a base for such a program. Several projects were being run independently in Siberia and the idea of putting them into a common framework, leveraging funds, exchanging data and information, planning measurements, etc., seemed rather appealing. They began promoting the idea to the US, European, and Russian scientists and agencies. Quite early, a decision was made that the Northern Eurasia program should encompass more than just Siberia or Russia. This was Don who delineated the present NEESPI geographic domain as the former Soviet Union, Mongolia, northern China, Scandinavia, and Eastern Europe.

It was not easy to start a program like NEESPI and to convert it into a large international, multi-institutional program with a broad scientific scope, supported by multiple agencies. However, in the spring of 2003, preliminary development of a NEESPI Science Plan in Suzdal, Russia, culminated in the review and approval of the plan by an external scientific panel in Yalta, Ukraine, that September, and the first NEESPI Science Team meeting was held in February 2006 in IIASA, near Vienna, Austria. Since then, NEESPI have held meetings on forests in the Far East, Siberia, and European Russia; on cold land regions in Alaska; on the carbon cycle in Germany and Russia; on the Baltic Sea basin in Poland; on dry lands in China, Kyrgyzstan, and Kazakhstan; on the Arctic zone in Finland; and on nonboreal Europe in Hungary and Ukraine. During the past five years, NEESPI has expanded tremendously. More than 150 projects supported by the US, Russian, Chinese, European Union, Japanese, and Canadian agencies have been conducted (half of them had been already completed) to address the NEESPI Science Plan objectives. Over 650 scientists from more than 200 institutions in 30 countries contributed or are contributing to the initiative. NEESPI has produced numerous tangible products, such as special issues of peer-reviewed journals based on the NEESPI sessions organized at American Geophysical Union and European Geosciences Union assemblies as well as on numerous NEESPI workshops, regional and thematic meetings, and summer schools. Several books based on NEESPI results have been (or are being) published. All of the above would not have been possible if not for Don's perseverance in dealing with NEESPI issues and his optimism and enthusiasm in reaching the initiative goals.

In 2006, Don was diagnosed with amyotrophic lateral sclerosis, often referred to as "Lou Gehrig's disease." In May 2007, Don joined the NEESPI team for the last time at the NEESPI Summit held in Helsinki. He gave a retrospective presentation of the history of NEESPI's inception including some great pictures highlighting the program's milestones and growing pains during its early development. All was very upbeat and humorous despite his condition. The NEESPI session at the European Geosciences Assembly in Vienna in April 2010 was dedicated to Don's memory. At that time, it was decided to dedicate this book to Don Deering, a scientist who was involved in Siberia projects before NEESPI's inception, made efforts in promoting measurements in Siberian forests by US scientists, and brought the rough idea of a Northern Eurasia program to fruition.

Acknowledgements

The studies in this volume have been supported through numerous grants, national and international programs and individual Institutions under the auspices of the Northern Eurasia Earth Science Partnership Initiative (NEESPI; http://neespi.org). Below we list the primary sources, programs and institutions that made this publication possible:

- The NASA Land Cover and Land Use Change Program and its dedicated investments into decade-long studies in Siberia (http://lcluc.nasa.gov), with its core started by late Don Deering - see Preface,
- Russian Academy of Sciences, particularly its multiannual Siberian Integrated Regional Study (SIRS; http://iopscience.iop.org/1748-9326/5/1/015007),
- NOAA Climate Program Office ((http://www.climate.noaa.gov) and the All-Russia Institute for Hydrometeorological Information of the Russian Federal Service for Hydrometeorology and Environmental Monitoring (http://www.meteo.ru) and their support of the regional data capacity building, dissemination, and climate change research in Siberia
- International Institute for Applied Systems Analysis, Laxenburg Austria, particularly its "Ecosystems Services and Management" Program (http://www.iiasa.ac.at//Research/ESM/index.html);
- European Commission (http://ec.europa.eu/index_en.htm), particularly its support of numerous International Association for the Promotion of Co-operation with Scientists from the New Independent States of the Former Soviet Union (INTAS) projects in Siberia and (within the 6th Framework Programme of the European Commission) the project "Enviro-Risks: Man-induced Environmental Risks: Monitoring, Management and Remediation of Man-made Changes in Siberia; http:// http://risks.scert.ru/), and
- The Russian Foundation for Basic Research (http://www.rfbr.ru/rffi/eng/) and its dedicated investments in environmental studies and capacity building efforts in Siberia.

The following external reviewers are acknowledged:

- Andrey P. Sokolov (Massachusetts Institute for Technology, Cambridge, USA;
- Elena Yu. Novenko, RAS Institute of Geography of the Russian Academy of Sciences, Moscow, Russia;
- Eugene L. Genikhovich, Voeikov Main Geophysical Observatory, St. Petersburg, Russia;
- Natalia N. Vygodskaya, A. N. Severtsov Institute of Ecology and Evolution of the Russian Academy of Sciences, Moscow, Russia.

Also, the Editors (P.Ya. and G.G) express special thanks to the Team of Springer Editors under the leadership of Mss. Hermine Vloemans and Raaj Vijayalakshmi for diligent work on this volume and their extraordinary patience while bringing it to fruition.

Contents

Chapter 1
Introduction: Regional Features of Siberia

**Pavel Ya. Groisman, Garik Gutman, Anatoly Z. Shvidenko,
Kathleen M. Bergen, Alexander A. Baklanov,
and Paul W. Stackhouse Jr.**

Abstract In this introduction chapter, we describe geographical, climatic, environmental, and demographic characteristics of Siberia and outline major problems dealt with in regional studies of this vast region including those important for the Global Earth System. The science questions, which are put in this chapter, are further addressed in detail throughout the book.

Siberia has increasingly been the focus of attention in the context of global change due to ongoing changes including increases in surface temperature, deepening of permafrost active layer depths, shifting vegetation zones, and declines in snow and sea ice extent (ACIA 2005; IPCC 2007). Numerous studies indicate that the global climate change signal in Siberia is pronounced and has already exceeded natural climate variability (NEESPI 2004).

P.Ya. Groisman (✉)
NOAA National Climatic Data Center, Asheville, NC, USA

State Hydrological Institute, St. Petersburg, Russia
e-mail: Pasha.Groisman@noaa.gov

G. Gutman
The NASA Land-Cover/Land-Use Change Program, NASA Headquarters,
Washington, DC, USA

A.Z. Shvidenko
Forestry Program, International Institute for Applied Systems Analysis, Laxenburg, Austria

Forestry Institute, Siberian Branch, Russian Academy of Science, Krasnoyarsk, Russia

K.M. Bergen
School of Natural Resources and Environment, University of Michigan, Ann Arbor, MI, USA

A.A. Baklanov
Danish Meteorological Institute, Copenhagen, Denmark

P.W. Stackhouse Jr.
NASA Langley Research Center, Hampton, VA, USA

P.Ya. Groisman and G. Gutman (eds.), *Regional Environmental Changes in Siberia and Their Global Consequences*, Springer Environmental Science and Engineering, DOI 10.1007/978-94-007-4569-8_1, © Springer Science+Business Media Dordrecht 2013

1.1   Siberia Defined

The historical name, Siberia, comes from "Sibir Khanate," a small Tatar kingdom founded in the fifteenth century and located in the middle reaches of Ob and Irtysh Rivers eastward of the Ural Mountains. The Khanate was conquered by Russian Cossacks at the end of the sixteenth century. In the following decades, Russian explorers, adventurists, and administrators moved eastward across northern Asia several thousand kilometers, establishing forts (*ostrogs*) and settlements on their way (e.g., Yakutsk, the current capital of the Sakha Autonomous Republic, was founded in 1632) and by 1639 reached the Pacific Ocean. Mapping and further colonization followed. However, it was not until the nineteenth century that large-scale agricultural development and industrial exploration of Siberia gained a significant foothold. Development at this time was predominantly focused in the south, along the Trans-Siberian Railroad which was begun in the late nineteenth century and completed in 1916. Over the past (twentieth) century including during the Soviet era (1922–1991), large reserves of oil, gas, gold, and other minerals as well as forest resources became the backbone of the region's economy. Still, today these resources contribute a disproportionally large (compared to the population fraction) input to the economy of the Russian Federation (1991–present). Severe climate conditions restrict the population growth in Siberia. As of 2009, only ~25% of 141 million of the Russian Federation population resided east of the Ural Mountains. In addition to climate-related conditions, socioeconomic changes of the past two decades took a heavy toll on the region's population (cf. Chap. 7 of this book).

With its 10 million km² of land, Siberia comprises almost 60% of the total land area of the Russian Federation, with the latter comprising approximately 11% of global terrestrial land area. Geographically, the area of Siberia extends from the Ural Mountains on the west to approximately the divide between the Arctic and Pacific watersheds on the east and north to south from the Arctic Ocean (~75°N) to the mountains and steppes bordering China, Mongolia, and the Central Asian states (~45°N). Siberia as defined in this book is comprised of the entire area of the contemporary Russian Federation Siberian Federal *Okrug* (Federal District), the Sakha Republic (Yakutia, administratively part of the Russian Far Eastern Federal *Okrug*), and the majority of the land area of the Urals Federal *Okrug*. Based on this definition, Siberia is comprised of the entire area of Russian Federation east of the Ural Mountains except for the monsoon climatic areas of the Russian Far East and Beringia (Fig. 1.1). The physical characteristics of this vast region are described below.

1.2   Geographic and Physical Characteristics

1.2.1   General Characteristics

Geographic and physical descriptions of Siberia can be found in many monographs (see, e.g., Lydolph 1977; Shahgedanova 2003; Balzter 2010). The Siberian landscape is large and geographically varied, and so it has typically been treated using regional

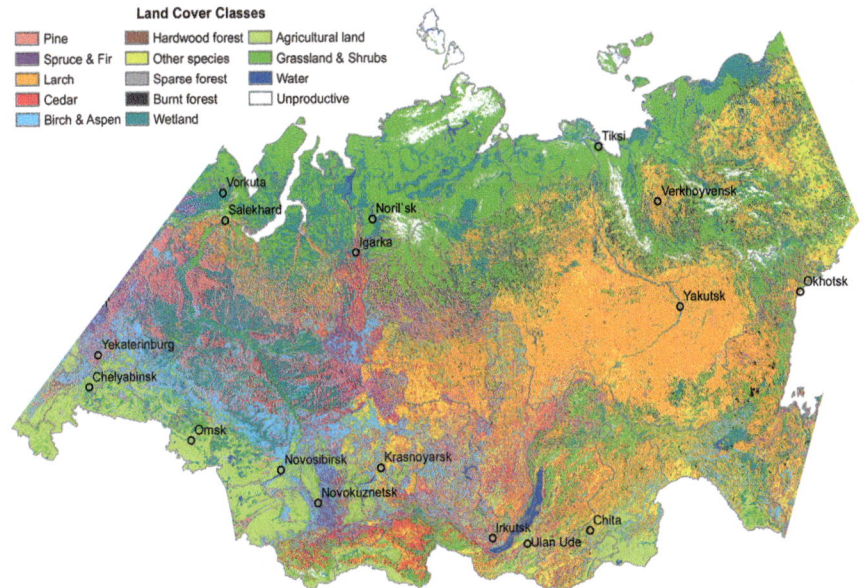

Fig. 1.1 The hybrid land-cover dataset of Siberia. This blended dataset was built as a synergy of remote sensing products, land statistics, and in situ measurements (Schepaschenko et al. 2010). The dataset contains the most accurate land-cover information for Siberia, updated for 2009. Resolutions and uncertainties of the dataset satisfy the requirements of a verified terrestrial ecosystems full greenhouse gas account (more, cf. Chap. 6). This land-cover dataset is also considered as a platform for an integrated observing system for the country

divisions based on a combination of geography and terrain. The geographic division of *West Siberia* is generally understood to extend from the Ural Mountains on the west to the Yenisei River on the east. Comprised primarily of the West Siberian Lowland, this is the vast basin of the Ob and Irtysh Rivers. Swamps and bogs are the dominant land cover; taiga forest grows in belts primarily along the many river courses which flood annually over large expanses. The geographic division of *East Siberia* is generally considered to extend east from the Yenisei over the Central Siberian Plateau to the divide between the Arctic and Pacific drainage basins. Here the terrain trends toward higher relief north- and eastward.

A significant diversity of climatic conditions and affiliated processes are found in Siberia. The large landmass of Siberia causes an extreme range in seasonal temperatures, especially in its central part, where the difference between summer and winter reaches a record high in comparison with any other nonmountainous climatic region on the planet. The important climate-forming factor of central Siberia is the Arctic Ocean. Cold and dry air masses may form over the Arctic in summer as well as in winter. Siberian rivers, in turn, discharge relatively warm water back to the Arctic Ocean, playing a critical role in the delicate energy and freshwater balances of the oceanic thermohaline circulation.

Ongoing global change is an integral and inherent feature of the dynamics of Siberian ecosystems. Major drivers which impact the Siberian ecosystems include (1) dramatically increasing temperature coupled with diverse regional trends of

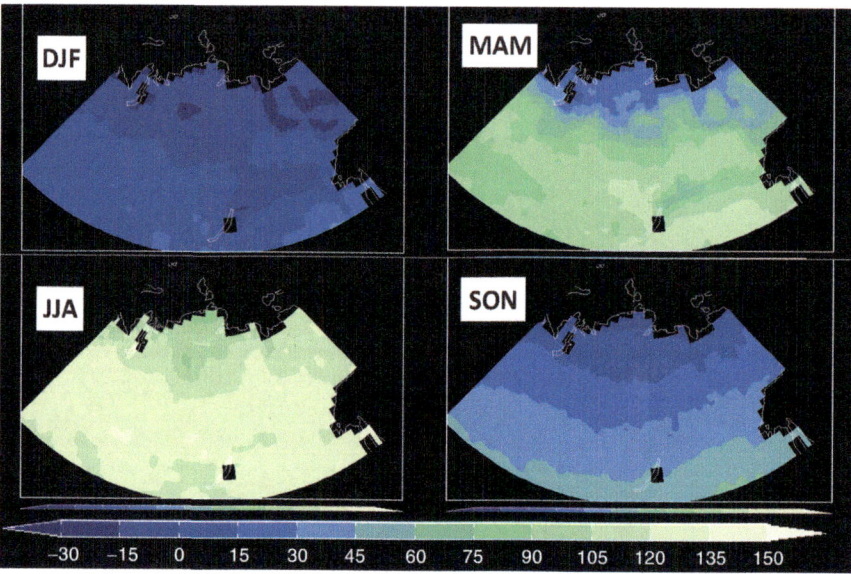

Fig. 1.2 Contemporary mean seasonal total net surface radiation budget (SRB; average for the 1884–2007 period) over Siberia as determined by the GEWEX SRB project (Gupta et al. 2006), W m⁻²

changes of precipitation; increasing aridity of climate was observed in vast continental regions; (2) increasing seasonal variability of weather; (3) changes in surface albedo due to regional variations in snow cover; (4) changes in disturbance regimes, basically in extent and severity of wildfire and insect outbreaks, following changing of the temporal dynamics of temperature and precipitation; (5) changes in hydrological regimes connected to permafrost thawing and land-use changes; and (6) increasing and mostly unregulated anthropogenic impacts (McGuire et al. 2006; Soja et al. 2007; Vygodskaya et al. 2007; Shvidenko 2009; Quegan et al. 2011).

1.2.2 Surface Energy Balance

The land surface is a focal area where (a) most of the solar energy absorption and a significant fraction of reflection occur, (b) a significant fraction of the outgoing long-wave radiation forms, and (c) atmospheric warming by turbulent and long-wave radiation fluxes is generated. The surface heat exchange defines temperatures at the surface and above and below ground, and changes the surface substance by snowmelt and evaporation as well as by providing the energy for the functioning of the biosphere. While at the top of the atmosphere above Siberia, the annual radiative budget is negative, the surface is a heat sink only in the winter season and for northern Siberia in the autumn season also (Fig. 1.2). The negative heat balance in Siberia is compensated by transport of latent heat from more temperate regions, a considerable part of which comes from northern Atlantic region via the westerly atmospheric circulation (Pavlov 1984).

Siberia is sparsely populated (and therefore sparsely covered by in situ observations) and has harsh weather conditions most of the year. Therefore, the use of remote sensing from space is a natural choice to document climate conditions in this part of the world. Moreover, the remote sensing products corroborate quite well with regionally averaged in situ data (when they exist and can be directly compared, cf. Chap. 3). Chapter 2 describes the available remote assets in detail.

1.2.3   Water Cycle

Advection processes (mostly westerlies, but Arctic effects are also present) modify the climate of Siberia, reduce its continentality, and are the source of water for interiors of the continent. Weather conditions favorable for cloud formation and precipitation are highly variable in time and space. Thus, atmospheric circulation is a major source of the variability in land surface processes. Siberia is "protected" by mountain ridges from the direct influx of water vapor from the tropics (Kuznetsova 1978; Shver 1976). Major sources of water vapor in that area include the Atlantic Ocean, the Arctic Ocean, and their coastal seas. Large interior lakes like the Caspian Sea, Baikal, and (up to the recent years) the Aral Sea also contribute. The advection by extratropical storms is the major means of moisture transport. This makes precipitation conditions of Siberia highly variable and very sensitive to circulation changes. Thus, relatively modest changes in the global circulation of the atmosphere and ocean may substantially affect climate and environmental conditions in Siberia.

Historically, small shifts in storm tracks resulted in enormous variations in water balance of interior lakes, in deep ground water circulation systems, and in corresponding shifts of the ecosystem boundaries. Paleoclimatic, archeological, and historical records indicate propagation of forested and steppe areas far southward into the desert and semidesert areas during "wet" epochs and their retreat and desertification during the prolonged periods of insufficient precipitation. These changes were quite swift (with a time scale of several decades) and were accompanied by prosperity and/or collapse of local agricultural and nomadic civilizations (Wigley et al. 1981; Lamb 1988; Gumilev 1990; Kaplin and Selivanov 1995; Pirazzoli 1996; Selivanov 2000).

Six of the ten great rivers of the Russian Federation are located in Siberia: the Ob, Irtysh, Yenisei, Angara, Lena, and Amur (Antipov et al. 2006). The Yenisei-Angara-Selenga Rivers, which drain northern Mongolia and Siberia northward toward the Arctic Ocean, comprise the world's fifth longest river system; the Ob-Irtysh is the seventh longest. The largest lake in Siberia (and the world's deepest and largest freshwater reservoir) is Lake Baikal (31,500 km^2 with a maximum depth of 1,680 m). The relatively flat and poorly drained west Siberian lowlands consist of over one million lakes along with significant groundwater resources and peatlands (Antipov et al. 2006). Details on the regional water cycle are given in Chaps. 3 and 4.

1.2.4 Carbon Cycle

In the biomass, soils, and peatlands of Siberia, boreal Russia holds one of the largest pools of terrestrial carbon (see Chap. 6). Because Siberia is located where some of the largest temperature increases are expected to occur under current climate change scenarios, stored carbon has the potential to be released with associated changes in disturbance regimes. Carbon cycle of Siberian land is basically driven by interconnection between net primary production and ecosystem's heterotrophic respiration and fluxes due to disturbances. A recent study that is based on a detailed full carbon account has estimated net ecosystem productivity at 671 ± 168 Tg C yr^{-1}. The emission caused by disturbances is estimated at 145 Tg C yr^{-1}. Taking into account wood decomposition (134 Tg C yr^{-1}), lateral fluxes (43 Tg C yr^{-1}), and consumption of plant products (105 Tg C yr^{-1}), the net ecosystem carbon balance has been estimated at 245 Tg C yr^{-1}. On average for the region, forest is estimated as a net sink (336 Tg C yr^{-1}), and the rest of land categories (except wetlands) – as a relatively small source (see more details in Chap. 6). The above results have been obtained for 2009. In addition, Russian land (including interim water reservoirs) has been estimated as a source of methane at 16.2 Tg C-CH$_4$ year^{-1}, a greenhouse gas 23 times more powerful than carbon dioxide. Of this total amount, 65% is delivered by wetlands. Assessment of carbon budget of forest ecosystems for 1990–2007 based on forest inventory data also estimated the region's forests as a sink of 255 ± 64 Tg C yr^{-1} and 264 ± 66 Tg C yr^{-1} for 1990–1999 and 2000–2007, respectively (Pan et al. 2011). Overall, these results are consistent taking into account that different definitions of forest (national and FAO's) have been used in the above studies.

Interannual variability of carbon emissions is basically defined by seasonal weather specifics and disturbances. While average values of net ecosystem carbon balance (NECB) for Russia as a whole usually vary in limits of 10–15%, the variability of emissions for individual regions and reasons (e.g., wildfires) could be substantial. For instance, vegetation fire carbon emissions varied from 50 (2000) to 231 (2003) Tg C yr^{-1} during 1998–2010 (Shvidenko et al. 2011). The large variability of direct fire fluxes has been reported by different authors (e.g., Soja et al. 2004; Van der Werf et al. 2006).

Gas locked inside Siberia's frozen soil and under its lakes has been seeping out since the end of the last ice age 10,000 years ago. But in the past few decades, as the Earth has warmed, the icy ground has begun thawing more rapidly, accelerating at a perilous rate the release of methane (cf. Forster et al. 2007). Furthermore, this permafrost thaw began affecting Siberian infrastructure (cf. Chaps. 5 and 7).

1.3 Human Geography

While Siberia's land area is larger than the conterminous United States, Siberia is inhabited by only an estimated 33 million people and has a population density of about 4.5 person per km^2 in the Siberian and Urals Federal *Okrugs* and 0.3 in the

Sakha Republic (Europa Publications 2010) – this is compared with mean population densities 8.3, 32, and 139 in the Russian Federation as a whole, the United States, and China, respectively. Siberia is more densely populated in the west and in the south than in the east and north. Overall, the human geography is also fragmented and concentrated in a handful of large cities, and much of the region is virtually uninhabited. Five of the 20 largest (by population) cities in the Russian Federation are in Siberia: Novosibirsk (1.4 million), Yekaterinburg (1.3 million), Omsk (1.1 million), Chelyabinsk (1.0 million), and Krasnoyarsk (0.9 million), and five others have populations over 0.5 million, although none can rival Moscow at 10.5 million (Europa Publications 2010). About 70% of Siberia's people live in these and other cities, and within cities, most of the population live in apartments. However, urban Siberians have long had a tie to the forest, and *dachas* (rural second dwellings with a small plot of land) have allowed many urban residents over the past decades to maintain connection with the forests and nature and to raise their own vegetables and fruits. Permanent inhabitants of rural areas live in small settlements and simple houses. In the nineteenth and twentieth centuries, some smaller settlements have been supported by forestry or other state operations. During several decades of the twentieth century, some of these settlements were hubs of transportation and supervision of convicts and exiles from the European part of Russia. The majority of the population of Siberia today is of Russian, Ukrainian, or other western descent. A minority of the current population are descendants of Mongol or Turkic peoples (e.g., Buryats, Yakuts) or northern indigenous peoples.

In terms of the geographic distribution of populations, in West Siberia, lowlands and permafrost inhibit settlement in large areas; primarily only the south has major cities (such as Omsk and Novosibirsk) and these are largely within the corridor of the Trans-Siberian Railroad (considered completed in 1916). The oil town of Khanty-Mansiysk is in north-central West Siberia, north of the historic industrial town of Tyumen. Novosibirsk is located on the Ob River in southern West Siberia and is the largest city in Siberia. Only the southern parts of West and East Siberia are suitable for agriculture and this area was settled starting in the eighteenth to the nineteenth century by farmers in search of available agriculture lands.

East Siberia, while containing more inhabitable uplands, is also sparsely settled, and much of the area is remote and permafrost-affected. Larger settlements in East Siberia are located in the southern taiga zone along the Yenisei and Angara Rivers. The city of Krasnoyarsk lies at the western doorway to East Siberia on the Yenisei. The hydropower center of Bratsk and the Bratsk reservoir is found to the east of Krasnoyarsk on the Angara and on the Baikal-Amur Mainline Railway (BAM), which diverges from the Trans-Siberian at Tayshet between Krasnoyarsk and Bratsk. Further to the east but also in the south and on the Trans-Siberian railway route are Lake Baikal and the city of Irkutsk. Irkutsk lies on the Angara just to the west of Lake Baikal and is the principal service center for the regions to the north and east. Mining, lumbering, and farming occur in the valley hinterlands of southern taiga cities in East Siberia such as Krasnoyarsk and Irkutsk. The Buryat Republic lies east of Lake Baikal and is home to the Buryats, Siberia's largest ethnic minority. In the Arctic part of East Siberia just to the east of the Yenisei as it makes its almost due northward journey is the industrial nickel-smelting city of Norilsk.

In the vast northeastern part of East Siberia and in the Sakha Republic, permafrost and remoteness again have mitigated against significant human settlement. An exception, the sizeable city of Yakutsk (population over 200,000), lies as a port on the Lena River more than halfway between its headwaters in the highlands west of Lake Baikal and the Arctic Ocean. Yakutsk has, among other things, the RAS Permafrost Research Institute for the study of built infrastructure on permafrost substrates. The smaller city of Lensk in The Sakha Republic lies west of Yakutsk. Formerly known for diamond processing, Lensk is now a key point on a new oil pipeline linking the oil fields of East Siberia with the Far East and China.

1.4   Natural Resources

Siberia is a veritable "storeroom" of natural resources for Russia. Its contents include almost the majority of Russian reserves of natural gas and oil as well as the majority of ores, metallic minerals, and precious minerals. Siberian forests are part of the "world's largest forest," the northern Eurasian boreal taiga that extends from the Pacific to Atlantic Oceans. The region has huge resources of renewable energy, e.g., installed capacity of hydroelectric power stations of the Angara-Yenisei basin comprises half of the capacity of all hydroelectric power stations of Russia.

The majority of the Russian Federation's known energy sector reserves are located in Siberia (and the Russian Far East), including 85% of its prospected gas reserves, 75% of its prospected coal reserves, and 65% of its prospected petroleum reserves (Oldfield 2006). Approximately one-third of the world's reserves of natural gas are located in Russia, and these are predominantly within Siberia. The Central Siberian Plateau is also exceptionally rich in ores and metallic minerals, containing some of the world's greatest deposits of manganese, lead, zinc, iron, platinum, nickel, palladium, cobalt, and molybdenum. As a proportion of the Russian Federation total reserves, Siberia has more than 90 and 75% of coal and lignite, respectively; more than 95% lead; approximately 90% molybdenum, platinum, and platinoids; 80% diamonds, 75% gold, 70% nickel and copper; 50% tin and zinc; etc. (Korytny 2009). The world's largest nickel deposits are those of the Norilsk site in Siberia, and the Norilsk industry currently produces over 20% of nickel and 40% of palladium globally. Siberia is also known for diamonds, gold, graphite, and semi-precious stones; gold is mined in the Sakha and Buryatia Republics, in the Krasnoyarsk region, and in other locations.

The forests of Siberia represent up to approximately 65% of the Russian Federation's forest area (Fomchenkov et al. 2003). Across the geographic regions of Siberia, there exists a north-south gradient, which patterns its primarily boreal forest resources and their suitability (or lack of) for timber exploitation, conversion to agriculture, or even susceptibility to fire (Antipov et al. 2006; Walter and Breckle 2002; also, see Chap. 6). Human management and logging of these forests over the past century is described in Chap. 7 of this book. In addition to their role as timber resources, Siberian forests sequester carbon and provide other sustenance for humans in rural areas (mushrooms, berries, pine nuts, game) plus habitat for fur-bearing animals and other biodiversity (Krankina and Dixon 1992). While boreal regions are generally not as biologically diverse as temperate or tropical biomes, Siberia has some of Earth's largest intact landscapes (critical for migratory herds such as caribou, *Rangifer tarandus*) and is habitat to internationally important species such as the Siberian tiger (*Panthera tigris altaica*), the snow leopard (*Panthera uncial*), and the Siberian crane (*Grus leucogeranus*; Dinerstein 1994).

Approximately 27% of the utilized arable farmland in the Russian Federation is in Siberia, with 7% and 20% located in the Urals and Siberian okrugs, respectively, and a negligible amount in the Sakha Republic (Oldfield 2006). Because of the short growing season in the north, most agricultural production in Siberia is concentrated in the southern part of the region in the forest-steppe and steppe zones. These zones are the regions of increased societal water demand (Vörösmarty et al. 2000). Unfortunately, most of agricultural fields and pastures in these zones are not irrigated and are prone to frequent droughts. Water availability has become a central issue for social and ecological sustainability.

Recent developments of industry in Siberia, man-made changes of environment and ecosystems, as well as ongoing and expected climate change generate many risks to the Siberian landscape. Current state and recent tendencies of dynamics of terrestrial ecosystems, particularly forests, are negatively impacted by increasing anthropogenic pressure and insufficient governance of natural resources in many large Siberian regions. Construction of hydroelectric power stations on large Siberian rivers leads to large land-use changes that substantially impact regional and local climates. The significant impacts of global change on the Siberian environment and human health, as well as on the social and economic safety of human well-being, are very likely in terms of short-term impacts and become crucial when assessing the long-term consequences. Overall, the region is one of the most vulnerable vast territories on earth and is indicated as a "hot spot" by the ESSP Global Carbon Project (2010).

1.5 Types of Environmental Changes

A prerequisite to understanding of human-driven change is knowledge of natural environmental dynamics; these are discussed in greater detail in Chap. 6. Up to the industrial era, humans minimally altered the Siberian environment through fishing, hunting, and limited permanent and shifting agriculture. Subsequent to industrialization, the human impact has been significantly greater, including fire, logging, pollution, and industrialized agriculture; these are discussed in greater depth in Chap. 7.

In general, natural dynamics of the boreal forests include gap dynamics, fire, wind throw, plant-animal interactions, and growth and succession. Fires occur largely in conifer and mixed forests, ranging in severity and size from small light surface fires to large stand-replacing crown fires (Conard and Ivanova 1997). The Siberian silk moth (*Dendrolimus superans sibiricus*) is one of the principal causes of disturbance in conifer forests (Kharuk et al. 2007). In the middle and southern taiga zones, following stand-replacing disturbance, the most widespread successional pathway is from young birch-aspen regeneration, through maturing deciduous forests, to mixed and conifer forest (Hytteborn et al. 2005; Schulze et al. 2005). However, in the larch-dominant (*Larix sibirica, L. gmelinii*) as well as in some Scotch pine-dominant (*Pinus sylvestris*) communities, sites often are regenerated directly to larch and pine, respectively (Kharuk et al. 2010).

Human-driven changes in forests include fire, logging, post-logging succession, and agriculture development and abandonment. Within Siberia, the three administrative units with the largest forest industries have been Irkutsk Oblast, Krasnoyarsk Kray, and Tomsk Oblast (these have been followed by Khabarovsk Kray in the Russian Far East). In the middle and southern taiga, logging has historically occurred most often as clear-cuts in conifer forests, with most sites left to natural regrowth and succession. Natural fire return cycles may be significantly modified by human-ignited fires (Achard et al. 2008; Korovin 1996). Establishment of agriculture has incurred large-scale conversion of temperate and southern boreal forests in the more arable southern portions of Siberia; however, more recently, abandonment of farming areas contributes to forest regrowth (Bergen et al. 2008).

Characteristic environmental changes associated with energy and mineral resources include development and growth of large urban centers (including many of Siberia's largest cities); establishment of large industrial complexes; mines and mining-dominated landscapes; air and water pollution from petrochemical and smelter industries; hydropower complexes including dams, reservoirs, and hydroelectric power stations; and transportation infrastructures. In West Siberia, the Khanty-Mansi region has historically contained 70% or more of Russia's developed oil fields, with the Samotlor field being the largest in Russia. Other regional cities developed large petrochemical industries (e.g., Tomsk, Tobol'sk, and Omsk). Large coal and iron ore deposits (used in steel production) are located in southeastern West Siberia in the 27,000 km^2 Kuznetsk Basin. Production of natural gas is concentrated near Urengoy (the world's second largest gas field) and the Yamal Peninsula in northern West Siberia. More recently, oil reserves have been explored in East Siberia including the Yurubchen-Tohomo oil fields (cf., synthesis case study in Chap. 7) north of Krasnoyarsk and a cluster of others further east and north of Lake Baikal. Because urbanization and industrialization require power, Siberia's great rivers have been harnessed for hydropower. The upper reaches of a number of rivers, including the Yenisei and Angara, have strong gradients and flow, making them especially conducive to development of hydropower.

Environmental changes of consequence to plant and animal biodiversity are generally similar to those of forests: conversion of steppe and forest lands to agriculture, fire, logging, insects, mining, and pollution. Additionally, there is the added threat of hunting and poaching to animal biodiversity. Given such threats, there are a number of protected area types in Siberia that are intended to limit human-driven environmental change: strict nature reserves (*zapovedniki*), national parks (*natsional'nye parki*), natural landmarks (*pamyatniki prirody*), natural reserves and hunting grounds (*zakazniki*), global heritage sites, wetlands, regional natural parks, and traditional nature management zones. As of 2001, in Siberia there were estimated to be 30 *zapovedniki* and nine *natsional'nye parki* out of about 135 total combined in Russia (Oldfield 2006), and the number of protected areas has been increasing since that time.

Requirements for large-scale agriculture (wheat and other grains, potatoes, grazing of sheep and cattle) include fertile soils, absence of significant permafrost or more moderate climate conditions, or both. These are met primarily in the steppe zone of southern Siberia; thus, this area has largely been converted to agriculture, as have floodplain valleys of rivers in the southern taiga zone. When abandoned, agricultural areas in the taiga zones may revert to forest (Krankina et al. 2005; Bergen et al. 2003, 2008; Chap. 7 synthesis case study). In more mountainous areas, pasture is also established on hillsides. In the northern taiga and Arctic zones, reindeer pastoralism and breeding have long been human subsistence activities and continue today with fairly minimal environmental consequences.

1.6 Impacts of Environmental Changes

The state and dynamics of ecosystems are a product of the sophisticated interplay and mutual conditionality of impacts, responses, and feedbacks of natural, economic, and social components; environment; and human society. An important feature of the region is the fragility of ecosystems, which in evolutionary terms developed under a rather stable and cold climate. Ecological thresholds and buffering capacity of ecosystems under rapid substantial warming have no precedent in recent history and are poorly understood. This generates major challenges for understanding the current and future state, vitality, and resilience of ecosystems of Northern Eurasia.

Global changes provide direct and indirect effects on Siberian ecosystems, and these interact with multiple natural and anthropogenic disturbances and other ecological processes. Some of these changes may be irreversible on century time scales and have the potential to cause rapid changes in the earth system (McGuire et al. 2006). The impacts of both global and regional change on ecosystems often cannot be understood within the simple "cause-and-effect" paradigm. Understanding and appropriate description of these changes require us to take into account many climate-forming factors of cosmophysical (including heliospheric), geospheric, biospheric, and anthropogenic origin, to determine not only changes of state of the climatic system but also evolution of these physical processes and phenomena, which may be regionally specific. Predicting the cumulative impacts of such complex interactions is difficult and usually requires an integrative modeling approach (Vygodskaya et al. 2007; Milne et al. 2009) that would combine different types of models (empirical, process-based, etc.) and involve different dimensions of the surrounding world – ecological, social, and economic.

Increasingly, Siberian ecosystems become socioecological systems taking into account the transformation of vegetation cover through development of previously untouched territories and the diverse destructive anthropogenic impacts on the environment and natural landscapes. For example, the following types of oil- and gas-related complex impacts on the environment, quality of life, and ecosystems are most dangerous: (1) infrastructure-related ecological impacts that lead to destruction of soil cover and changes of hydrological regimes; (2) ordinary ecological impact caused by pollutants during extraction and transport of oil and gas; such impacts cannot be eliminated by current technologies of oil and gas exploration and extraction; (3) extraordinary ecological impacts caused by accidents and technogenic catastrophes; and (4) impacts on ecosystems by population outside of industrial areas (Sibgatulin et al. 2009).

Methods practiced today of industrial exploitation of northern territories in the region are wasteful and often provide extremely negative impacts on the environment and ecosystems. For example, the Siberian Federal *Okrug* (including part of West and the entire central Siberia) is second in the Russian Federation in terms of total emissions from stationary sources (5.8 million tons in 2007 that is 28% of the total emissions of the country; cf. Fig. 1.3). An abundance of large industrial enterprises is a specific feature of the region. For instance, the company "Norilsk Nickel" that emits about 2 million tons of pollutants per year (mostly sulfur dioxide) generated about 2 million ha of technogenic desert during the last 40 years, of which about 500 thousand ha were forests (this is addressed in detail in Chap. 8). In regions of intensive oil and gas extraction of West Siberia, (1) up to 35,000 breaks of oil pipelines occur annually; of this number, about 300 accidents are officially registered with oil spills >10 000 tons each, (2) the tundra surface is destroyed for more than 15%, and (3) physical destruction of natural landscapes exceeded 30% of the total area of the territories of middle and southern taiga. Utilization and use of oil casinghead gas is unsatisfactory. By different estimates, from 15 to 25 billion m^3 of such gas are burnt in flares annually. The Government Commission on Fuel and Energy reported that the amount of extracted casinghead gas in 2007 comprised 61.2 billion m^3 with 16.7 billion m^3 burnt in flares (Kryukov and Tokarev 2010). The quality of river water, specifically in industrial areas and southern regions with high population density, does not correspond to the norms of water use for drinking and fisheries. The governance of natural resources (in particular, forests) and the control of natural resource management are below any acceptable levels.

The level of atmospheric pollution and soil contamination in major regions of intensive oil and gas extraction substantially exceeds acceptable limits (see Chap. 8 for details). Soil pollution and water contamination are widespread in some regions and high in industrial populated territories. For instance, the following share of area which did not correspond to requirements of the health code of populated areas was observed in central Siberia: Minusinsk district 71.4%, Krasnoyarsk 61.1%, Norilsk 12.3% (but 83.0% in 2006), etc. (Kaliagina 2009). The impacts of toxic anthropogenic water contamination, the decline of the human immune system, increasing stress, impacts of many negative social phenomena connected, among others, to intensification of migration processes, will likely accelerate the negative impacts of climatic change on the standards of life and health of the population, as well as enforcing undesirable feedbacks.

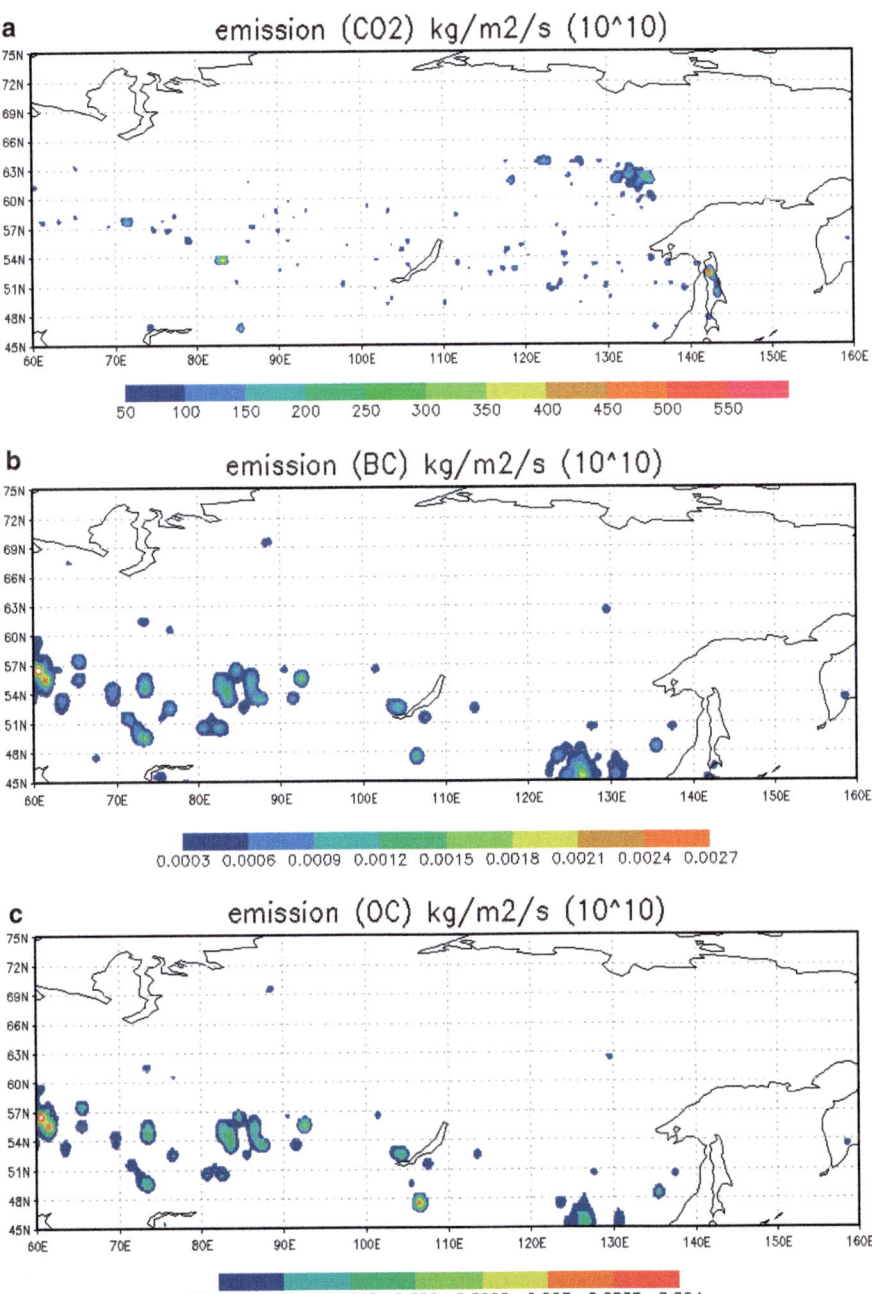

Fig. 1.3 Emissions of main pollutants in Siberia based on different global emission inventories (Extracted by A. Zakey, see Chap. 8)

A major present-day paradigm of the interaction of humanity and nature in the contemporary world is the transition to sustainable development. One of the most important prerequisites of sustainable development is maintenance of regional stability of the biosphere (in particular, the problem of balancing major biogeochemical cycles within ecological regions). In Russia, this transition is declared to be a cornerstone of national and regional policy of natural resources management. However, the reality is far from such declarations. The ecological and environmental situation in large regions of Northern Eurasia could be characterized as an ongoing global ecological crisis initiated by the unregulated anthropogenic pressure on nature and the explosive increase of fossil fuel consumption. In all, this results in the decreasing quality of major components of the environment – air, water, soil, and vegetation. Siberia is a typical and illustrative example of such negative processes.

1.7 Structure of the Following Book Chapters

Above, we briefly described several features of Siberia that make it a special and important area of the globe. This is a region with severe continental climate that:

- Is harsh as well as sensitive to changes
- Is quite dry and exercises controls both on the Arctic freshwater budget (and thus to the global thermohaline circulation) and on the regional ecosystem well-being and is capable to quite quickly change the land cover from taiga to steppe or even to semidesert
- Has a unique storage of carbon in the frozen ground and across the expansive taiga zone that can change (be burned, or simply be released from thawing permafrost) creating potential positive feedback to the global Earth system
- Has experienced dramatic changes in human-driven resource exploitation and management over the past century.

Chapter 2 contains descriptions of the tools (information systems and their content) accumulated for the Siberian domain in its national institutions as well as abroad in Europe, Japan, and the United States. In the following six chapters, we assess environmental changes in Siberia. We start with a description of the regional climate changes and their projections (Chap. 3), paying a particular attention to changes in the water cycle of the region (Chap. 4) and cryosphere impact on the regional infrastructure (Chap. 5). Changes in the ecosystems and carbon cycle of Siberia are described in Chap. 6. Human dimensions of environmental change in the context of land-cover/land-use change during Soviet, early post-Soviet, and emerging eras are the subject of Chap. 7. Atmospheric pollution, soil and water contamination issues are addressed in Chap. 8. Chapter 9 provides the summary of findings described in the previous chapters.

References

Achard F, Eva HD, Mollicone D, Beuchle R (2008) The effect of climate anomalies and human ignition factor on wildfires in Russian boreal forests. Philos Trans R Soc B Biol Sci 363(1501): 2331–2339

ACIA (2005) Arctic climate impact assessment scientific report. Cambridge University Press, New York, 1042 pp

Antipov AN, Korytny LM, Plyusnin VM (2006) Siberia and its geographical research. In: Antipov AN (ed) Geography of Siberia. Research India Publications, Delhi, pp 3–36

Balzter H (ed) (2010) Environmental change in Siberia. Springer, Dordrecht/Heidelberg/London/New York. doi:10.1007/978–90–481–8641–9, 282 pp

Bergen KM, Conard SG, Houghton RA, Kasischke ES, Kharuk VI, Krankina ON, Ranson KJ, Shugart HH, Sukinin AI, Treyfeld RF (2003) NASA and Russian scientists observe land-cover and land-use change and carbon in Russian forests. J For 101(4):34–41

Bergen KM, Zhao T, Kharuk V, Blam Y, Brown DG, Peterson LK et al (2008) Changing regimes: forested land cover dynamics in central Siberia 1974 to 2001. Photogramm Eng Remote Sens 74:787–798

Conard SG, Ivanova GA (1997) Determining effects of area burned and fire severity on carbon cycling and emissions in Siberia. Clim Chang 1(2):197–211

Dinerstein E (1994) An emergency strategy to rescue Russia's biological diversity. Conserv Biol 8(4):934–942

Europa Publications (2010) Russian Federation. In: Eastern Europe, Russia and Central Asia. Europa Publications Limited, London, pp 352–354

Fomchenkov VF, Sdobnova VV, Danilov NK, Danilova SV, Kurdina GV, Beljakova TF (2003) Forest fund of Russia (Data of State Forest Account). All-Russia Research Institute of Forestry and Mechanization, Moscow, 640 pp

Forster P, Ramaswamy V, Artaxo P, Berntsen T, Betts R, Fahey DW, Haywood J, Lean J, Lowe DC, Myhre G, Nganga J, Prinn R, Raga G, Schulz M, Van Dorland R (2007) Changes in atmospheric constituents and in radiative forcing. In: Solomon S, Qin D, Manning M, Chen Z, Marquis M, Averyt KB, Tignor M, Miller HL (eds) Climate change 2007: the physical science basis. Contribution of working group I to the fourth assessment report of the Intergovernmental Panel on Climate Change. Cambridge University Press, Cambridge/New York

Gumilev LN (1990) Ethnogenesis and the Earth's biosphere. Gidrometeoizdat, Moscow, 528 pp (in Russian)

Gupta SK, Stackhouse PW Jr, Cox SJ, Mikovitz JC, Zhang T (2006) 22-year surface radiation budget data set. GEWEX News 16(4):12–13

Hytteborn H, Maslov AA, Nazimova DI, Rysin LP (2005) Boreal forests of Eurasia. In: Anderssen F (ed) Ecosystems of the World: coniferous forests. Elsevier, Amsterdam, pp 23–99

IPCC (2007) Climate change 2007: working group I report "The Physical Science Basis", contribution of working group I to the fourth assessment report of the Intergovernmental Panel on Climate Change, 2007. In: Solomon S, Qin D, Manning M, Chen Z, Marquis M, Averyt KB, Tignor M, Miller HL (eds). Cambridge University Press, Cambridge/New York

Kaliagina LV (2009) Economic regulation of ecological land-use in Krasnoyarsk kray. In: Vaganov E (ed) Resource economics, environmental economics and climate change – 2009. Siberian Federal University, Krasnoyarsk, pp 325–340

Kaplin PA, Selivanov AO (1995) The flood that was, that is and that will be. Science Russ 2:16–23

Kharuk VI, Ranson KJ, Fedotova EV (2007) Spatial pattern of Siberian silkmoth outbreak and taiga mortality. Scand J For Res 22(6):531–536

Kharuk VI, Ranson KJ, Dvinskaya ML (2010) Wildfire dynamics in Mid-Siberian larch dominated forests. In: Balzter H (ed) Environmental change in Siberia. Springer, London, pp 83–100

Korovin GN (1996) Analysis of the distribution of forest fires in Russia. In: Goldammer JG, Furyaev VV (eds) Fire in ecosystems of Boreal Eurasia. Kluwer, Boston, pp 112–128

Korytny LM (2009) Actual tasks of geographical resource management. In: Vaganov E (ed) Resource economics, environmental economics and climate change – 2009. Siberian Federal University, Krasnoyarsk, pp 359–366

Krankina ON, Dixon RK (1992) Forest management in Russia – challenges and opportunities in the era of Perestroika. J For 90(6):29–34

Krankina ON, Sun G, Shugart HH, Kasischke E, Kharuk VI, Bergen KM, Masek JG, Cohen WB, Oetter DR, Duane MV (2005) Northern Eurasia: remote sensing of boreal forest in selected regions. In: Gutman G, Janetos AC, Justice CO, Moran EF, Mustard JF, Rindfuss RR, Skole D, Turner BL, Cochrane MA (eds) Land change science: observing, monitoring, and understanding trajectories of change on the earth's surface. Kluwer Academic Publishers, Dordrecht, pp 123–138

Kryukov VA, Tokarev AN (2010) Evolution of oil resource management in Russia. J Sib Fed Univ Hum Soc Sci 6(3):864–890

Kuznetsova LP (1978) Water vapor migration over the USSR territory. Nauka, Moscow, 92 pp (in Russian)

Lamb HH (1988) Weather. Climate and Human Affairs, Routledge/London, 438 pp

Lydolph PE (1977) Climate of the Soviet Union. Elsevier, Amsterdam/Oxford/New York, 443 pp

McGuire AD, Chapin RS III, Wirth C, Apps M, Bhatti J, Callaghan T, Christiansen TR, Clein JS, Fukuda M, Onuchin A, Shvidenko A, Vaganov E (2006) Responses of high latitude ecosystems to global change: Potential consequences for the climate system. In: Canadell JG, Pataki DE, Pitelka LF (eds) Terrestrial ecosystems in a changing world. Springer, Berlin/Heidelberg/New York, pp 297–310, Chapter 24

Milne E, Aspinall RJ, Veldkamp TA (2009) Integrated modelling of natural and social systems in land change science. Landsc Ecol 24:1145–1147

NEESPI 2004: The Northern Eurasia Earth Science Partnership Initiative (NEESPI) executive overview, NEESPI Science Plan Development Team, 20 pp. Available at http://neespi.gsfc.nasa.gov/science/ExecutiveSummary15W.pdf

Oldfield JD (2006) Russian nature: exploring the environmental consequences of societal change. Ashgate, Burlington

Pan Y, Birdsey RA, Fang J, Houghton R, Kauppi P, Kurz WA, Phillips OL, Shvidenko A, Lewis SL, Canadell JG, Ciais P, Jackson RB, Pacala S, McGuire AD, Piao S, Rautiainen A, Sitch S, Hayes D, Wayson C (2011) A large and persistent carbon sink in the world forests. Science 333:988–993. doi:10.1126/science.1201609

Pavlov AV (1984) Energy exchange in the landscape sphere of the Earth. Nauka, Novosibrsk, 256 pp

Pirazzoli PA (1996) Sea-level changes: the last 20,000 years. Wiley, Chichester, 212 pp

Quegan S, Beer C, Shvidenko A, McCallum I, Handoh IC, Peylin P, Rödenbeck C, Lucht W, Nilsson S, Schmullius C (2011) Estimating the carbon balance of central Siberia using a landscape-ecosystem approach, atmospheric inversion and dynamic global vegetation models. Glob Chang Biol 17:351–365

Schepaschenko D, McCallum I, Shvidenko A, Fritz S, Kraxner F, Obersteiner M (2010) A new hybrid land cover dataset for Russia: a methodology for integrating statistics, remote sensing and in-situ information. J Land Use Sci. doi:10.1080/1747423X.2010.511681 (Published online 22 Dec 2010)

Schulze ED, Wirth C, Mollicone D, Ziegler W (2005) Succession after stand replacing disturbances by fire, wind throw, and insects in the dark taiga of Central Siberia. Oecologia 146(1):77–88

Selivanov AO (2000) Nature, history, culture: environmental aspects of ethnic cultures of the world. Moscow, GEOS, 324 pp (in Russian)

Shahgedanova M (2003) The physical geography of Northern Eurasia. Oxford University Press, Oxford, 571 pp

Shver TsA (1976) Atmospheric precipitation over the USSR territory (in Russian). Gidrometeoizdat, Leningrad, 302 pp

Shvidenko A (2009) Terrestrial ecosystems of Northern Asia, global change and post Kyoto developments. In: Vaganov EA (ed) Resource economics, environmental economics and climate change-2009, conference proceedings, Krasnoyarsk, 1–7 July 2009, 667–678

Shvidenko A, Schepaschenko D, Vaganov EA, Sukhinin A, Maksyutov SS, McCallum I, Lakida I (2011) Impact of vegetation fires in Russia on ecosystems and global carbon budget. Proceedings of the Russian Academy of Sciences (Doklady Earth Sciences) 441(Part 2):1678–1682

Sibgatulin VG, Simonov KV, Peretokin SA et al (2009) Estimates of geodynamics for urbanized areas. Montane Inform – Analyt Bull 18(12):51–55 (in Russian)

Soja AJ, Cofer WR III, Shugart HH, Sukhinin AI, Stackhouse PW Jr, McRae DJ, Conard SG (2004) Estimating fire **emissions** and disparities in boreal Siberia (1998 through 2002). J Geophys Res 109:D14S06. doi:10.1029/2004JD004570

Soja AJ, Tchebakova NM, French NHF, Flannigan MD, Shugart HH, Stocks BJ, Sukhinin AI, Parfenova EI, Chapin Iii FS, Stackhouse PW Jr (2007) Climate-induced boreal forest change: predictions versus current observations. Glob Planet Chang 56:274–296

The Global Carbon Project (2010) GEO carbon strategy. Document is available at: http://www. globalcarbonproject.org/global/pdf/GEO_CARBONSTRATEGY_20101020.pdf

Van der Werf GF, Randerson JT, Giglio L, Collatz GJ, Kasibhalts PS, Arellano Jt AF (2006) Interannual variability of global biomass burning emissions from 1997 to 2004. Atmos Chem Phys Discuss 6:3175–3226

Vörösmarty CJ, Fekete BM, Meybeck M, Lammers RB (2000) Global system of rivers: its role in organizing continental land mass and defining land-to ocean linkages. Global Biogeochem Cycles 14:599–621

Vygodskaya NN, Groisman PY, Tchebakova NM, Kurbatova JA, Panfyorov O, Parfenova EI, Sogachev AF (2007) Ecosystems and climate interactions in the boreal zone of northern Eurasia. Environ Res Lett 2. doi:10.1088/1748-9326/2/4/045033

Walter H, Breckle SW (2002) Walter's vegetation of the earth: the ecological systems of the geo-biosphere, 4th edn. Springer, Berlin, 527 pp

Wigley TML, Ingram MJ, Farmer J (eds) (1981) Climate and history. Cambridge University Press, Cambridge, 456 pp

Chapter 2
Development of Information-Computational Infrastructure for Environmental Research in Siberia as a Baseline Component of the Northern Eurasia Earth Science Partnership Initiative (NEESPI) Studies

Evgeny P. Gordov, Keith Bryant, Olga N. Bulygina, Ivan Csiszar, Jonas Eberle, Steffen Fritz, Irina Gerasimov, Roman Gerlach, Sören Hese, Florian Kraxner, Richard B. Lammers, Gregory Leptoukh, Tatiana V. Loboda, Ian McCallum, Michael Obersteiner, Igor G. Okladnikov, Jianfu Pan, Alexander A. Prusevich, Vyacheslav N. Razuvaev, Peter Romanov, Hualan Rui, Dmitry Schepaschenko, Christiane C. Schmullius, Suhung Shen, Alexander I. Shiklomanov, Tamara M. Shulgina, Anatoly Z. Shvidenko, and Alexander G. Titov

Abstract This chapter provides a brief description of the information resources currently supporting environmental studies of Siberia including key references and points of contact. It describes environmental, hydrological, and meteorological datasets available for Siberia as well as the tools developed to organize and seamlessly deliver these data to the international research community for studying regional environmental and climatic dynamics of the ongoing global changes. Three-hour and daily datasets of major meteorological characteristics measured at the Siberian

E.P. Gordov (✉) • I.G. Okladnikov • T.M. Shulgina • A.G. Titov
Siberian Center for Environmental Research and Training and Institute of Monitoring of Climatic and Ecological Systems SB RAS, Tomsk, Russian Federation
e-mail: gordov@scert.ru

K. Bryant • I. Gerasimov • J. Pan • H. Rui
Adnet and NASA Goddard Space Flight Center, Greenbelt, MD, USA

O.N. Bulygina • V.N. Razuvaev
Russian Research Institute of Hydrometeorological Information – World Data Centre (RIHMI-WDC), Obninsk, Russian Federation

I. Csiszar
NOAA NOAA/NESDIS/STAR, Camp Springs, MD, USA

J. Eberle • R. Gerlach • S. Hese • C.C. Schmullius
Department of Earth Observations, Institute for Geography, Friedrich-Schiller-University, Jena, Germany

P.Ya. Groisman and G. Gutman (eds.), *Regional Environmental Changes in Siberia and Their Global Consequences*, Springer Environmental Science and Engineering, DOI 10.1007/978-94-007-4569-8_2, © Springer Science+Business Media Dordrecht 2013

weather stations and relevant metadata sets are the first tangible resources available to the researchers. However, most of the Siberian territory is sparsely populated and the observational networks that provide regional in situ observations are also sparse. Therefore, other information resources described below are based upon, or include as their integral part, remote sensing and model output data. These resources are (a) land information system for Siberia that includes cartographical materials, data of different inventories and surveys, diverse databases of in situ measurements and remote sensing products, and numerous auxiliary models for assessment of relevant biophysical indicators of Siberian ecosystems; (b) remote sensing Earth observation products and tools for data search, data access, data visualization, and analysis over Siberia; and (c) a suite of online systems to monitor, process, visualize, analyze, and access Earth science remote sensing products and regional climatic and meteorological geospatial datasets, as well as a variety of geospatial data on climate, climate forecast, hydrology, hydrological forecast, environmental remote sensing, socioeconomic information, etc.

2.1   Introduction

This chapter describes IT tools developed to support studying Northern Eurasia regional environment dynamics under Global Change impact. While developed for the entire Northern Eurasia, here we focus on those tools that are most relevant to studies in Northern Asia (Siberia). The approaches used and tools developed are based on modern tendencies in Earth and Space Science Informatics (ESSI) and

S. Fritz • F. Kraxner • I. McCallum • M. Obersteiner
Forestry Program, International Institute for Applied Systems Analysis, Laxenburg, Austria

R.B. Lammers • A.A. Prusevich • A.I. Shiklomanov
Water Systems Analysis Group, Complex Systems Research Center, Institute for the Study of Earth, Ocean, and Space, University of New Hampshire, Morse Hall, Durham, NH, USA

G. Leptoukh
NASA Goddard Space Flight Center, Greenbelt, MD, USA

T.V. Loboda
Department of Geography, University of Maryland, College Park, MD, USA

P. Romanov
NOAA Cooperative Remote Sensing Science and Technology Center (NOAA-CREST), City College of City University of New York, Maryland, MD, USA

D. Schepaschenko
Forestry Program, International Institute for Applied Systems Analysis, Laxenburg, Austria

Moscow State Forest University, Mytischi, Moscow Reg., Russian Federation

S. Shen
NASA Goddard Space Flight Center, George Mason University, Greenbelt, MD, USA

A.Z. Shvidenko
Forestry Program, International Institute for Applied Systems Analysis, Laxenburg, Austria

Forestry Institute, Siberian Branch, Russian Academy of Science, Krasnoyarsk, Russian Federation include such conventional tools as GIS-based information systems and data centers of instrumental meteorological observations as well as such novel tools as thematic web portals comprising features of web-based information-computational systems with GIS functionalities.

Earth system science research often demands multidisciplinary approaches involving a number of collaborating institutions from different research disciplines spread around the globe. Historically, within each discipline, specific data formats, tools, and community standards evolved to meet their individual requirements. The heterogeneity of these community-specific practices has been identified as one impediment to interdisciplinary collaboration and exchange of data and tools, and mitigation through interoperability is mostly employed as a concept to overcome such heterogeneity (Bigagli et al. 2006). Interoperability is achieved by implementing standardized interfaces encapsulating, and thus retaining, the underlying domain specifics (Ramamurthy 2006). These interfaces are typically provided as web services following a service-oriented architecture (SOA) approach. Within the environmental and Earth science community, standards of the Open Geospatial Consortium (OGC), the International Organization for Standardization (ISO), or the World Wide Web Consortium (W3C) are increasingly adopted. Whereas the primary goal of such infrastructures is to provide access to disparate data resources, they should further enable researchers to integrate and federate data and computing resources into their workflow or application (Hey and Trefethen 2005). Another important issue is integration of web services with modeling frameworks. The approach has been previously implemented for specific projects in the fields of climatology (Bernholdt et al. 2005), atmosphere (Gordov et al. 2006), hydrology (Kumar et al. 2006), oceanography (Allen et al. 2008), and global ecosystems (Best et al. 2007).

This chapter has the following structure. Section 2.1 describes the state of the art in developing information-computational infrastructure of Siberia Integrated Regional Study and its key element – the CLimate and Environment Analysis and Research System (CLEARS). The ability to produce a highly detailed (both spatially and thematically) land-cover/land-use dataset over Russia, by combining existing datasets into a hybrid information system, is demonstrated in Sect. 2.2. The system of storage and acquisition of land-cover data over the Russian Federation is described in Sect. 2.3. The Siberian Earth System Science Cluster (SIB-ESS-C) has been established at the Friedrich Schiller University (Jena, Germany) as a spatial data infrastructure aimed at providing researchers with a focus on Siberia with the technical means for data discovery, data access, data publication, and data analysis. It is described in Sect. 2.4. The SIB-ESS-C infrastructure will be further developed as one of the nodes in a network of similar systems for the NEESPI region (e.g., NASA GES-DISC (Goddard Earth Sciences Data and Information Services Center) Interactive Online Visualization ANd aNalysis Infrastructure (Giovanni) developed at the NASA Goddard Space Flight Center). The capabilities of Giovanni are presented in Sect. 2.5. This is a mature web-based application that provides a simple and intuitive way to visualize, analyze, and access a vast amount of Earth science data online without necessarily having to download them. Giovanni is one of the major components of the NASA NEESPI Data Center (http://neespi.gsfc.nasa.gov), where remote sensing and modeled data from different resources are preprocessed and archived in a cohesive, well-architected data management system with convenient data access. Giovanni-NEESPI is a portal that focuses on atmospheric, land surface, and cryospheric products for Northern Eurasia in support of the NEESPI projects. Novel web-based services for spatially distributed Earth system, climate, and hydrologic data are described in Sect. 2.6. The Rapid Integrated Mapping System (RIMS) develops further integration of web services with modeling frameworks that allow streamlined interaction and management of large and diverse data holdings in project specific environments. It is aimed at visualization, data exploration, querying, manipulation, and arbitrary calculations with any gridded or vector polygon dataset loaded into the system. The system was adopted for hydrology and climatology studies in the NEESPI region and has a potential for applications in other fields of Earth sciences.

In spite of the fact that these tools are aimed at targeted region applications, all of them have quite generic character and might be used for environmental investigations beyond the Northern Eurasia domain.

2.2 Earth and Space Science Information for Regional Studies: Towards Distributed Information-Computational Infrastructure to Support Monitoring, Modeling, and Analysis of Global Change Impact in Siberia

Distributed information-computational infrastructure is a prerequisite to successful NEESPI development. A key feature of this infrastructure is that it should provide for chosen thematic domains organization of data and knowledge exchange between the various elements in order to form a distributed collaborative information-computational environment supporting investigations in multidisciplinary area of Earth regional environment studies. It appears that using of GIS combined with computational resources required to support modern models and sharing of huge data archives is not very promising. Therefore, an approach relying mainly upon web technologies potential and information systems employing the three-level model – data/metadata, computation, and knowledge levels (De Roure et al. 2001) – was chosen for development of the information-computational infrastructure in Siberia (Gordov 2004a, b; Gordov et al. 2006, 2007). The web-based environmental information systems (with accompanying applications) representing distributed collaborative information-computational environment to support multidisciplinary investigation of Northern Eurasia form a powerful tool for better understanding of the interactions between the ecosystem, atmosphere, and human dynamics for the vast Siberia region under the impact of global climate change. This activity is stimulated by strong demand of the NEESPI community investigating dynamics of Northern Eurasia environment under Global Change impact, and in turn, its development allows us to achieve significant progress in reaching the NEESPI scientific plan objectives. Being generic, the activity should provide researchers with a reference, open platform (portal plus tools) that may be used, adapted, enriched, or altered on the basis of the specific needs of particular applications in different regions as well.

Due to development of powerful software, easily deployed on web servers, such as Mapserver (http://mapserver.org/), GeoTools (http://www.geotools.org/), etc., web based data visualization, generation, processing and analysis recently became a new approach for working with georeferenced data. Usage of Web-GIS technologies, particularly Web Map Service (WMS) and Web Feature Service (WFS) protocols, is intended for realization of the following system functionality: scaling of graphical processing results, geographical region selection, representation of different cartographical information using layers, and providing access to information about geographical object selected by user. However, in this research area, there is still a lack of powerful tools combining various capabilities to process, analyze, and visualize multiple-source data collections using unified web interface for integrated study of global and regional climate change. To illustrate the state of the art and tendencies in developing of information-computational Web-GIS systems for Siberia, CLEARS, which is aimed at regional climatic and meteorological datasets processing and visualization, is described in this section.

2.2.1   General CLEARS Description

Using experience obtained during system's prototype development (http://climate. risks.scert.ru/, Gordov et al. 2007; Okladnikov et al. 2008), general requirements to its final version were elaborated. First of all, verified series of long-term regular observations of meteorological characteristics are used for research. Computational elements of the system provide a wide set of procedures for complex analysis of meteorological and climatic characteristic behaviors both at global and regional scales. Processing results are represented both in graphical form as colored data fields, time series, or plots, and as a file in selected format for possible following usage. Technically, the information-computational Web-GIS system consists of four key parts (Fig. 2.1):

- Structured archives of georeferenced geophysical data accompanied by corresponding metadata
- Computational kernel consisting of an independent modules set implemented in ITTVIS IDL (Interactive Data Language, [http://www.ittvis.com/ProductServices/IDL.aspx])
- Web portal providing data interchange for internal web applications, connections with web services, and access to metadata collections
- Graphical user interface (GUI)

In order to provide the system with data, regional and global geophysical datasets were and still are collected and stored on a disk array directly connected to a high-performance server. Since data collected by different organizations and institutes differ by parameter set, file formats, etc., automatic organization and conversion to NetCDF or HDF5 file formats were performed. High-resolution data field reconstruction is performed by statistical methods and modern numerical model WRF (http://www.wrf-model.org/), which includes assimilation algorithms

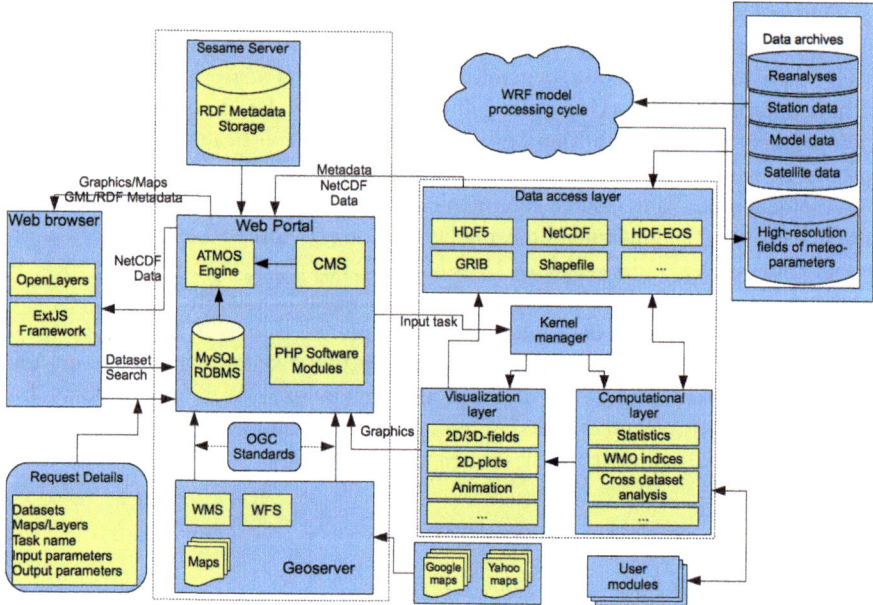

Fig. 2.1 General architecture of the system

for observation data inside a selected region. Users of this system use a web browser as a standard client which is present on every modern workstation. GUI for administration and operation of the information-computational system gives the user tools to generate a task for georeferenced data processing. A relevant processing task and visualization parameters are selected by a system's user via a graphical interface inside a web browser window. The system functionality includes calculation of basic statistics such as extremes, mean values, and standard deviation; counting of days where a characteristic value lies in a specified region; calculation of correlation and linear regression coefficients, trends, and climate change indices (Peterson 2005); and processing results to a user through the GUI as graphics and links to files. Annotation, storage, organization of effective semantic search, and access to geophysical datasets are among the most frequently requested tasks. Web applications developed at SCERT provide these services too (Titov 2006; Titov et al. 2009, 2010).

2.2.2 Datasets

The following archives of georeferenced data are currently accessible by the Web-GIS system for processing and visualization: the first and second editions of Reanalysis of National Center for Atmospheric Research (NCAR)/National Centers for Environmental Prediction (NCEP), JRA-25 Reanalysis of Japan Meteorological Agency (JMA)/Central Research Institute of Electric Power Industry (CRIEPI),

ERA-40 Reanalysis of European Centre for Medium-Range Weather Forecasts (ECMWF), and National Oceanic and Atmospheric Administration (NOAA)– Cooperative Institute for Research in Environmental Sciences (CIRES) Twentieth Century Global Reanalysis. Also Landsat 4–7, Global Land Survey (GLS) and Moderate Resolution Imaging Spectroradiometer (MODIS) satellite data archives (mosaics for 1975, 1990, 2000, and 2005 for Northern Eurasia http://glovis.tsc.ru/), and observations from the former USSR meteorological network are available. These datasets are stored as structured file archives on the high-performance storage system.

2.2.3 Current State of the CLEARS Development

The information-computational Web-GIS system architecture and structure of geophysical data archives were developed. Aforementioned archives of georeferenced data (cf., Sect. 2.1.2) were collected and prepared for use, and methods for their preprocessing were selected and developed (Gordov et al. 2007; Okladnikov et al. 2008; Titov 2006; Titov et al. 2009, 2010). A basic set of georeferenced maps (including maps of land cover, nature ecosystems, and NDVI) was prepared for use by a Web-GIS service. Later on, the set of available data will be expanded. In particular, one of the applications of the system is a feature assessment and analysis of dynamics and interrelations of key meteorological characteristics. This means that a computational kernel task manager and a set of 28 computational modules for basic statistical analysis and calculation of climate change indices selected by the Commission of Climatology, Climate Variability and Predictability Project, and the Joint WMO–IOC Technical Commission for Oceanography and Marine Meteorology (CCl/CLIVAR/JCOMM) Expert Team on Climate Change Detection and Indices (ETCCDI) (http://cccma.seos.uvic.ca/ETCCDMI/indices.shtml) were developed (cf. an example in Fig. 2.2). Finally, kernel graphical modules for visualization of processing results, writing them into Encapsulated Postscript, GeoTIFF and ESRI Shapefile formats, and a special module are ready to provide the user with cartographical legends by corresponding WMS request. GIS functionality integrated into the CLEARS web portal forms a basis for its further development.

2.2.4 Summary and Outlook

The information-computational system CLEARS outlined above has been used in Siberia Integrated Regional Study (SIRS, Gordov and Vaganov 2010) aimed at investigation of Global Change impact on the region and its interrelations with the Global Earth system. First results can be found in Gordov et al. (2011a, b). However, recent progress in meteorological modeling and data assimilation techniques has led to the appearance of new Reanalysis projects, which generate huge datasets of

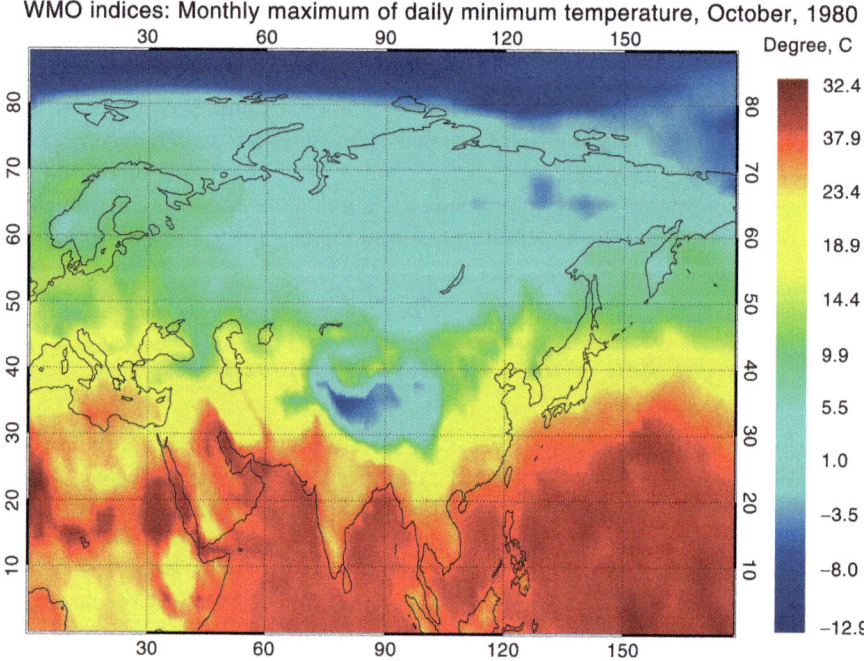

Fig. 2.2 Climate change index "monthly maximum of daily minimum temperature," October, 1980

high-resolution meteorological fields (cf., Saha et al. 2010). Since volume of these global datasets is near 100 Tb, traditional in-house processing of those is becoming obsolete and one has to rely upon only thematic information-computational web systems employing relevant distributed data storages. It should be mentioned that at the time of this chapter preparation, such distributed Data Center for Siberia was only at the planning stage.

2.3 Development of the Land Information System for Russia: A Methodology for Integrating Statistics, Remote Sensing, In Situ Information, and Models

2.3.1 Objectives and Rationale

Required in biospheric studies across the globe, land-cover information provides the foundation for environmental monitoring and diverse ecosystem studies (e.g., land-use competition, food security). In recent years, major advances have taken place and researchers now have several global 1-km products available to choose from, each with approximately 20 thematic classes (McCallum et al. 2006). Additionally, a 500-m global product exists from the Moderate Resolution Imaging Spectro-radiometer (MODIS) (Friedl et al. 2002), and a 300-m global product (GlobCover) from the Medium Resolution Imaging Spectrometer (MERIS) has recently been made available (Bicheron et al. 2008). Besides these global datasets, continental and regional products exist around the globe (Bartalev et al. 2003). However, for many applications at regional and continental scales, especially those relating to biogeochemical cycles, these products in their current form are limited. In particu-lar, practically, all existing land-cover datasets temporally and thematically inhibit progress in accurate modeling of the terrestrial biosphere. Temporally, the majority of datasets are restricted to a certain time period with no updates planned (MODIS being the exception). Thematically, the lack of detail among the land-cover classes identified prevents the use of valuable inventory and statistical data. Tree species, in particular, are difficult to classify from remote sensing data alone and therefore the majority of land-cover products produced to date stop short of delineating tree species and related attributes. However, forest science has accumulated many data and semiempirical models of forest growth and productivity. These models contain such attributes as tree species, age, site index, etc., but cannot be readily combined with remote sensing products.

There are evident difficulties in providing land-cover information for vast, mostly sparsely populated territories of Siberia. In spite of availability of new information tools and technologies, current knowledge of land cover and state of terrestrial eco-systems in this region is not satisfactory. Official statistics (like *State Land and Forest Accounts*) systematically report obsolete data of unknown certainty for such a large region of Russia as Siberia. At the same time, the region is represented by substantial area of rapid changes (burnt forest area, industrially transformed territo-ries, etc.). Such a situation leads to a need of development of information products which would satisfy a definite level of details and quality. The latter is defined by requirements of the problems which should be resolved. Here (and in Chap. 6) an information product is presented which has been developed based on requirements for a terrestrial ecosystems verified full carbon account at a national level (Shvidenko et al. 2010). The major idea used was to develop an Integrated Land Information System (ILIS) for the entire Russian Federation for 2009, which would combine all relevant information sources of information (Fig. 2.3). The ILIS includes all avail-able cartographical materials (digitized maps of vegetation, land use–land cover, soil, landscapes, administrative maps, etc.), data of different inventories and surveys (*State Land Account, State Forest Account*, fire maps, etc.), diverse databases of *in situ* measurements (live biomass, net primary production, heterotrophic respiration, and many others), spatially distributed climate data, different remote sensing products, and numerous auxiliary models for assessment of relevant biophysical indicators of ecosystems. All this information is organized around a hybrid land cover at the resolution 1 km².

Development of a methodology to create a hybrid land-cover dataset for large territory as an information base for a terrestrial biota full carbon account necessi-tates the detailed quantification of land classes (e.g., for forests – dominant species, age, growing stock, net primary production, etc.). The major idea of the hybrid

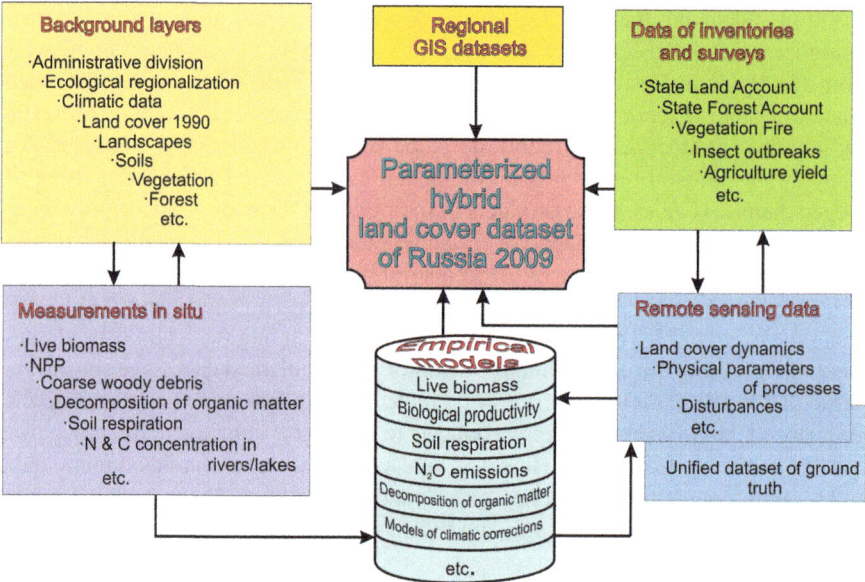

Fig. 2.3 Structure of integrated land information system

dataset approach involves integration of all relevant information to explore syner-gies, in particular the merging and harmonization of land and forest inventories, ecological monitoring data, remote sensing data, and in situ information. The base year was adopted as 2009 owing to available data sources. Additionally, this meth-odology seems suitable for application at the global level, dependent upon the avail-ability of required input data.

2.3.2   Datasets

Initial efforts involved a survey of the available data to determine which datasets would be suitable for inclusion in this process (Table 2.1). All spatial datasets collected for the study were processed in a Geographic Information System (GIS) and converted to 1-km raster resolution.

Remote sensing datasets: Two land-cover datasets were basically used: the global land cover 2000 (Bartholome and Belward 2005) and MODIS land cover. GLC2000 seems to be the most reliable product produced by regional experts (Bartalev et al. 2003). MODIS on the other hand provides the most up-to-date information. The MODIS Vegetation Continuous Fields (VCF) product is an annual representation of percent tree, herbaceous/shrublands and barren cover for each pixel (Hansen et al. 2002). The VCF provides the necessary flexibility, allowing us to prioritize the assignment of statistical data to land-cover data. The MODIS net primary production

Table 2.1 Datasets used in creation of the hybrid Russian land-cover dataset

Dataset	Resolution	Date	Reference
Remote sensing			
Global land cover (GLC2000)	1 km	2000	Bartholome and Belward (2005)
MODIS land cover	300 m	2008	Friedl et al. (2002)
Vegetation continuous fields (forest and herbaceous)	500 m	2008	Hansen et al. (2002)
MODIS NPP		2008	Running et al. (2004)
Vegetation fire (AVHRR and LANDSAT)	35 m and 1 km	2000–2009	Sukhinin (2008)
GIS			
Soil	1:2.5 Mil	1988	Dokuchaev Soil Science Institute, Moscow
Administrative regions	1:2.5 Mil	2009	Stolbovoi and McCallum (2002) updated
Forest enterprises	1:2.5 Mil	2009	IIASA in-house data base
Vegetation	1:4 Mil	1990	Stolbovoi and McCallum (2002)
Bioclimatic zones	1:4 Mil	1990	Stolbovoi and McCallum (2002)
Rivers/lakes and roads/ railways	1:1 Mil	1990	IIASA in-house data base
Statistics			
State land account	81 Regions	2009	FACRE'RF (2009)
State forest account	1585 Forest Enterprises	2003–2009	Russian Forest Information Service Roslesinforg (http:// www.roslesinforg.ru)
Disturbances in forests	81 regions	1991–2009	FFS'RF (2009)

(NPP) product (Running et al. 2004) presents up-to-date estimation of vegetation productivity and was used as an auxiliary information for design of hybrid land cover. Natural disturbance plays a large role in shaping the landscape of Northern Eurasia. In particular, wildfire is responsible for large areas of annual land-cover change and needs to be included in such a dataset. Wildfire data were acquired based on the Advanced Very High Resolution Radiometer (AVHRR) (hot spots) with control of burnt area by the LANDSAT Thematic Mapper (Sukhinin 2008). The dataset contains burnt area and the date of fire for each 1-km^2 pixel. These data are assumed substantially more reliable than official fire statistics (Shvidenko and Goldammer 2001).

GIS datasets. The first key GIS dataset used to assign data from different statistical inventories – the administrative region coverage – contains 81 regions. The original dataset dates from 1993 (Stolbovoi and McCallum 2002) and was updated for year 2009. The second key dataset in the assignment of statistics is the Forest Enterprise dataset for year 2005. This was created at IIASA with the aid of hardcopy and digital products, with a total of about 1,600 polygons.

A *Soil database* was one of the key components for selecting the appropriate area of arable land and wetlands. It contains a total of 292 unique soil types across the country with 21,988 polygons. The digitized soil map was developed by

V.V. Dokuchaev Soil Science Institute (Moscow) in 1996 based on a hard copy of the soil map of Russia, 1:2.5 million scale (Fridland 1989).

A *Vegetation dataset* was also utilized to provide broad vegetation classes and bioclimatic zones (derived from the dataset titled Vegetation of the former USSR), produced at a scale of 1:4 million (Stolbovoi and McCallum 2002). The dataset includes georeferencing of 101 vegetation classes (e.g., "spruce, fir-spruce and spruce-fir forest with mosaic grass-low bush and grass-spruce cover" or "northern semishrub and bunchgrass steppe"). Bioclimatic Zones were derived from the vegetation database. A total of 8 zones (polar desert; tundra; forest–tundra, northern and sparse taiga; middle taiga; southern taiga; temperate forests; steppe; deserts and semideserts) were identified.

In order to account for data not captured in the above datasets due to small areas of individual polygons, but which could have substantial areas aggregated by administrative regions (e.g., linear features, small water bodies, harvested areas), we relied on the Russian 1:1 million Planimetric dataset to account for "Virtual polygons." This dataset is based on original cartographic work from the State Committee for Geology and Cartography, USSR, and other sources. All existing railway lines and roads were buffered with 15 m, creating 30-m-wide polygons. Due to the enormous number of small rivers, only rivers >10 km in length were considered and buffered 20 m to create 40-m-wide polygons. Lakes smaller in size than 400 ha were also included. All of these *virtual polygons* (not presented on the map, but taken into account for the area balance and further calculations) were then tabulated per administrative region.

Data of different inventories and statistics: The State Forest Account (SFA, currently named Forest Cadaster) is the only source of aggregated forest inventory data for Russia. The last available accounts (http://www.roslesinforg.ru) date from 2003–2008. It contains statistics for approximately 1,600 forest enterprises. The SFA data contains areas and growing stock by dominant forest species distributed by age, site index, and relative stocking. There are approximately 50 sets of records for each enterprise on average. More information is available online (Shvidenko et al. 2007). The State Land Account (SLA) is provided annually by the State Committee of Land Resources of Russia based on land statistics. Originally, the SLA is provided by administrative districts (about 3,000 for Russia) and contains areas by (approximately 50) land classes. Publicly available data are by administrative regions. The SLA contains a two-dimensional official Russian land-use and land-cover hierarchical classification. The latter includes 7 primary land-use categories: (1) agricultural lands; (2) populated areas; (3) lands for industry, energy, transport, communications, aerospace activities, defense, etc.; (4) special protective territories; (5) forest fund; (6) water fund; and (7) state reserves. Land-cover classes are defined by their dominant use and are based on natural and historical characteristics. They include agricultural classes (arable, fallow, forage production (hayfields and pastures), and perennial vegetation) and nonagricultural land classes (lands under surface water including bogs; forest lands and lands under tree and shrub vegetation; built-up land; lands under roads; disturbed land – mining operations, earthmoving, etc.; and other land – ravines, sand, dumps, etc.) (FACRE'RF 2009).

Table 2.2 Set of parameters for comparison statistics and spatial information

Parameter name	Statistics (s)	Remote sensing/GIS (t)
Land cover	Land use, tree species	GLC2000; MODIS land cover
Tree stocking	Relative stocking	VCF trees
Site quality	Site index	Zone; soil
NPP	Ground NPP	MODIS NPP

2.3.3   Methodology of Land-Cover Assignment

This assignment was performed on a per-pixel (1-km) basis across the entire country. The distribution of land surface by land classes is provided based upon the relevant combination of remote sensing products, GIS data, and statistics from different sources, applying the general principle that the most accurate and updated information has priority in assignment. The quantitative correspondence of statistics (forest and land account) and spatial (remote sensing, GIS) data – agreement index (S_{ts}) for each pixel-pair (grid of territory (t) and statistics record (s)) within the territory unit (forest enterprise, administrative region) – were calculated:

$$S_{ts} = \frac{1}{q} \left(\sum_{j=1}^{q} (x_{tj}^{norm} - x_{sj}^{norm})^2 \right)^{1/2}, \text{ where } q - \text{number of parameter (Table 2.2)};$$

$x_{tj}^{norm}, x_{sj}^{norm}$ – normalized value of parameter j for territory pixel t and j; and

$$x_{j}^{norm} = \frac{x_j - x_{j\min}}{x_{j\max} - x_{j\min}}, \text{ where } x_{j\max}, x_{j\min} \text{ – maximum and minimum values of}$$

parameter j within the certain area (forest enterprise, administrative unit).

Data on a nominal scale, i.e., GLC land-cover classes, were ranked with respect to a certain vegetation class in the statistics. For example, the most suitable GLC class for pine forest is "Tree Cover, Needle-leaved, Evergreen"; thus, this would receive the highest rank. Pine forest could also fit into "Tree Cover, Mixed Leaf Type" or might be found in the "Tree Cover, Broadleaved, Deciduous" GLC class (and would receive a lower rank). In accordance with SFA instructions, mixed forest is considered as coniferous if the coniferous growing stock is more than 50 %. The resultant agreement index S varies from 0 to 1. It can be interpreted as a distance between objects (grid of territory and statistics record) within the space of parameters. The lower the index value, the higher agreement of initial datasets and more suitable is the current piece of territory for the given statistical data. This method corresponds to the methods of land suitability assessment (Dokuchaev 1951) accepted in 1976 by Food and Agriculture Organization (FAO).

The final stage involved the optimization of distribution statistics data on the territory based on the agreement index results. Each forest and land account record in the statistics was assigned to the most suitable grid within each forest enterprise.

The major land categories of land-cover assignment include *forest, agriculture, wetlands, open woodland, burnt area, shrub and grassland, water,* and *unproductive*. Detailed description of the assignment can be found at the IILC web site (www.iiasa.ac.at/Research/FOR/hlc).

2.3.4   Results and Discussion

Application of the methodology applied has resulted in the new hybrid land-cover/land-use map of Siberia at 1-km resolution (Fig. 1.1). A total of six major land-cover types were identified, namely, forest, agriculture, wetlands, shrubs/grasses, water, and unproductive land. These are further subdivided into the following classes: forest – each grid links to the SFA database (the SFA data contains areas and growing stock by dominant forest species distributed by age, site index, and relative stocking) containing 78,639 records; agriculture – 6 classes, parameterized by 81 administrative units; wetlands – 8 classes, parameterized by 83 zones/regions; and shrub/grassland – 58 classes, parameterized by 321 zones/regions.

A principal problem of merging and harmonizing substantially different datasets deals with the compatibility of definitions and classification schemes used. In Russia, many important definitions do not correspond to those agreed internationally, and inconsistencies between national definitions of different datasets (e.g., *State Forest Account* and *State Land Account*) are common. Additionally, in situ indicators (e.g., by forest inventory) and those delivered by remote sensing are often not compatible. Therefore, decision rules and regional empirical models were used for harmonizing the definitions if relevant. Evaluation of the Russian hybrid land-cover dataset is rather difficult, owing to the fact that the majority of available material of relevance has been used in the creation of the dataset itself. In addition, the resultant coverage is highly detailed (i.e., forest species at 1-km resolution), making comparison to existing coarser datasets somewhat irrelevant. Therefore, the levels of confidence in the assignments were based on the assessed agreement between the input datasets (Schepaschenko et al. 2010). It was found that substantial part of the country (52 % of area) infers a high degree of confidence in the agreement among the remote sensing products and the in situ statistics. A vegetated GLC class, combined with maximum VCF classes, was matched with corresponding statistical data. Another 42 % of the territory has lower agreement between initial datasets. Coniferous forest could mix with deciduous, or forest with shrubs. For example, we assign "forest" to the GLC class "shrub cover" (not forest, but still vegetation) with VCF trees exceeding 50 %. High disagreement has only about 5 % of area. This is mostly sparse vegetation with typical forms of disagreement such as wetland–grassland in West Siberia and forest–tundra ecotones in East Siberia and the Far East. In GLC, it is indicated as "bare areas", but VCF shows herbaceous cover in the range of 61–98 %. In accordance with the SLA data, we assigned "pasture" to this area. This class also appears in mountains and on the outskirts of large cities. Forest statistics incompleteness appears on about 1 % of area. The remote sensing products report there forest area, but the SFA contains far fewer forests. Most of such areas in Siberia are in the north (outside of managed forest area). Some information on detailed description of development of diverse layer of biophysical and other indicators included in the ILIS (live biomes, NPP, HR, coarse woody debris among many others) can be found in Shvidenko and Nilsson (2003) and Shvidenko et al. (2005, 2007, 2008, 2010).

2.3.5 Summary and Outlook

There is a critical need for accurate land-cover information for resource assessment, biophysical modeling, greenhouse gas studies, and for the estimation of possible terrestrial responses and feedbacks to climate change (Frey and Smith 2007). Over the past two decades, evidence has accumulated of significant contributions of extratropical Northern Hemisphere land areas to the global uptake of anthropogenic CO_2 (Schimel et al. 2001). Without accurate baseline measures of crucial datasets such as land use and land cover, we will have little hope of monitoring the effects of global change on vegetation over large regions. A variety of global information sets exist with data on land use and land cover, although none of these alone satisfy the requirements of the above mentioned aims. The IIASA study (cf., Fig. 1.1) demonstrated the ability to produce a highly detailed (both spatially and thematically) land-cover/land-use dataset over Russia – the largest country on the planet – by combining existing datasets into a hybrid information system. The new land-cover product uses the advantages of all sources, supplies up-to-date geographically explicit and well-parameterized information, thus allowing for reduction of source's uncertainty.

The IIASA approach briefly described above is flexible and allows the inclusion of additional existing or newly created datasets in the future (i.e., elevation, lidar biomass, and more). The main advantage of the methodology is the ability to link on-ground data and models to the remote sensing products. Adoption of such techniques by FAO might provide a useful enhancement to their provision of global forest statistics.

2.4 Russian Baseline Datasets for Climatic Studies Over Siberia

2.4.1 Introduction

In this section, the system of acquisition and storage of in situ meteorological observations over the Russian Federation is briefly described. It changed several times for the period of its existence primarily due to the introduction of new methods of

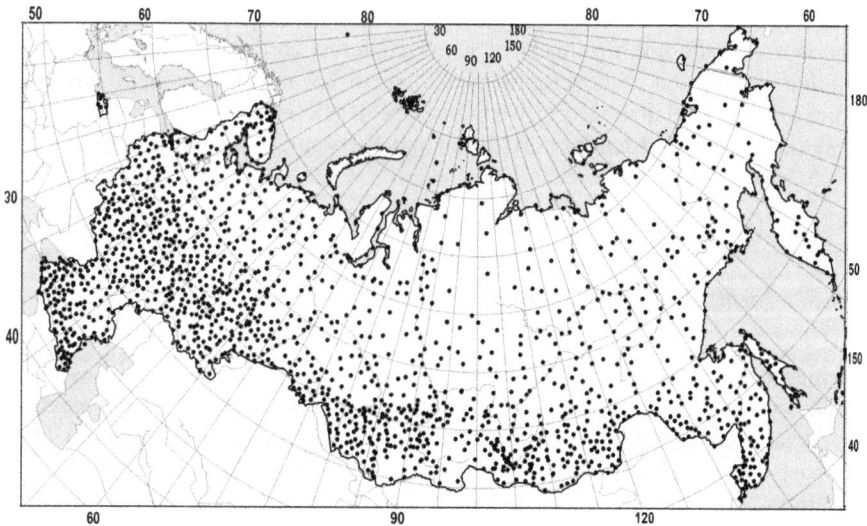

Fig. 2.4 Network of meteorological stations over the Russian Federation in 2010 (Sherstukov et al. 2007, updated)

automatic data processing and the improved methods of data storage. Historically, transfer of data from an out-of-date carrier to more advanced medium occasionally leads to data omission, and the problem of data recovery still exists and receives much attention. Currently, the Russian State Data Fund stores its data on present-day carriers (large-volume data cartridges) and regularly updates the data carriers. Meteorological in situ observations comprise about 80 % of the Fund archives. These are of prime interest for scientific and application research, including that of regional climate change over Northern Eurasia.

In Russia, regular large-scale meteorological observations started in 1881. However, these observations were conducted even earlier at individual sites. For example, the first meteorological observations in St. Petersburg were performed in 1734. No more than 100 stations are available in Russia whose observation series are as long as 120 years and over. The network of Russian meteorological stations is quite dense, except for hard-to-reach regions in Siberia and circumpolar areas (Fig. 2.4). At present, 1,627 points of meteorological observations are in operation. A benchmark climatic network contains 454 stations. These are generally long-series representative stations with a complete observation program that covers a meteorologically homogeneous territory. The number of full-size active meteorological stations in Russia changed in time with the maximum of their number during the 1961–1980 period (about 2,300 stations). In the 1990s, the number of active stations decreased substantially. Since 1966, observations at meteorological stations have been conducted every 3 h. All meteorological elements included in the Global Climate Observation System (GCOS) are measured. Gradually, the content

Table 2.3 List of meteorological elements included in datasets

Three-hour datasets	Daily datasets
Air temperature	Mean daily air temperature
Water vapor pressure	Maximum air temperature
Dew-point temperature	Minimum air temperature
Relative humidity	Daily precipitation
Sea level pressure	Snow depth
Station level pressure	Snow coverage
Air pressure tendency	Characteristics of site
Visibility	Minimum of relative humidity
Total and low cloud amount	Minimum of surface temperature
Cloud genera for each cloud layer	Wind speed maximum
Height of cloud base	Atmospheric phenomena
Wind speed and wind direction at 10 above the ground	Atmospheric phenomena duration
Precipitation	Daily total and low cloud amount
Present and past weather information	Sunshine duration
Surface temperature and state of the ground	
Atmospheric phenomena	

of the datasets held in the State Data Fund has increased. For example, since 1977, measurements of soil temperatures at different depths have been archived, and since 1984, the same has been done with automatic recorder data.

2.4.2 Data and Access

In situ meteorological elements observed at the Russian meteorological stations are listed in Table 2.3. A detailed description of measurement techniques and changes in the appropriate procedures for each of the stations can be found in the documents held at meteorological stations, regional administrations of the Hydrometeorological Service and RIHMI-WDC.

Special attention at the State Data Fund is given to the preparation of the surface air temperature and atmospheric precipitation data. At each meteorological station, surface air temperature is measured at an observation hour. Additionally, daily maximum and minimum temperatures are determined. In the 1891–1935 period, measurements were made three times a day, at 07.00 a.m., 01.00 p.m., and 09.00 p.m., local astronomic time. In the 1936–1965 period, a nocturnal observation hour was added and measurements started to be made four times a day, at 01.00 a.m., 07.00 a.m., 01.00 p.m., and 07.00 p.m., local astronomic time. From 1966 up to the present, measurements have been made every 3 h, Moscow decree time (since 1993, Greenwich Mean Time). These changes of time are to be taken into account in analyses of climatic changes over the nation (cf., Brown 2000).

A procedure of atmospheric precipitation measurement was also modified several times. Primarily, it was connected with the change in the number of precipitation observations during the day, the rain gauge design, and the observational practice (cf., Groisman and Rankova 2001). Precipitation was generally measured twice a day, in the morning and evening observation hours. For a number of stations in European Russia, however, precipitation was measured every 6 h. The design of the rain gauge and observing practices also changed more than once. The most significant change occurred in or around 1953 and in 1966. The users must be aware of these inhomogeneities in observed precipitation data if they are going to study climatic changes of this element.

The main objective of the State Data Fund is to archive, preserve, and disseminate observed geophysical data. Upon user request, RIHMI-WDC is providing information on meteorological conditions recorded at observation hours by national meteorological stations. Currently, the data archiving is made through the automated data collection system developed by RIHMI-WDC. Monthly observation data packages arrive at RIHMI-WDC from each meteorological station in the 4-month delayed mode due to a necessity of preliminary data processing and quality control in regional meteorological centers of the Russian Hydrometeorological Service. The data archiving system is designed to keep data in the formats in which they arrived at the State Data Fund throughout the time of its existence. The Fund's datasets include every change in data processing and observation procedures, changes in data formats, etc. When necessary, the Fund's data can be reprocessed on demand into specialized baseline datasets for various scientific and applied problems, for example, to study climate change.

The major requirement for meteorological data in the NEESPI framework is data presentation in the form of time series for main meteorological elements. Special attention is to be given to analyzing the homogeneity of data series. Broken homogeneity of data, which may be caused by station relocation, change of instruments, and modification of data processing or observation procedure, can lead to false information on climate change. At present, by using the Fund's data, RIHMI-WDC is actively involved in creating specialized datasets for climate research. Prepared are datasets containing mean monthly values of air temperature, atmospheric precipitation, air pressure, water vapor pressure, and sunshine duration from 518 stations that are rather equally located over Russia. In particular, daily-resolution data on air temperature and precipitation, as well as soil temperature at standard depths and snow characteristics (snow depth and amount of snow covering the near-station area) are collected and are available via the RIHMI-WDC web service (Razuvaev and Bulygina 2008; Bulygina et al. 2009). All the datasets in the ASCII format can be found on the RIHMI-WDC site (http://meteo.ru/english/climate/cl_data.php). Relevant descriptions contain information on data sources, quality control methods used, and essential metadata (coordinates, elevation relocations, and instrumentation). Specialized technology on the RIHMI-WDC server (Veselov et al. 2000) provides access to a dataset, data retrieval for the stations of users' interest, data scanning, and data copying. Such specialized datasets are constantly being created on demand. Access to specialized climatic datasets for a worldwide international research community allows new important information on the state of the Global Earth system (Bulygina et al. 2007). Therefore, the prepared datasets have been already widely used in climate monitoring, climate change research, and analysis of extreme climatic events.

2.4.3 Summary

The above description of the Russian baseline meteorological datasets shows that, firstly, those should be converted from the data archiving format to the climate data format (time series of data for individual stations) and relevant metadata sets should be created. These datasets have been and will be prepared by RIHMI-WDC for broad dissemination among the international research community (Sherstukov et al. 2007; Razuvaev and Bulygina 2009).

2.5 The Siberian Earth System Science Cluster: A Data Discovery, Access, and Analysis System for Siberia

2.5.1 Objectives and Rationale

There are currently 137 active satellite missions attributed to Earth science, Earth observation, remote sensing, or meteorology (only civil, commercial, governmental use) collecting terabytes of data daily (USC 2010). Although the ability to collect and store vast amounts of data has been improved considerably thanks to the general development in information technology, the capabilities to process, analyze, and exploit these data volumes are lacking behind, and scientists need further assistance from data mining and visualization tools (Compieta et al. 2007; Hey and Trefethen 2005). There are a number of national and international initiatives where researchers and computer scientists collaborate under the e-Science or Digital Earth paradigm to develop the required tools and services as part of cyberinfrastructures, e-Infrastructures, spatial data infrastructures, GEOSS, or the grid (Baru et al. 2009; Craglia et al. 2008; Foresman 2008; GEO 2005; Hey and Trefethen 2005; Kindermann et al. 2007; Yang and Raskin 2009). It should be noted that these terms are not synonymous because of different applications, scale, actors, and target users (Craglia et al. 2008), although they still share some common goals (e.g., to provide tools for data discovery and access) and face similar challenges (e.g., large data volume).

The objective of the Siberian Earth System Science Cluster (SIB-ESS-C) established at the Friedrich Schiller University (Jena, Germany) is to facilitate environmental research and Earth system science in the Asian part of Russia using remote sensing products. SIB-ESS-C forms an element of the web-based infrastructure for data generation, dissemination, archiving, and analysis of remote sensing products over this part of the globe. It is focused on Earth observation data and services for scientific users but also for decision makers and the general public enabling its users to search, access, evaluate, and analyze distributed data resources. The aim of this section is to provide an overview of the concept of SIB-ESS-C in the context of spatial data infrastructures utilized in scientific research. The description is addressing potential users and collaborators describing the capabilities and services of SIB-ESS-C.

2.5.2   Cluster Architecture and Data Discovery Services

The overall design philosophy of SIB-ESS-C follows three major principles: (a) adhere to standards to ensure interoperability, (b) utilize components that are well established in the Earth science, Earth observation, and GIS communities, and (c) implement free and open source software components whenever possible. The first and second principle are in-line with the GEOSS (Global Earth Observation System of Systems) approach to establish a "system of systems" consisting of distributed but interconnected Earth observation systems (GEO 2005). SIB-ESS-C is developed as one node in such a global framework. Consequently, SIB-ESS-C resources and components are interoperable and registered at the GEOSS portal to raise awareness and foster its widespread use.

The technical implementation of SIB-ESS-C follows a service-oriented architecture (SOA) approach, and the components for data discovery, access, and analysis are implemented as web services. These services as well as the communication between them are based on international standards and specifications by the Open Geospatial Consortium (OGC), the International Organization for Standardization (ISO), or the World Wide Web Consortiums (W3C). Adhering to these standards ensures interoperability and boosts exchange with similar systems by broadening the availability and accessibility of data and applications. Data access through the SIB-ESS-C Portal is provided free of charge, but data users are requested to credit the respective data creator and cite the appropriate publications. In its current development stage, the SIB-ESS-C architecture comprises components for data discovery, data access, and data analysis, as well as a web portal for direct user interaction (Fig. 2.5). Further details on the individual components are given in the sections below.

The SIB-ESS-C infrastructure has been designed to act as a data provider including its own catalogue and respective user interface for data dissemination and discovery. The key component of the data discovery use case is the catalogue at the service tier capable of publishing metadata but also functioning as a mediator between different metadata models on both the server and the client side. This additional service tier provides flexibility and interoperability in heterogeneous environments customary to interdisciplinary research. The SIB-ESS-C catalogue implementation builds on GI-cat, an active open source software project (Bigagli et al. 2004; Nativi et al. 2009). It enables users to perform queries on external catalogues and in turn allows other registries to harvest information of all services federated into GI-cat. On the

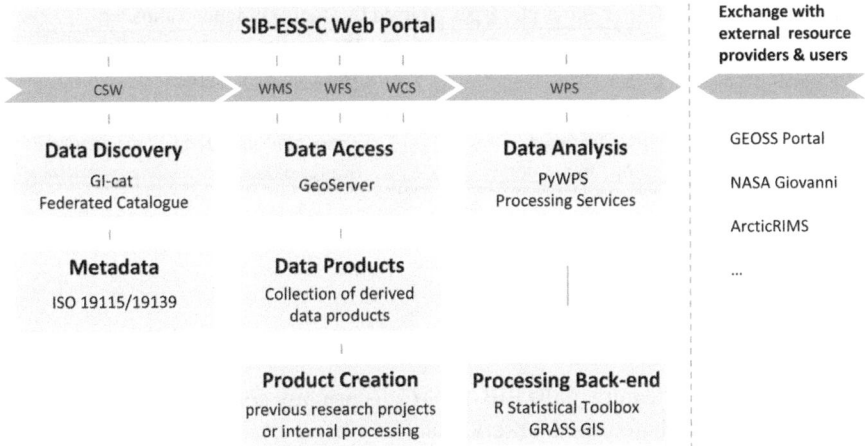

Fig. 2.5 SIB-ESS-C architecture, its components, and interfaces

client side, the SIB-ESS-C GI-cat instance provides a standard OGC CSW/ISO interface (OGC 2007a) plus an extended version specifically targeted to the SIB-ESS-C web portal.

2.5.3 Data Access Services

The SIB-ESS-C infrastructure provides access to data products created during a number of research projects over the last decade derived from multiple Earth observation sensors (e.g., land cover, biomass, forest disturbances, water bodies, phenology), resulted from model experiments (e.g., carbon emissions from forest fires), and supplementary datasets (e.g., "Land Resources of Russia"; Stolbovoi and McCallum 2002). A detailed description of these datasets is available at http://www.sibessc.uni-jena.de/.

The SIB-ESS-C infrastructure offers two different methods for data access. First, each single dataset, along with supplemental information, is packaged into a compressed archive (gzip) which can be downloaded individually through the SIB-ESS-C web portal in the original format chosen by the data creator. The second method, following the service-oriented architecture (SOA) approach, introduces a service layer between the client and the data level and enables clients to retrieve portions of a server's data holdings based on spatial constraints and other criteria. Thus, users retrieve only data they really need and not files of arbitrary size defined by a provider. Within this second method, SIB-ESS-C offers three different access services: through Web Coverage Services (OGC 2005, 2007b), through Web Feature Services (OGC 2005b, 2006), and/or as Web Map Services (OGC 2002). To provide these three services, an instance of GeoServer (http://geoserver.org/) has been implemented. The data served by GeoServer had been further processed to improve performance of the web services (e.g., reformatting, reprojecting, tiling, and overviews).

Table 2.4 SIB-ESS-C time-series plot types grouped by number of input parameters

Single parameter plots	Dual parameter plots
2D time-averaged grid	2D difference grid
Time series, area averaged	2D zonal statistics grid
Seasonal times series, area averaged	Time-series difference, area averaged
Time-series decomposition, area averaged	Scatter plots
Histogram	

2.5.4 Data Analysis Services

This is an advanced feature of the SIB-ESS-C. Its primary goal is the spatiotemporal exploration of distributed Earth observation time-series data. Spatially, Earth observation data consists of a multitude of single measurements leading to a large spatial coverage and high-information density. Visualizing and exploring the spatial and temporal characteristics of these remote sensing time-series datasets typically requires aggregation in time, space, or both. The SIB-ESS-C analysis tools contain both ways of aggregation depending on the visualization type selected. These can be two-dimensional (2D), map-like visualizations of a user specified area of interest that are averaged across the temporal range selected. Additionally, SIB-ESS-C offers visualization along a timeline that is averaged in space according to the user-defined area of interest. A list of plot types available from SIB-ESS-C is given in Table 2.4 along with a few sample plots in Fig. 2.6 selected to demonstrate some of the capabilities of these Data Analysis Services.

The implementation of the analysis component adopts the overall design strategy of the SIB-ESS-C infrastructure by utilizing standards, open source technology, and well-established tools and methods. The creation of graphs is based on a client-server architecture using standard compliant interfaces for interaction. The selection of the input data sources by the user is supported by the web portal and the catalogue search functionality implemented therein. After successful execution, the service returns a graphics file to the portal. For time-series plots also, an XML file with actual data values can be retrieved, allowing the clients to perform customized output rendering as needed.

2.5.5 SIB-ESS-C Web Portal

SIB-ESS-C contains a web-based client for human interaction with the web services described earlier in this chapter. It is the central access point to all services provided by SIB-ESS-C servers but also integrates federated services from external resource providers if those services comply with OGC standards. The web portal offers functionalities for data search, data access, data visualization, and data analysis. The default interface of the web portal shows a map view on the central panel

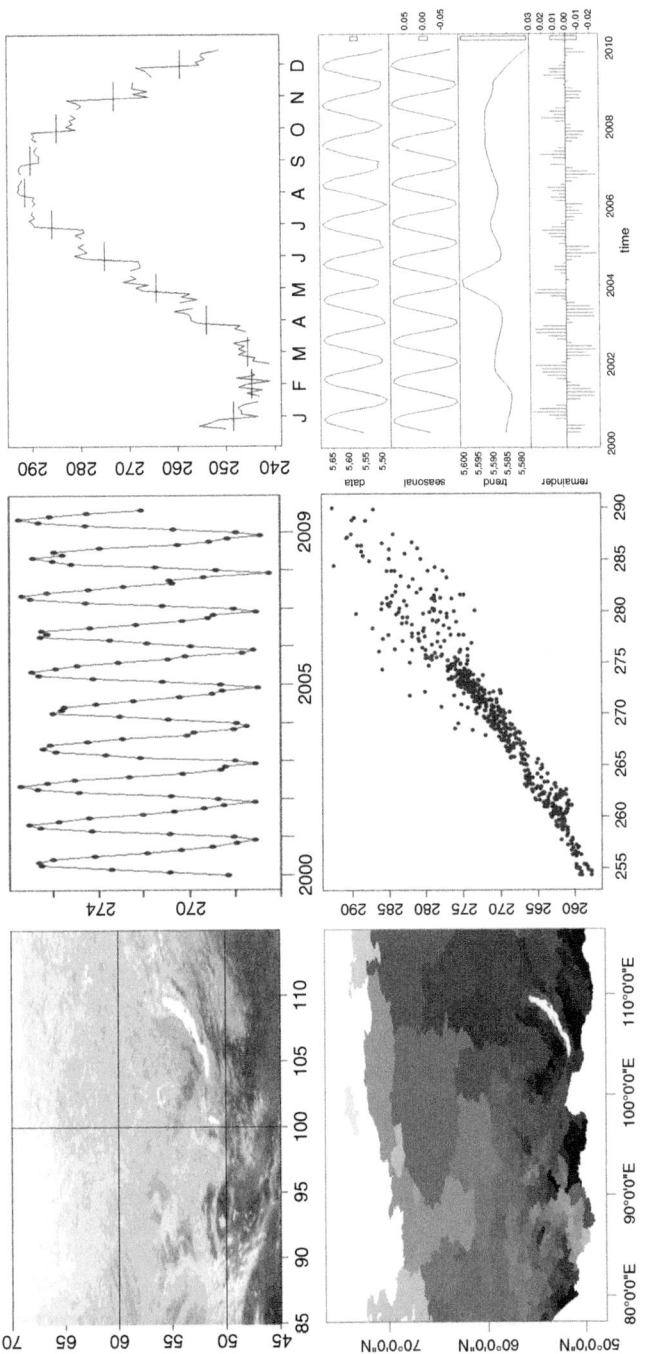

Fig. 2.6 Samples of SIB-ESS-C data analysis plots

collocated by a layer tree on the left and legend panel on the right. On top of the map panel, common navigation controls (e.g., zoom, pan) as well as functions for feature selection and retrieval of feature attributes are available. The Google Maps API (http://code.google.com/intl/de-DE/apis/maps/index.html) has been integrated as a background layer because of its rich content (satellite imagery, road/rail network, place names, etc.), the free and straightforward implementation, and the well-designed cartographic presentation. These advantages outweigh the drawback of the typical northern latitude distortions introduced by the Mercator projection used by Google. All other layers added to the map are reprojected to this coordinate system. Besides the map view, users may form a search request specified by entering keywords, a time span, or a spatial extent. The interactive nature of the web portal also allows drawing a selection box in the map view and transferring the coordinates to the search form. Other actions include direct data access (e.g., download) and adding datasets to the map view for further analysis. An advanced feature of the web portal is the timeline view of the search results, which is particularly useful for users seeking time-series data. The portal integrates standard compliant web services, has a modular, extendible architecture, and builds on existing free and open source toolkits. All these tools and features have an objective of optimizing the user interaction with SIB-ESS-C web services.

2.5.6   Summary and Outlook

The above conceptual overview of the Siberian Earth System Science Cluster and the described capabilities and services available shows that it is quite a powerful tool yet facilitating distributed multidisciplinary research in Siberia. With its standard OGC web services for data discovery, data access, and data analysis, the system can be integrated into a broader context, such as GEOSS. Future developments of SIB-ESS-C will focus on improving performance and usability, extending data holdings by in-house processing of relevant Earth observation products, extending the analysis and visualization capabilities, web service orchestration, and interoperability experiments with similar systems and services.

2.6   Giovanni Data and Information System for the NEESPI Domain

2.6.1   Introduction

The Goddard Interactive Online Visualization ANd aNalysis Infrastructure (Giovanni) system (http://giovanni.gsfc.nasa.gov/) is an online visualization and analysis system developed by NASA at Goddard Earth Sciences Data and Information Services

Center (GES DISC). It provides a simple and intuitive way to visualize, analyze, and access vast amount of Earth science data without having to download the data (Acker and Leptoukh 2007; Berrick et al. 2009). The system consists of the following components: web interfaces, back end data processing software, image renders, and a portal generator. The portal generator is able to create customized Giovanni portals based on scientific needs by selecting desired analysis functions and parameters of one or more satellite instruments or numerical models from the Giovanni database. Giovanni can display images and provides capability for downloading images in PNG, and KML for Google Earth presentation as well as the corresponding data in several formats (ASCII, HDF, netCDF). The data processing and image generation are done on the fly according to selected options. Other features include:

- Data lineage which presents metadata and supporting documentation, i.e., data provenance, such as dataset's structure, parameters, format, spatial extent, fill values, temporal coverage, source data, and processing approach used in filtering, masking, regridding, aggregating, image rendering, and other Giovanni processing steps.
- Machine-to-machine gateway via Web Mapping Service (WMS) and Web Coverage Service (WCS) protocols. It can act as WMS or WCS, thus allowing any GIS clients to add layers or get subsetted data from Giovanni. In addition, the system can act as a client by accessing remotely located data via WCS or WMS.
- Input data formats: HDF-4, HDF-5, HDF-EOS, netCDF, GRIB, and binary.
- Access input data from local and various remote systems via ftp, OPeNDAP, and WCS protocols.

2.6.2   Giovanni-NEESPI

The Northern Eurasia Earth Science Partnership Initiative (NEESPI) Data Center at GES DISC is a NASA-funded project that focuses on collecting satellite remote sensing data and providing tools, information, and services in support of NEESPI scientific objectives (Leptoukh et al. 2007). The data can be accessed online through anonymous ftp, or the advanced data search and order system, called Mirador, which uses keywords to find data quickly in a Google-like interface. In addition to traditional data access tools, NASA NEESPI data center provides web-based online visualization and analysis tools. The data from different satellite sensors and numerical models have been processed and integrated into Giovanni system so that they can be used easily to perform online analysis, such as seasonal, interannual variations and events, as well as intercomparison or relationship studies of multiple parameters from different sensors or models.

The current Giovanni system consists of more than 35 Giovanni portals serving various missions, projects, and communities. Giovanni-NEESPI is a customized Giovanni portal that integrates atmospheric, land surface, and cryospheric data products from a number of sensors over Northern Eurasia in support of the NEESPI. The Giovanni-NEESPI allows analyzing a single parameter, comparing two similar parameters from different sources, and analyzing a relationship between two parameters. The available data analysis features are summarized as following:

(a) *Single parameter analysis*:

- Latitude–longitude area plots of time-averaged parameters
- Time-series plots of area-averaged parameters
- Latitude/longitude–time Hovmöller diagram
- Animations of consecutive latitude–longitude area plots

(b) *Multiparameter analysis*:

- Latitude–longitude area plots of overlain time-averaged parameters
- Time-series plots of multiple parameters
- Time series of two-parameter difference
- Latitude–longitude area plot of two-parameter differences
- Scatter plots with regression statistics
- Temporal correlation maps

Two Giovanni-NEESPI portals are in operation: monthly and daily. The monthly Giovanni-NEESPI contains monthly products from multiple sensors: the Moderate Resolution Imaging Spectroradiometer (MODIS) Terra; MODIS Aqua; the Advanced Microwave Scanning Radiometer-Earth Observing System (AMSR-E); the Atmospheric Infrared Sounder (AIRS); Global Precipitation Climatology Project (GPCP); the National Environmental Satellite, Data, and Information Service (NESDIS)/the Interactive Multisensor Snow and Ice Mapping System (IMS); and the Goddard Chemistry Aerosol Radiation and Transport (GOCART) model. The parameters in the monthly portal have been grouped into three: atmosphere, land surface, and cryosphere. The daily Giovanni-NEESPI portal contains daily products from MODIS Terra, MODIS Aqua, GPCP, AIRS, and GOCART model. Both portals contain products of $1° \times 1°$ horizontal resolution. Table 2.5 lists parameters, instrument name, temporal coverage, and status of products in Giovanni-NEESPI.

2.6.3 Summary

Giovanni at NASA GES DISC is a user-friendly online visualization and analysis system. The customized Giovanni portals for NEESPI enable one to explore satellite remote sensing data from multiple sensors and modeled data for Northern Eurasia by performing preliminary statistical analysis and conducting multidiscipline parameter relationship studies associated with land-cover/land-use changes and climate change. The website for NASA NEESPI Data Center and for Giovanni system can be found at http://disc.gsfc.nasa.gov/neespi and http://giovanni.gsfc.nasa.gov, respectively.

Table 2.5 Parameters provided by the Giovanni-NEESPI system

Group	Parameter name	Sensor/model name	Available since	Status Monthly	Daily
Atmosphere	Aerosol optical depth at 0.55 μm	MODIS-Terra	2000.02	OPS	OPS
		MODIS-Aqua	2002.07		
	Atmospheric water vapor (QA-weighted)	MODIS-Terra.	2000.02	OPS	OPS
		MODIS-Aqua	2002.07		
	Aerosol small mode fraction	MODIS-Terra,	2000.02	OPS	OPS
		MODIS-Aqua	2002.07		
	Cloud fraction	MODIS-Terra,	2000.02	OPS	OPS
		MODIS-Aqua	2002.07		
	Cloud optical depth	MODIS-Terra,	2000.02	OPS	OPS
		MODIS-Aqua	2002.07		
	Cloud effective radius	MODIS-Terra,	2000.02	OPS	OPS
		MODIS-Aqua	2002.07		
	Cloud top pressure	MODIS-Terra,	2000.02	OPS	OPS
		MODIS-Aqua	2002.07		
	Cloud top temperature	MODIS-Terra,	2000.02	OPS	OPS
		MODIS-Aqua	2002.07		
	Column amount ozone	Aura OMI	2004.08	N/A	OPS
	UV aerosol index	Aura OMI	2004.08	N/A	OPS
	GPCP precipitation	GPCP Derived	1979.01	OPS	OPS
	Dust aerosol column optical depth 550 nm (total, coarse, fine)	GOCART	2000.01	OPS	OPS
	Black carbon column optical depth	GOCART	2000.01	OPS	OPS
	Particulate organic matter column optical depth	GOCART	2000.01	OPS	OPS
	Sulfate aerosol column optical depth	GOCART	2000.01	OPS	OPS
Land surface	Cloud and overpass corrected fire pixel count	MODIS-Terra	2000.11	OPS	N/A
		MODIS-Aqua	2002.07		
	Overpass corrected fire pixel count	MODIS-Terra	2000.11	OPS	N/A
		MODIS-Aqua	2002.07		
	Mean cloud fraction over land for fire detection	MODIS-Terra	2000.11	OPS	N/A
		MODIS-Aqua	2002.07		
	Mean fire radiative power	MODIS-Terra	2000.11	OPS	N/A
		MODIS-Aqua	2002.07		
	Enhanced vegetation index (EVI)	MODIS-Terra	2000.02	OPS	N/A
		MODIS-Aqua	2002.07		
	Normalized difference vegetation index (NDVI)	MODIS-Terra	2000.02	OPS	N/A
		MODIS-Aqua	2002.07		
	Land surface temperature (daytime)	MODIS-Terra	2000.03	OPS	N/A
	Land surface temperature (nighttime)	MODIS-Terra	2000.03	OPS	N/A
	Soil moisture	AMSR-E	2002.10	OPS	N/A
	Surface air temperature	AIRS	2002.08	OPS	OPS
	Surface skin temperature	AIRS	2002.08	OPS	OPS
Cryosphere	Ice occurrence frequency	NESDIS/IMS	2000.01	OPS	N/A
	Snow occurrence frequency	NESDIS/IMS	2000.01	OPS	N/A

Note: *OPS* operational, *N/A* no data available

2.7 RIMS: An Integrated Mapping and Analysis System with Application to Siberia

2.7.1 Objectives and Rationale

The Rapid Integrated Mapping System (RIMS) has been developed at the University of New Hampshire to provide web interfacing and services to a wide range of geospatial data that relates to Earth system modeling, observation, and monitoring. It is predominantly a data-oriented system interfacing digital content of the datasets rather than their image representation similar to data viewing web tools such as Google Maps/Earth applications (Butler 2006). In addition to mapping, major web services of the RIMS for digital data content interfacing include (a) access to the raw data values for each pixel of the displayed map, (b) time-series maps and graphs, (c) access to the spatial- and temporal-aggregated subdatasets (DataCube), (d) data manipulation tools (the Data Calculator), (e) point/station data access, and (f) dataset search tools which are all described below. The system is designed to work with near-real-time data flow where new temporal data are generated by continuous operational computer clusters chained in data mining $\rightarrow$ processing $\rightarrow$ modeling $\rightarrow$ data product delivery in a similar manner as described in (Rodell et al. 2004). RIMS is used to serve gridded vector and point/station data pages on the NEESPI project web site (http://NEESPI.sr.unh.edu) at UNH for the research community, education, and public outreach. Some conceptual ideas of the system have been adopted from an earlier application of these concepts, ArcticRIMS (http://NEESPI.sr.unh.edu), which has been used for the study of water cycle in Arctic watersheds (Rawlins et al. 2010; Shiklomanov and Lammers 2009). The RIMS/NEESPI data pool is composed of a variety of climate, climate forecast, hydrology, hydrology forecast, remote sensing, human dimension, and other themes that presently comprise about 2,000 individual single layer (e.g., elevation) and time series (e.g., daily runoff) datasets, and 4 climate and hydrology station/point network datasets. In addition to NEESPI, the RIMS framework has been applied to other research projects such as Global Hydrology (Wisser et al. 2009), human impact on fresh water hydrology (Vorosmarty et al. 2010), Earth Atlas (http://earthatlas.sr.unh.edu), and others. These make RIMS an important resource for research and knowledge exploration in the field of Earth system science and for the application to the NEESPI domain, requiring the analysis of spatial data.

2.7.2 RIMS Software Design and Components

The conceptual software design chart is given on Fig. 2.7. Metadata of RIMS raw data holdings are composed in two local databases. The first represents a database of all gridded and vector polygon datasets in the system and the second a database

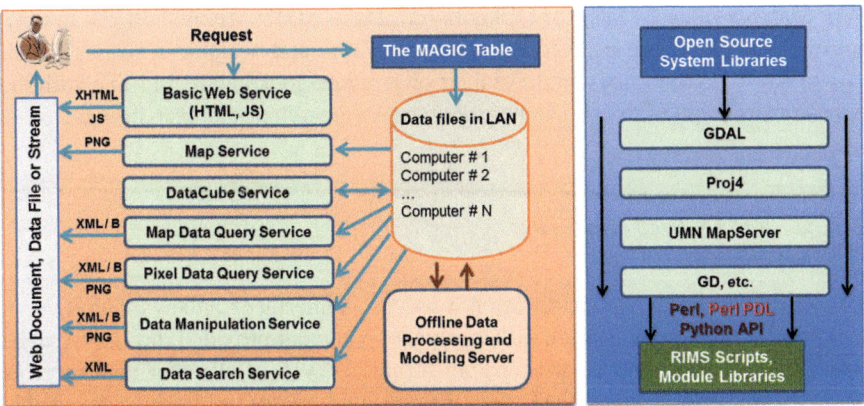

Fig. 2.7 Conceptual software design for the RIMS. A web user generates a sequence of requests to the RIMS which are evaluated, processed and assembled to a document, graphics, or data file utilizing a number of stand-alone services that use the same pool of raw data which, in turn, has all its metadata summarized in the Manipulation and Geographic Inquiry Control (MAGIC) Table

of point/station datasets/databases. These two Manipulation and Geographic Inquiry Control (MAGIC) Tables are essential to control all data flows within the RIMS. This includes all essential metadata for each dataset in the system to allow for manipulation and visualization on the fly without modification to the original data files. It is also a tool for straightforward and rapid mounting of new datasets to the system. Each record/row of this database refers to an individual dataset that has about 48 entries/columns of the dataset metadata and attributes. In addition to information about data file location, data layer pointers, time-series range, geographical projection, and some attributes, it controls data display via color palette and on-the-fly unit conversion.

RIMS design and its core functionality are based on existing web and GIS technologies, data access interfaces, and database software that are installed as modular system libraries on a cluster of server computers that use only Open Source (http://www.opensource.org) or Public Licensed software (http://mapserver.org/, http://www.libgd.org, http://www.gdal.org/, http://www.gdal.org/ogr/, http://trac.osgeo.org/proj, and http://www.postgresql.org) and some other system libraries. Application programming interfaces (API) are used to access these libraries to run a number of server-side online and offline applications. Online server scripts work in conjunction with client-side applications to support all RIMS web services, while offline applications automatically or manually aggregate source data to a variety of temporal resolutions and/or areal polygon statistics (e.g., per country, per watershed averages) that we refer as a DataCube aggregation. RIMS client-side web components (Kandaswamy et al. 2006) are developed on DHTML and AJAX technologies (Goodman 2002), XML data streaming (Harold and Means 2004) and

Web Map Services (WMS) for GIS data (cf., Davis 2007; Mitchell 2005). In turn, these components are used to build customizable and scalable web pages that utilize RIMS web services but have their own unique look and feel in-line with the appearance design of the site home page for the research or outreach project.

2.7.3   RIMS Web Services

The RIMS client-side web application works as an interface to many other services (cf., Fig. 2.8). This application utilizes dynamic content provided through Apache web server. RIMS applications for map service as well as all other data services listed below are built over GDAL and ORG system libraries so that they are capable to work with virtually all possible data file formats for single and multilayer gridded and polygon vector data (Warmerdam 2008). Another functional advantage of utilizing the aforementioned system libraries is independence of the source data projections and spatial resolutions. Both of these use the original data files in their original formats from project collaborators or unrestricted external web resources. All RIMS data maps served via the web are built on-the-fly and merged with choices of base layers such as administrative coverage, cartographic and digital elevation modeled (DEM) river networks. Client-side map navigation has zoom-in and zoom-out click and drag tools, map panning, back and forward map history buttons, map resizing, Lon/Lat reading, index map, and choices for data interpolation (nearest neighbor, bilinear, spline and cubic spline) and shading to allow user control to vary and improve data visualization for any dataset (Fig. 2.8).

The *DataCube and time series* service combines mapping or data query in three virtual dimensions of options for (a) scaled time aggregation, (b) nonscaled or climatological time aggregation, and (c) polygon aggregation. On the server side, (Data Server in Fig. 2.7) the *Map Data Query* Service allows generation of digital 2D arrays of data values for every pixel on the map by reprojecting and resampling the original data to the requested map projection, extent, and resolution. These data can be delivered to the RIMS client application for instant data value query for each map pixel or could be used by the Data Manipulation Service and the Data Calculator client application. If needed, the latter can call the Map Data Query Service multiple times for each date layer in a time-series dataset within a requested date range to generate a 3-dimensional array of data where the third dimension is date/time. *Data Manipulation Service* (DMS) allows the user to perform interactively mathematical, statistical, custom functions, and/or logical statements/queries on a pixel or polygon (area integral) level over given map extent with any number of datasets and time-series date layers and returns the calculation results in any combination of the following formats: (1) maps, (2) data value frequency histograms, (3) georeferenced data files, (4) time-series date vector of this polygon data (spreadsheet format), (5) graph of polygon data, and (6) statistical moments of the resulting data. Presently, the RIMS uses the DMS for the web application named as the "Data Calculator" that replicates some basic functionality of a GIS spatial analysis. *The Data Calculator*

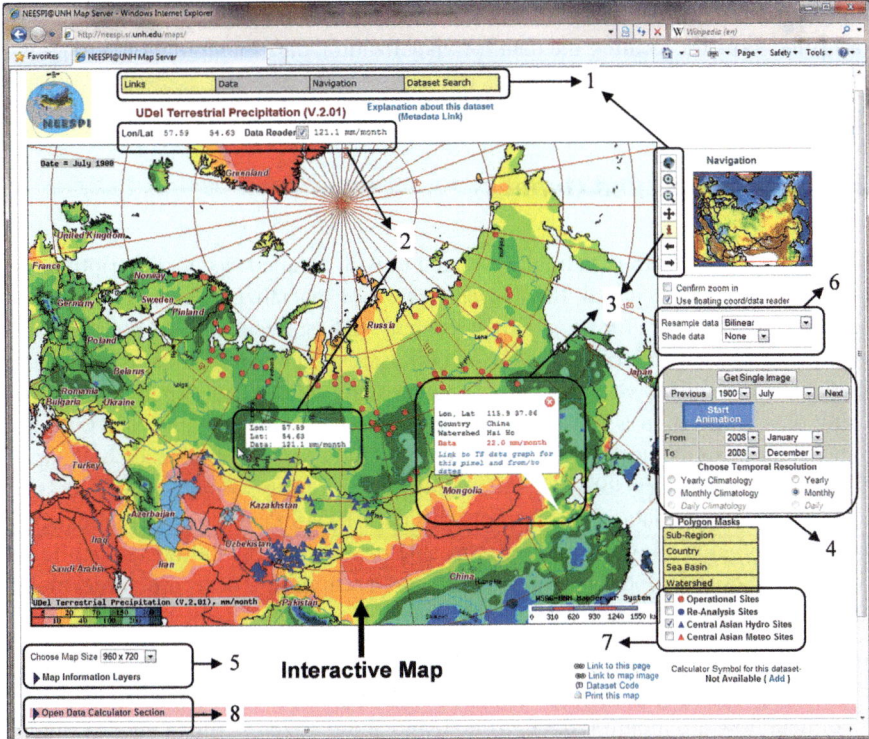

Fig. 2.8 Web client application of the RIMS customized for NEESPI site with adaptive dynamic reprojection (always North up). Keys – *1* basic data and map navigation tools include drop-down menu for data search/selection, side bar with zoom/pan, back/forward map history tools for spatial/data navigation, metadata link, etc.; *2* interactive coordinate and map data value reader for mouse over locations (loads map digital data to the browser); *3* pixel query tool (i-tool) gets coordinates, country, watershed, and map data value in a callout frame that has a link to time-series graphs and spreadsheet data for this pixel as well as prebuild polygon aggregations where this pixel present; *4* time-series navigation tool allows user to choose and display maps/animations for a given date(s) and switch temporal resolutions for the selected dataset (see DataCube concept section); *5* map size and base layer choices that has DEM river network option; *6* data interpolation and shading tools allow smoothed/interpolated visual interpretations; *7* point/station data list: selected datasets (stations) are displayed on the map with clickable symbols that open station pages in a separate browser window; *8* fold-out section to run the Data Calculator application to perform mathematical and logical functions over gridded or vector datasets to build custom maps, graphs, and spatial queries (see Data Calculator description below)

can use a single data layer or a list of time-series data layers for array specific operators such as *min()*, *max()*, or others. With these tools, a user can easily explore the difference between any two or more datasets, convert units, find an area where some conditions are true or false, etc.

Presently the RIMS data system at the Water Systems Analysis Group at the University of New Hampshire, USA accounts for over 2,000 Earth sciences-related datasets, and the Dataset Search Service (DSS) allows for the development of web or in-line tools for locating the data for visualization, data manipulation, modeling, or any other use.

2.7.4   Summary and Outlook

The summary of RIMS data holdings together with detailed system description can be found at the RIMS portal. RIMS *offers* a suite of specialized web and in-line GIS services and tools for climate and hydrological studies. It can be used both globally and in regional applications such as Northern Eurasia. The motivation to implement them as web services was driven by the increasing difficulties in maintaining and using data archives in a distributed workstation environment. Numerous GIS and data analysis tools suitable for hydrological studies are available in both open source and proprietary software suites; however, offering seamless data access to constantly updated data is increasingly difficult. RIMS *demonstrates* that existing web technologies allow complex server-side processing capabilities streamlining the centralization of data management and archiving. There are three major advantages offered by the RIMS:

- RIMS is directly integrated into a hydrological modeling framework that combines data mining, model runs, and data delivery to end users such that all stages of data flow are monitored, controlled, and summarized within a single computational environment.
- RIMS contains a set of web-based and in-line research and data analysis services and tools that could be used for virtually all scientific gridded, vector, and point (station) datasets. These, in turn, can be extended using the DataCube aggregation functions. The research and data analysis tools are built over specially designed, reusable data query, visualization/mapping/graphing, data delivery, and manipulation web services. An example of advanced web application is the *Data Calculator* that replicates many features of offline GIS data analysis tools, e.g., ESRI ArcInfo.
- RIMS includes hydrology-specific web tools for data query that are built around digital river networks. These tools allow users to analyze and/or summarize data on an individual watershed polygon basis or any upstream area from selected coordinates/grid points.

2.8   Brief Summary

This chapter provided a brief description of the NEESPI information resources currently supporting thematic investigations of Siberia. It shows that these resources form a significant segment of information-computation infrastructure required to study this region in its interrelationships with the Global Earth system. Technical details of systems only mentioned in this chapter have been or will be published in

Table 2.6 NEESPI/Siberia information resources

Information resource	Functions	URL and contact person
CLEARS CLimate and Environmental Analysis and Research System	Web–GIS system for processing, visualization and analysis regional climatic and meteorological geospatial datasets (major reanalyses and observations datasets)	http://clears.imces.ru system prototype http://climate.risks.scert.ru/ Dr. Alexander Titov titov@scert.ru
ILIS Integrated Land Information System	Land information system for Russia including cartographical materials, data of different inventories and surveys, diverse databases of *in situ* measurements, spatially distributed climate data, different remote sensing products, and numerous auxiliary models for assessment of relevant biophysical indicators of ecosystems	www.iiasa.ac.at/Research/ FOR/hlc Dr. Dmitry Shchepashchenko schepd@iiasa.ac.at
RIHMI-WDC Russian Hydrology and Meteorology Institute-World Data Center	Three-hour and daily datasets of major meteorological characteristics measured at Russia weather stations and relevant metadata sets prepared upon request	http://meteo.ru/english/ climate/cl_data.php Dr. Olga Bulygina bulygina@meteo.ru
SIB-ESS-C Siberian Earth System Science Cluster	Earth observation products: data search, data access, data visualization and analysis	http://www.sibessc.uni-jena.de/ Dr. Roman Gerlach roman.gerlach@uni-jena.de
Giovanni Goddard Interactive Online Visualization ANd aNalysis Infrastructure	Online system to visualize, analyze, and access vast amount of Earth science data	http://disc.gsfc.nasa.gov/neespi http://giovanni.gsfc.nasa.gov Dr. Gregory Leptoukh[a] gregory.g.leptoukh@nasa.gov Dr. Suhung Shen suhung.shen-1@nasa.gov
RIMS Rapid Integrated Mapping System	Online services (modeling, visualization, analysis) to modeling, observation, and monitoring geospatial data (variety of climate, climate forecast, hydrology, hydrology forecast, remote sensing, human dimension, etc.)	http://NEESPI.sr.unh.edu http://RIMS.unh.edu/ Dr. Alexander Prussevich, alex.proussevitch@unh.edu

[a]Dr. Gregory Leptokh (NASA Goddard Space Flight Center), the creator and long-term leader of GIOVANNI Project, passed away on January 12, 2012

thematic journals. Here we only underline the fact that all developments are in-line with progress of informatics, and in spite of the fact that the systems are oriented towards needs of researchers dealing with Siberia studies, they can be easily adapted to needs of different regions. Table 2.6 summarizes brief information about the systems, relevant web links, and contacts, which might be useful to potential users.

References

Acker J, Leptoukh G (2007) Online analysis enhances use of NASA earth science data, EOS. Trans Am Geophys Union 88:14

Allen G, Bogden P, Creager G, Decate C, Jesch C, Kaiser H, MacLaren J, Perrie W, Stone G, Zhang X (2008) Towards an integrated GIS-based coastal forecast workflow. Concurr Comput Pract Exper 20:1637–1651

Bartalev SA, Belward AS, Erchov DV, Isaev AS (2003) A new SPOT4-VEGETATION derived land cover map of Northern Eurasia. Int J Remote Sens 24:1977–1982

Bartholome E, Belward AS (2005) GLC2000: a new approach to global land cover mapping from earth observation data. Int J Remote Sens 26:1959–1977

Baru C, Chandra S, Lin K, Memon A, Youn C (2009) The GEON service-oriented architecture for Earth Science applications. Int J Digit Earth 2(1 supp 1):62–78

Bernholdt D, Bharathi S, Brown D, Chanchio K, Chen D, Chervenak A, Cinquini L, Drach B, Foster I, Fox P, Garcia J, Kesselman C, Markel R, Middleton D, Nefedova V, Pouchard L, Shoshani A, Sim A, Strand G, Williams D (2005) The earth system grid: supporting the next generation of climate modeling research. Proc IEEE 93(3):485–495. doi:10.1109/JPROC.2004.842745

Berrick SW, Leptoukh G, Farley JD, Hualan R (2009) Giovanni: a web service workflow-based data visualization and analysis system. IEEE Trans Geosci Remote Sens 47(1):106–113

Best BD, Halpin PN, Fujioka E, Read AJ, Qian SS, Hazen LJ, Schick RS (2007) Geospatial web services within a scientific workflow: predicting marine mammal habitats in a dynamic environment. Ecol Inf 2(3):210–223

Bicheron P, Defourny P, Brockmann C, Schouten L, Vancutsem C, Huc M, Bontemps S, Leroy M, Achard F, Herold M, Ranera F, Arino O (2008) Globcover: products description and validation report, MEDIAS-France

Bigagli L, Nativi S, Mazzetti P, Villoresi G (2004) GI-Cat: a web service for dataset cataloguing based on ISO 19115. In: Proceedings of the 15th international workshop on database and expert systems applications (DEXA'04), Zaragoza, Spain, 30 August–3 September 2004. IEEE Computer Society Press, Los Alamitos, pp 846–850

Bigagli L, Nativi S, Mazzetti P (2006) Mediation to deal with information heterogeneity – application to Earth System Science. Adv Geosci 8:3–9

Brown RD (2000) Northern Hemisphere snow cover variability and change, 1915–97. J Climate 13:2339–2355, http://dx.doi.org/10.1175/1520–0442(2000)013≤2339:NHSCVA≥2.0.CO;2

Bulygina ON, Razuvaev VN, Korshunova NN, Groisman PYa (2007) Climate variations and changes in extreme climate events in Russia. Environ Res Lett 2, 045020:7

Bulygina ON, Razuvaev VN, Korshunova NN (2009) Changes in snow cover over Northern Eurasia in the last few decades. Environ Res Lett 4, 045026:6. doi: 10.1088/1748–9326/4/4/045026

Butler D (2006) Virtual globes: the web-wide world. Nature 439:776–778

Compieta P, Di Martino S, Bertolotto M, Ferrucci F, Kechadi T (2007) Exploratory spatio-temporal data mining and visualization. J Vis Lang Comput 18(3):255–279

Craglia M, Goodchild MF, Annoni A, Camara G, Gould M, Kuhn W, Mark D, Masser I, Maguire D, Liang S, Parsons E (2008) Next-Generation Digital Earth – a position paper from the Vespucci Initiative for the Advancement of Geographic Information Science. Int J Spatial Data Infrastruct Res 3:146–167

Davis S (2007) GIS for web developers: adding where to your web applications. The Pragmatic Bookshelf, Releigh/Dallas, 275 pp

De Roure D, Jennings N, Shadbolt N (2001) A future e-science infrastructure. Report commissioned for EPSRC/DTI Core e-Science Programme

Dokuchaev VV (1951) Steppe transformation. Papers on soil study and land evaluation (1888–1900). Russian Academy of Science, Moscow (in Russian)

FACRE'RF (2009). State (national) report about the state and use of lands of Russian Federation in 2008. Federal Agency of Real Estate Cadastre Report, Moscow, Russia, 220 pp (in Russian)

FFS'RF (2009) Major indicators of forestry activities. Federal Forest Service, Moscow, Russia (in Russian)

Foresman TW (2008) Evolution and implementation of the Digital Earth vision, technology and society. Int J of Digit Earth 1(1):4–16

Frey KE, Smith LC (2007) How well do we know northern land cover? Comparison of four global vegetation and wetland products with a new ground-truth database for West Siberia. Global Biogeochem Cycles 21:GB1016. doi:10.1029/2006GB002706

Fridland VM (1989) Soil map of the USSR. Committee on Cartography and Geodesy, Moscow, Russia (in Russian)

Friedl MA, McIver DK, Hodges JCF, Zhang XY, Muchoney D, Strahler AH, Woodcock CE, Gopal S, Schneider A, Cooper A, Baccini A, Gao F, Schaaf C (2002) Global land cover mapping from MODIS: algorithms and early results. Remote Sens Environ 83:287–302

GEO (2005) Global earth observation system of systems (GEOSS) 10-year implementation plan. ESA Publications Division, Noordwijk

Goodman D (2002) Dynamic HTML: the definitive reference. O'Reilly, Sebastopol, 1401 pp

Gordov EP (2004a) Computational and information technologies for environmental sciences. Comput Technol 9(1):3–10 (in Russian)

Gordov EP (2004b) Modern tendencies in regional environmental studies. Geogr Nat Resour Special Issue 11–18 (in Russian)

Gordov EP, Vaganov EA (2010) Siberia integrated regional study: multidisciplinary investigations of the dynamic relationship between the Siberian environment and global climate change. Environ Res Lett 5:6

Gordov EP, Lykosov VN, Fazliev AZ (2006) Web portal on environmental sciences "ATMOS". Adv Geosci 8:33–38

Gordov EP, Okladnikov IG, Titov AG (2007) Development of elements of a web-based information-computational system for studies of regional environment processes. Comput Technol 12(Spec. issue 3):20–29 (in Russian)

Gordov EP, Bogomolov VYu, Genina EYu, Shulgina TM (2011a) Analysis of regional climatic processes in Siberia: approach, data and some results. Vestnik NGU Seria Informatsionye teknologii 9(Issue 1):56–66 (in Russian)

Gordov EP, Genina EYu, Shulgina TM (2011b) Climate change induced dynamics of bioclimatic indices for Siberia territory. Boreal forests in a changing world: challenges and needs for actions. In: Proceedings of the international conference, Sukachev Institute of Forest SV RAS, Krasnoyarsk, pp 216–219

Groisman PYa, Rankova EYa (2001) Precipitation trends over the Russian permafrost-free zone: removing the artifacts of pre-processing. Int J Climatol 21:657–678

Hansen MC, DeFries RS, Townshend JRG, Sohlberg R, Dimiceli C, Carroll M (2002) Towards an operational MODIS continuous field of percent tree cover algorithm: examples using AVHRR and MODIS data. Remote Sens Environ 83:303–319

Harold ER, Means WS (2004) XML in a nutshell. O'Reilly, Sebastopol, 689 pp

Hey T, Trefethen AE (2005) Cyberinfrastructure for e-science. Science 308(5723):817–821

Kandaswamy G, Fang L, Huang Y, Shirasuna S, Marru S, Gannon D (2006) Building web services for scientific grid applications. IBM J Res Dev 50(2/3):249–260

Kindermann S, Stockhause, M, Ronneberger K (2007) Intelligent data networking for the earth system science community. In: German e-science conference, Baden-Baden

Kumar SV, Peters-Lidard CD, Tian Y, Houser PR, Geiger J, Olden S, Lighty L, Eastman JL, Doty B, Dirmeyer P, Adams J, Mitchell K, Wood EF, Sheffield J (2006) Land information system: an interoperable framework for high resolution land surface modeling. Environ Model Software 21(10):1402–1415

Leptoukh G, Csiszar I, Romanov P, Shen S, Loboda T, Gerasimov I (2007) NASA NEESPI data center for satellite remote sensing data and services. Environ Res Lett 2, 045009. doi:10.1088/1748-9326/2/4/045009

McCallum I, Obersteiner M, Nilsson S, Shvidenko A (2006) A spatial comparison of four satellite derived 1 km global land cover datasets. Int J Appl Earth Observ Geoinf 8:246–255

Mitchell T (2005) Web mapping illustrated. O'Reilly, Sebastopol, 349 pp

Nativi S, Bigagli L, Mazzetti P, Boldrini E, Papeschi F (2009) GI-Cat: a mediation solution for building a clearinghouse catalog service, pp 68–74

OGC (2002) Web map service implementation specification, Version 1.1.1. OGC 01-068r3

OGC (2005) Web feature service implementation specification, version 1.1.0. OGC 04–094

OGC (2006) Corrigendum for the OpenGIS web feature service (WFS) implementation specification, Version 1.1.0. OGC 06-027r1

OGC (2007a) OpenGIS® catalogue services specification 2.0.2 – ISO metadata application profile. OGC 07–045

OGC (2007b) Web coverage service (WCS) implementation specification, version 1.1.1c1. OGC 07–067r2

Okladnikov IG, Titov AG, Melnikova VN, Shulgina TM (2008) Web-system for processing and visualization of meteorological and climatic data. Comput Technol 13(Spec. issue 3):64–69

Peterson TC (2005) Climate change indices. WMO Bull 54(2):83–86

Ramamurthy MK (2006) A new generation of cyberinfrastructure and data services for earth system science education and research. Adv Geosci 8:69–78

Rawlins MA, Steele M, Holland M, Adam J, Cherry J, Francis J, Groisman P, Hinzman L, Huntington T, Kane D, Kimball J, Kwok R, Lammers RB, Lee C, Lettenmaier D, McDonald K, Podest E, Pundsack J, Rudels B, Serreze M, Shiklomanov AI, Skagseth O, Troy T, Vorosmarty CJ, Wensnahan M, Wood EF, Woodgate R, Yang D, Zhang K, Zhang T (2010) Arctic system for freshwater cycle intensification: observations and expectations. J Climate 23:5715–5737

Razuvaev VN, Bulygina ON (2008) New version of the data set In: Daily temperature and precipitation data for 223 USSR stations. GC41A-0677, AGU 2008 fall meeting, 15–19 December 2008, San Francisco

Razuvaev VN, Bulygina ON (2009) Baseline climatological data sets for Eastern Europe area. In: Groisman P, Ivanov S (eds) Regional aspects of climate-terrestrial-hydrological interactions in non-boreal Eastern Europe. Springer, Dordrecht, pp 17–22

Rodell M, Houser PR, Jambor U, Gottschalck J, Mitchell K, Meng CJ, Arsenault K, Cosgrove B, Radakovich J, Bosilovich M, Entin JK, Walker JP, Lohmann D, Toll D (2004) The global land data assimilation system. Bull Am Meteorol Soc 85:381–394

Running SW, Nemani RR, Heinsch FA, Zhao M, Reeves M, Hashimoto H (2004) A continuous satellite-derived measure of global terrestrial primary production. Bioscience 54:547–560

Saha S, Moorthi S, Pan H-L et al (2010) The NCEP climate forecast system reanalysis. Bull Am Meteorol Soc 91:1015–1057

Schepaschenko D, McCallum I, Shvidenko A, Fritz S, Kraxner F, Obersteiner M (2010) A new hybrid land cover dataset for Russia: a methodology for integrating statistics, remote sensing and in-situ information. J Land Use Sci 6(4):245–259, TLUS-2009–0005.R1

Schimel DS, House JI, Hibbard KA, Bousquet P, Ciais P, Peylin P, Braswell BH, Apps MJ, Baker D, Bondeau A, Canadell J, Churkina G, Cramer W, Denning AS, Field CB, Friedlingstein P, Goodale C, Heimann M, Houghton RA, Melillo JM, Moore B, Murdiyarso D, Noble I, Pacala SW, Prentice IC, Raupach MR, Rayner PJ, Scholes RJ, Steffen WL, Wirth C (2001) Recent patterns and mechanisms of carbon exchange by terrestrial ecosystems. Nature 414:169–172

Sherstukov BG, Razuvaev VN, Bulygina ON, Groisman PYa (2007) NEESPI Science and Data Support Center for Hydrometeorological Information in Obninsk, Russia. Environ Res Lett 2, 045010:2

Shiklomanov AI, Lammers RB (2009) Record Russian river discharge in 2007 and the limits of analysis. Environ Res Lett 4:045015. doi:10.1088/1748-9326/4/4/045015

Shvidenko A, Goldammer JG (2001) Fire situation in Russia. Int Fire News 24:41–59

Shvidenko A, Nilsson S (2003) A synthesis of the impact of Russian forests on the global carbon budget for 1961–1998. Tellus B Chem Phys Meteorol 55:391–415

Shvidenko A, McCallum I, Nilsson S (2005) Data, results and assessment of full greenhouse gas accounting for the major GHG's for 2002/2003. IIASA, Laxenburg

Shvidenko A, Schepaschenko D, McCallum I, Nilsson S (2007) Russian forests and forestry (online). International Institute for Applied Systems Analysis and the Russian Academy of Science. Available from: http://www.iiasa.ac.at/Research/FOR/forest_cdrom. Accessed 27 Feb 2012

Shvidenko AZ, Schepashchenko DG, Vaganov EA, Nilsson S (2008) Net primary production of forest ecosystems of Russia: a new estimate. Doklady Earth Sci 421:1009–1012 (in Russian)

Shvidenko A, Schepaschenko D, McCallum I, Nilsson S (2010) Can the uncertainty of full carbon accounting of forest ecosystems be made acceptable to policymakers? Clim Chang 103:137–157

Stolbovoi V, McCallum I (2002) Land resources of Russia (online). IIASA & RAS. Laxenburg, Austria. Available from: http://www.iiasa.ac.at/Research/FOR/

Sukhinin AI (2008) Satellite data on wild fire (online). IIASA & RAS. Laxenburg, Austria. Available from: www.iiasa.ac.at/Research/FOR/forest_cdrom/index.html

Titov AG (2006) Metadata RDF-schema for meteorology and climate. In: Gordov EP (ed) Observations, modeling and information systems for environment study. Publishing house of Tomsk CNTI, Tomsk, pp 58–61

Titov A, Gordov E, Okladnikov I, Shulgina T (2009) Web-system for processing and visualization of meteorological data for Siberian environment research. Int J Digit Earth 2(1 suppl 1):105–119

Titov AG, Gordov EP, Okladnikov IG (2010) Semantic Web technique usage in information-computational system for environmental data analysis. Vestnik NGU Series Information Technologies 8(1):60–67

Union of Concerned Scientists (USC) (2010) USC satellite database 7-1-10. http://www.ucsusa.org/nuclear_weapons_and_global_security/space_weapons/technical_issues/ucs-satellite-database.html. Accessed 22 Sept 2010

Veselov VM, Pribylskaya IR, Proskurin VI (2000) Aisori system as a standard tool to handle archived hydrometeorological data. Proc RIHMI-WDC 166:8–25

Vorosmarty CJ, McIntyre PB, Gessner MO, Dungeon D, Prusevich AA, Green P, Glidden S, Bunn SE, Sullivan CA, Reidy Liermann C, Davies PM (2010) Global threats to human water security and river biodiversity. Nature 467:555–561. doi:10.1038/nature09440

Warmerdam F (2008) The geospatial data abstraction library. In: Hall GB, Leahy MG (eds) Open source approaches in spatial data handling. Springer, Berlin/Heidelberg, pp 87–104

Wisser D, Fekete BM, Vorosmarty CJ, Schumann AH (2009) Reconstructing 20th century global hydrography: a contribution to the Global Terrestrial Network- Hydrology (GTN-H). Hydrol Earth Syst Sci 14:1–24

Yang C, Raskin R (2009) Introduction to distributed geographic information processing research. Int J Geogr Inf Sci 23:553–560

Chapter 3
Climate Changes in Siberia

**Pavel Ya. Groisman, Tatiana A. Blyakharchuk, Alexander V. Chernokulsky,
Maksim M. Arzhanov, Luca Belelli Marchesini, Esfir G. Bogdanova,
Irena I. Borzenkova, Olga N. Bulygina, Andrey A. Karpenko,
Lyudmila V. Karpenko, Richard W. Knight, Vyacheslav Ch. Khon,
Georgiy N. Korovin, Anna V. Meshcherskaya, Igor I. Mokhov,
Elena I. Parfenova, Vyacheslav N. Razuvaev, Nina A. Speranskaya,
Nadezhda M. Tchebakova, and Natalia N. Vygodskaya**

Abstract This chapter provides observational evidence of climatic variations in Siberia for three time scales: during the past 10,000 years, during the past millennium prior to instrumental observations, and for the past 130 years during the period of large-scale meteorological observations. The observational evidence is appended with the global climate model projections for the twenty-first century based on the most probable scenarios of the future dynamics of the major anthropogenic and natural factors responsible for contemporary climatic changes. Historically, climate of Siberia varied broadly. It was both warmer and colder than the present. However, during the past century, it became much warmer; the cold season precipitation north of 55°N increased, but no rainfall increase over most of Siberia has occurred. This led to drier summer conditions and to increased possibility of droughts and fire weather. Projections of the future climate indicate the further temperature increases, more in the cold season and less in the warm season, significant changes in the

P.Ya. Groisman (✉)
NOAA National Climatic Data Center, Asheville, NC, USA

State Hydrological Institute, St. Petersburg, Russian Federation
e-mail: pasha.groisman@noaa.gov

T.A. Blyakharchuk
Tomsk State University, Tomsk, Russian Federation

A.V. Chernokulsky • M.M. Arzhanov • A.A. Karpenko • V.Ch. Khon • I.I. Mokhov
A.M. Obukhov Institute of Atmospheric Physics, Russian Academy of Sciences,
Moscow, Russian Federation

L.B. Marchesini
Forest Ecology Laboratory, University of Tuscia, Viterbo, Italy

E.G. Bogdanova • A.V. Meshcherskaya
Voeikov Main Geophysical Observatory, St. Petersburg, Russian Federation

P.Ya. Groisman and G. Gutman (eds.), *Regional Environmental Changes in Siberia
and Their Global Consequences*, Springer Environmental Science and Engineering,
DOI 10.1007/978-94-007-4569-8_3, © Springer Science+Business Media Dordrecht 2013

hydrological cycle in Central and southern Siberia (summer dryness), ecosystems' shifts, and changes in the permafrost distribution and stability. Observed and projected frequencies of various extreme events have increased recently and are projected to further increase. While in the north of Siberia, contemporary models predict warmer winters at the end of the twenty-first century and paleoreconstructions hint to warmer summers compared to the present warming observed during the period of instrumental observations. These three groups of estimates are broadly consistent with each other.

3.1 Paleoclimatic Evidence of Climatic Changes in Siberia

3.1.1 Late Glacial and Holocene Paleoclimatic Reconstructions in Siberia

Table 3.1 presents a partition of the Holocene by periods used below. It is close to standard definitions (cf. Khotinsky 1977) but has small regional specifics for Siberia in order to accommodate time partitioning used by major research groups who conducted research in the area. Time boundaries of periods of Holocene in Table 3.1 and in text are given in uncalibrated years before present (uncal. yr BP), and only in special cases they are in calibrated thousand years before present (kyr BP). The difference between these two time scales is described in Velichko (2010).

Data of pollen analysis of peat and lake sediments accompanied by radiocarbon dating and, where possible, by macrofossil data are a strong instrument for study of past changes of vegetation and climates. A close connection of vegetation types with climatic conditions is expressed especially well in Siberia. Therefore, paleopalyno-logical data can be used for qualitative and quantitative reconstructions of change of

I.I. Borzenkova • N.A. Speranskaya
State Hydrological Institute, St. Petersburg, Russian Federation

O.N. Bulygina • V.N. Razuvaev
Russian Research Institute of Hydrometeorological Information – World Data Centre
(RIHMI-WDC), Obninsk, Russian Federation

L.V. Karpenko • E.I. Parfenova • N.M. Tchebakova
V.N. Sukachev Institute of Forest, Siberian Branch of Russian Academy of Sciences,
Krasnoyarsk Akademgorodok, Russian Federation

R.W. Knight
North Carolina State University, Raleigh, NC, USA

G.N. Korovin
Center for Problems of Forest Ecology and Productivity, Moscow, Russian Federation

N.N. Vygodskaya
A.N. Severtsov Institute of Ecology and Evolution, Russian Academy of Sciences,
Moscow, Russian Federation

Table 3.1   Major Holocene periods, their abbreviations, and time frames according to the generalization of Khotinsky (1977) used throughout this chapter

Period name	Abbreviation	Time boundaries of periods (in uncal. kyr BP)
Alleröd	AL	11.7–11.0
Younger Dryas stage	YD	11.0–10.3
Preboreal period	PB	10.3–9.1
Boreal period	BO	9.1–8.0
Atlantic period	AT	8.0–4.5
Subboreal period	SB	4.5–2.5
Subatlantic period	SA	2.5–present

paleoclimate long before the instrumental era of climate observations began. To the end of twentieth century, quite a number of pollen diagrams supplied by radiocarbon dates have been published for Siberia providing information about local and regional fluctuations of Siberian climate.

The first fundamental compilation of paleopalynological data for all Northern Eurasia was performed by N. A Khotinsky (1977) who found that changes of moistening of the subcontinent were very diverse while changes of temperature were mostly synchronous. He revealed three main types of climate dynamic in Northern Eurasia after termination of last Glacial period:

1. Atlantico-continental type (the Russian Plain) with two thermal maxima in Atlantic and Subboreal periods and hygrotic phase after Atlantic period
2. Continental type (most of Siberia) with two thermal maxima and wet phase in Boreal and Atlantic periods
3. Oceanic (Pacific) type (Kamchatka Peninsula and Sakhalin Island) with two thermal maxima in Boreal and Atlantic periods and a wet phase after Boreal time.

By Khotinsky reconstruction, Atlantico-continental and oceanic types of dynamic are similar by the same tendency of development of moistening of climate, while continental and oceanic types of dynamic have a similar wave pattern of temperature change. He found that with increase of moistening and warming of climate in Siberia in the Boreal period, in Atlantico-continental and oceanic areas, the climate was relatively cold and dry (Russian Plain) or dry and warm (Pacific Ocean coast). On the contrary, an increase of moistening in the oceanic areas of Northern Eurasia in the second half of the Holocene was accompanied by a decrease of moistening of climate in Siberia. "Boreal phenomenon" by the opinion of Khotinsky (1977) was caused by decreased glaciation of the Arctic seas and had favored penetration of warm sea currents from the Atlantic Ocean to the Barents Sea and formation of intensive northwestern transfer of wet air masses which passed from the north of Fennoscandia and increased the precipitation over Siberia. In the Atlantic period, the final disappearance of Scandinavian Ice Sheet and destruction of associated European High led to formation of direct transfer of air masses from Atlantic eastward. This effect was stronger than at present time due to weakening of Siberian High.

Cooling and increasing of continentality of climate of Siberia beginning from Subboreal time is explained by increasing of glaciation of Arctic sea basins, strengthening of Siberian High, and weakening of transportation of western wet air masses to the east. Therefore, during the Holocene in Siberia, the main changes in climate and landscape were probably caused by global factors. Below, we outline major features of dynamics of the climatic changes over Siberia using a wide body of paleoclimatic reconstructions for the past 11 kyr BP (Andreev et al. 1989, 1997, 2001, 2002; Andreev and Klimanov 1991; Andreev and Siegert 2002; Belorusova et al. 1987; Belov et al. 2006; Belova et al. 1982; Bezrukova et al. 2005; Blyakharchuk 1989, 2003; Blyakharchuk and Sulerzhitsky 1999; Blyakharchuk et al. 2004, 2007; Borisova et al. 2005, 2011; Borzenkova and Zubakov 1984; Burashnikova et al. 1982; Demske et al. 2005; Firsov et al. 1974; Glebov et al. 1996; Kaplina and Lozhkin 1982; Kaplina and Chekhovski 1987; Karpenko 2006; Khotinsky 1977, 1984, 1989; Khotinsky and Klimanov 1985; Kind and Leonov 1982; Kirpotin et al. 2003; Klimanov 1976, 1984, 1994; Klimanov and Sirin 1997; Koshkarova 1989, 2004; Koshkarova and Koshkarov 2004; Kutafiyeva 1975; Laukhin 1994; Levina and Orlova 1993; Levkovskaya et al. 1970; Lister and Sher 1995; Lozhkin et al. 1995; Lozhkin and Vazhenin 1987; L'vov and Blyakharchuk 1982; Mangerud et al. 1974; MacDonald et al. 2000; Mironenko and Savina 1975; Naurzbayev et al. 2002; Nikol'skaya 1980; Orlova and Panychev 1989; Peteet et al. 1998; P'yavchenko 1983; Savina and Khotinsky 1982, 1984; Shichi et al. 2009; Shnitnikov 1951; Shumilova 1962; Tarasov et al. 1997, 2009; van Geel et al. 2004; Velichko 1989, 2010; Velichko et al. 1992, 1997; Volkova and Klimanov 1988; Volkova and Levina 1985; Yamskikh 1995; Yamskikh et al. 1981; Zubakov 1986; Zubakov and Borzenkova 1990; Zubareva 1987).

3.1.2   Compilation of Paleoclimatic Reconstructions for Siberia by Periods of Holocene

3.1.2.1   Late Glacial, and Preboreal (PB)

In Taymyr Lowland, a warming of Alleröd (11,700–11,000 uncal. yr BP or 13.3–12.8 kyr BP) was characterized by mean July temperatures 1.5 °C warmer than today and mean January temperatures 1 °C lower than today, and average annual temperature was close to modern values. The reason of such strong summer warming these authors explain by more continental climate due to the northerner Arctic Ocean coastline compared to its present position.

Palynological data from Khomustakh Lake in central Yakutia show that this territory was covered by shrub lands of shrub and dwarf *Betula* and of steppe plant communities with *Artemisia*, *Poaceae*, *Chenopodiaceae* species, and different herbs. Forest plant communities existed only in the flood plains as islands of steppified (i.e., with a large number of steppe species in ground cover) forests with *Larix* and *Betula*.

For central and southern Yakutia, reconstructed Alleröd July temperatures were by 1.5 to 2 °C below the present values, January by 2 to 5 °C, and the mean annual ones by 3 to 4 °C below the present values. In West Siberia, this period corresponds to the Taymyr warming when arboreal (tree) vegetation moved considerably northward. This warming occurred at the same time when rapid mammoth fauna extinction occurred in northeastern Russia and on the Taymyr Peninsula. In western Beringia, *Larix dahurica* first appeared (ca. 11,600 BP), and in eastern Beringia, forests with *Populus* appeared in the end of the Alleröd period. In southern Siberia in Baikal region in glacial landscape dominated by the cold and drought-resistant steppe and tundra communities during Alleröd warming, trees quickly spread out of their glacial refugia occupying up to ca. 25 % of the area. A spreading of complex vegetation of forest islands of spruce and larch together with steppe vegetation was marked in the south of Western Siberia. At the same time in Altai region, Alleröd warming did not cause afforestation due to extremely dry climatic conditions.

During the cooling of the Younger Dryas stage (11,000–10,300 uncal. yr BP or 12.8–11.6 kyr BP), the temperatures were 3–4 °C cooler and precipitation about 100 mm less than present in Taymyr Lowland. In the Putorana Plateau (Taymyr), mountain glaciation reactivated, the southern boundary of ground glaciers shifted southward to 56°N, and an arctic desert landscape predominated on the Taymyr Peninsula. In general due to Younger Dryas cooling in Northern Siberia, the arboreal vegetation retreated southward by more than 700–800 km. In central and southern Yakutia during Younger Dryas, cooling steppe vegetation dominated in landscape, and forest and shrub areas decreased considerably. Paleoclimatic reconstructions show that during this cooling, the mean summer temperature dropped by 3 °C and that of January by 6–7 °C; the precipitation was 150 mm less than present. The latest publication for Baikal region indicates the replacement of boreal woodland by shrubby tundra communities (decrease of woody cover percentages from ca. 25 % during AL to below 10 %) during Younger Dryas cool interval. But in most cases, YD cooling had little or no effect on vegetation in southern Siberia due to arid conditions of previous stage.

The first Holocene warming in West Siberia took place in the first half of the Preboreal (PB), and severe continental climatic conditions with widespread permafrost changed to a milder climate. As a result of this, tundra–steppe vegetation of periglacial areas was replaced by spruce and larch forest-tundra with shrub understory. The thawing and lowering of permafrost started to spread quickly. The further warming of climate promoted the spread of spruce and larch forests along the river valleys and on interfluve areas. Climatic conditions became much warmer than in late glacial time but did not reach the present level.

In the second half of the Preboreal (PB) in the area of the present forest zone of West Siberia, the climatic conditions became drier which was favorable for the spreading of birch forest-steppe. Spreading of relatively thermophilous water plants (*Myriophyllum*) in the south of the forest zone of West Siberia and the first appearance of *Abies sibirica* in taiga zone provides evidence of a comparatively warm climate. At the same time, in the eastern part of West Siberia adjacent to Yenisei River at about 9,500 year BP, the forest plant communities spread in the zone of present tundra on Cape Karginsky and in the south in the area of the present southern taiga zone during PB forest-steppe changed to birch forest and then to dark coniferous forest due to decreasing climate continentality and increasing moisture.

In the north of East Siberia in early Holocene, with climate warming about 10,000 year BP, tundra–steppe vegetation changed to *Betula nana – Discheikia fruticosa* shrub tundra. *Larix* appeared in the area at 9,400 year BP and disappeared after 2,900 year BP. The beginning of the Holocene here was marked by two coolings at 10,500 and 9,600 year BP. *Artemisia–Chenopodiaceae* plant communities disappeared from vegetation cover of central Yakutia about 9,800 year BP, when forests with *Larix* and tree species of *Betula* (*Betula pendula* + *B. alba*) with additions of spruce (*Picea obovata*) spread widely in the region.

In southern Siberia in Baikal region, the onset of the Holocene interglacial conditions is marked by the increase in woody cover to above 27 % at 10,300 uncal. yr BP (11.5 kyr BP) caused by significant climate amelioration (e.g., increase in T_w winter, T_c summer, and P_{ann} nual). Since that time, boreal forest became a major feature of the landscape around Baikal Lake although it took about 1 kyr until forest coverage in the area reached maximum at the beginning of Atlantic period (6,100 uncal. yr BP). In the Altai Mountains, quick afforestation of landscaped by closed forests of *Pinus sibirica* and *Larix sibirica* with admixture of *Picea obovata* and *Abies sibirica* took place later at the beginning of Boreal period.

During PB, the deviations of July and January temperatures, as well as of precipitation, regularly diminish from west to east, January's deviations being of greater magnitude than those of July.

3.1.2.2 Boreal Period (BO)

During the Boreal period, January temperature anomalies divided the former USSR territory into two regions at the latitude 60°N across the European Russia and then southward to Lake Baikal in Siberia. North of this boundary, the winter January temperature was warmer and annual precipitation higher than it is today. South of this boundary, a cooling was noted. In summer (July), the temperatures in the north of West Siberia (up to 60°N) and in East Siberia were lower than they are now. Elsewhere, a warming occurred in the northeastern maritime regions. Total annual precipitation was 25 mm greater than present in the northern and northeastern (Penzhina Delta) areas of the former USSR, less than present south of 60° and in Yakutia.

Increase of temperature and precipitation took place in the mid-Boreal at 8,300–8,500 year BP and promoted the widespread expansion of *Picea obovata* in the north of Western Siberia. Spruce and larch forests advanced in the tundra zone. In West Siberia in present taiga zone, the *Larix–Picea* forests alternated with birch forest-steppe widespread, but the type of BO latitudinal zonation was different from present.

In the area of West Siberia adjacent to Yenisei River in the southern taiga zone in BO (unlike to PB), the climate was warmer and drier than present (the amount of precipitation was 40–50 mm less). Reconstructed annual air temperatures exceeded today's values by 0.6 and 3.5 °C, respectively. Latitudinal zones were shifted to the north by 4–5° from their present position. The process of podzolization was interrupted by formation of steppe soils in the zone of present southern taiga reflecting drier climate during BO with a short period moisture increase in the middle of BO when wetter climate was observed in the taiga zone.

To the east, a BO temperature optimum was most explicit in the ice-free regions, e.g., in Taymyr Peninsula, at the maritime lowlands of Yakutia and in the northeastern regions of the former USSR. During this warming, the northernmost position (up to 76°N) of the tree-birch boundary was recorded over the entire Holocene. Its remnants (fossil trunks) are discovered even on the Arctic islands. The July temperature was 3–4 °C above the present, and annual precipitation was 100–150 mm higher. The temperature in the north of the Taymyr Peninsula was even higher: 5–6 °C above the present. In forest zone of Eastern Siberia, BO was the warmest and most humid period of the Holocene. Occurrence of single *Picea sibirica* trees, shrub birches (*Betula nana, B. exilis, B. humilis*), and mountain pine (*Pinus pumila*) in Kheta and Khatanga river valleys suggests a considerable climate warming. January and July air temperatures were 6 °C and 1–2 °C higher, respectively, and precipitation was 50–100 mm greater as compared to the present.

For central Yakutia, the boreal thermal maximum was observed at 8,370 BP during which a portion of spruce in vegetation cover considerably increased. During the following "Novosanchugovskoye" cooling (8,200 year BP), spruce forests degraded, and the role of herbal and shrub plant communities increased. The next warming in central Yakutia took place ca.8000 year BP and led to an increased area of larch-birch forests and a degradation of shrub formations (Andreev et al. 1989). Southward in the Baikal area and in Altai Mountains, forests consisting of *Pinus sibirica, Abies sibirica, Larix sibirica, Betula pendula*, and *Picea obovata* dominated in moderately continental and humid climate ca. 9,300–8,500 year BP.

During the maximum Boreal warming, the earlier established regular decrease in the deviations of temperature and precipitation from the west to the east became less prominent.

3.1.2.3 Atlantic Period (AT)

The climatic optimum of the Holocene is best expressed in West Siberia at 5,500–6,000 year BP being warmer and wetter than the twentieth century climate. Signs of Atlantic climatic warming can be observed especially clear in the north (Fig. 3.1). Thermokarst processes, the result of which was the formation of alas depressions, developed on the islands of the Arctic Basin. It caused a new expansion of forests into the tundra zone on Cape Karginsky.

Current landscapes of northern West Siberia adjacent to the Arctic can be considered as analogs of the landscapes in northwestern Yakutia in AT. In Holocene climatic optimum, most of the North Siberian Lowland was covered by *Larix* forest,

Fig. 3.1 Larch wood remnants on the sandy beach of Pionerskaya rivulet, the Khatanga River. The trunk was cut for the needs of dendroclimatological analysis (Picture by M.M. Naurzbayev)

with tree birches (*Betula pendula* and *B. alba*) and *Picea obovata* occurring as minor components in its southern part. Past southern tundra climatic characteristics were reconstructed for as far back as 6,000–6,500 years BP and show that at that time air temperatures of January and July were 2–4°C and 1–2°C higher, respectively, than the present and precipitation exceeded the present mean values by 50–150 mm. Larch wood findings (Fig. 3.1) suggest that July isotherm deviation from the current values was at least 8–10°C.

In the forest zone of West Siberia between 55° and 65°N, the annual temperatures exceeded present by 1–1.5°, and precipitation was 25–50 mm greater. These conditions promoted widespread *Picea obovata/Betula pendula* forests with *Pinus sibirica* in the central part of the West Siberian Plain but in its southwestern part were spread *Betula pendula/Pinus sylvestris* forests with an admixture of dark-coniferous (*Abies sibirica, Picea obovata*) and broad-leaved species (*Tilia cordata* and *Ulmus*). The role of *Abies* forests increased across all taiga zones of West Siberia. Vegetation zones advanced 200–300 km to the north of their present position.

The southern boundary of the forest zone in West Siberia was shifted northward in West Siberia in the AT climatic optimum. This shift is indicated by presence of a second humus horizon in the relic soils of the southern taiga. In the steppe zone, the great decrease in the size of Lake Chany as a result of drying was dated as 5,530 year BP. Steppe communities spread into the present forest-steppe zone. Precipitation change was not the same in high and low latitudes. In the steppe zone, precipitation was less than present. In forest-steppe zone during Atlantic period, intervals of warming coincided with a moister climate. Paleoclimatic curves of the forest zone (northern and southern taiga) demonstrate warmer and moister climatic conditions in AT with a short period of cooling in the middle of AT.

In southern taiga of West Siberia adjacent to Yenisei River, according to carpological data, forests of *Pinus sylvestris, Picea obovata, Abies sibirica,* and *Pinus sibirica* were spread. Also, a large number of water plants were found in the pollen spectra of this region which is indicative of the optimal heat/moisture ratio. In the north, the duration of frost-free and vegetative periods increased by 20–25 days and by 60 days, respectively. In the south, vegetative period increased by 30–40 days compared with the present.

To the east in Central Siberia on the interfluves of the Nizhnyaya Tunguska and Podkamennaya Tunguska rivers, a dark-coniferous taiga (*Picea obovata, Larix sibirica, Pinus sibirica*) predominated in western regions, and light *Larix sibirica* forests with *Pinus sibirica* and *Picea obovata* occupied eastern regions. Climate conditions improved considerably in East Siberia at AT. Northern taiga which occupied the Tunguska Plateau, Central Yakutian Plain, and Vilyuy and Lena Plateaus in the Boreal time was replaced by middle taiga, and middle taiga which had occupied Angara and Lena-Angara Plateaus was replaced by southern taiga with an admixture of broad-leaved species (*Tilia, Ulmus,* etc.) in some regions.

In central Yakutia, coniferous forests with *Picea obovata* increased by area. The data of archaeological site Belkashi (in the middle reaches of the Aldan river) provide evidence that first light *Larix* forests with *Betula* and *Pinus sylvestris* were established on sand soils in the mid-Holocene time.

Pollen diagrams from Yana River lowlands in Arctic Yakutia show that larch (*Larix dahurica*) forests with shrub alder (*Duschekia fruticosa*) and dwarf birch (*Betula exilis*) dominated the area during the last 6,400 years BP. Climate reconstructions at the Arctic coast of Yakutia show that the warmest time was between 6,000 and 4,500 year BP.

Picea obovata and *Pinus sylvestris* domination in forest stands in early AT in central Yakutia indicated that climate was considerably warmer than before. During the maximum of this warming, $7,880 \pm 120$ BP, average July, annual, and January temperatures were 0.1–1, 1, and 1–1.5 °C lower, respectively, and precipitation was also somewhat lower as compared to current values. Humid and warm climate of the late AT (average July and January temperatures exceeding current values by 1–2 and 2 °C, respectively) vastly promoted an extension of *Pinus sylvestris* stands with an admixture of *Picea obovata*. At that time, *Pinus sylvestris* forests were widespread across Yakutia.

Southward in the Baikal region, mixed *Pinus sibirica/Abies sibirica* and *Larix/Betula* stands with a minor component of *Picea obovata* were widely spread in the first half of AT, whereas the second half of this period was characterized by well-developed *Pinus sibirica* stands with minor components of *Picea obovata* and *Abies sibirica*, and mixed *Larix/Betula* and *Betula* stands with *Abies sibirica* and *Pinus sibirica* as minor components. The most favorable conditions ($T_{July} = 17$–18 °C, $T_{January} = -19$ °C, and $P_{annual} = 500$–550 mm) for the taiga growing in Baikal region occurred at 9.800–6,400 uncal. year BP. In Altai Mountains, amelioration of climate was marked by maximal spreading of *Abies sibirica* in mountain forests and by spreading of tree vegetation in high mountain areas contemporary covered by treeless alpine vegetation.

In East Siberia, Monserud et al. (1995) reconstructed late Atlantic (6,000–4,600 year BP) and present climates applying their Siberian vegetation model (Tchebakova et al. 1994) to the paleovegetation map of Khotinsky (1984) and the present vegetation map of Isachenko et al. (1988), respectively. They found that January temperature was 5.6 °C warmer than present, July temperature was only 0.6 °C warmer, and precipitation was 150 mm greater in the mid-Holocene than present. Larger precipitation and winter warming favored less continentality and a thicker active layer promoting the deep penetration of dark-needled species (*Picea obovata, Pinus sibirica*) into East Siberia where currently only one permafrost resistant *Larix dahurica* (*L. gmelini + L. cajanderii)* survives. In the Holocene optimum, the mean global surface air temperature was 1.0–1.2 °C higher than its values in the middle of the twentieth century, which exactly corresponds to the Northern *Hemisphere surface air temperature anomaly in the past decade* (cf. Lugina et al. 2006 updated; Fig. 3.3). At high latitudes of Asia, the temperature increase was maximal in winter (+4°), whereas in the south, the positive deviations decreased, and in the latitudinal belt from 40° to 50°N mostly cooling took place.

3.1.2.4   Subboreal Period (SB)

The SB climate in West Siberia was unstable but mostly cooler than that of AT. The northern boundary of the forest zone retreated to the south by almost 2° latitude. Period around 4,500 year BP was a time of dry cooling. It was indicated by the renewed regression of lakes in the steppe and forest-steppe zones. After this dry period, the climate became wet and cool, causing the afforestation of forest-steppe and partly of the steppe zone. The cooling stimulated the freezing of some wet mires in the middle taiga and formation of palsa bogs in the northern taiga and forest-tundra. In forest zone of West Siberia, relatively thermophyllous *Abies* forests retreated to the south to the position of the present southern taiga. The widespread expansion of *Pinus sibirica* began in central areas of West Siberia in the second half of the Holocene. Continentality of climate increased. In forest-steppe, unstable climatic conditions were documented with fluctuations of from more cool and dry conditions to more warm and dry intervals.

The forests of West Siberia adjacent to Yenisei River were mainly represented by *Pinus sibirica* and mixed *Pinus sylvestris/Pinus sibirica* stands. In East Siberia, climatic conditions in SB became more severe on the Taymyr Peninsula causing forest-tundra migration southward and forest degradation. In Ary-Mas forest, *Picea obovata* and tree-like birch (*Betula pendula and B. alba*) became extinct from *Picea obovata/Larix* open forests, and only *Larix dahurica* remained; peat development decreased considerably across this region. Temperature maximum of a lesser magnitude than in AT took place 3,500 years BP.

In Yakutia, a decrease in *Picea obovata* forest extent indicates climate cooling: surface air temperatures were ~1 °C lower, and precipitation was somewhat lower than the present. The regional climate somewhat warmed in the second half of SB, which can be noticed from a decrease in shrub birch (*Betula exilis, Betula humilis*) extent and partial recovery of *Picea obovata* forest range.

In Central Siberia, average July, January, and annual temperatures exceeded present values by 0.5, 2.4, and 1.6 °C, respectively, and the frost-free period was 35 days longer (Karpenko 2006). The Minusinsk Hollow climate in the second half of SB (3,650–3,200 BP) was dry and warm, and tall grass/sedge and pigweed/woodworm steppe plant communities were abundant. In Baikal region, *Pinus sylvestris*, *Pinus sibirica*, and *Larix* forests were common, but unlike the Minusinsk Hollow, decreases in temperature and precipitation were documented in the Baikal Lake region. There is a hypothesis that in the Lake Baikal region, the Holocene precipitation and tree cover dynamics may be linked to shifts in intensity of the Pacific monsoon. Cooling and drying of climate in SB was marked by retreating of forest vegetation from high mountain areas of southeastern Altai and decreasing of role of *Abies sibirica* in mountain forests of central and northern Altai.

3.1.2.5 Subatlantic Period (SA)

During SA in West Siberia, cold phases took place at 2,800, 1,700, and 500 year BP. The following dry events with maxima at 2,500–3,000, 1,500, and 600–800 year BP were noticed by afforestation of mires in the southern taiga zone and by regression of lakes in the forest-steppe and steppe zones. In the opinion of Khotinsky (1989), the centennial climate oscillations during the last millennium indicated a tendency in which warming in humid areas coincided with stages of aridification in the steppe zone, and cooling in humid areas coincided with wetter conditions in arid zones. This is confirmed by pollen diagrams and paleoclimatic curves of taiga and forest-steppe zones. On an interdecadal time scale, volcanic activity was among the important factors controlling the last millennium climate (cf. Sect. 3.1.4).

During SA, most of Siberian contemporary landscapes had been established including Western Siberia, Central Siberia, and Eastern Siberia. In Western Siberia, the northern boundary of forest zone retreated to the south reaching contemporary position. In SA, it was formed mostly by birch-thinned forests with admixture of *Picea obovata*. In northern and middle taiga zone of Western Siberia, *Pinus sibirica* forests played a leading role, but in southern taiga, *Abies sibirica* and *Picea obovata* with admixture of tree *Betula* and *Pinus sibirica* predominated. *Pinus sylvestris* occupied sandy soils and widespread on oligotrophic bogs, as well in "belt forests" of forest-steppe zone.

In the Taymyr Peninsula, *Larix* occurred within its present range, and near the Khatanga settlement, climate was close to what it is now. Forests of Evenkia were dominated by *Picea obovata*, mixed *Pinus sylvestris/Larix*, and *Betula/Larix* stands in early SA, and the regional climate was cooler and more continental than present. In the second half of SA, larch forests were gradually acquiring their present look. Average January and annual air temperatures were 5 and 2.5 °C higher, respectively, than present. In Yakutia, in SA, a few intervals of warming of smaller amplitude than in the mid-Holocene optimum at 2,000 and 1,000 year BP were reported. The vegetation cover of central Yakutia was similar to present: *Larix* and *Betula* were widely spread, with *Pinus sylvestris* being common on sandy soils. A marked climate cooling (with average July, annual, and January temperatures about 1, 1.5, and 2 °C lower, respectively, than present) occurred ca. 1,500 BP with lower precipitation by about 50 mm. Climate became warmer (Medieval climatic optimum) ca. 1,000 BP with average surface air temperatures higher by 0.5–1 °C and precipitation 25 mm higher than present.

During SA, the Baikal region was dominated by dark-needled forests which suggest a higher climate humidity and decreasing continentality. The forest cover of Kas-Dubches interfluves of West Siberia adjacent to Yenisei River was dominated by *Pinus sylvestris* and *Pinus sibirica*, whereas *Betula, Picea obovata*, and *Abies sibirica* ceased to be among the forest components. July temperature was close to the present, while January and annual temperatures were 0.9 and 1.4 °C higher. The area received 60-mm less precipitation, and the frost-free period lasted 38 days longer than now. In the Minusinsk Hollow, warm temperatures and alternating humid periods promoted wormwood/herb steppe to spread, and the foothill slopes were covered by mixed *Betula/Pinus sylvestris* forest-steppe in early SA. The end of SA here was characterized by warm and humid climate. In the Altai Mountains, vegetation cover was similar to those formed in SB. Mountains having a great variety of environments evolved from complex topography may serve as a good model for reconstructions of both paleovegetation and climates. Below, we describe in more details paleoreconstructions in one of such regions.

3.1.3 Regional Case: Paleoreconstructions of Vegetation and Climate in the Altai–Sayan Mountains, Southern Siberia, Throughout the Holocene

Two quantitative methods were used to reconstruct paleoenvironments and vegetation in the Altai–Sayan mountains during the Holocene. Firstly, nine retrospective pollen diagrams from lake and peat deposits across the study area were constructed (Blyakharchuk et al. 2004, 2007, 2008). The pollen spectra were examined for five time slices of the Holocene: 10,000, 8,000, 5,300, and 3,200 years BP (before present) and the present. The "biomization" method of Prentice et al. (1996) was applied to the fossil pollen data to reconstruct the site paleovegetation. Thereafter, the montane bioclimatic model, MontBioCliM (Tchebakova et al. 1994, 2009b), was inversely used to convert site paleovegetation into site paleoclimates: average July temperatures and annual precipitation. Climate change was evaluated as the differences between the July temperatures and annual precipitation totals between contemporary and paleoclimates in each of nine sites for each of five selected Holocene time slices. Thereafter, these differences were scaled down to 1-km grid cells and added to current July temperature and annual precipitation surfaces which were mapped using thin plate smoothing splines (cf. Hutchinson and de Hoog 1985) on the base map at a resolution of 1 km. Thus, detailed paleo July temperature and annual precipitation maps were obtained for each time slice.

Pollen-based reconstructions of vegetation patterns during the Holocene across the Russian part in the Altai–Sayan mountains included the following (Tchebakova et al. 2009b):

Before 10,000 BP, Late Glacial Time. Most of the Altai–Sayan mountains were covered by treeless steppe and tundra vegetation. Only in the Kuznetsk Alatau Mountains were there islands of spruce and larch forests spread along river valleys and around lakes.

10,000 BP, Early Holocene. In the Kuznetsk Alatau Mountains, mountain tundra dominated. Intermountain depressions were Covered by cold steppe with islands of spruce forests. In central Altai and southwestern Tuva, vast areas were occupied by grass *Artemisia* steppe with subalpine shrubs upslope. Small islands of forests started to spread. Steppes in the southern part in the mountains were drier compared to those in the northern part. Vegetation-based climate reconstructions show that in the Altai–Sayan mountains, the early Holocene ca. 10,000 BP was cold and dry. July temperature anomalies with respect to the contemporary climate were greatly negative −2 to −5 °C 10,000 BP.

8,000 BP, BO. The Kuznetsk Alatau was covered by thick dark conifer forests with *Abies sibirica* and with tall herbs and ferns in the ground layer. The lower Batenev Ridge of the Kuznetsk Alatau was covered by forests with *Pinus sibirica* and *Picea obovata* which stretched to lower elevations along river valleys. In the drier central Altai, *Larix sibirica–Pinus sibirica* forests with admixture of *Abies sibirica* were observed. On the western ridges of Altai, the portion of *Pinus sibirica* increased in the forests. At the upper elevations, mountain tundra and meadows developed. Lower than 1,800-m real steppe occurred. Forests of *Larix sibirica* mixed with *Picea obovata* were spread further to the dry southeastern Altai and southwestern Tuva. In those forests, *Pinus sibirica* was also found. The climate in the period around 8,000 BP was warm and moist; July anomalies were positive (2–4 °C).

5,300 BP, AT. Vegetation of the Kuznetsk Alatau did not change much compared to 8,000 BP, but in the forests of the Batenev Ridge, *Picea obovata* disappeared. Also from the forests of the central Altai, *Abies sibirica* almost disappeared. The steppe vegetation of intermountain depressions became more xerophytic with a greater role played by *Artemisia*. July anomalies were positive (2–4 °C) as were at 8,000 BP. For comparison, in the Minusink depression located in the north of the mountains, Koshkarova (2004) based on the macrofossil analysis reconstructed July temperature as being 3 °C higher in 7,000–5,800 BP.

3,200 BP, SB. Dark conifer "chern" forests retreated from central areas of Kuznetsk Alatau to its western macroslope. In the central Altai, vegetation cover was similar to that which occurred in 5,300 BP. In the lower mountains, adjacent to Altai and West Sayan, the role of *Betula pendula* increased. The climate ca. 3,200 BP was cooler and dryer than the present. July temperature anomalies with respect to the contemporary climate were negative −2 to −3 °C. Koshkarova (2004) reconstructed July temperature 1 °C lower in 3,000–2,400 BP, and Zubareva (1987) reconstructed July temperature 2 °C lower about 3,000 BP than today.

Vegetation change across the Altai–Sayan mountains reflects a relatively rapid climatic change from dry and cold in the early Holocene to wet and warm from 8,000 BP to 5,300 BP. Thereafter, a gradual change to a more continental and less humid climate took place in the second half of the Holocene. A more pronounced climate and vegetation change occurred in the leeward southeastern and eastern parts over the Altai–Sayan mountains and a less pronounced change on windward western macroslopes. Such changes can be explained by the weakened Atlantic cyclone activities in the Subboreal over Northern Eurasia (Velichko et al. 1997; MacDonald et al. 2000; Blyakharchuk et al. 2004).

The reconstructed paleovegetation and paleoclimates of the Holocene described above mostly agreed with previous findings of other paleoecologists (cf. Blyakharchuk et al. 2004, 2007, 2008). However, the current reconstruction of the 5,300 BP climate shows wetter and warmer climate compared to the results of other authors (e.g., at 5,000–6,000 BP by Wu and Lin 1988, at 5,400 BP by Herzschuh et al. 2004, at 5,300 BP by Yamskikh et al. 1981, and at 5,200 BP by Savina and Koshkarova 1981). The 3,200 BP climate reconstructed in the above analysis was cooler and dryer than the present climate. This result corresponds well with the reconstructions of Herzschuh et al. (2004) at 3,100 BP and Savina and Koshkarova (1981) at 3,200 BP, although Wu and Lin (1988) reconstructed the 3,000 BP climate as cooler and wetter than the present climate.

The mid-Holocene is frequently hypothesized to be an analog of future climate warming. However, reconstructed as warm and moist in Siberia, the mid-Holocene climate differed from the projected warm and dry climate of the twenty-first mid-century. Future vegetation at 2050 was predicted to be dissimilar to any of the paleovegetation reconstructed at 8,000 BP and 5,300 BP (Tchebakova et al. 2009b).

3.1.4   Climatic Change During the Past 1,000 Years from Noninstrumental Data

Global warming reported from instrumental observations that span the past 130 years increased our interest in detailed information (with annual resolution) about the preinstrumental period regional and global temperature changes at the millennium time scale. This information includes dendroclimatological data on tree rings width and density (particularly at the boundaries of the forest zone in high latitudes and/or in the mountains), cf. D'Arrigo et al. 2006; Luterbacher et al. 2004; Wilson et al. 2007; Khantemirov 1999; Vaganov et al. 1996, stalagmites' core measurements in caves (Smith et al. 2006), and isotopes measurements in ice cores in Greenland (www.ncdc.noaa.gov/paleo/icecore). Figure 3.2 shows time series of surface air temperature (anomalies from the present-day mean values) derived from dendro-climatological information across Siberia with 1-year time step. It compiles the work of D'Arrigo et al. (2006), Jones and Mann (2004), Osborn and Briffa (2006), Luterbacher et al. (2004), and Wilson et al. (2007). Comparison of these annual time

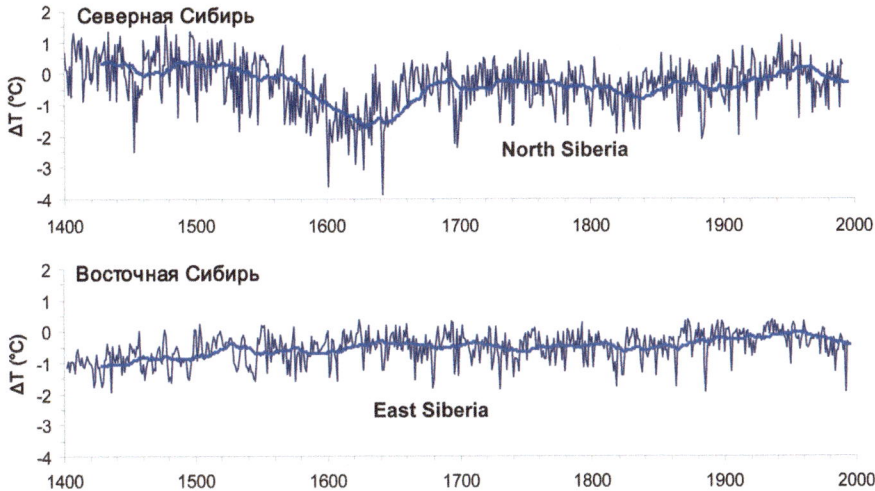

Fig. 3.2 Annual surface air temperature anomalies from the long-term mean contemporary values for two regions of Siberia based upon dendroclimatological information (available at ftp://ftp.ncdc.noaa.gov/pub/data/paleo/treering/records). *Blue lines* represent 30-year running average values

series with catalog of major volcanic eruptions in the past centuries (Borzenkova and Brook 1989; Borzenkova 1992) shows that many distinctive features of abrupt temperature changes in Northern Siberia corroborate sufficiently well with the dates of major volcanic eruptions. For example, temperature decreases documented by dendroclimatic data began nearly simultaneously with a sequence of eruptions of Kuwae volcano located in the tropical Pacific Ocean (Vanuatu). The most significant eruption of this volcano occurred in 1452/1453 AD that was the greatest sulfate volcanic eruption during the past 700 years (Gao et al. 2006). The cooling in the early 1600 s is recorded in all temperature reconstructions worldwide and is related to catastrophic eruption in Peru of Huaynaputina volcano (1600/1601) that was the strongest in the past 500 years. Thereafter, between 1600 and 1690 AD, there were 12 powerful volcanic eruptions in the tropics which are known to have a stronger impact on the temperature in both hemispheres than eruptions in the polar regions (Khmelevtsov 1986).

Filtration of these (and other) temperature time series for the past millennium shows that (in addition to the latest greenhouse-induced warming) 50 % of the climate variability can be ascribed to interannual variability. Interdecadal variations are responsible for approximately 30 % and century long-term variations for another 20 % of variance of these time series (Borzenkova et al. 2011). An assumption that the interdecadal variability is mostly related to impact of the volcanic activity is supported with model calculations in Crowley (2000) and Crowley and Berner (2001). In these studies, it was shown that about 30 % of the global temperature variance during the past 1,000 years can be described by known changes in stratospheric aerosol loading due to volcanic activity.

3.2 Observational Evidence of Contemporary Climatic Changes in Siberia

3.2.1 Meteorological Network in Siberia

Meteorological network in Siberia was established in the early nineteenth century (e.g., Irkutsk in 1830; Verkhoyansk in 1869; Yakutsk in 1829; Tomsk in 1837; Berezov in 1832; Yehiseysk in 1853; Krasnoyarsk in 1838; Turukhansk in 1877; Barnaul in 1838; Chita in 1828, and Nerchinskiy Zavod in 1839). It is commonly accepted that the temperature time series over the region (as well as over the entire Russia) can be assessed since ca. 1881 when the national standards of temperature measurements were adopted (Vannari 1911; Vinnikov et al. 1990). Since the early 1880 s and up to date, the meteorological network over Siberia has been sparse (Sherstyukov et al. 2007; Chap. 2) but is sufficient to deliver the pattern of mean monthly/seasonal temperature anomalies as well as the regionally averaged time series for the past 130 years (Fig. 3.3). This occurs due to a large radius of spatial correlation of the temperature field over the expansive interior of the Northern Eurasia, especially during the cold season that can reach more than 3,000 km (Gandin et al. 1976). It would be enough to say that in winter the positive spatial correlation of temperature in Barnaul (in the foothills of Altai Mountains of the Central Siberia) and Moscow is still statistically significant.

3.2.2 Temperature Changes

Figure 3.3 shows that Siberia is a region of the Northern Hemisphere with the largest temperature changes within the Northern Hemisphere (1.39 °C/100 years). Changes here are higher than over Northern Eurasia, Northern Asia (1.29 °C/100 years), over the Arctic (1.28 °C/100 years), or over the entire hemisphere (0.77 °C/100 years). Very high variability (even in the annual means that are mostly controlled by the cold season temperatures) is an important feature of Siberian climate. In winter, the major source of heat in Siberia is not a "reliable and stable" solar input but atmospheric advection that brings (or does not bring) warmer air masses from outside. Therefore, there are no physical reasons to expect that the amplitude of the temperature anomalies in this part of the world will be reduced in the foreseen future, and contemporary climate models' projections (e.g., those that explore the pattern of the greenhouse gases-induced global warming) suggest that the systematic temperature changes in Siberia (trends) will remain among the largest over the globe.

3.2.3 Arctic Sea Ice Changes Northward of Siberia

An overview of contemporary Arctic sea ice changes and what these changes mean for hemispheric climate are given by Kwok and Untersteiner (2011). The data on sea ice extent are routinely provided at http://nsidc.org/data/seaice_index/. This informa-

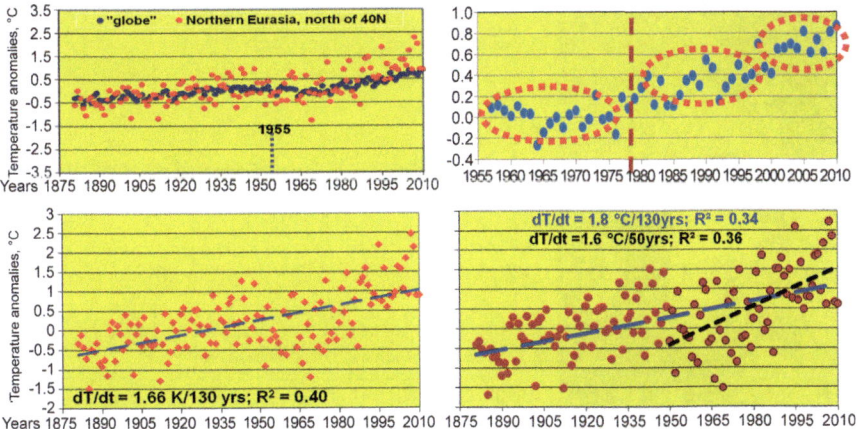

Fig. 3.3 *Upper panels. Left.* Annual surface air temperature area averaged over most of the "globe" (60°S to 90°N) and Northern Eurasia (linear trends are 0.91 °C/130 years and 1.5 °C/130 years, respectively; Lugina et al. 2006, updated). *Right.* Global surface air temperature anomalies (90°S to 90°N) from the same source. *Vertical dashed line* indicates the year when the projection of the contemporary global warming was made by Budyko and Vinnikov (1976). *Bottom panels.* Annual surface air temperature anomalies for Asia north of 40°N (*left*) and Siberia (*right*; a region outlined in Fig. 1.1). All anomalies are calculated as deviations from the long-term mean values for the 1951–1975 period

tion and the information about sea ice thickness that now combines the submarine sonar data and satellite ice monitoring show a dramatic decrease of the Arctic ice cover in the last three decades. These changes are among the most important indications of the ongoing global climatic change and impact upon the energy and water vapor supply to the polar atmosphere particularly in the end of the warm season when other factors (e.g., solar radiation) are weaker and circulation factors (e.g., westerlies) are affected themselves by a reduction of meridional temperature gradient (cf. Groisman and Soja 2009). Figure 3.4 shows also that most of the Arctic sea ice changes in the end of the warm season occur exactly offshore of the Eurasian continent. This is due to a peculiarity of the Arctic Ocean circulation (cf. Proshutinsky et al. 2007; Kwok and Untersteiner 2011). Changes in the Arctic exercise control on energy and water vapor transport into the interior of Eurasia. In Siberia in the cold season, this development manifests itself by increasing interannual temperature variability (ACIA 2005), frozen precipitation increase (Rawlins et al. 2010), and increase in snow depth and snow water equivalent (Bulygina et al. 2009, 2011, Fig. 3.5).

3.2.4 Changes in Precipitation and Snow Cover

Analyses of Siberian precipitation (e.g., Groisman 1981; NCDC 2005; Rawlins et al. 2010) show that while in the second half of the twentieth century the annual precipitation totals became noticeably higher than in the first half (by ~10 %),

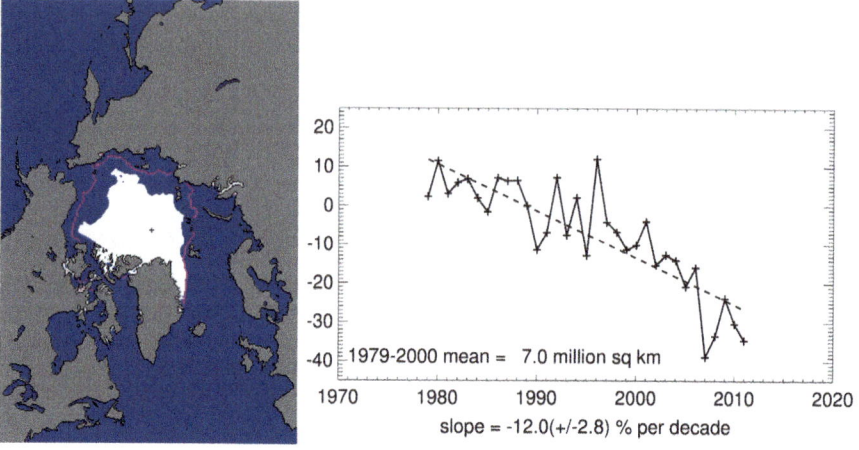

Fig. 3.4 *Left*. Northern Hemisphere sea ice extent as of mid-September 2011; magenta line, a median for the 1979–2000 period; (*right*) September 1979–2011 anomalies. Archive of the US National Snow and Ice Data Center (http://www.nsidc.org/data/seaice_index/index.html)

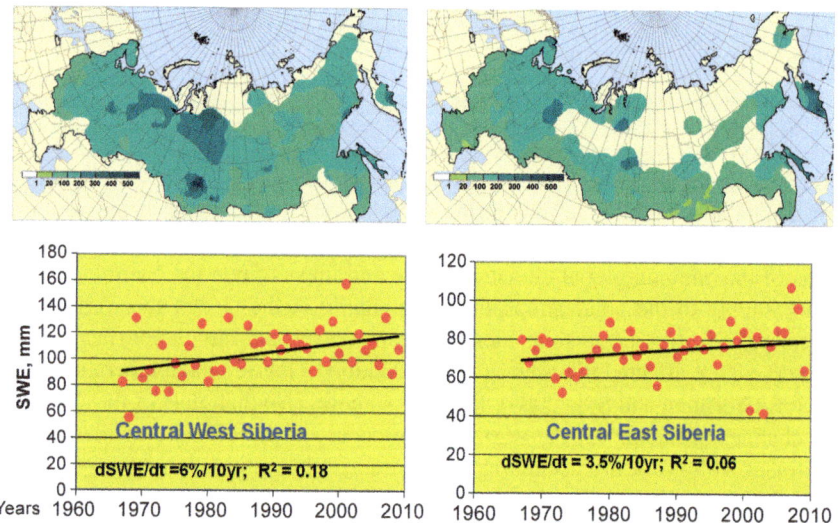

Fig. 3.5 *Upper panel*. Mean maximum snow water equivalent, mm, along the snow surveys in the forested (*left*) and open ("field," *right*) areas. Mean values for the 1966–2009 period. *Bottom panel*. Changes in snow water equivalent over Northern Siberia along the "field" snow survey routes (approximately 55–65°N lat. belt)

the consequent changes became subtle. Annual totals as well as the number of days with precipitation in Siberia have not further changed during the following decades, but in the warm season (while there were no systematic changes in precipitation totals), a significant redistribution by intensity among the days with precipitation occurs: the number of days with heavy rainfall and very heavy rainfall has increased

(Sect. 3.2.6.2). Recently, analyses of precipitation in the cold season over Siberia revealed an increase in the cold season precipitation over most of Siberia (Rawlins et al. 2010).

There are seemingly conflicting reports about the snow cover changes over Siberia (ACIA 2005; Brown and Mote 2009; Groisman et al. 1994, 2006; Shmakin 2010; Bulygina et al. 2009, 2010, 2011; Callaghan et al. 2011). However, we cannot attribute the discrepancies to the lack of data. The in situ reports of snow on the ground when area-averaged over Russia and/or Siberia reasonably well reproduce the mean values, interannual variability, and trends of snow cover extent delivered by NOAA satellites (Groisman et al. 2006). The reported differences mostly arose from different characteristics that were assessed and different tendencies in these characteristics. Snow cover extent over Siberia did not appreciably change during winter (defined as the Dec.–March season) but significantly retreated in spring–early summer, from April through June (Robinson et al. 1993; Groisman et al. 1994, 2006; Bulygina et al. 2009; Arndt et al. 2010).

Satellites cannot yet deliver reliably the snow depth and snow water equivalent over Siberia, and the airborne products (e.g., Carroll and Carroll 1989) are costly and not used on the large uninhabited areas of North Asia. Therefore, only in situ data of meteorological networks serve as basis for assessment of changes in these characteristics (e.g., NCDC 2010). Analyses of the snow depth data show that particular maximum snow depth and the number of days with snow depth above 20 cm over most of Siberia is increasing (Bulygina et al. 2007, 2009). To assess the maximum snow water equivalent of snowpack, it is possible to use the snow surveys data (Bulygina et al. 2010). The analysis of these data shows that over most of Siberia since 1966 snow water equivalent has increased (Fig. 3.5) or did not change even while the duration of period with snow on the ground shrunk mostly due to the earlier snowmelt (Bulygina et al. 2011).

Significant climatic changes over Siberia in the twentieth century have been reflected in many atmospheric and terrestrial variables. While changes in surface air temperature and precipitation are most commonly addressed in the literature (cf. above), changes in their derived variables (variables of economic, social, and ecological interest based upon daily temperatures and precipitation) have received less attention. The list of these variables (indices) includes frequency of extremes in precipitation and temperature; frequency of thaws; heating degree days; growing season duration and degree days; sum of temperatures above/below a given threshold; duration of the frost-free period; day-to-day temperature variability; precipitation frequency and duration of prolonged no-rain periods; and various drought and water deficit indices such as Nesterov (NI), Zhdanko (ZhI), Keetch-Byram (KBDI), Palmer Drought Severity (PDSI), and other indices (e.g., Nesterov 1949; Zhdanko 1965; Palmer 1965; Keetch and Byram 1968; Mescherskaya and Blazhevich 1997). In practice, these and other indices are often used instead of "raw" temperature and precipitation values for numerous applications that include modeling of crop yields, prediction and planning for pest management, plant-species development, greenhouse operations, food processing, heat oil consumption in remote locations, electricity sales, heating system design, power plant construction, energy distribution, reservoir operations, floods, and forest fires. These indices provide measures for the analysis of changes

Table 3.2 Mean duration of the growing season (*GS*) and growing degree days (*GDD*) in Siberia south of the Arctic Circle and the rates of its change during the 1966–2009 period

Region	Growing season, days	GC change, days $(10 \text{ year})^{-1}$	GS change, % $(10 \text{ year})^{-1}$	GDD change, % $(10 \text{ year})^{-1}$
West Siberia, north of 55°N	135	3.5	2.6	3.9
West Siberia, south of 55°N	155	3.4	2.2	2.8
Central Siberia, north of 55°N	125	*1.0*	*0.8*	2.8
Central Siberia, south of 55°N	145	2.3	1.6	3.1

All trend estimates except that for GS over Central Siberia north of 55°N are statistically significant at the 0.01 level. Linear trends of the regional GDD describe from 22 % (southern West Siberia) to 32 % (southern East Siberia) of the time series variances

that might impact agriculture, energy, and ecological aspects of Siberia. In the remaining sub-sections of Sect. 3.2, we present information about changes in some of these characteristics during the past several decades.

3.2.5 Changes in the Seasonal Cycle that Impact Ecosystems and/or Human Activity

3.2.5.1 Growing Season Duration and Degree Days for Different Species

Growing season in Siberia is short, and therefore, temperature changes that began occurring here in the warm season in the last several decades dramatically increased its duration (ACIA 2005). Table 3.2 provides estimates of its duration (defined as a period with mean daily surface air temperature being stably above 5 °C) and changes for the 1966–2009 period. In the regions where energy supply to the biosphere is a restricting factor for bioproductivity (e.g., over most of Siberia), growing season duration (GS) itself, while being important, does not directly affect the net ecosystem productivity (NEP). Annual growing degree days (GDD, defined as the sum of temperatures above a given threshold specific for vegetation type/species) represent a better temperature characteristic that correlates closer to NEP (cf. Chap. 6). Changes in GDD shown in the last column of Table 3.2 (where we selected 5 °C as this threshold) for Siberia are much more prominent, and their relative changes (in %) during the past four decades reach two-digit numbers. There is also a second component that controls NEP in Siberia: water availability that should be sufficient for vegetation growth. If we assume that for current ecosystems the water demand from the atmosphere (i.e., needs in precipitation) is balanced by evapotranspiration during the warm season, then a two-digit GDD increase will (a) increase this water demand proportionally and/or (b) cause internal changes in the ecosystems composition that is more congruent to the new regional thermal regime. We have not observed noticeable changes in the warm season precipitation (Sect. 3.2.4), and the earlier spring snow cover retreat just exacerbates the water demand in the warm season. Further consequences of these temperature regime changes are discussed in Sect. 3.2.6.

Table 3.3 Mean duration of the heating season (HS) and heating degree days (HDD) in Siberia south of the Arctic Circle and the rates of its change during the 1966–2009 period

Region	Heating season, days	HS change, days (10 year)$^{-1}$	HDD change, °C (10 year)$^{-1}$	% (10 year)$^{-1}$
West Siberia, north of 55°N	326	−2.0	−154	−2.3
West Siberia, south of 55°N	308	−2.7	−174	−2.9
Central Siberia, north of 55°N	339	−1.4	−131	−1.5
Central Siberia, south of 55°N	333	−3.6	−147	−2.1

All trend estimates except that for HS in West Siberia north of 55°N are statistically significant at the 0.01 level or above (in West Siberia north of 55°N the trend estimate is statistically significant only at the 0.1 level)

Table 3.4 Mean duration of the no-frost period (NFP) in Siberia south of the Arctic Circle and the rates of its change during the 1966–2009 period

Region	No-frost period, days	NFP change, days (10 year)$^{-1}$	NFP change, % (10 year)$^{-1}$
West Siberia, north of 55°N	110	2.3	2.1
West Siberia, south of 55°N	128	3.4	2.6
Central Siberia, north of 55°N	90	2.0	2.2
Central Siberia, south of 55°N	107	2.8	2.6

All trend estimates are statistically significant at the 0.01 level or above (except West Siberia north of 55°N where the trend estimate is statistically significant at the 0.05 level)

3.2.5.2 Heating Season Duration and Degree Days

Heating degree days are the sum of positive mean daily temperature (T_{mean}) anomalies from the base temperature $(T_{base} - T_{mean})_+$. For our calculations shown, we used T_{base} equal to 18 °C (a compromise between 65 °F routinely used in the United States and 17 °C used in Norway). Heating degree days closely correlate to energy consumption for heating and have numerous other practical implications (Guttman and Lehman 1992). Over Siberia, significant reductions in heating degree days are observed (8–12 % per 44 years; Table 3.3). This indicates that there have been reduced heating costs in relative terms.

3.2.5.3 Frost-Free Season

The length of the frost-free (no-frost) period is among the most carefully monitored variables in the continental climate of Siberia. Table 3.4 provides climatology and regional trends for this variable. The regionally averaged duration of the frost-free period varies from less than 100 days in Central Siberia to more than 4 months in steppes of southern West Siberia and has increased by ~10 % during the past four decades over the entire region. It is interesting to note the increase in the frost-free period in Central Siberia north of 55 N (by ~9 days during the 1966–2009 period) where the duration of growing season changed insignificantly.

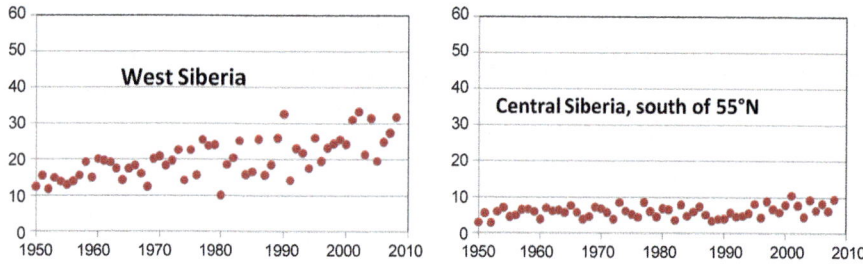

Fig. 3.6 Annual number of days with thaw in West Siberia and in Central Siberia south of 55°N

3.2.6 Changes in Extreme Events Frequency and Intensity

3.2.6.1 Days with Thaws

We define the days with thaw as the days with mean daily temperature greater or equal to −2 °C while snow on the ground is greater or equal to 5 cm (Brown 2000). These days, when occurring during the cold season, dramatically change snow cover properties and cause earlier gradual snowmelt (and the following runoff). Figure 3.6 shows that while in West Siberia the number of days with thaw nearly tripled (following the same pattern as in northern Europe, cf., ACIA 2005; Groisman et al. 2011) farther in the continental interior, the switch between warm and cold seasons is swift, and thaws remain quite infrequent (Fig. 3.6). Nevertheless, their number in the south of Central Siberia has increased during the past decades by 2d (60 year)$^{-1}$, and this increase was statistically significant at the 0.05 level.

In Siberia, due to a northward direction of major rivers, an earlier freshet runoff causes higher peak streamflow in the downstream areas where meltwater arrives before regional melt occurs (cf. Chap. 4, Shiklomanov and Lammers 2009). Moreover, an earlier freshet runoff reduces the amount of the accumulation in the cold season snowfall that remains available for evapotranspiration in the following short vegetation season.

3.2.6.2 Heavy and Very Heavy Rain Events

Siberia south of the Arctic Circle is a cold dry continental region with seasonal precipitation maximum in summer months and with a large areas (e.g., over most of Yakutia) where annual precipitation totals do not exceed 300 mm (Korzun et al. 1974; Shver 1976; Shahgedanova 2003). Nevertheless, on average meteorological stations of Siberia report approximately 105 nonzero precipitation events above 0.5 mm. Additionally, there is a substantial number of very light precipitation events especially in the cold season including those with only a trace daily amount (i.e., less than 0.05 mm day^{-1}). A sufficiently large number of nonzero daily precipitation

Table 3.5 A few characteristics of mean annual measured precipitation at meteorological stations of Siberia south of the Arctic Circle (climatology for the 1966–1995 period when the observational practice and rain gauge design have not substantially changed)

Region	Mean number of wet days (with $P > 0.5$ mm)	Mean daily precipitation in wet days, mm	Thresholds (mm) that define heavy (upper 5th percent) and very heavy (upper 1st percent) daily rain events	
			Heavy	Very heavy
West Siberia, north of 55°N	117	3.8	6	16
West Siberia, south of 55°N	96	3.8	6	14
East Siberia, north of 55°N	112	3.5	6	14
East Siberia, south of 55°N	83	5.3	7	20
Siberia south of the Arctic Circle	105	4.0	6	16

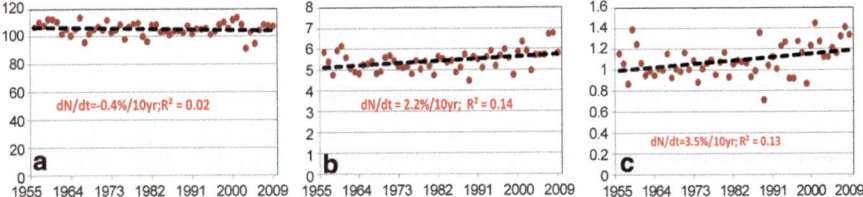

Fig. 3.7 Annual number of days with precipitation above 0.5 mm (**a**), in the upper 5 % (**b**), and 1 % (**c**) of daily rain events area-averaged over Siberia south of the Arctic Circle

events and modest annual precipitation totals hint that daily precipitation totals are quite low. Moreover, a rainy day with precipitation above 15 mm represents a sufficiently rare event that is reported at Siberian stations on average only once per year (the upper 1 % of precipitation events, Table 3.5) and in mountainous regions of southern Siberia 2–3 times per year.

Firstly, from observations (Karl and Knight 1998; Easterling et al. 2000; Groisman et al. 1999, 2005) and then from theoretical considerations (Karl and Trenberth 2003; IPCC 2007; Trenberth 2011), it became clear that the contemporary process of global warming has one distinctive feature: heavy and very heavy precipitation in the extratopical regions has been increasing. In Siberia, this tendency became apparent during the past five decades (Fig. 3.7). Those hydrologists whose major concern is the regional water budget may not be interested in changes of a relatively small fraction of precipitation totals reported in this Figure (upper 1 % of precipitation events contribute a one-digit fraction to the annual totals). However, those whose concern is the frequency of very heavy rainfall (e.g., for disaster control and water resources management) may take notice of the observed changes reported in Fig. 3.7. This information may help them to better operate in the present, already changed climatic conditions, and begin preparations for the future when (according to model projections) the occurrence of these rain events will be more frequent (Sect. 3.3).

3.2.6.3   Prolonged No-Rain Periods

It has been shown that with an increase or decrease of total precipitation, dispropor-tionate changes occur in the upper end of the precipitation frequency distribution in many regions of the extratropical land areas (Groisman et al. 1999, 2005) including Siberia (Sect. 3.2.6.2). These upward trends are primarily a warm season phenom-enon when the most intense rainfall events typically occur. The tendencies, which emerged during the past several decades with a disproportional increase in precipi-tation coming from intense rain events, should lead to discontinuities in the parallel increase/decrease of both total precipitation and precipitation frequency. For the Asian part of Russia, this discontinuity was first reported by Sun and Groisman (2000). They found a simultaneous decrease in the number of days with measurable precipitation and an increase in the number of days with heavy showers, "livni," or (according to the definition of the Russian Meteorological Service) a day with daily totals above 20 mm. If continued, this decrease in precipitation frequency may lead to an increase in the frequency of another potentially hazardous type of extreme event: prolonged periods without precipitation (even when the mean seasonal rainfall totals increase). This development is already occurring during the past several decades over a significant part of North America (Groisman and Knight 2007, 2008) and Europe (Zolina et al. 2009, 2010). The approach developed by Groisman and Knight (2008) was applied to precipitation data over Russia to test how the frequency of the prolonged no-rain periods during the warm season changed during the past several decades. This approach requires the use of a sufficiently dense network of meteorological stations with serially complete daily rainfall and could not be applied to northern regions of Russia. However, over its southern parts in Europe (not shown) and Asia (Fig. 3.8), the frequency of occurrence of prolonged no-rain episodes (30 days and above) in the warm season has increased. While over the European part of Russia, this increase occurred on the background of a general increase of climate humidity, in southern Siberia, the prolonged no-rain periods became more frequent (Groisman et al. 2009, Fig. 3.8), but no discernable rainfall increase was observed. Here, higher severity of agricultural droughts and a general increase in the "fire weather" conditions were reported (Mescherskaya and Blazhevich 1997, updated; Groisman et al. 2007).

Next section (Sect. 3.2.6.4) is focused on meteorological conditions associated with higher than usual chances to ignite fire and on the changes of the occurrence of such conditions in Siberia.

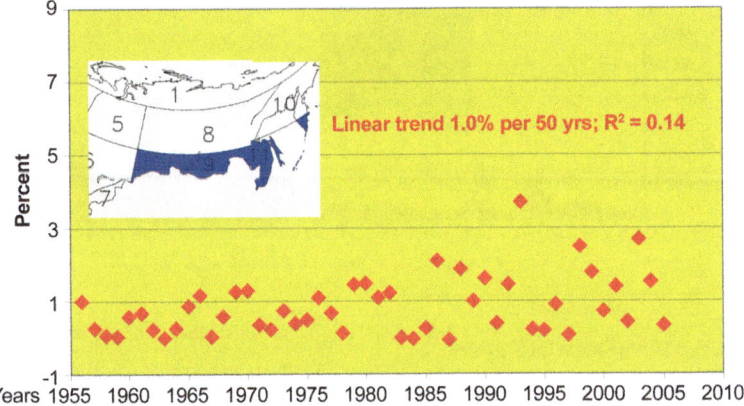

Fig. 3.8 Changes in the occurrence of prolonged no-rain episodes (30 days and above) in the warm season over the Asian part of Russia south of 55°N. Methodology of estimation of these events was described in Groisman and Knight (2008)

3.2.6.4 "Fire" Weather and Droughts

Each summer, all climatic zones in Siberia (from steppe to tundra) suffer from numerous naturally caused fires that are difficult to fight due to their remote locations. Among meteorological variables that affect the potential fire danger are surface air temperature, soil moisture, humidity deficit, probability of lightning, and atmospheric stability (Zhdanko 1965; Keetch and Byram 1968; Turner and Lawson 1978; Gillett et al. 2004). To characterize the level of potential fire danger, numerous indices have been suggested. An overview of these indices is provided in Groisman et al. (2007) where their trends over the Russian Federation were assessed for the past 100 years up to year 2001. It was shown that no matter which forest fire index was used, all these indices report an increase in the potential forest fire danger (so-called fire weather) over broad areas of Siberia south of the Arctic Circle. Updated statistics of actual areas consumed by forest fires and updated time series of these fire weather indices show that this tendency of the north Asia drying does not change and even strengthens (cf. Fig. 3.9; Korovin and Zukkert 2003 updated; Conard et al. 2002 updated).

Summarizing their findings for Siberia, Groisman et al. (2007) stated three lines of evidence:

- Up to two-digit (%) increases in temperature derivatives (e.g., the "warm season" duration and sum of temperatures above the phenologically important thresholds of 5 and 10 °C), which suggest that evapotranspiration may increase
- Earlier snowmelt and more frequent thaws which suggest that more cold season precipitation may go into runoff and become unavailable for vegetation in the warm season
- A moderate (or none) increase in precipitation but a larger increase in thunderstorm activity (thus, lightning frequency may also be on ascent), which suggests that a larger fraction of warm season precipitation may also go into runoff

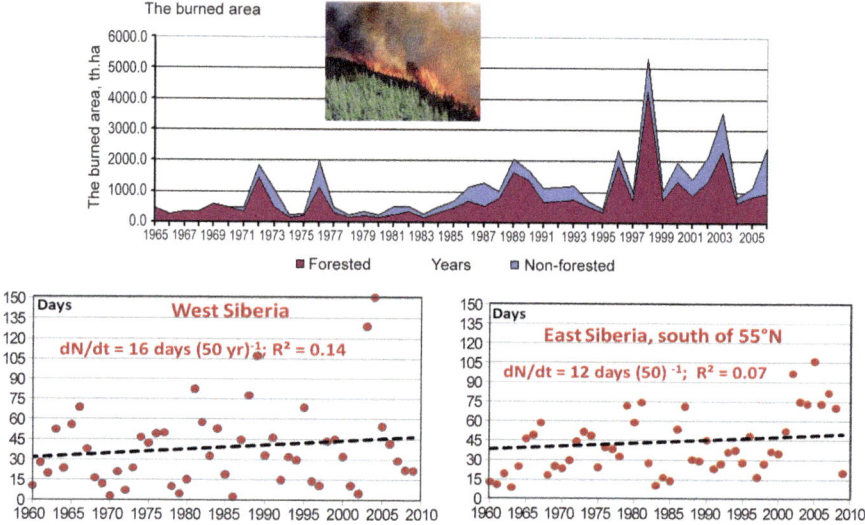

Fig. 3.9 *Top*. Dynamics of burned area over protected territory of Russia (Korovin and Zukkert 2003, updated), presented for the entire Russian Federation. These areas are mostly observed over its Asian part where widespread forest fires cannot be quickly extinguished. *Bottom*. Annual number of days with local Keetch-Byram Drought Indices (*KBDI*) values above their upper 10th percentile for West Siberia and southern part of East Siberia (Updated from Groisman et al. 2007)

Based upon the above factors, Groisman et al. (2007) concluded that there is an increase of the possibility of drier summer conditions in Siberia. Contemporary observations (Meshcherskaya and Blazhevich 1997 updated; Korovin and Zukkert 2003 updated; Mokhov et al. 2003; Dai et al. 2004; Figs. 3.8 and 3.9) further support the observational evidence of the ongoing tendency for "summer dryness" in Siberia. The detrimental consequences of this tendency are an increase in potential forest fire danger (that leads to more frequent forest fires) and an increase in the probability of extremely dry weather conditions (droughts) discussed in Chaps. 4, 6, and 7. Theoretical projections of increasing summer dryness in the interior of the continents as a result of the CO_2-induced global warming were first suggested 30 years ago (Manabe et al. 1981). Further GCM's simulations that are in line with these conclusions are provided in Sect. 3.3.

3.3 Projected Climatic Changes in Siberia: Assessment Based upon an Ensemble of Global Climate Models

3.3.1 Introduction

Climate warming that has already been observed is expected to accelerate further in the twenty-first century (Climate Change 2007; ACIA 2005; Eliseev et al. 2007; Mokhov and Khon 2002a, b; Mokhov et al. 2003). According to estimations from global climate models (GCM), surface air temperature (SAT) can increase up to 3 °C (median; for some scenarios models give an upper range of the temperature increase up to 6 °C) compared to the end of twentieth century (Climate Change 2007) because of anthropogenic emissions of greenhouse gases. The increase of SAT over land area of the Northern Hemisphere is expected to be more than twice that of the global rate (Climate Change 2007; Russian Assessment Report 2008). Dramatic climate change in the northern regions particularly in Siberia is likely caused by the amplification of the greenhouse warming by the ice-albedo feedback (Serreze and Francis 2006; Serreze et al. 2009). The previous Section provides more details on the observed climate changes in Siberia. Next, the projected climate changes are presented.

During the last decade, a GCMs ensemble approach was proposed (Tebaldi and Knutti 2007; Climate Change 2007; Meleshko et al. 2004, 2008a, b). This approach assumes that climate model biases are random with respect to the choice of particular models within a given ensemble. Proper statistical treatment of this model ensemble, in principle, allows one to remove biases and arrive at a more confident projection. At present, the most frequently used ensemble of climate models includes the models involved in phase 3 of the Coupled Model Intercomparison Project (CMIP3, also called AR4 ensemble after their wide usage in preparation of the Fourth Assessment Report issued by the Intergovernmental Panel on Climate Change) (Meehl et al. 2007). While individual models entering this ensemble are not strictly mutually statistically independent (Jun et al. 2008) and do not necessarily span the whole possible uncertainty range (Allen and Ingram 2002), they do form the best contemporary available ensemble of climate models. As pointed out by Govorkova et al. (2008), the CMIP3 models ensemble has performed well for estimating climate over Northern Eurasia. Thus, this ensemble was used below to estimate projections for climate changes in Siberia to the end of twenty-first century. Brief information about CMIP3 models is given in Table 3.6. Simulations with the moderate scenario SRES A1B for the anthropogenic emissions (Nakicenovic et al. 2000) were used.

3.3.2   Projected Changes in the Temperature Regimes

Figure 3.10 shows changes of annual- and seasonal-mean surface air temperature (SAT) in Siberia for the period 2081–2100 compared with 1980–2000 according to the CMIP3 ensemble mean. It should be noted that all individual GCMs show increase of SAT in twenty-first century. The ensemble mean changes of SAT are higher than one standard deviation among different GCMs. These changes are statistically significant at the 0.01 significance level using t-test (it is 0.05 for some regions for seasonal mean). It is expected the increase of annual-mean SAT will be 3–5 °C to the end of the twenty-first century. Furthermore, the largest changes are expected in the polar regions, mostly because of winter-mean changes (up to 7–9 °C). At the same time, in summer, the largest changes are expected in southern regions. These changes in general are less than in winter (about 3–5 °C). Maximum

Table 3.6 CMIP3 model information

Model short name	Research center	Atmosphere/ocean model resolution	Applied models				
			SAT/TCF	T_{max}/T_{min}	Precipitation	Runoff (Amur, Lena, Ob, Yenisei)	Permafrost
BCCR BCM 2.0	Bjerknes Centre for Climate Research, University of Bergen, Norway	T63 L31/(0.5–1.5)° × 1.5° L35	×	×	×		
CCMA CGCM 3.1	Canadian Centre for Climate Modeling and Analysis, Victoria, BC, Canada	T63 L31/0.9° × 1.4° L29	×		×	A, L, O, Y	×
CNRM CM 3.0	Centre National de Recherches Meteorologiques, Meteo–France, Toulouse, France	T63 L45/0.5–2° × 2° L31	×		×	A, L, Y	
CSIRO MK 3.0	Commonwealth Scientific and Research Organization Atmospheric Research, Melbourne, Australia	T63 L18/0.8° × 1.9° L31	×	×	×	A, O, Y	
CSIRO MK 3.5d			×	×	×	A, Y	
GFDL CM 2.0	US Department of Commerce/ National Oceanic and Atmospheric Administration/ Geophysical Fluid Dynamics Laboratory, Princeton, NJ, USA	2.0° × 2.5° L24(0.3–1)° ×	×		×	A, L, O, Y	
GFDL CM 2.1		1° L50	×		×	A, L, O, Y	×
GISS AOM	National Aeronautics and Space Administration/Goddard Institute for Space Studies, New York, USA	3° × 4° L12/3° × 4° L16	×	×	×	A, L, O, Y	
GISS model EH		4° × 5° L20/4° × 5° L13	×		×	A, L, Y	×
GISS model ER			×		×	A, L, O, Y	
IAP FGOALS-g1.0	National Key Laboratory of Numerical Modeling for Atmospheric Sciences and Geophysical Fluid Dynamics/Institute of Atmospheric Physics, Beijing, China	T42 L26/1° × 1° L30	×		×	A, L, O, Y	
INGV SXG	National Institute of Geophysics and Volcanology, Bologna, Italy	T106 L19/2° × 2° L31	×				

Model	Institution	Resolution				Codes	
INM CM 3.0	Institute of Numerical Mathematics, Moscow, Russia	4°×5° L21/2°×2.5° L33	×			A, L, Y	×
IPSL CM 4	Institut Pierre Simon Laplace, Paris, France	2.5°×3.75° L19/(1–2)° × 2° L31	×	×	×	A, L, Y	×
MIROC 3.2	Center for Climate System Research, Tokyo/National Institute for Environmental Studies, Ibaraki/Frontier Research Center for Global Change, Kanagawa, Japan	T40 L20/(0.5–1.4)° × 1.4° L43	×	×	×	A, L, Y	×
MIUB ECHO-G	Meteorological Institute of the University of Bonn, Bonn, Germany/Meteorological Research Institute of the Korea Meteorological Administration, Korea	T30 L19/0.5–2.8° × 2.8° L20	×		×	A, L, Y	
MPI-OM ECHAM 5	Max Planck Institute for Meteorology, Hamburg, Germany	T63 L31/1.5°×1.5° L40	×		×	A, L, Y	×
MRI CGCM2.3.2	Meteorological Research Institute, Tsukuba, Ibaraki, Japan	T42 L30/(0.5–2.0)° × 2.5° L23	×			A, L, O, Y	×
NCAR CCSM 3	National Center for Atmospheric Research, Boulder, CO, USA	T85 L26/(0.3–1)° × 1° L40	×	×	×	A, L, Y	×
NCAR PCM 1	National Center for Atmospheric Research, Boulder, CO, USA	T42 L26/(0.5–0.7)° × 1.1° L40	×		×	O	×
UKMO HadCM 3	Hadley Centre for Climate Prediction and Research/	2.5°×3.8° L19/1.5°× 1.5° L20	×			A, L, O, Y	
UKMO HadGEM 1	Met Office, Exeter, Devon, UK	1.3°×1.9° L38/ (0.3–1.0)° × 1.0° L40	×			A, L, Y	×

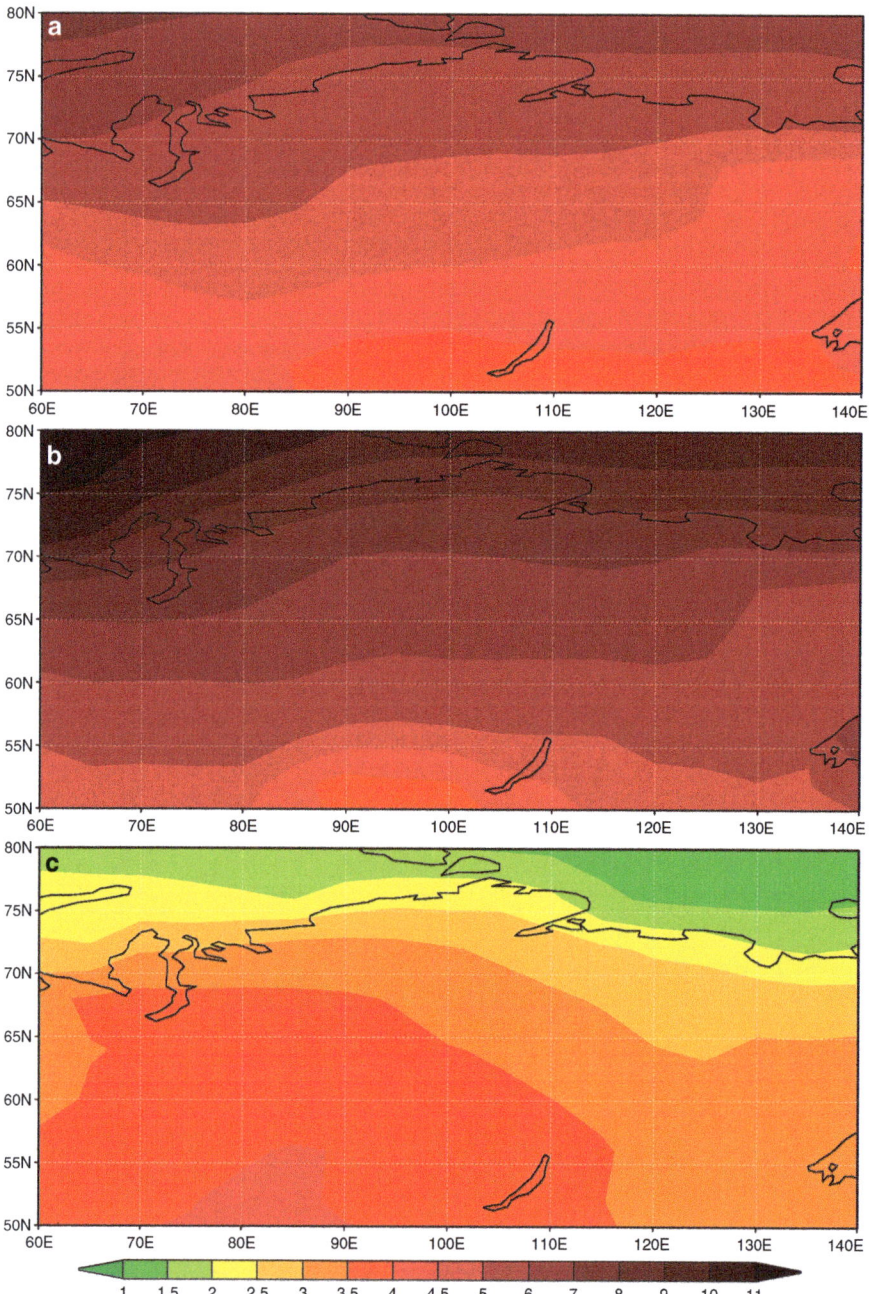

Fig. 3.10 Model-derived surface air temperature changes for the period 2081–2100 compared with 1980–2000 for annual mean (**a**), winter mean (**b**), and summer mean (**c**)

SAT (T_{max}) and minimum SAT (T_{min}) in winter will increase by 1–2 °C to the end of the twenty-first century in southern Siberia and by 2–3 °C in Northern Siberia (not shown). In summer, T_{max} and T_{min} will increase by 1–1.5 °C in southern Siberia and by less than 0.5 °C in Northern Siberia. Changes of T_{max} and T_{min} have no statistical significance for all seasons.

Changes in diurnal temperature range (DTR) are also expected in the twenty-first century. These changes are strongly anti-correlated with cloudiness changes. In Fig. 3.11, changes of total cloud fraction (TCF) over Siberia are shown for the end of the twenty-first century compared with the end of the twentieth century from the CMIP3 ensemble mean (Table 3.6). In general, the increase of annual-mean TCF is expected in Northern Siberia (up to 0.05), and the decrease of annual-mean TCF is expected in southern Siberia (up to 0.02). These changes are likely associated with changes in cyclonic activity (Mokhov et al. 2009). According to model simulations, the frequency of cyclones would increase in the twenty-first century in northern regions (north of 60°N) and decrease in southern regions (south of 60°N) (Yin 2005; Mokhov et al. 2009). Opposite changes are also expected in summer and in winter. In winter, overall increase of TCF is expected over Siberia (except southeast region) with largest changes in northern regions (up to 0.1). These changes could lead to reduction of DTR. In summer, DTR could increase due to overall decreasing of TCF especially over the central and southern part of Siberia where TCF is expected to decrease by 0.02–0.05. In general, changes of TCF in both seasons would exacerbate an expected warming. It is worth noting that changes of TCF have no statistical significance for all seasons.

3.3.3 Projected Changes in the Hydrological Regime

Alongside the temperature increase, significant changes in the hydrological regime are observed in Siberia, particularly runoff increase (for more details, see Sect. 3.2 and Chap. 4 of this book). According to model simulations, there is an expected further increase of runoff of Siberian rivers in the twenty-first century (Mokhov and Khon 2002a, b; Mokhov et al. 2003; Shiklomanov and Shiklomanov 2003; Meleshko et al. 2004, 2008b; Nohara et al. 2006; Kattsov et al. 2007;Khon and Mokhov 2012) along with an increase of the number of extreme precipitation events (Mokhov et al. 2005b, 2006; Khon et al. 2007). In general, GCMs are able to reproduce the hydrological regime of Siberia quite well when compared to observational data (Meleshko et al. 2004; Govorkova et al. 2008; Khon and Mokhov 2012). As pointed out by Khon and Mokhov (2012), GCMs reproduce well the precipitation regime and runoff for main Siberian rivers on an annual basis (Fig. 3.12). At the same time, most GCMs produce a 1-month shift of the maximum runoff to earlier months (Fig. 3.13).

3.3.3.1 Projected Changes in Precipitation and Snow Cover

Figure 3.14 shows spatial distribution of changes of precipitation rate, intensity (c, d), and probability for winter and summer for the end of the twenty-first century compared

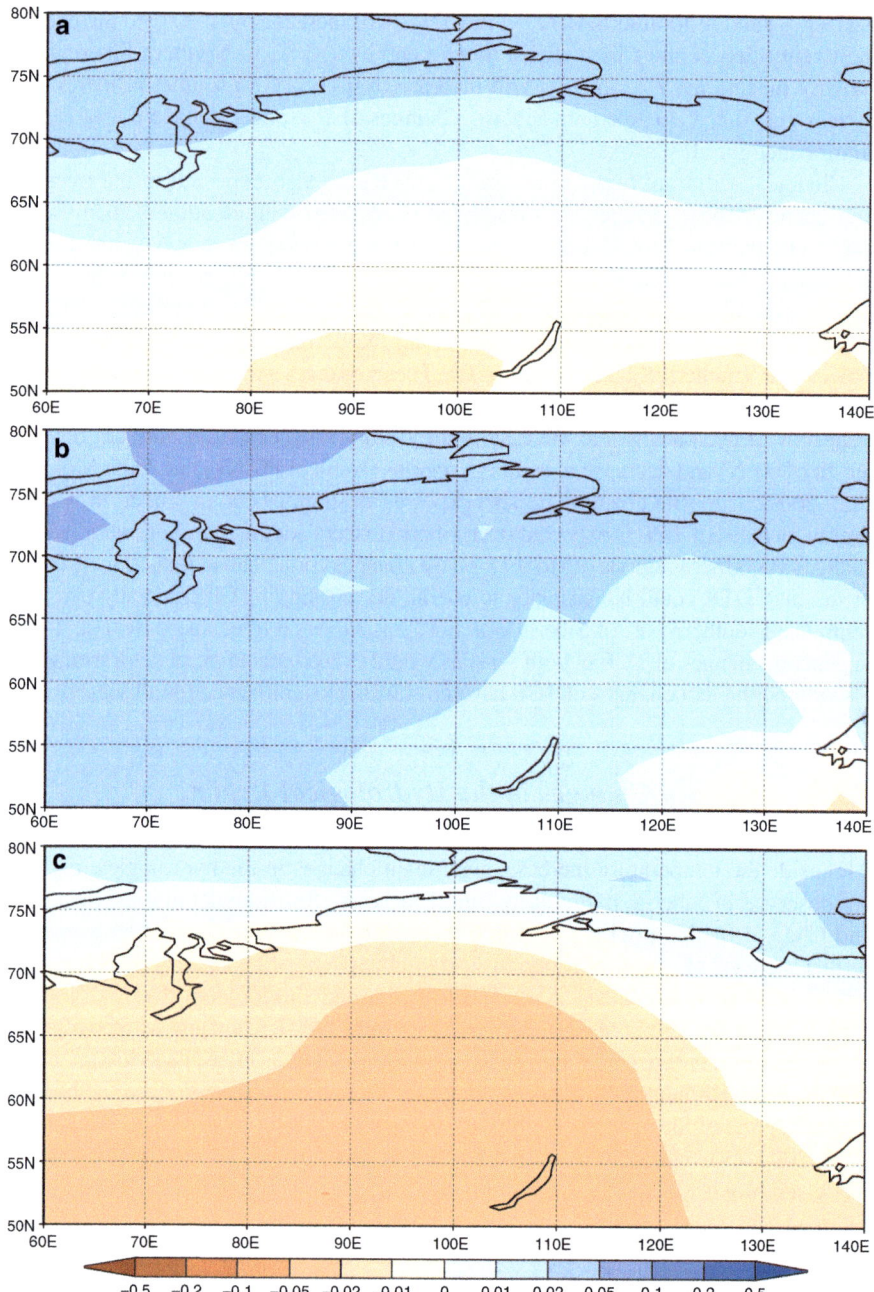

Fig. 3.11 Model-derived total cloud fraction changes for the period 2081–2100 compared with 1980–2000 for annual mean (**a**), winter mean (**b**), and summer mean (**c**)

R, mm/year

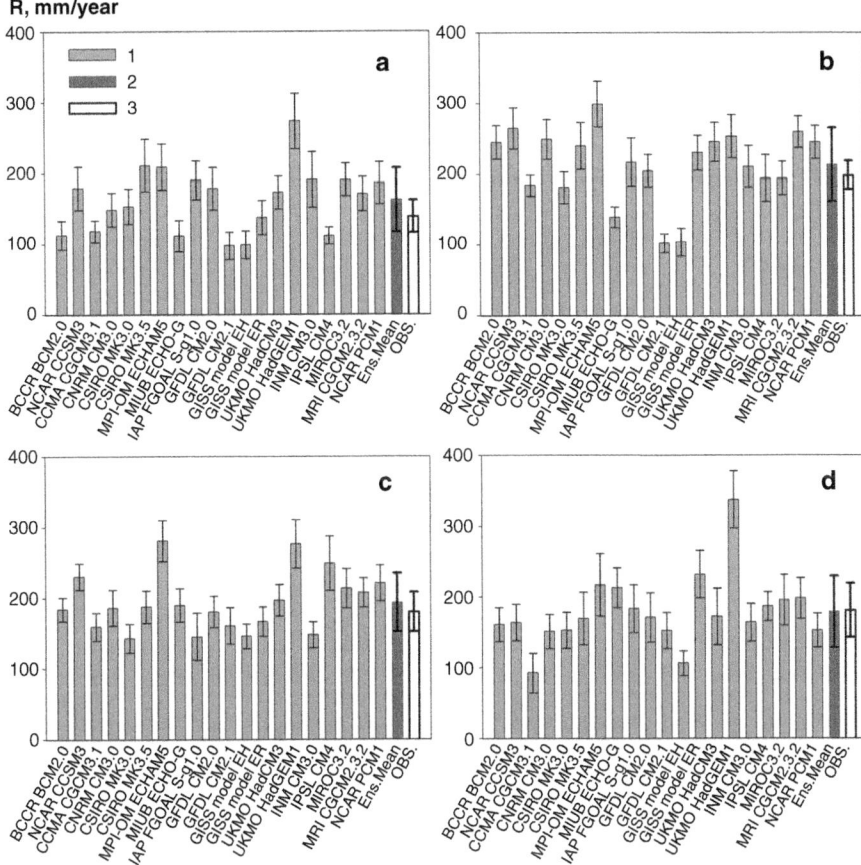

Fig. 3.12 Annual-mean runoff (mm/year) and its standard deviation for (**a**) Ob', (**b**) Yenisei, (**c**) Lena, and (**d**) Amur rivers according to different models *1*, ensemble mean *2*, and observational data *3* from Climate Research Unit (New et al. 2000)

with the end of the twentieth century. The largest increase of precipitation rate is expected in winter for all river basins especially over northeastern part of Eurasia (up to 40 % for Yenisei, Lena, and Amur river basins). This will lead to an additional amount of snow over most of Siberia. Therefore, increasing of snow water equivalent, which is noted over Siberia during the last decades (cf. Sect. 3.2), is expected to continue up to the end of the twenty-first century. In summer, changes of precipitation rate are not particularly strong. The same is true for spring and autumn. The intensity of precipitation is increased for the entire year with maximum changes in summer. The changes of probability of precipitation are different for summer and winter. In particular, the frequency of precipitation increases in winter and decreases in summer. Thus, in winter, precipitation is becoming more frequent and heavy. In summer, the overall precipitation rate will change slightly, but precipitation events will become less frequent and more intensive.

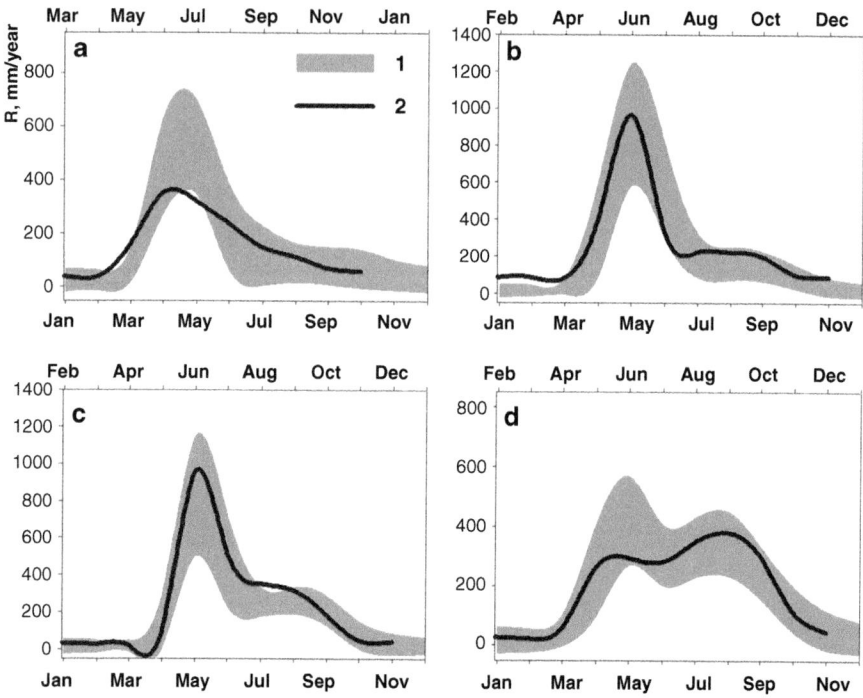

Fig. 3.13 Runoff annual cycle (mm/year) for (**a**) Ob′, (**b**) Yenisei, (**c**) Lena, and (**d**) Amur rivers according to ensemble mean (*1, lower panels*) and observational data CRU (*2, upper panels*)

3.3.3.2   Projected Runoff Changes

The precipitation increase will lead to runoff increase for main Siberian rivers. Changes of the annual-mean runoff for Ob, Yenisei, Lena, and Amur rivers in the twenty-first century are depicted in Fig. 4.25 (next chapter). All GCMs show an increase of annual-mean runoff. It is up to 15 % for Ob River, 20 % for Yenisei River, and 25 % for Lena River. For Yenisei and Lena rivers, these changes are larger than the interannual standard deviation. Model-derived changes for the Amur River runoff are smaller than for other rivers. Some models show a little decrease for the Amur River runoff. In general, these changes are not statistically significant. In Fig. 3.15, changes of the seasonal-mean runoff for different rivers are depicted. Because of the accumulation of additional water in winter, the highest increase of the runoff is expected in spring for most of Siberian rivers, especially for Lena and Yenisei rivers. The summer runoff will be decreased for Western and Central Siberian rivers (Ob and Yenisei rivers), and it will be increased for Eastern Siberian rivers (Lena and Amur rivers). In Chap. 4, analysis based on ensemble with 8 GCMs provides more detailed information on the Siberian rivers' runoff.

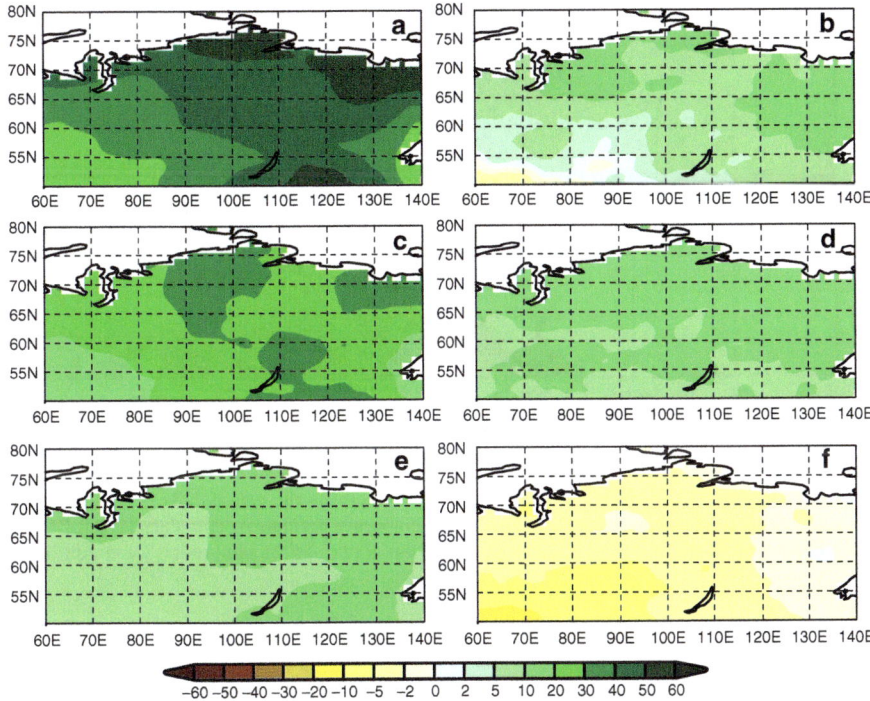

Fig. 3.14 Projected precipitation rate changes (**a, b**), precipitation intensity changes (**c, d**), and precipitation probability changes (**e, f**) for winter (**a, c, e**) and for summer (**b, d, f**) for the end of the twenty-first century compared with the end of the twentieth century according to model ensemble mean

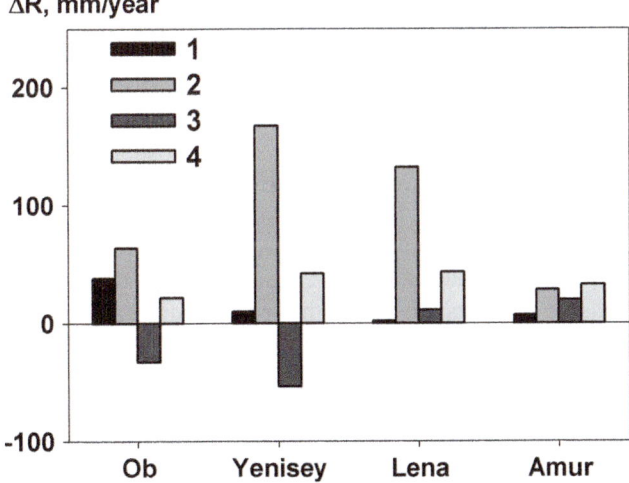

Fig. 3.15 Runoff changes (mm/year) for Ob′, Yenisei, Lena, and Amur rivers to the end of twenty-first century according to model ensemble mean for winter *1*, spring *2*, summer *3*, and autumn *4*

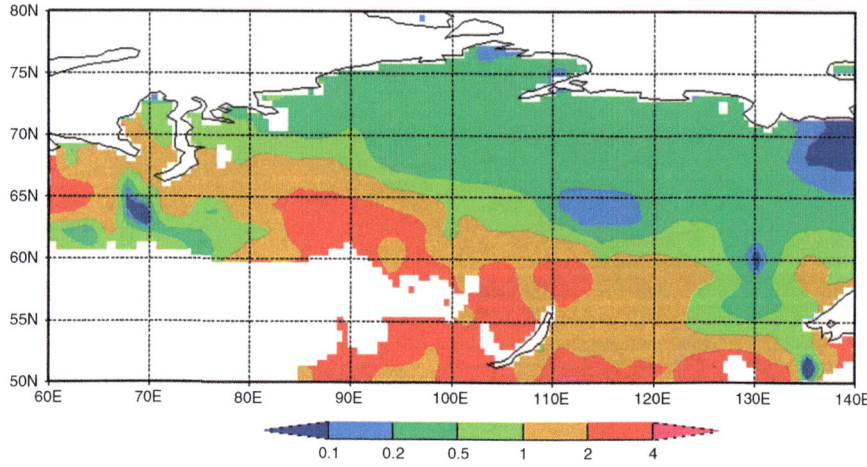

Fig. 3.16 Model-derived changes of permafrost table depth for the period 2081–2100 compared with 1980–2000. *Shaded area* corresponds to contemporary permafrost boundary (From Brown et al. 1998. revised February 2001. Circum-Arctic map of permafrost and ground-ice conditions)

3.3.4 Projected Changes in the State of Permafrost

Dramatic changes of SAT over Siberian region lead to changes of soil temperature, particularly in permafrost areas. Active layer thickness increase and permafrost thawing will cause surface subsidence and destructive geocryological processes evolution including soil flow, thermokarst, and thermoerosion processes (Mazhitova and Kaverin 2007). Modeling is the main approach to investigate climate impacts on permafrost and to make projections for permafrost changes in the twenty-first century. Various scientific groups' models simulate processes in permafrost in different ways. Anisimov (2009) suggested using a stationary model with information provided about soil temperature based on surface temperature and thermal conductivity coefficients. Another approach is to use a dynamical one-dimensional heat transfer model. In particular, this is implemented in the permafrost model of the University of Alaska Fairbanks (Tipenko et al. 2004; Marchenko et al. 2008).

Soil model of A.M. Obukhov Institute of Atmospheric Physics (Arzhanov et al. 2008) was used to estimate projected climate impacts on permafrost area over Siberia in the twenty-first century. In particular, assessment of seasonal active layer thickness and talik's depth (talik is part of thawed ground in the permafrost area) was performed. The main atmospheric variables including SAT, precipitation, incoming solar radiation, and air moisture came from 10 GCMs (see Table 3.6). Figure 3.16 shows the maximum changes of permafrost table depth (PTD) for the last 20 years of the twenty-first century compared with the last 20 years of the twentieth century from the ensemble mean. The major changes of PTD are noted over the permafrost southern boundaries mostly because of talik's depth increase. According to model-based projections, there is expected to be an increase of PTD by 2–3 m in West

Siberia and in Zabaikalie region (to the east of Lake Baikal). The increase of PTD by 0.5–1.5 m is expected to the mid-twenty-first century (cf. also Anisimov 2009).

These changes are confirmed by observational data showing an increase of talik's depth by 0.6–0.7 m for the last 20–30 years (Oberman 2008). The expected changes of permafrost conditions are less in the northern regions. In particular, there can be a PDT increase by no more than 0.2–0.3 m (Fig. 3.16). Permafrost condition changes can lead to forcing processes with negative consequences for economics as well as natural ecosystems of Siberia (Shiklomanov et al. 2008; Romanovsky et al. 2010; Anisimov 2009; Nelson et al. 2002). For more details on permafrost changes impacts, see Sect. 4.8 and Chaps. 5 and 7 of this book.

3.3.5   Projected Changes in Extreme Climate and Weather Events

Temperature changes can be explained by changes in mean values as well as changes in variance. Model projections show that rising temperature will be driven by both causes (Houghton 2009). These could lead to reduction of events with record cold weather and increasing of days with extremely hot weather. According to Meleshko et al. (2008b), the major changes will occur in wintertime. In particular, the number of days with extreme low temperature in Northern Siberia is expected to decrease by 6–8 days by the middle of the twenty-first century (Meleshko et al. 2008b). The number of days with minimum temperature lower than 0 °C could decrease by 10–15 days by the middle of the twenty-first century (Meleshko et al. 2008b) and by 15–20 days to the end of the twenty-first century. In summer, the number of days with extremely high temperature is expected to increase up to 10–20 over Northern Siberia (Russian Assessment Report 2008). In addition, the duration of periods with record hot weather is also expected to increase. These hot weather duration periods will increase by 2–4 days over most of Siberia (Russian Assessment Report 2008).

There is a strong likelihood that mutual changes of temperature and precipitation regimes over Siberia will lead to increases of probability and intensity of extreme climate and weather events such as droughts and forest fires (Mokhov et al. 2005a; Malevskii-Malevich et al. 2007; Shkol'nik et al. 2008; Mokhov and Chernokulsky 2010), which has been confirmed in the observations during recent decades (Sect. 3.2.6.4). As pointed out by Mokhov et al. (2005a), an increase of temperature by 1 °C could lead to an increase of drought areas up to 10 %. Assessments with Regional Climate Model of Main Geophysical Observatory (Shkol'nik et al. 2008; Mokhov and Chernokulsky 2010) show that overall temperature increase during summer over most of Siberia with prolongation of period without rain will lead to an increase in severity of fire danger indices. The major increase is expected over Eastern Siberia and Zabaikalie to the middle of the twenty-first century, mostly because of changes in precipitation regime (Mokhov and Chernokulsky 2010). Because of earlier snowmelt, it is expected that the duration of fire danger weather could be prolonged by 1–1.5 months in southern Siberia. Numerous studies (e.g., Soja et al. 2007; Vygodskaya et al. 2007) show that climate changes in Siberia, particularly increasing of forest fires in southern regions and warming in northern regions, could lead to the boreal forest zone changes. It is expected that the climate-induced ecosystem zones will shift to the north.

3.4 Energy and Water Exchange in Ecosystems of Central Siberia

Annual and seasonal dynamics of energy and water exchange are controlled by both the physical properties of the environment and the biological properties of ecosystems. Climatic conditions influence ecosystem energy and mass (water and CO_2) exchange and carbon balance. Ecosystems in turn influence surface climate, transforming components of radiation, energy, and water balances. Thus, feedback mechanisms between the environment and the biosphere are being formed. In Central Siberia, these feedbacks may reverse some of the climatic changes and exercise an impact on the global earth system (Groisman and Bartalev 2007).

Energy and water fluxes in various ecosystems (tussock tundra, pine forest, Sphagnum bog, true steppe) in Central Siberia, totals for the growing season (May–September) of 1998–2000, were analyzed from eddy covariance measurements taken during the EuroSiberian Carbon Flux project (1998–2000) and the TCOS-Siberia project (2002–2004). Energy and water balance components were summarized and are presented in Table 3.7. In this table, components in similar ecosystems from different geographical regions in Siberia extracted from previous publications are also presented for comparison.

Components of the energy balance in a pine forest and a Sphagnum bog located as close as 500 m to each other differed due to the type of underlying surface. Due to the difference in the albedo of the forest (lower albedo, 10 %) and the bog (higher albedo, 15 %), vegetation radiation balance and air temperature were higher above the forest than above the bog (Table 3.7). During the growing season, sensible and latent heat fluxes in these ecosystems had different signs. In the forest, sensible heat exceeded latent heat by a factor of 1.3. In the bog, latent heat was threefold greater than sensible heat because available water did not limit evaporation. In the tussock tundra, for the second, warm part of the growing season, latent heat was 1.3 times greater than sensible heat. In the steppe, sensible heat was greater in the first half of the growing season, but with increasing rain during the summer, latent heat prevailed sensible heat resulting in total 710 MJ m^{-2} of latent heat versus 410 MJ m^{-2} of sensible heat (Table 3.7).

The difference in the surface roughness and available moisture of these two ecosystems initiated differences in the Bowen ratio, β. The average daily Bowen ratio was always below 1.0, varying between 0.2 and 0.7 in different bog ecosystems (Schulze et al. 1999), and was 0.6 in the assessed Sphagnum bog (Table 3.7). The transparent pine forest with its large gaps in forest cover consumed more energy for sensible heat and less energy for evaporation compared to other ecosystems, thus

Table 3.7 Summary of annual and seasonal energy and water balance (May–September) in major Central Siberian ecosystems: a pine forest (1998–2000), a Sphagnum bog (1998–2000), a steppe (2002–2004), a tussock tundra (2002–2004), and a larch permafrost forest (1998–2006) derived from eddy covariance measurements[a] (Schulze et al. 1999; Tchebakova et al. 2002; Kurbatova et al. 2002; Corradi et al. 2005; Belelli Marchesini 2007; Ohta 2010) and annual energy balance in analogous Siberian ecosystems from gradient measurements[b] (Pavlov 1984; Martyanova 1970)

Geography, an ecosystem type	Energy balance components, MJ m^{-2}				Water balance components, mm		
	Net radiation	Sensible heat	Latent heat	Bowen ratio	Rain	Evaporation	Soil water and runoff
Central Siberia[a]							
Pine forest	1,415	615	465	1.4	235	195	+40
Sphagnum bog	1,330	210	680	0.6	235	235	0
True steppe	1,480	410	710	0.8	315	310	+5
Tussock tundra	625	170	220	0.6	50	100	−50
Larch forest in central Yakutia					190	196	−6
Central Yakutia[b]							
Larch forest	1,740	610	930	0.7			
Pine forest	1,780	650	1,000	0.8			
Stony tundra	1,440	800	465	1.7			
Transbaikaliye steppe[b]	1,718	760	834	0.9			

resulting in a higher Bowen ratio (Table 3.7). In early spring, before photosynthesis commenced, the monthly Bowen ratio was 2–3 on average and could reach 8–10 on some hot days (Tchebakova et al. 2002). In the neighboring fir–spruce and birch ecosystems, in moister habitats of the Yeniseisky range, β varied around 1 and was less than 1, respectively (Roser et al., 2002). In open larch ecosystems on permafrost in Yakutia, β was as high as 10–20 in early spring before leaves came out then decreased downward to ~1 at the peak of the growing season (Ohta 2010). Wilson and Baldocchi (2000) noted that in evergreen conifer forests, the Bowen ratio steadily decreased in the seasonal course with sensible heat fluxes exceeding latent heat fluxes ($\beta > 1$) from spring to the midsummer and then remained around 1 or lower till fall. In comparison, in deciduous broadleaf and conifer (*Larch spp.*) forests and meadows, latent heat exceeded sensible heat by two to three times once photosynthetic activity began and remained high until the leaves fell.

In dry climate of Yakutia, the 6-year water balance (soil moisture + runoff) was slightly negative. Therefore, in the absence of permafrost (which thaws in summer and provides water in addition to rain), moisture would not be sufficient to support tree growth, and steppe landscapes would rather dominate there (Pozdnyakov 1993). In the tussock tundra, evaporation exceeded rain by twofold due to unlimited water to evaporate in the floodplain on permafrost. In the pine forest, water discharge from runoff and soil water was 20 % of rain. In the neighboring bog, the water balance was null for these 3 growing seasons although some discharge from creeks was available. In the steppe, the water balance was slightly positive providing some soil moisture for shallow-rooted grass to prevail (Table 3.7).

In Siberia, the climate at the end of the twenty-first century is predicted by GCMs to be much warmer and dryer in terms of the ratio between heat and water available for evaporation (cf. Sect. 3.3; Vygodskaya et al. 2007; Tchebakova et al. 2009a). Even if the predicted precipitation increase of 10–20 % occurs (or decrease in some areas) in Siberia (IPCC 2007), this increase would not be large enough to balance temperature increase and would actually cause a greater evaporation deficit. This means that there would be a greater difference between potential and actual evaporation creating an unfavorable condition for plants and especially trees to grow. Thus, the energy balance would be partitioned in favor of sensible flux that would exceed latent flux resulting in an increasing Bowen ratio. To survive unfavorable conditions, plants would have to close stomata. If these conditions would last for long (years), first, plant production would drop like it happened in Europe in 2003 (Ciais et al. 2005), and then plants of a given vegetation type would die out and be replaced by a more drought-resistant vegetation type. So climate warming would initiate such effects as vegetation shifts (Tchebakova et al. 2009a, b), an increased potential for severe and large fires (Flannigan et al. 2005) that affect forest structure and functions (Amiro et al. 2006), a shorter snow cover period (Chapin et al. 2005; Bulygina et al. 2009), etc., which feedback to each other and all together would change surface albedo. Albedo change would in turn affect surface energy balance resulting in both positive and negative feedbacks to the climate system accelerating/slowing down climate warming. Chapter 6 further addresses dynamics of this ecosystem development in Siberia.

3.5 Brief Summary

During the past 10,000 years, climate of Siberia varied broadly. It was both warmer and colder than the present. In high latitudes, the warmer climates were mostly accompanied by higher humidity, and in the south of Siberia, they could be both wet and dry in different warm periods of the past.

During the past century, the climate of Siberia became much warmer (cf. Fig. 3.3), and snow data hint that the cold season precipitation north of 55°N increased. At the same time, we do not observe rainfall increase over most of Siberia that, in conjunction with temperature increase, increase in duration of the warm season, and thus, a general increase in water demand for evapotranspiration led to drier summer conditions, prolonged "no-rain" situations, and increased possibility of manifestations of extreme dry events (droughts, fire weather, and northward advance of steppe into the forest zone and forest into the tundra zone).

Projections of the future climate for the end of the twenty-first century indicate the further temperature increases, more in the cold season (by up to +8 °C in high latitudes) and less in the warm season (by up to +5 °C in southern Siberia). For precipitation, there is a projected increase but only in the cold season. There is likely to be a further reduction of the difference of "precipitation minus evapotranspiration" $(P-E)$

Table 3.8 Estimates and projections of the mean surface air temperature changes over Siberia in the "climatic optimum" of the Holocene (paleoclimatic reconstructions; Klimanov and Novenko 2010; Velichko et al. 1992), during the period of instrumental observations (past 130 years, e.g., Fig. 3.3), and as projected by the contemporary GCMs (Sect. 3.3.2)

| Period | Mean temperature changes, °C | | | Comments |
	Annual	Winter	Summer	
5,500–6,000 BP	1–4	1–4	0–5	Compared to the mid of the twentieth century
Past 130 years	1.8	2.4	1	Regional linear trends, °C per 130 years
2,081–2,100	3–5	3–8	1–2	Compared with the 1980–2000 period

toward more negative values that may lead to (a) significant changes in the hydrological cycle in Central and southern Siberia (Chap. 4), (b) significant ecosystems' shifts and changes in the regional carbon cycle (Chap. 6), and (c) changes in the permafrost distribution and stability (Sect. 3.3.3) with potential of detrimental impact on the regional infrastructure (cf. Chaps. 5 and 7).

Observed (Sect. 3.2.6) and projected (Sect. 3.3.4) frequencies of various extreme events have increased recently and are projected to further increase.

Taking into account the expected further global and regional warming in Siberia, we focus on this feature of Siberian climate and put together in Table 3.8 a few most prominent general numbers that characterize three (past, present, and projected) warming periods in the Holocene. Comparison among the lines of this table shows that:

(a) The projected summer warming is already close to that observed, but the winter projections are much higher, probably due to the positive feedbacks (mostly related to the future snow cover retreat) that have not yet materialized over Siberia on a full scale.

(b) Changes in the climatic optimum of the Holocene (while its spatial distribution and radiation forcing were very different from the present) were quite homogeneous spatially in the cold season (mostly +2 to +3 °C) but had a strong meridional gradient in the warm season (from less than +1° in the southern Siberia to more than +4 °C in the Taymyr Peninsula).

Warm period of the past was accompanied with drying in the south and more humid climatic conditions in the north of Siberia that manifested itself in vegetation shifts. The same pattern is projected for the future (cf. Vygodskaya et al. 2007). While in the north of Siberia, contemporary models predict warmer winters in the end of the twenty-first century and paleoreconstructions hint to warmer summers compared to the present warming observed during the period of instrumental observations, these three groups of estimates are broadly consistent with each other. In fact, we can even explain the discrepancies with present climate changes by their transitional nature. The entire summer warming of the past 130 years occurred only in the past two decades, and the vegetation and permafrost (and therefore, positive climatic feedbacks associated with them) had not yet fully reacted to this warming.

Furthermore, the multiannual Arctic sea ice extent prevailing in the past 130 years will not be a factor in decreasing the future winter warming at the end of the twenty-first century.

References

ACIA (Arctic Climate Impact Assessment) (2005) Chapter 2 "Arctic climate system and its global role". In: Arctic climate impact assessment, "impact of a warming arctic". Cambridge University Press, Cambridge, 144 pp

Allen M, Ingram W (2002) Constraints on future changes in climate and the hydrologic cycle. Nature 419(6903):224–232

Amiro BD, Orchansky AL, Barr AG, Black TA, Chambers FS, Chapin FS, Goulden ML, Litvak M, Liu HP, McCaughey JH, McMillan A, Randerson JT (2006) The effect of post-fire stand age on the boreal forest energy balance. Agr Forest Meteorol 140:41–50

Andreev AA, Klimanov VA (1991) Vegetation history and climate changes in the interfluve of the rivers Ugra and Yakonit (the Southern Yakutia) in Holocene. J Bot 76(3):334–351 (in Russian)

Andreev AA, Klimanov VA, Sulerzhitsky LD (1997) Younger Dryas pollen records from central and southern Yakutia. Quat Int 41/42:111–117

Andreev AA, Klimanov VA, Sulerzhitsky LD, Khotinsky NA (1989) Chronology landscape-climatic changes in Central Yakutia during the Holocene. In: Khotinsky NA (ed) Paleoclimate of the late glacial – Holocene time. Nauka, Moscow, pp 116–121

Andreev AA, Klimanov VA, Sulerzhitsky LD (2001) Vegetation and climate history of the Yana River lowland, Russia, during the last 6400 yr. Quat Sci Rev 20:259–266

Andreev AA, Siegert C (2002) Late pleisticene and holocene vegetation and climate on the Taymyr lowland, Northern Siberia. Quat Res 57:138–150

Andreev AA, Siegert C, Klimanov VA, Derevyagin AY, Shilova G, Melles M (2002) Late Pleistocene and Holocene vegetation and climate on the Taymyr Lowland, Northern Siberia. Quat Res 57:138–150

Anisimov OA (2009) Stochastic modelling of the active layer thickness under conditions of the current and future climate. Kriosfera Zemli 13(3):36–44

Arndt DS, Baringer MO, Johnson MR (eds) (2010) State of the climate in 2009. Bull Am Meteor Soc 91(6):S1–S224

Arzhanov MM, Eliseev AV, Demchenko PF, Mokhov II, Khon VCh (2008) Simulation of thermal and hydrological regimes of Siberian river watersheds under permafrost conditions from reanalysis data. Izvestiya Atmos Ocean Phys 44(1):83–89

Belelli Marchesini L (2007) Analysis of the carbon cycle of steppe and old field ecosystems of Central Asia. PhD thesis, University of Tuscia, Viterbo Italy, 227 pp. http://dspace.unitus.it/handle/2067/540

Belorusova ZhM, Lovelius NV, Ukraintseva VV (1987) Regional features of nature variations on Taimyr in the Holocene. Bot J 72:610–618 (in Russian)

Belov AV, Bezrukova EV, Sokolova LP, Abzaeva AA, Letunova PP, Fisher EE, Orlova LA (2006) Vegetation of Pribaykal'ye as indicator of global and regional changes of environmental conditions of Northern Asia in Late Cenozoic. Geogr Nat Resour 27(3):5–18 (in Russian)

Belova VA, Barysheva YeM, Kol'tsova VG, Kutaph'eva TK, Nikol'skaya MV, Savina LN (1982) Vegetation cover of Eastern Siberia in Holocene. In: Logachev NA (ed) Late Quaternary and Holocene of the southern past of East Siberia. Academy of Sciences of the USSR, Siberian branch, Institute of the Earth's crust, Novosibirsk, pp 64–70 (in Russian)

Bezrukova EV, Abzaeva AA, Letunova PP, Kulagina NV, Veershinin KE, Belov AV, Orlova LA, Danko LV, Krapivina SV (2005) Post-glacial history of Siberian spruce (*Picea obovata*) in the Lake Baikal area and the significance of this species as a paleoenvironmental indicator. Quat Int 136:47–57

Blyakharchuk TA (1989) History of vegetation and climate of south-east of West Siberia in the Holocene (according to macrofossil and spore-pollen analyses of peat deposits). PhD thesis, Tomsk, 226 pp (in Russian)

Blyakharchuk TA (2003) Four new pollen sections tracing the Holocene vegetational development of the southern part of the West Siberian Lowland. The Holocene 13(5):715–731

Blyakharchuk TA, Sulerzhitsky LD (1999) Holocene vegetation and climatic Change in the forest zone of West Siberia according to pollen records from the extrazonal palsa bog Bugristoe. The Holocene 9(5):621–628

Blyakharchuk TA, Wright HE, Borodavko PS, van der Knaap WO, Ammann B (2004) Late Glacial and Holocene vegetational changes on the Ulagan high-mountain plateau, Altai Mountains, southern Siberia. Paleogeogr Paleoclimatol Paleoecol 209:259–279

Blyakharchuk TA, Wright HE, Borodavko PS, van der Knaap WO, Ammann B (2007) Late Glacial and Holocene vegetational history of the Altai Mountains (southwestern Tuva Republic, Siberia). Paleogeogr Paleoclimatol Palynol 245:518–534

Blyakharchuk TA, Wright HE, Borodavko PS, van der Knaap WO, Ammann B (2008) The role of pingos in the development of the Dzhangyskol lake-pingo complex, central Altai Mountains, southern Siberia. Paleogeogr Paleoclimatol Palynol 257:404–420

Borisova OK, Zelikson EM, Kremenetski KV, Novenko EYu (2005) Landscape and climatic changes in West Siberia in Late glacial and Holocene based upon a new palynological data. Trans Russ Acad Sci Ser Geogr 6:38–49 (in Russian)

Borisova OK, Novenko EYu, Zelikson EM, Kremenetski KV (2011) Lateglacial and Holocene vegetational and climatic changes in the southern taiga zone of West Siberia according to pollen records from Zhukovskoye peat mire. Quat Int 237:65–73. doi:10.1016/j.quaint.2011.01.015

Borzenkova II (1992) Climatic changes during the Cenozoic era. Gidrometeoizdat, Sankt-Petersburg, 247 pp (in Russian)

Borzenkova II, Brook SA (1989) On impact of volcanic eruptions on the climate changes in the late-glaciate period through Holocene. Trans State Hydrol Inst 347:40–56 (in Russian)

Borzenkova II, Zubakov VA (1984) The climate optimum of the Holocene as a model of the beginning of the 21st century. Meteorol Hydrol 8:69–77 (in Russian) (Eng. trans. Russ Meteorol Hydrol, 1984, No. 8)

Borzenkova II, Zhiltsova HL, Lobanov VA (2011) Ice cores data and tree-ring temperature reconstructions as sources of climate variations over the historical time. Ice and snow 2. "Nauka", M (in Russian)

Brown J, Ferrians OJ Jr, Heginbottom JA, Melnikov ES (eds) (1998) Circum-Arctic map of permafrost and ground-ice conditions. U.S. Geological Survey in Cooperation with the Circum-Pacific Council for Energy and Mineral Resources, Washington, DC. Circum-Pacific Map Series CP-45, scale 1:10,000,000, 1 sheet. Available at: http://nsidc.org/data/docs/fgdc/ggd318_map_circumarctic/index.html

Brown RD (2000) Northern hemisphere snow cover variability and change, 1915–1997. J Climate 13:2339–2355

Brown RD, Mote PhW (2009) The response of northern hemisphere snow cover to a changing climate. J Climate 22:2124–2145. doi:10.1175/2008JCLI2665.1

Budyko MI, Vinnikov KYa (1976) Global warming. Soviet Meteorol Hydrol 7:16–26

Bulygina ON, Razuvaev VN, Korshunova NN, Groisman PYa (2007) Climate variations and changes in the climate extreme events in Russia. Environ Res Lett 2, 045020:7. doi:10.1088/1748-9326/2/4/045020

Bulygina ON, Razuvaev VN, Korshunova NN (2009) Changes in snow cover over Northern Eurasia in the last few decades. Environ Res Lett 4. doi:10.1088/1748-9326/4/4/045026

Bulygina ON, Ya P, Groisman VNRazuvaev, Radionov VF (2010) Snow cover basal ice layer changes over Northern Eurasia since 1966. Environ Res Lett 5:10. doi:10.1088/1748-9326/5/1/015004

Bulygina ON, Groisman YaP, Razuvaev VN, Korshunova NN (2011) Changes in snow cover characteristics over Northern Eurasia since 1966. Environ Res Lett 6, 045204: 10. doi:10.1088/1748-9326/6/4/045204

Burashnikova TA, Muratova MV, Suetova IA (1982) Climatic model of territory of Soviet Union during the Holocene optimum. In: Velichko AA (ed) Development of nature of USSR in the late Pleistocene and the Holocene. Nauka, Moscow, pp 245–251 (in Russian)

Callaghan TV, Johansson M, Brown RD, Groisman PYa, Labba N, Radionov V, Contributors (2011) Changing snow cover and its impacts. In: Snow, Water, Ice and Permafrost in the Arctic (SWIPA), 59 pp. AMAP Report to the Arctic Council [Available at http://amap.no/swipa/]

Carroll SC, Carroll TR (1989) Effect of uneven snow cover on airborne snow water equivalent estimates obtained by measuring terrestrial gamma radiation. Water Resour Res 25(7): 1505–1510

Chapin FS III, Sturm M, Serreze MC, McFadden GP, Key JR, Lloyd AH, McGuire AD, Rupp TS, Lynch AH, Schimel JP et al (2005) Role of land-surface changes in arctic summer warming. Science 310:657–660

Ciais P, Reichstein M, Viovy N, Granier A, Ogee J, Allard V, Aubinet M, Buchmann N, Bernhofer C, Carrara A et al (2005) Europe-wide reduction in primary productivity caused by the heat and drought in 2003. Nature 437:529–533

Conard SG, Sukhinin AI, Stocks BJ, Cahoon DR, Davidenko EP, Ivanova GA (2002) Determining effects of area burned and fire severity on carbon cycling and emissions in Siberia. Clim Change 55:197–211

Corradi C, Kolle O, Walter K, Zimov SA, Schulze E-D (2005) Carbon dioxide and methane exchange of a north-east Siberian tussock tundra. Glob Chang Biol 11:1910–1925

Crowley TJ (2000) Causes of climate change over the past 1000 years. Science 289:270–277

Crowley TJ, Berner RA (2001) CO_2 and climate change. Science 392:870–872

D'Arrigo R, Wilson R, Jacoby G (2006) On the long-term context for late twentieth century warming. J Geophys Res 111:D03103. doi:10.1029/2005JD006352

Dai A, Trenberth KE, Qian T (2004) A global dataset of Palmer Drought Severity Index for 1870–2002: relationship with soil moisture and effects of surface warming. J Hydrometeorol 5:1117–1130

Demske D, Heumann G, Granoszewski W, Nita M, Mamakowa K, Tarasov PE, Oberhansli H (2005) Late glacial and Holocene vegetation and regional climate variability evidenced in high-resolution pollen records from Lake Baikal. Glob Planet Change 46:255–279

Easterling DR, Evans JL, Groisman PYa, Karl TR, Kunkel KE, Ambenje P (2000) Observed variability and trends in extreme climate events: a brief review. Bull Am Meteor Soc 81:417–425

Eliseev A, Mokhov I, Karpenko A (2007) Influence of direct sulfate-aerosol radiative forcing on the results of numerical experiments with a climate model of intermediate complexity. Izvestiya Atmos Ocean Phys 42(5):544–554

Firsov LV, Troitski SL, Levina TP, Nikitin VP, Panychev VP (1974) Absolute age and first standard pollen diagram of Holocene peat section in northern Siberia. Bull Comm Investig Quat Period 41:121–127 (in Russian)

Flannigan MD, Logan KA, Amiro BD, Skinner WR, Stocks BJ (2005) Future area burned in Canada. Clim Chang 72:1–16

Gandin LS, Celnai R, Zakhariev VE (eds) (1976) Statistical structure of meteorological fields. Az Orszagos Meteorologiai Szolgalat, Budapest, 364 pp [In Russian and Hungarian, resumes in German]

Gao C, Robock A, Self S, Witter JB, Steffenson JP, Clausen HB, Siggaard-Andersen M, Johnsen S, Mayewski PA, Ammann C (2006) The 1452 or 1453 A.D. Kuwae eruption signal derived from multiple ice core records: greatest volcanic sulfate event of the past 700 years. J Geophys Res 111:D12107. doi:10.29/2005JD006710

Gillett NP, Weaver AJ, Zwiers FW, Flannigan MD (2004) Detecting the effect of climate change on Canadian forest fires. Geophys Res Lett 31:L18211. doi:10.1029/2004GL020876

Glebov FZ, Karpenko LV, Klimanov VA, Mindeeva TN (1996) Paleoecological analysis of peat section in Ob'-Vasyugan inter fluver area. Sib Ecol J 6:497–504 (in Russian)

Govorkova VA, Kattsov VM, Meleshko VP, Pavlova TV, Shkol'nik IM (2008) Climate of Russia in the 21st century. Part 2. Verification of atmosphere – ocean general circulation models CMIP3 for projections of future climate changes. Russ Meteorol Hydrol 33(8):467–477

Groisman PYa, Bartalev SA (2007) Northern Eurasia earth science partnership initiative (NEESPI): science plan overview. Glob Planet Change 56(3–4):215–234

Groisman PYa, Knight RW (2007) Prolonged dry episodes over North America: new tendencies emerging during the last 40 years. Adv Earth Sci 22(11):1191–1207

Groisman PYa, Knight RW (2008) Prolonged dry episodes over the conterminous United States: new tendencies emerging during the last 40 years. J Climate 21:1850–1862

Groisman PYa (1981) Empirical estimates of relationship between processes of global warming (cooling) and precipitation on USSR territory. Izw Acad Sci USSR Ser Geogr 1981(5):86–95 (in Russian, in English in Soviet Geography)

Groisman PYa, Soja AJ (2009) Ongoing climatic change in Northern Eurasia: justification for expedient research. Environ Res Lett 4:7. doi:10.1088/1748–9326/4/4/045002

Groisman PYa, Karl TR, Knight RW, Stenchikov GL (1994) Changes of snow cover, temperature, and radiative heat balance over the northern hemisphere. J Climate 7:1633–1656

Groisman PYa et al (1999) Changes in the probability of heavy precipitation: important indicators of climatic change. Clim Chang 42(1):243–283

Groisman PYa, Knight RW, Easterling DR, Karl TR, Hegerl GC, Razuvaev VN (2005) Trends in intense precipitation in the climate record. J Climate 18:1343–1367

Groisman PYa, Knight RW, Razuvaev VN, Bulygina ON, Karl TR (2006) "State of the ground" rarely used characteristic of snow cover and frozen land: climatology and changes during the past 69 years over Northern Eurasia. J Climate 19:4933–4955

Groisman PYa et al (2007) Potential forest fire danger over Northern Eurasia: changes during the 20th century. Glob Planet Change 56(3–4):371–386

Groisman PYa et al (2009) The Northern Eurasia earth science partnership: an example of science applied to societal needs. Bull Am Meteorol Soc 90:671–688

Groisman PYa, Gutman G, Reissell A (2011) Chapter 1. Introduction: climate and land-cover changes in the arctic. In: Gutman G, Reissell A (eds) Arctic land cover and land use in a changing climate: focus on Eurasia, vol VI. Springer, Amsterdam, 306 pp

Guttman NB, Lehman RL (1992) Estimation of daily degree-days. J Appl Meteorol 31:797–810

Herzschuh U, Tarasov P, Wunnemann B, Hartmann K (2004) Holocene vegetation and climate of the Alashan Plateau, NW China, reconstructed from pollen data. Paleogeogr Paleoclimatol Paleoecol 211:1–17

Houghton J (2009) Global warming. The complete briefing. Cambridge University Press, Cambridge, 438 pp

Hutchinson MF, de Hoog FR (1985) Smoothing noisy data with spline functions. Numerische Mathematik 47:99–106

IPCC (2007) In: Solomon S, Qin D, Manning M, Chen Z, Marquis M, Tignor KBM, Miller HL (eds) Climate change 2007: the physical science basis. Contribution of Working Group I to the fourth assessment report of the Intergovernmental Panel on Climate Change. Cambridge University Press, Cambridge/New York, 996 pp

Isachenko AG, Shlyapnikov AA, Robozertseva OD, Filipetskaya AZ (1988) The landscape map of the USSR. General Ministry of Geodesy and Cartography of the USSR, Moscow (4 plates)

Jones PD, Mann ME (2004) Climate over past millennia. Rev Geophys 42(2):RG2002. doi:10.1029/2003RG000143

Jun M, Knutti R, Nychka D (2008) Spatial analysis to quantify numerical model bias and dependence: how many climate models are there? J Am Stat Soc 103(483):934–947

Kaplina TN, Chekhovski (1987) Reconstruction paleogeographical situation in the maritime lowland of Yakutia during the Holocene climatic optimum. In: Pokhilainen VP (ed) Quaternary period of the North-Eastern Asia. SVKNII DVO AN SSSR, Magadan, pp 145–151 (in Russian)

Kaplina TN, Lozhkin AV (1982) The history of Holocene vegetation development on coastal low-lands of Jakutia. In: The development of nature in the USSR territory in the Late Pleistocene-Holocene. Nauka, Moscow, pp 207–220

Karl TR, Knight RW (1998) Secular trends of precipitation amount, frequency, and intensity in the United States. Bull Am Meteorol Soc 79:231–241

Karl TR, Trenberth KE (2003) Modern global climate change. Science 302:1719–1723. doi:10.1126/science.1090228

Karpenko LV (2006) Reconstructing quantitative indices of climate and vegetation successions of the Sym-Dubches interfluve for the Holocene. Geogr Nat Resour 2006(2):77–82

Kattsov VM, Walsh JE, Chapman WL, Govorkova VA, Pavlova T, Zhang X (2007) Simulation and projection of arctic freshwater budget components by the IPCC AR4 global climate models. J Hydrometeorol 8:571–589

Keetch JJ, Byram GM (1968) A drought index for forest fire control. U.S.D.A. Forest Service Research Paper SE-38, 35 pp. Available from: http://www.srs.fs.fed.us/pubs/

Khantemirov RM (1999) Tree-ring reconstruction of summer temperatures in the North of Western Siberia over the past 3248 years. Sib Ekolog J 6(2):185–191 (in Russian)

Khmelevtsov SS (ed) (1986) Volcanoes, stratospheric aerosol, and the Earth climate. Gidrometeoizdat, Leningrad, 256 pp (in Russian)

Khon VCh, Mokhov II (2012) The hydrological regime of large river basins in Northern Eurasia in the XX-XXI centuries. Water Resour 39(1):1–10

Khon VCh, Mokhov II, Roeckner E, Semenov VA (2007) Regional changes of precipitation characteristics in Northern Eurasia from simulations with global climate model. Glob Planet Change 57:118–123

Khotinsky NA (1977) The Holocene of northern Eurasia. Nauka, Moscow, 200 p (in Russian)

Khotinsky NA (1984) Holocene vegetation history. In: Velichko AA, Wright HE Jr, Barnosky CW (eds) Late Quaternary environments of the Soviet Union. University of Minnesota Press, Minneapolis, pp 179–200

Khotinsky NA (1989) Disputable questions of reconstruction and correlation of paleoclimates of the Holocene. In: Khotinsky NA (ed) Paleoclimates of the late Glacial and the Holocene. Nauka, Moscow, pp 12–17 (in Russian)

Khotinsky NA, Klimanov VA (1985) Radiocarbon age and climatic conditions of development of palsa mires in Nadym-Kasym inteffluve area in Holocene. In: Voprosy ekologii rastenii bolot, bolotnukh mestoobitanii I torfyanykh zalezhei. Karelian Branch of RAS, Petrozavodsk, pp 132–140

Kind NV, Leonov BN (eds) (1982) The anthropogen of Taimyr. Paleobotanical and paleoclimatical reconstructions. Nauka, Moscow, 183 p (in Russian)

Kirpotin SN, Blyakharchuk TA, Vorob'ev SN (2003) Dynamics of subarctic tussock mires in West Siberian Lowlands as an indicator of the global climatic changes. News of the Tomsk State University, Series in Biological Sciences, pp 122–134 (in Russian)

Klimanov VA (1976) To the method of reconstruction of quantitative parameters of past climate. Vestnik MGU Ser Geogr 2:92–98

Klimanov VA (1984) Paleoclimatic reconstructions based on the information statistical method. In: Velichko AA, Wright HE Jr, Barnosky CW (eds) Late Quaternary environments of the Soviet Union. University of Minnesota Press, Minneapolis, pp 297–303

Klimanov VA (1994) The climate of the Northern Eurasia in the Alleröd time interval. Rep Acad Nauk Russ 339:533–537 (in Russian)

Klimanov VA, Novenko EYu (2010) Climatic conditions in Northern Eurasia during the Holocene optimum. In: Velichko AA (ed) Atlas-monograph "evolution of landscapes and climates of Northern Eurasia". Late Pleistocene – Holocene – elements of prognosis. Issue 3. GEOS, Moscow, 220 pp

Klimanov VA, Sirin AA (1997) Dynamics of peat accumulation in northern Asia during last 3000 years. Dokl Akad Nauk 354:683–686 (in Russian)

Korovin GN, Zukkert NV (2003) Climatic change impact of forest fires in Russia. In: Danilov-Danilyan VI (ed) Climatic change: view from Russia. TEIS Publishers, Moscow, pp 69–98, 416

Korzun VI, Sokolov AA, Budyko MI, Voskresensky KP, Kalinin GP, Konoplyantsev AA, Korotkevich ES, L'vovitch MI (eds) (1974) Atlas of world water balance. USSR National Committee for the International Hydrological Decade. English translation. UNESCO, Paris, 35 pp + 65 maps

Koshkarova BL (1989) Holocene climatic changes in the area of Yenisei River (according to paleo-ecological data). In: Khotinsky NA (ed) Paleoclimates of the late Glacial and the Holocene. Nauka, Moscow, pp 96–98 (in Russian)

Koshkarova VL (2004) Evolution of vegetation and climate changes in the Minusa basin throughout the Holocene (according to paleocarpology data). Geogr Nat Resour 2:84–89

Koshkarova VL, Koshkarov AD (2004) Regional signatures of changing landscape and climate of northern central Siberia in the Holocene. Russ Geol Geophys 45(6):672–685

Kurbatova J, Arneth A, Vygodskaya NN, Kolle O, Varlagin AB, Milyukova IM, Tchebakova NM, Schulze E-D, Lloyd J (2002) Comparative ecosystem–atmosphere exchange of energy and mass in a European Russian and a central Siberian bog I. Interseasonal and interannual variability of energy and latent heat fluxes during the snow free period. Tellus 54B:497–513

Kutafiyeva TK (1975) Vegetation history in the interstream between the rivers Low and Podkamennaya Tunguska in the Holocene. In: Savina LN (ed) History of the Siberian forests in the Holocene. Forest Institute USSR Academy of Sciences Publishers, Krasnoyarsk, pp 72–95 (in Russian)

Kwok R, Untersteiner N (2011) The thinning of Arctic sea ice. Phys Today 41(4):36–41

Laukhin SA (1994) Evolution landscape-vegetation zonality of the north-eastern part of Asia over the Pleistocene. Rept Acad Nauk Russia 338:683–686 (In Russian)

L'vov YuA, Blyakharchuk TA (1982) Paleobotanical studies of turf in the center of Tunguska meteorite fallout. In: Meteorite and meteor studies. Nauka Publishing House, Novosibirsk, pp 89–99

Levina TP, Orlova LA (1993) The Holocene climatic cycles in the south of West Siberia. Geol Geophys 3:38–55 (in Russian)

Levkovskaya GM, Kind NV, Zavelski FS, Firsov VS (1970) Absolute chronology of peat deposits in the area of Igarka and partition of the Holocene in West Siberia. Bull Comm Investig Quat period 39:94–101

Lister AM, Sher AV (1995) Ice cores and mammoth extinction. Nature 378:23–24

Lozhkin AV, Vazhenin LN (1987) Vegetation evolution of the maritime Lowland of Kolyman in the early Holocene. In: Pokhilainen VP (ed) Quaternary period of the North-Eastern Asia. SVKNII DVO AN SSSR, Magadan, pp 135–144 (in Russian)

Lozhkin AV, Anderson PM et al (1995) New palynological and radiocarbon data about evolution the vegetation of West Beringia during the Late Pleistocene and Holocene. In: YuM Bychkov, Lozhkin AV (eds) History of climate and vegetation in Beringia during the Late Cenozoic. North East Interdisciplinary Research Institute, Far East branch, Russian Academy of Sciences, Magadan, pp 5–24 (in Russian)

Lugina KM, Groisman PYa, Vinnikov KYa, Koknaeva VV, Speranskaya NA (2006) Monthly surface air temperature time series area-averaged over the 30-degree latitudinal belts of the globe. In: Trends online: a compendium of data on global change. Carbon Dioxide Information Analysis Center, Oak Ridge National Laboratory, U.S. Department of Energy, Oak Ridge, Tennessee, pp 1881–2005. doi: 10.3334/CDIAC/cli.003

Luterbacher J, Dietrich D, Xoplaki E, Grosjean M, Wanner H (2004) European seasonal and annual temperature variability, trends, and extremes since 1500. Science 303(5663):1499–1503

MacDonald G et al (2000) Holocene treeline history and climate change across Northern Eurasia. Quat Res 53:302–311

Malevskii-Malevich SP, Mol'kentin EK, Nadezhina ED, Semioshina AA, Sall' IA, Khlebnikova EI, Shklyarevich OB (2007) Analysis of changes in fire-hazard conditions in the forests in Russia in the 20th and 21st centuries on the basis of climate modeling. Russ Meteorol Hydrol 32(3):154–161

Manabe S, Wetherald RT, Stouffer RJ (1981) Summer dryness due to an increase of atmospheric CO_2 concentration. Clim Chang 3:347–386

Mangerud J, Andersen ST, Berglund BE, Donner JJ (1974) Quaternary stratigraphy of Norden, a proposal for terminology and classification. Boreas 3:109–128

Marchenko SS, Romanovsky VE, Tipenko G (2008) Numerical modeling of spatial permafrost dynamics in Alaska. In: Proceedings of the ninth international conference on permafrost, 29 June–3 July 2008, Fairbanks, Alaska. Institute of Northern Engineering, University of Alaska, Fairbanks

Martyanova GN (1970) Turbulent heat and water exchange in steppes of Transbaikalia. In: Bachurin GV, Nechaeva EC, Snytko VA (eds) Proceedings of the conference on topological heat, water and matter exchange in geosystems. Institute of Geography, Sib Branch, USSR Academy of Science, Irkutsk, pp 9–12

Mazhitova GG, Kaverin DA (2007) Thaw depth dynamics and soil surface subsidence at a circumpolar active layer monitoring (CALM) site, the European North of Russia. Kriosfera Zemli 11(4):20–30

Meehl GA, Covey C, Taylor KE, Delworth T, Stouffer RJ, Latif M, McAvaney B, Mitchell JFB (2007) THE WCRP CMIP3 multimodel dataset: a new era in climate change research. Bull Am Meteor Soc 88:1383–1394

Meleshko VP, Govorkova VA, Kattsov VM, Malevskii-Malevich SP, Nadezhina ED, Sporyshev PV, Golitsyn GS, Demchenko PF, Eliseev AV, Mokhov II, Semenov VA, Khon VC (2004) Anthropogenic climate change in Russia in the twenty-first century: an ensemble of climate model projections. Russ Meteorol Hydrol 29(4):22–30

Meleshko VP, Kattsov VM, Mirvis VM, Govorkova VA, Pavlova TV (2008a) Climate of Russia in the 21st century. Part 1. New evidence of anthropogenic climate change and the state of the Art of its simulation. Russ Meteorol Hydrol 33(6):341–350

Meleshko VP, Kattsov VM, Govorkova VA, Sporyshev PV, Shkol'nik IM, Shneerov BE (2008b) Climate of Russia in the 21st century. Part 3. Future climate changes calculated with an ensemble of coupled atmosphere-ocean general circulation CMIP3 models. Russ Meteorol Hydrol 33(9):541–552

Meshcherskaya AV, Blazhevich VG (1997) The drought and excessive moisture indices in a historical perspective in the principal grain-producing regions of the former Soviet Union. J Climate 10:2670–2682

Mironenko ON, Savina LN (1975) On the vegetation history of Central Siberia at its northern boundary. In: History of Siberian forest, Siberian branch of the Russian Academy of Science. V.N. Sukachev Inst. of Forest, Krasnoyarsk, pp 37–59 (In Russian)

Mokhov II, Chernokulsky AV (2010) Regional model assessments of forest fire risks in the Asian part of Russia under climate change. Geogr Nat Resour 31(2):165–169

Mokhov I, Khon V (2002a) Hydrological regime in basins of Siberian rivers: model estimates of changes in the 21st century. Russ Meteorol Hydrol 27(8):77–93

Mokhov I, Khon V (2002b) Model scenarios of changes in the runoff of Siberian rivers in the 21st century. Doklady Earth Sci 383(3):329–332

Mokhov I, Semenov V, Khon V (2003) Estimates of possible regional hydrologic regime changes in the 21st century based on global climate models. Izvestiya Atmos Ocean Phys 39(2):130–144

Mokhov II, Dufresne J-L, Le Treut H, Tikhonov VA, Chernokulsky AV (2005a) Changes in drought and bioproductivity regimes in land ecosystems in regions of northern Eurasia based on calculations using a global climatic model with carbon. Doklady Earth Sci 405(6):810–814

Mokhov II, Semenov VA, Khon VCh, Roeckner E (2005b) Extreme precipitation regimes in northern Eurasia in the 20th century and their possible changes in the 21st century. Doklady Earth Sci 403(5):767–770

Mokhov II, Semenov VA, Khon VCh, Roeckner E (2006) Possible regional changes in precipitation regimes in northern Eurasia in the 21st century. Water Resour 33(6):702–710

Mokhov II, Chernokul'skii AV, Akperov MG, Dufresne J-L, Le Treut H (2009) Variations in the characteristics of cyclonic activity and cloudiness in the atmosphere of extratropical latitudes of the northern hemisphere based from model calculations compared with the data of the reanalysis and satellite data. Doklady Earth Sci 424(1):147–150

Monserud RA, Tchebakova NM, Kolchugina TP, Denissenko OV (1995) Change in phytomass and net primary productivity for Siberia from the Mid-Holocene to the present. Glob Biogeochem Cycle 9:213–226

Nakicenovic N, Davidson O, Davis G, Grubler A, Kram T, Lebre La Rovere E, Metz B, Morita T, Pepper W, Pitcher H, Sankovski A, Shukla P, Swart R, Watson R, Dadi Z (2000) Emissions scenarios, special report of Working Group III of the Intergovernmental Panel on Climate Change. Cambridge University Press, Cambridge, 599 pp

National Climatic Data Center (NCDC) (2005) Data documentation for dataset 9813. Raw, homogenized, and bias-free daily precipitation for the former USSR and Russia, 14 pp [Available at http://www1.ncdc.noaa.gov/pub/data/documentlibrary/tddoc/td9813.pdf]

National Climatic Data Center (NCDC) (2010) DSI-9808 "Snow depth and ice crust at snow courses of the Russian Federation during the 1966–2009 period". Data set description is available from NOAA National Climatic Data Center, Asheville, NC

Naurzbaev MM, Vaganov EA, Sidorova OV, Schweingruber FH (2002) Summer temperatures in eastern Taimyr inferred from a 2427-year late-Holocene tree-ring chronology and earlier floating series. Holocene 12:727–736

Nelson FE, Anisimov OA, Shiklomanov NI (2002) Climate change and hazard zonation in the Circum-Arctic permafrost regions. Nat Hazard 26:203–225

Nesterov VG (1949) Forest fire potential and methods of its determination. Goslesbumizdat Publishing House, Moscow (in Russian)

New M, Hulme M, Jones P (2000) Representing twentieth-century space-time climate variability. Part II: development of 1901–96 monthly grids of terrestrial surface climate. J Climate 13: 2217–2238

Nikol'skaya MB (1980) Paleobotanic characteristics of the Upper Pleistocene and Holocene deposits on the Taymyr. In: Volkova VS (ed) Paleopalinologia Sibiri. Nauka, Moscow pp 97–111 (In Russian)

Nohara D, Kiton A, Hosaka M, Oki T (2006) Impact of climate change on river runoff. J Hydrometeorol 7:1076–1089

Oberman N (2008) Contemporary permafrost degradation of European North of Russia. In: Proceedings of the ninth international conference on permafrost, vol 2, June 29–July 3, Fairbanks, Alaska, pp 1305–1310

Ohta T (2010) Hydrological aspects in a Siberian larch forest. In: Osawa A et al (eds) Permafrost ecosystems: Siberian larch forests. Springer, Dordrecht, pp 245–269

Orlova LA, Panychev VA (1989) Radiocarbon dating of podzolic soils with second humus horizon. In: Regional stratigraphy of Siberia and Far East. Nauka, Novosibirsk, pp 125–135 (in Russian)

Osborn TJ, Briffa KR (2006) The spatial extent of 20th century warmth in the context of the past 1200 years. Science 311(5762):841–844

Palmer WC (1965) Meteorological drought. Research Paper No. 45. U.S. Weather Bureau. [NOAA Library and Information Services Division, Washington, DC 20852]

Pavlov AV (1984) Energy exchange in the landscape sphere of the Earth. Nauka, Novosibirsk, 256 pp (in Russian)

Peteet D, Andreev A, Bardeen W, Mistretta F (1998) Long-term Arctic peatland dynamics, vegetation, and climatic history of the Pur-Taz regions, West Siberia. Boreas 27:115–126

Pozdnyakov LK (1993) Forestry on permafrost. Nauka, Novosibirsk, 192 pp (In Russian)

Prentice IC, Guiot J, Huntley B, Jolly D, Cheddadi R (1996) Reconstructing biomes from palaeoecological data: a general method and its application to European pollen data at 0 and 6Ka. Clim Dyn 12:185–196

Proshutinsky A, Ashik I, Hakkinen S, Hunke E, Krishfield R, Maltrud M, Maslowski W, Zhang J (2007) Sea level variability in the Arctic Ocean from AOMIP models. J Geophys Res 112:C04S08. doi:10.10292006JC003916

P'yavchenko NI (1983) About age of peat deposits and Holocene change of vegetation in the south of West Siberia. Bull Comm Investig Quat period 52:164–170 (in Russian)

Rawlings MA, 29 Co-Authors (2010) Analysis of the arctic system for freshwater cycle intensification: observations and expectations. J Climate 23:5715–5737

Robinson DA, Dewey KF, Heim R Jr (1993) Global snow cover monitoring: an update. Bull Am Meteorol Soc 74:1689–1696

Romanovsky VE, Drozdov DS, Oberman NG, Malkova GV, Kholodov AL, Marchenko SS, Moskalenko NG, Sergeev DO, Ukraintseva NG, Abramov AA, Gilichinsky DA, Vasiliev AA (2010) Thermal state of permafrost in Russia. Permafr Periglac Process 21:136–155

Roser C, Montagnani L, Schulze ED, Mollicone D, Kolle O, Meroni M, Papale D, Belelli Marchesini L, Federici S, Valentini R (2002) Net CO_2 exchange rates in three different successional stages of the "Dark Taiga" of central Siberia. Tellus B 54:642–654. doi: 10.1034/j.1600-0889.2002.01351.x

Russian Assessment Report (2008) In: Meleshko VP, Semenov SM (eds) Russian assessment report on climate change and its impact on the Russian territory. Technical summary. VNIIGMI-MCD, Moscow, 92 pp (in Russian)

Savina SS, Khotinsky NA (1982) Zonal method of reconstruction of paleoclimates of the Holocene. In: Velichko AA (ed) Development of nature on the territory of USSR in the late Pleistocene and the Holocene. Nauka, Moscow, pp 179–186 (in Russian)

Savina SS, Khotinsky NA (1984) Holocene paleoclimatic reconstructions based on the zonal method. Late Quaternary environments of the Soviet Union. In: Velichko AA, Wright HE Jr, Barnosky CW (eds) Late Quaternary environments of the Soviet Union. University of Minnesota Press, Minneapolis, pp 287–296

Savina LN, Koshkarova VL (1981) In: Yamskikh AF (ed) Natural conditions of the Minusink basin. Krasnoyarsk Pedagogical Institute, Krasnoyarsk, pp 101–110

Schulze E-D, 21 Co-Authors (1999) Productivity of forests in the Eurosiberian boreal region and their potential to act as a carbon sink – a synthesis. Glob Change Biol 5:703–722

Serreze MC, Francis JA (2006) The arctic amplification debate. Clim Chang 76(3–4):241–264

Serreze MC, Barrett AP, Stroeve JC, Kindig DN, Holland MM (2009) The emergence of surface-based Arctic amplification. The Cryosphere 3:11–19

Shahgedanova M (2003) The physical geography of northern Eurasia. Oxford University Press, Oxford, 571 pp

Sherstyukov BG, Razuvaev VN, Bulygina ON, Groisman PYa (2007) NEESPI Science and Data Support Center for Hydrometeorological Information in Obninsk, Russia. Environ Res Lett 2, 045010:2. doi:10.1088/1748–9326/2/4/045010

Shichi K, Takahara H, Krivonogov SK, Bezrukova EV, Kashiwaya K, Takehara A (2009) Vegetation and climate records for the last 50 kir from Lake Kotokel, the middle Lake Baikal area, East Siberia. Quat Int 205:98–110

Shiklomanov AI, Lammers RB (2009) Record Russian river discharge in 2007 and the limits of analysis. Environ Res Lett 4. 045015 (9 pp) doi: 10.1088/1748–9326/4/4/045015

Shiklomanov IA, Shiklomanov AI (2003) Climatic change and dynamics of river discharge into the Arctic Ocean. Water Resour 30(6):593–601

Shiklomanov NI, Nelson FE, Streletskiy DA, Hinkel KM, Brown J (2008) The circumpolar active layer monitoring (CALM) program: data collection, management, and dissemination strategies. In: Kane DL, Hinkel KM (eds) Proceedings of the ninth international conference on permafrost, vol 2, Fairbanks Institute of Northern Engineering, University of Alaska Fairbanks, Fairbanks, Alaska, June 29–July 3 2008, pp 1647–1652

Shkol'nik IM, Mol'kentin EK, Nadezhina ED, Khlebnikova EI, Sall IA (2008) Temperature extremes and wildfires in Siberia in the 21st century: the MGO regional climate model simulation. Russ Meteorol Hydrol 33(3):135–142

Shmakin AB (2010) Climatic characteristics of snow cover over North Eurasia and their change during the last decades. Ice Snow 1(1):43–57

Shnitnikov AV (1951) Variability of solar radiation during historical epoch on the base of it's Earth's manifestations. Bull Comm Investig Sun 7(21):47–70 (in Russian)

Shumilova LV (1962) Botanical geography of Siberia. Tomsk University Press, Tomsk, 439 pp (in Russian)

Shver TsA (1976) Atmospheric precipitation over the USSR territory (in Russian). Gidrometeoizdat, Leningrad, 302 pp

Smith CL, Baker A, Fairchild IJ, Frisia S, Borsato A (2006) Reconstructing hemispheric-scale climates from multiple stalagmite records. Int J Climatol 26(10):1417–1424

Soja AJ, Tchebakova NM, French NH, Flannigan MD, Shugart HH, Stocks BJ, Sukhinin AI, Parfenova EI, Chapin FS III (2007) Climate-induced boreal forest change: predictions versus current observations. Glob Planet Change 56:274–296

Sun B, Groisman PYa (2000) Cloudiness variations over the former Soviet Union. Int J Climatol 20:1097–1111

Tarasov PE, Jolly D, Kaplan JO (1997) A continuous Late Glacial and Holocene record of vegetation changes in Kazakhstan. Palaeogeogr Palaeoclimatol Palaeoecol 136:281–292

Tarasov PE, Bezrukova EV, Krivonogov SK (2009) Late Glacial and Holocene changes in vegetation cover and climate in southern Siberia derived from a 15 kyr long pollen record from Lake Kotokel. Clim Past 5:285–295

Tchebakova NM, Monserud RA, Nazimova DI (1994) A Siberian vegetation model based on climatic parameters. Can J Forest Res 24:1597–1607

Tchebakova NM, Kolle O, Zolotukhine D, Lloyd J, Arneth A, Parfenova EI, Schulze E-D (2002) Annual and seasonal dynamics of energy and mass exchange in a middle taiga pine forest. In: Pleshikov FI (ed) Forest ecosystems of the Yenisei meridian. SB RAS Publishers, Novosibirsk, pp 252–264

Tchebakova NM, Parfenova E, Soja A (2009a) Effects of climate, permafrost and fire on vegetation change in Siberia in a changing climate. Environ Res Lett 4. 045013 (9 pp) doi:10.1088/1748–9326/4/4/045013

Tchebakova NM, Blyakharchuk TA, Parfenova EI (2009b) Reconstruction and prediction of climate and vegetation change in the Holocene in the Altai-Sayan Mts, central Asia. Environ Res Lett 4. 045025 (11 pp) doi:10.1088/1748–9326/4/4/045025

Tebaldi C, Knutti R (2007) The use of the multi-model ensemble in probabilistic climate projections. Philos Trans R Soc Ser A 364(1857):2053–2075

Tipenko GS, Marchenko S, Romanovsky VE, Groshev V, Sazonova T (2004) Spatially distributed model of permafrost dynamics in Alaska. Presented at the AGU Fall Meeting. San Francisco, CA, 13–17 Dec 2004. EOS 85:C12A-02

Trenberth KE (2011) Changes in precipitation with climate change. Clim Res 47:123–138. doi:10.3354/cr00953

Turner JA, Lawson BD (1978) Weather in the Canadian forest fire danger rating system. A user guide to national standards and practices. Environment Canada, Pacific Forest Research Centre, Victoria, BC. BC-X-177

Vaganov EA, Shiyatov SG, Mazepa VS (1996) Dendroclimatic study in Ural-Siberian subarctic. Nauka, Novosibirsk, 246 pp

van Geel B, Bokovenko NA, Burova ND, Chugunov KV, Dergachev VA, Dirksen VG, Kulkova M, Nagler A, Parzinger H, van der Plicht J, Visiliev SS, Zaitseva GI (2004) Climate change and the expansion of the Scythian culture after 850 BC: a hypothesis. J Archaeol Sci 31(12):1735–1742

Vannari PI (1911) Meteorological networks in Russia and other countries. Issue of Meteorological Papers in the Memory of the Chief of the Meteorological Committee of Imperator Russian Geographical Society A.I. Voeikov 1:51–64 (in Russian)

Velichko AA (1989) The Holocene like element of global natural processes. In: Velicko AA (ed) Paleoclimates of the late Glacial and the Holocene. Nauka, Moscow, pp 2–12 (in Russian)

Velichko AA (ed) (2010) Atlas-monograph "evolution of landscapes and climates of Northern Eurasia". Late Pleistocene – Holocene – elements of prognosis. Issue 3. GEOS, Moscow, 220 pp

Velichko AA, Klimanov VA, Borzenkova II (1992) Climates between 6,000 and 5,500 yr BP. In: Frenzel B, Pecsi M, Velichko AA (eds) Atlas of paleoclimates and paleoenvironments of the Northern Hemisphere (Late Pleistocene – Holocene). Geographical Research Institute/Gustav Fischer Verlag, Budapest/Stutgart, p 65, 69, 73, 77, 137–139

Velichko AA, Andreev AA, Klimanov VA (1997) Climate and vegetation dynamics in the tundra and forest zone during the late Glacial and Holocene. Quat Int 41(42):71–96

Vinnikov KYa, Groisman PYa, Lugina KM (1990) The empirical data on modern global climate changes (temperature and precipitation). J Climate 3(6):662–677

Volkova VS, Klimanov VA (1988) Palynology and climate of West Siberia during main thermal maximums of the Holocene (8500, 5500, 3500 yr. BP). In: Shatski SB (ed) Microfissils and stratigraphy of Mesozoic and Cenozoic of Siberia. Nauka, Novisibirsk, pp 91–99

Volkova VS, Levina TP (1985) The Holocene like model for investigation of interglacial epoch of West Siberia. In: Khlonov AF (ed) Palynostratigraphy of the Mesozoic and Cenozoic of Siberia. Nauka, Novosibirsk, pp 74–84 (in Russian)

Vygodskaya NN, Groisman PYa, Tchebakova NM, Kurbatova JA, Panfyorov O, Parfenova EI, Sogachev AF (2007) Ecosystems and climate interactions in the boreal zone of Northern Eurasia. Environ Res Lett 2, 045033: 7. doi:10.1088/1748–9326/2/4/045033

Wilson KB, Baldocchi DD (2000) Seasonal and interannual variability of energy fluxes over a broadleaved temperate deciduous forest in North America. Agr Forest Meteorol 100:1–18

Wilson R, D'Arrigo R, Buckley B, Büntgen U, Esper J, Frank D, Luckman B, Payette S, Vose R, Youngblut D (2007) A matter of divergence: tracking recent warming at hemispheric scales using tree ring data. J Geophys Res 112:D17103. doi:10.1029/2006JD008318

Wu X, Lin Z (1988) In: Zhang J (ed) The reconstruction of climate in China for historical times. Science Press, Beijing, pp 114–128

Yamskikh AF, Sulerzhitsky LD, Abramova GM, Zubareva GY (1981) Paleogeographic conditions of forming of the Bolshoi Kemchug River Valley. In: Yamskikh AF (ed) Natural conditions of the Minusink Hollow. Krasnoyarsk Pedagogical Institute Press, Krasnoyarsk, pp 101–112

Yamskikh GY (1995) Holocene vegetation and climate of the Minusinsk Hollow. University of Krasnoyarsk Press, Krasnoyarsk, 180 pp

Yin JH (2005) A consistent poleward shift of the storm tracks in simulations of 21st century climate. Geophys Res Lett 32:L18701

Zhdanko VA (1965) Scientific basis of development of regional scales and their importance for forest fire management. In: Melekhov IS (ed) Contemporary problems of forest protection from fire and firefighting. Lesnaya Promyshlennost' Publishers, Moscow, pp 53–86 (in Russian)

Zolina OG, Simmer C, Belyaev K, Kapala A, Gulev S (2009) Improving estimates of heavy and extreme precipitation using daily records from European rain gauges. J Hydrometeorol 10:701–716

Zolina O, Simmer C, Gulev SK, Kollet S (2010) Changing structure of European precipitation: longer wet periods leading to more abundant rainfalls. Geophys Res Lett 37:L06704. doi:10.1029/2010GL042468

Zubakov VA (1986) Global climate events of the Pleistocene. Gidrometeoizdat, Leningrad, 287 pp (in Russian)

Zubakov VA, Borzenkova II (1990) Global paleoclimate of the late cenozoic. Ser Dev Paleontol 12:456. Elsevier, Amsterdam etc.

Zubareva GYu (1987) Change of paleoclimate of south-Minusinsk hollow in the late Holocene. In: Yamskikh AF (ed) Paleogeography of the Central Siberia. Krasnoyarsk Pedagogical Institute Press, Krasnoyarsk, pp 41–64

Supplementary References

Girs AA (1974) Macrocirculation method of long term forecasts. Gidrometeoizdat, Leningrad, 488 pp (in Russian)

Huntington TG (2006) Evidence for intensification of the global water cycle: Review and synthesis. J Hydrol 319(1–4):83–95

Intergovernmental Panel on Climate Change (IPCC) (1990) In: Houghton JT, Jenkins GJ, Ephraums JJ (eds) Climate change. The IPCC scientific assessment. Cambridge University Press, New York, 362 pp

Intergovernmental Panel on Climate Change (IPCC) (1996) In: Houghton JT, Meira Filho LG, Callendar BA, Harris N, Kattenberg A, Maskell K (eds) Climate change 1995: the science of climate change. The second IPCC scientific assessment. Cambridge University Press, New York, 572 pp

Intergovernmental Panel on Climate Change (IPCC) (2000) In: Houghton JT, Ding Y, Griggs DJ, Noguer M, van der Linden PJ, Dai X, Maskell K, Johnson CA (eds) Climate change 2001: the scientific basis. Contribution of Working Group 1 to the third IPCC scientific assessment. Cambridge University Press, Cambridge/New York, 881 pp

Intergovernmental Panel on Climate Change (IPCC) (2007) In: Solomon S, Qin D, Manning M, Chen Z, Marquis M, Averyt KB, Tignor M, Miller HL (eds) Climate change 2007: the physical science basis. Contribution of Working Group I to the fourth assessment report of the intergovernmental panel on climate change. Cambridge University Press, Cambridge/New York, 996 pp

Manabe S, Wetherald RT, Milly PCD, Delworth TL, Stouffer RJ (2004) Century-scale change in water availability: CO_2-quadrupling experiment. Clim Chang 64:59–76

Shkolnik IM, Meleshko VP, Pavlova TV (2000) Hydrodynamical limited area model for climate studies over Russia. Russ Meteorol Hydrol 4:32–49

Shkolnik IM, Meleshko VP, Gavrilina VM (2005) Validation of the MGO regional climate model. Russ Meteorol Hydrol 1:9–19

Shkolnik IM, Molkentin EK, Nadezhina ED, Khlebnikova EI, Sall IA (2008) Temperature extremes and wildfires in Siberia in the 21st century: MGO regional climate model simulation. Russ Meteorol Hydrol 3:5–15

Chapter 4
Hydrological Changes: Historical Analysis, Contemporary Status, and Future Projections

Alexander I. Shiklomanov, Richard B. Lammers, Dennis P. Lettenmaier,
Yuriy M. Polischuk, Oleg G. Savichev, Laurence C. Smith,
and Alexander V. Chernokulsky

Abstract This chapter looks at several aspects of the hydrological regime across Siberia using long-term historical data and model simulation results to provide a better understanding of ongoing changes and future directions. It begins with a survey of the major components of water balance: river flow, precipitation, and evapotranspiration. This is followed by the primary focus on the Siberian river systems with emphasis on annual variability and the anomalously high river discharge in 2007, the seasonality of river flow with increases in winter discharge, and changes in magnitude of minimum river flow and the temporal shifts in maximum river flow. Other components related to the river systems are also explored, including the thermal regime showing a lack of widespread evidence for increasing river temperature while the ice cover over the major rivers is decreasing in terms of both the duration of ice cover

---

A.I. Shiklomanov (✉) • R.B. Lammers
Water Systems Analysis Group, Complex Systems Research Center, Institute for the Study
of Earth, Ocean, and Space, University of New Hampshire, Morse Hall, Durham, NH, USA
e-mail: alex.shiklomanov@unh.edu

D.P. Lettenmaier
Department of Civil, and Environmental Engineering, University of Washington,
Seattle, WA, USA

Y.M. Polischuk
Department of Ecology and Landuse Yugra State University,
Khanty-Mansiysk, Russian Federation,

O.G. Savichev
Department of Hydrogeology, Tomsk Polytechnic University,
Tomsk, Russian Federation

L.C. Smith
Department of Geography, University of California-Los Angeles, Los Angeles, CA, USA

A.V. Chernokulsky
A.M. Obukhov Institute of Atmospheric Physics, Russian Academy of Sciences,
Moscow, Russian Federation
e-mail: a.chernokulsky@ifaran.ru

P.Ya. Groisman and G. Gutman (eds.), *Regional Environmental Changes in Siberia* 111
and Their Global Consequences, Springer Environmental Science and Engineering,
DOI 10.1007/978-94-007-4569-8_4, © Springer Science+Business Media Dordrecht 2013

and ice thickness. Related hydrological conditions (e.g., groundwater hydrology) demonstrate an increase in both levels and temperatures; however, there is evidence for some local decreases in groundwater level. Additionally, increases in groundwater runoff from the taiga zone are observed. Total thermokarst lake area is changing, depending on the landscape zone. Northern zones of tundra are gaining lake area, while the southern tundra and taiga regions are losing lake area. This chapter concludes with a look at possible future changes in the region's hydrology. River discharge in the major Siberian watersheds is expected to rise, and this result is consistent across a majority of the global climate models' projections for the twenty-first century.

Many aspects of Siberian hydrology related to monitoring, interactions between water and carbon cycles, land cover, and land use impacts were discussed in the recently published monograph *Eurasian Arctic Land Cover and Land Use in a Changing Climate* (Gutman and Reissell 2011). In this chapter, we focus on a more detailed analysis of changes in the hydrological regime across Siberia using historical data and modeling results to provide better understanding of ongoing changes and future directions.

4.1 Introduction

There is mounting evidence that Siberia is experiencing an unprecedented degree of environmental change in various components of the water cycle. These include loss of permafrost (Yang et al. 2002), changing river flows (Shiklomanov et al. 2000; Shiklomanov and Shiklomanov 2003, Peterson et al. 2002; Shiklomanov and Lammers 2009), lengthened ice-free periods in lakes and rivers (Magnuson et al. 2000), disappearance of lakes (Smith et al. 2005), reductions in snow cover (Armstrong and Brodzik 2001), and melting of glaciers (Steffen et al. 2004). Ongoing and projected changes in the water cycle of the region are beginning to interfere with the regional human environment and could have direct and immediate implications for the economy, subsistence, and social life (see Chaps. 5 and 7 in this book).

River flow is an integrated characteristic that reflects numerous environmental processes at play over the upstream drainage area. River runoff also plays a significant role in the freshwater budget of the Arctic, accounting for about 2/3 of the freshwater flux to the Arctic Ocean. Ocean salinity and sea ice formation in turn are critically affected by this river input. Changes in the freshwater flux to the Arctic Ocean can exert significant control over global ocean circulation by influencing the North Atlantic deepwater formation (Manabe and Stouffer 1994; Rahmstorf 2002).

Eurasia contributes 75 % of the total terrestrial runoff to the Arctic Ocean (Shiklomanov et al. 2000), and three of the four major Arctic rivers are in Siberia. The combined river discharge from the six largest Russian north-flowing rivers (N. Dvina, Pechora, Ob, Yenisei, Lena, and Kolyma) has shown an increase of 7 % over the period 1936–1999 (Peterson et al. 2002) (Fig. 4.1). More recent estimates have shown that this increase has accelerated in the twenty-first century (Shiklomanov 2010; Shiklomanov and Lammers 2009). More detailed analysis of changes in annual river discharge for large Eurasian rivers is given in Sect. 4.3.

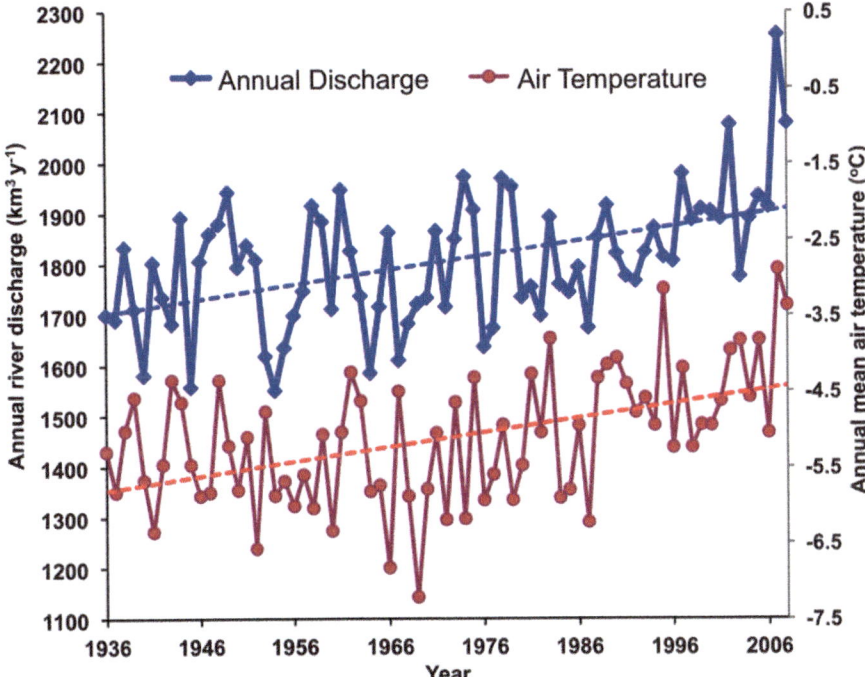

Rivers: Ob', Yenisey, Lena, Severnaya Dvina, Pechora, Kolyma

Fig. 4.1 Total annual river discharge to the Arctic Ocean from the six largest rivers in the Eurasian Arctic for the period 1936–2007 (*blue line*) (Shiklomanov and Lammers 2009) and mean annual air temperature aggregated for the same basins from University of Delaware observational climate data (Willmott and Matsuura 1995, 2001) for 1936–2008 (*red line*). Least squares linear trends are shown as *dashed lines*

4.2   Changes in Water Balance Components

The physical mechanisms driving observed runoff changes are not yet completely understood. Several research projects focused on understanding hydrological change in the pan-Arctic region were funded by the Freshwater Integration study – a program initiated in 2002 by the US National Science Foundation as a contribution to the Study of Environmental Arctic Change (SEARCH). However, comprehensive analyses of water balance components (Serreze et al. 2006; Rawlins et al. 2006b, 2010; Shiklomanov et al. 2007), human impacts (Adam et al. 2007; Yang et al. 2004), and hydrological modeling experiments (Rawlins et al. 2003, 2006a) suggested that multiple factors, rather than one unique cause, lie behind the observed discharge increase in the Eurasian pan-Arctic. For instance, there is some evidence for increased precipitation over the terrestrial Arctic area, but there are also indications that increased precipitation may be in part balanced by higher evapotranspiration due to an increasing air temperature (see Sect. 3.2.2). McClelland et al. (2004) evaluated

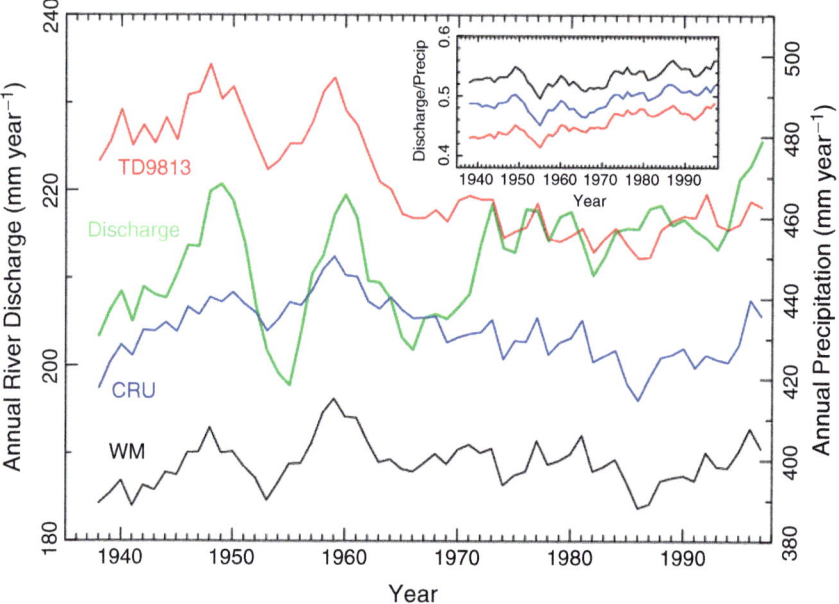

Fig. 4.2 Five-year running means of spatially averaged river discharge (mm year⁻¹) and precipitation (mm year⁻¹) across the six largest Eurasian river basins from 1936 to 1999. *Inset* shows the ratio of annual discharge to annual precipitation. TD9813 is NCDC's Dataset 9813, Daily and Sub-daily Precipitation for the Former USSR; WM is Arctic terrestrial air temperature and precipitation from University of Delaware (Willmott and Matsuura 1995); *CRU* is Climate Research Unit data V.2 (Mitchell et al. 2004) (Source: Rawlins et al. 2006b)

potential causes of the increasing Eurasian river discharge due to river impoundments, thawing permafrost and fire. They found that none of these could contribute to the large observed changes in the river systems and concluded that the cause must be precipitation change. Comprehensive analyses of precipitation and river runoff have been carried out using long-term precipitation data sets (Pavelsky and Smith 2006; Rawlins et al. 2006b; Shiklomanov et al. 2007) and river discharge from R-ArcticNet (Lammers et al. 2001; Shiklomanov and Lammers 2009). Rawlins et al. (2006b) analyzed trends and correlations between annual river discharge for the six largest Russian Arctic rivers and annual precipitation derived from three different data sets (Fig. 4.2). They found that the correlation coefficient between annual precipitation and discharge over the period 1936–1970 was around 0.55 but significantly lower thereafter, when discharge increases occurred despite precipitation declines. However, the relationship between annual runoff and precipitation over calendar year includes accumulated winter precipitation, which roughly corresponds to spring snow storage and in turn, is strongly related to spring and summer runoff. Therefore, annual precipitation trends may not be well related to annual runoff. For instance, despite a strong rainfall decline after the 1970s, declines in snowfall were moderate (it should be noted that measurements of solid precipitation are much more error-prone than are liquid precipitation measurements). Furthermore, there have been important spatial variations in snowfall trends. Snowfall increased from 1936 to the late 1950s across north-central Eurasia, an area where rainfall decreases were most prominent (Rawlins et al. 2006b). The mean maximum snow water equivalent has tended to increase across Central and South Siberia over the last 50 years (see Fig. 3.6). However, due to large spatial variations, an integrated signal across the northern Eurasian drainage basins yielded no significant change in interannual precipitation. A similar conclusion about the general inconsistency between runoff and precipitation was reached by Berezovskaya et al. (2004), who suggested that either existing precipitation products were of insufficient quality to capture high-latitude precipitation changes or some other processes are responsible.

An analysis of trends (1936–2001) in the water budget components across 56 subbasins of the six Eurasian river systems was carried out at the University of New Hampshire to identify those regions where runoff changes were consistent with trends in precipitation (P) and evapotranspiration (ET). Using the 25-km × 25-km EASE grid river network from ArcticRIMS (http://RIMS.unh.edu), river discharge data from R-ArcticNet (http://RArcticNet.sr.unh.edu), precipitation from University of Delaware (Willmott and Matsuura 2001), and estimated potential evapotranspiration (Hamon 1963), all 56 subbasins were tested to identify those regions most responsible for observed increase in total discharge to the Arctic Ocean and to map spatial patterns of major water budget components over the period 1936–2001. Runoff changes were found to have a complex spatial distribution, with greatest increasing trends in the northern latitudes probably responsible for most of the observed increase in total discharge to the ocean (Fig. 4.3). The sign of precipitation and runoff trends across the region is consistent; however, the rate of runoff increase in the northern regions is at least double the precipitation change. Thus, even if the entire increase in annual precipitation went into runoff change, it would explain less than half of the observed runoff, suggesting that other causes of runoff change are also under way. In contrast to observed runoff changes, potential evapotranspiration changes, which mostly reflect changes in air temperature, appear to be negligible across the region. Furthermore, the greatest discrepancies among water budget components are found in northern areas underlain by permafrost (Fig. 4.3). This supports, at least indirectly, the idea that permafrost thawing and a deepening of the active layer may be providing some amount of additional water release to river networks, leading to the observed increases in annual river discharge.

4.3   River Discharge Variability

The spatial distribution of runoff across Siberia is highly heterogeneous. The lowest values (less than 10 mm/year) are found in the steppe regions of West Siberia, and the highest values (exceeding 600 mm/year) are found in the mountainous regions of Siberia (Lammers et al. 2001).

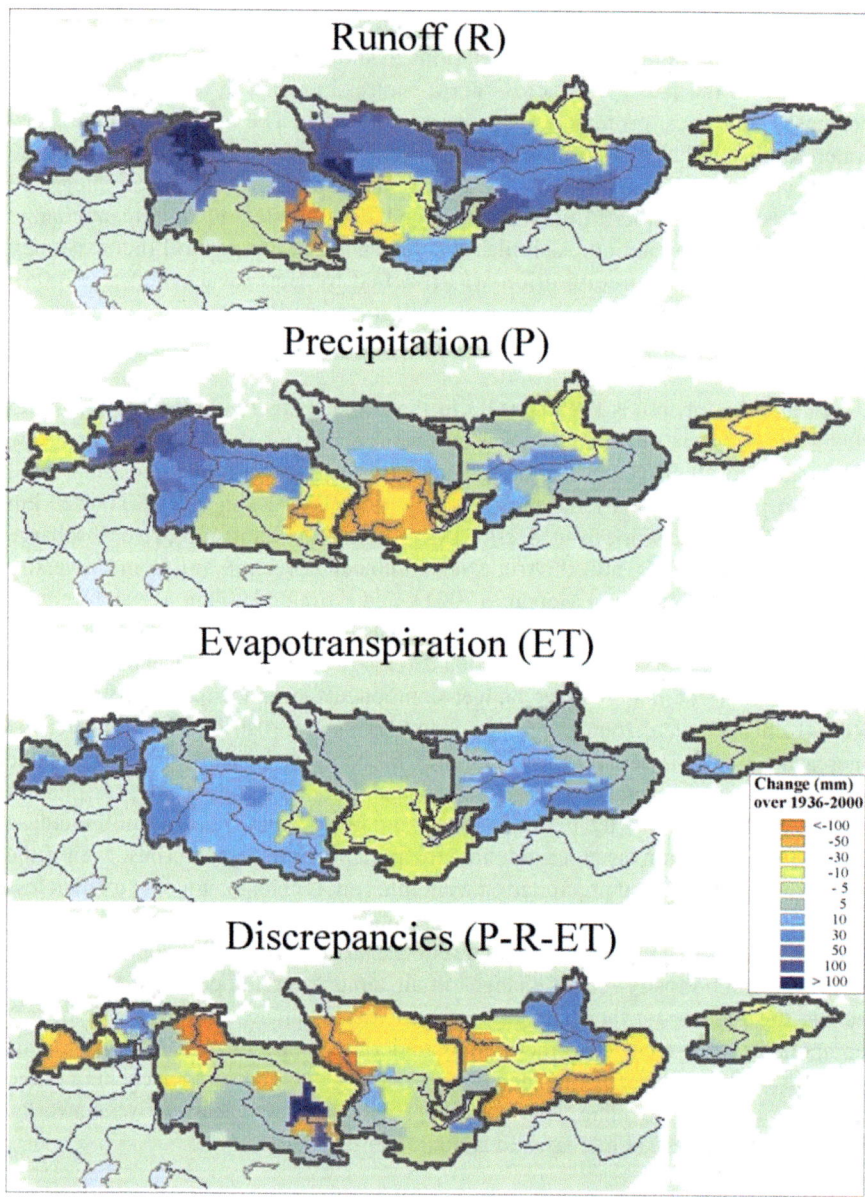

Fig. 4.3 Annual changes in major components of water balance (1936–2000) for hydrological subbasins located across the six largest Eurasian Arctic rivers computed based on a slope of linear trend line computed from observed annual runoff records for 56 gauges (*R*), annual precipitation (*P*) aggregated for these basins based on University of Delaware climate fields (Willmott and Matsuura 2001) and potential evapotranspiration, (*ET*) estimated using gridded air temperature data from University of Delaware (Willmott and Matsuura 2001) (Source: A. Shiklomanov, AGU presentation, 2008)

Rivers flowing into the Arctic Ocean are notable for their very low winter runoff, high spring peak flows, and rain-induced floods in the summer–fall period. Their degree of seasonality depends on climatic conditions, land cover, permafrost extent, and various effects of natural and artificial runoff regulation. Snowmelt-derived runoff may account for up to 80 % of the annual total in regions with a strong continental climate and continuous permafrost, such as the northern parts of Central and Eastern Siberia. At the same time, snowmelt runoff is about 50 % of the annual total in the northern parts of European Russia and Western Siberia (AARI 1985). Most Eastern Siberian rivers flowing through the continuous permafrost zone with drainage areas less than 100,000 km^2 have practically no runoff during winter due to their limited inflow of groundwater (ACIA 2005).

The most complete time series of discharge into the Arctic Ocean is available for large Russian rivers. From 1936–2008, the mean annual river discharge for the six largest Russian rivers flowing to the Arctic Ocean was about 1,800 km^3 year^{-1} (Fig. 4.1). Analysis of discharge variability for individual river basins over this time period shows significant positive trends (ranging from 11 to 21 % with 5 % significance level) for the Yenisei, Lena, Pechora, and Severnaya Dvina basins and no trend for the Kolyma and Ob basins. Even higher rates of river discharge increases are observed over the period 1980–2008 when all rivers show accelerating discharge increases (Fig. 4.4). For this more recent time period, the total annual discharge for these six rivers averaged 100 km^3 year^{-1} (~6 %) higher than the 1936–1979 period. This difference is greater than the mean annual discharge for the Kolyma River.

2007 saw a massive flux of freshwater from the Russian land surface of 2,250 km^3 year^{-1} (Fig. 4.1), about 25 % above the long-term mean. This was greater than the total annual flow from the Ob basin, the 8th largest basin in the world by drainage area (Vörösmarty et al. 2000) and approximately equal to the total estimated mean annual runoff from Greenland (Mernild et al. 2009). 2007 was also record high-flow runoff year in the Pechora and Yenisei basins, and discharges were above the 85th percentiles for the Ob, Lena, and Kolyma Rivers. Of the six largest basins, only the Severnaya Dvina experienced 2007 flows near the long-term mean. Except for the Kolyma (which has a negligible long-term trend), the 2007 river discharge anomalies were directionally consistent with observed linear discharge trends from 1936 to 2006 (Fig. 4.5). However, the magnitudes of 2007 anomalies significantly exceeded the long-term changes for all rivers (Shiklomanov and Lammers 2009).

Maximum Eurasian river discharges in 2007 corresponded to minimum sea ice extent in the Arctic Ocean. Long-term river discharge and sea ice have a negative correlation ($r = -0.7$), suggesting that both rivers and sea ice respond to changes in large-scale hemispheric climate patterns in similar ways (Rawlins et al. 2009) and that an increasingly ice-free summer in the Arctic Ocean may have contributed to wetter conditions on land via atmospheric moisture transport from open sea areas. This latter hypothesis can be tested through analysis of precipitation data. Over the periods 1980–2007, basin-aggregated precipitation for six river basins showed a weak correlation with river discharge ($r = 0.42$), with a general correspondence in the linear trend direction. The mean runoff ratio (runoff/precipitation) aggregated over these six basins was about 45 %. This is relatively low compared with other

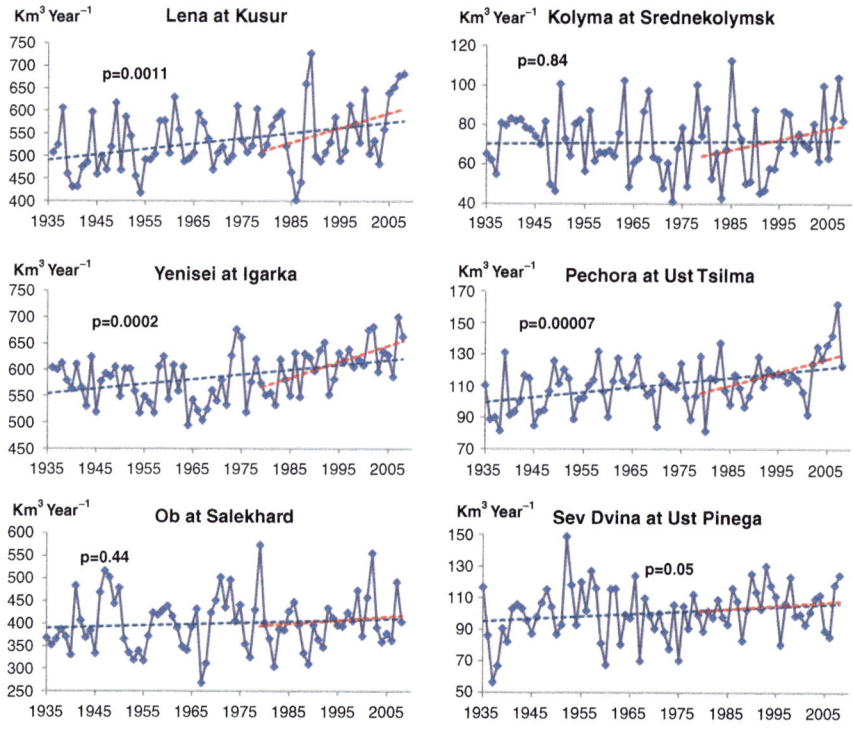

Fig. 4.4 Annual discharge of the six largest Russian rivers flowing to the Arctic Ocean for 1936–2008. *Dashed lines* show linear trends for 1936–2008 (*blue*) and 1980–2008 (*red*) (Data provided by the State Hydrological Institute, Russia)

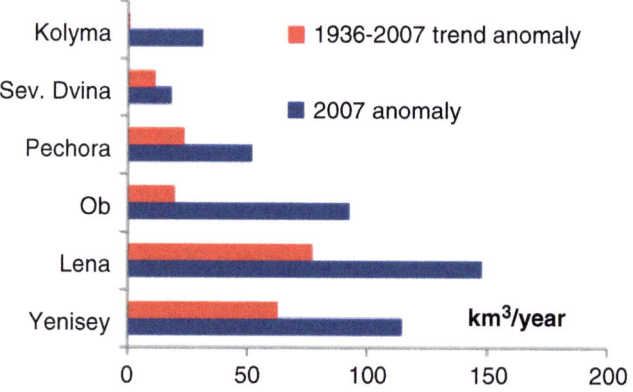

Fig. 4.5 Contributions of major Russian Arctic drainage basins to the 1936–2007 trend (*red bars*) and the 2007 anomaly (*blue bars*) (Source: Shiklomanov and Lammers 2009)

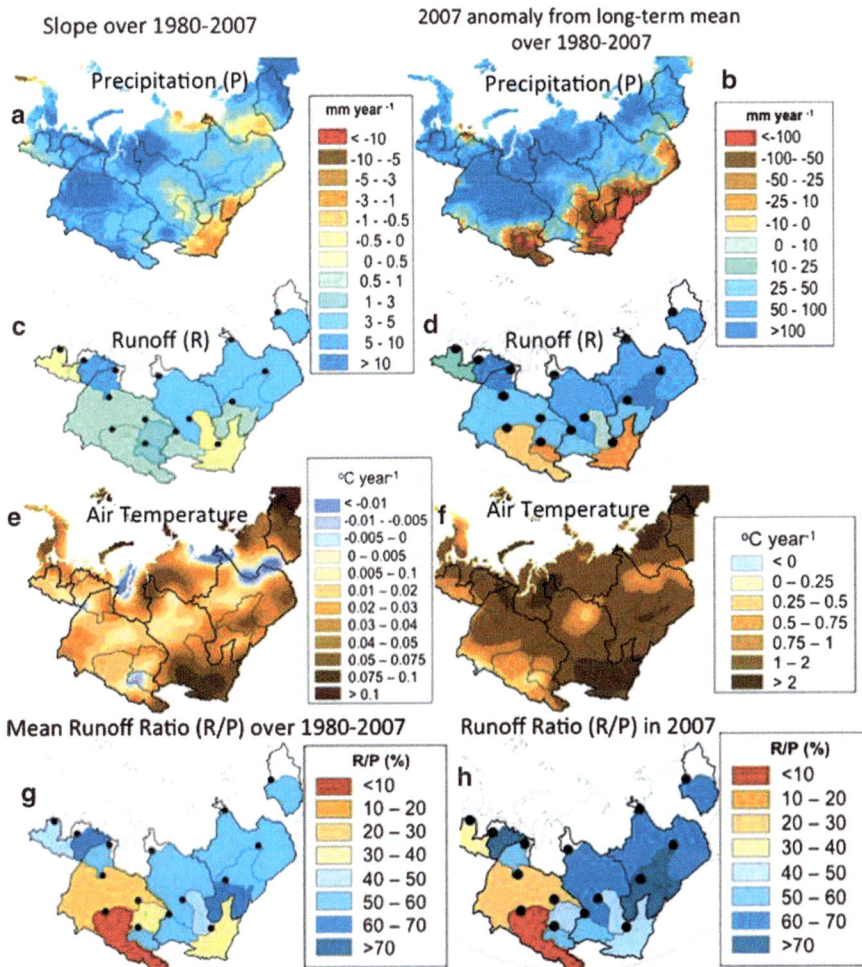

Fig. 4.6 Maps of long-term trend (*left column*) and 2007 anomalies relative to the long-term trend (*right column*) of precipitation (*P*), river runoff (*R*), and air temperature, along with mean runoff ratio (*R/P*) over the period 1980–2007 (*panel g*) and for 2007 (*panel h*). River runoff *R* and *R/P* ratios were calculated over 15 interstation regions, as shown by *black* and *gray border lines* with *circles* representing the gauge locations (Table 4.1). Note different *color* scales in the *left* and *right* columns (Source: Shiklomanov and Lammers 2009)

river basins in the pan-Arctic domain and is probably due to extensive dryland areas in the southern Ob and Yenisei basins. To investigate the spatial characteristics of hydroclimatological anomalies of 2007 relative to long-term variations in river discharge, 15 gauges were used (1 gauge each for Sev. Dvina, Pechora, and Kolyma; the Ob, Yenisei, and Lena were further disaggregated into 12 subbasins) along with spatially aggregated precipitation and air temperature. Precipitation anomalies (Fig. 4.6b) show that much of the observed river discharge increase to the ocean for

2007 originated in the northern parts of the drainage basins despite decreased anomalies in the southern headwaters of the largest basins (Fig. 4.6d). The annual trends from 1980 to 2007 (Fig. 4.6a, c) display similar patterns to the 2007 precipitation and discharge anomalies, further supporting the suggestion that 2007 was effectively a higher-magnitude expression of an ongoing pattern of change over the last 28 years. It is also important to note that 2007 was the warmest year in the last six decades for the Yenisei, Lena, and Kolyma basins, which together contain about 90 % of the total permafrost extent in the six basins (Fig. 4.6f). The warm 2007 conditions could therefore have intensified permafrost thaw, contributing additional thawed water to total runoff. Fyodorov-Davydov et al. (2008) showed that during warm summer, thawing of ice-rich transient layer located below active layer (upper-layer thawing and freezing annually) can contribute as much as 25 % of total precipitation to the runoff. Indeed, the runoff ratio (runoff/precipitation) (Fig. 4.6h) was significantly higher in 2007 than the long-term mean over 1980–2007 (Fig. 4.6g) for all subbasins located inside the three permafrost-dominated watersheds. All subbasins where the 2007 runoff anomaly exceeded the precipitation anomaly by >25 mm (bold font in Table 4.1) were located in the transition zone between non-permafrost and permafrost regions, where permafrost degradation is expected to have a greater impact from global warming (Stanilovskaya et al. 2008; Marchenko et al. 2007). Thus, it appears that increased precipitation across the northern part of these drainage basins combined with deeper active-layer thawing could be the trigger to the record 2007 Eurasian river discharge to the Arctic Ocean (Shiklomanov and Lammers 2009).

Any analysis of the long-term hydrologic variability of large Siberian rivers must consider the effects of reservoir construction, especially during the second half of the twentieth century. The total reservoir storage capacity in these river basins is 532 km^3, with an effective volume (active storage) of 232 km^3. Most big Siberian reservoirs (seven) and the largest (Bratskoje) are located in the Yenisei River basin. Total storage capacity of reservoirs in this basin is 423 km^3 and active storage is 172 km^3. There are three large reservoirs with total storage capacity 59 km^3 and active storage of 36 km^3 in the Ob River basin. Both the Lena and Kolyma Rivers have only one reservoir each in their basins with a storage capacity exceeding 1 km^3.

A comprehensive analysis of the influences of reservoirs and other human activity on the Ob and Yenisei Rivers was performed by the State Hydrological Institute of Russia during the 1970s and 1980s, during investigations of the potential effect of water transfers from the Ob and Yenisei basins to the Aral Sea basin (Shiklomanov and Markova 1987). The most significant decline in annual discharge at the downstream gauges due to human activity was about 3 % for the Ob River. The main cause of this decrease was consumptive water use for irrigation in China, Kazakhstan, and Russia. The effects of water management on annual river discharge at the downstream-most gauges of the Yenisei, Lena, and Kolyma did not exceed ±1 %. However, during periods of reservoir infilling, the declines in annual river discharge were much greater. Similar results were obtained using different methods including hydrological modeling by Yang et al. (2004), Adam et al. (2007), and Shiklomanov and Lammers (2009). Based on an analysis of "naturalized" and observed river flow data, Shiklomanov and Lammers (2009) showed that changes in the annual

Table 4.1 Change in runoff, precipitation, air temperature, and runoff ratio over 1980–2007 and their anomalies in 2007 for 12 subbasins in Yenisei, Lena, and Ob watersheds and 3 river basins

Code	Name	Drainage area (km²)	Permafrost (%)	Runoff (mm year⁻¹) Change 1980–2007	Runoff (mm year⁻¹) 2007 anomaly	Precipitation (mm year⁻¹) Change 1980–2007	Precipitation (mm year⁻¹) 2007 anomaly	Air temperature (°C) Change 1980–2007	Air temperature (°C) 2007 anomaly	Runoff ratio (%) Change 1980–2007	Runoff ratio (%) 2007 anomaly
1801	Kolyma–Srednekolymsk	361,000	100	47.8	82.2	80.2	67.3	1.33	1.45	57	68
3029	Lena–Krestovskoe	440,000	63	12.1	**47.2**	−16.3	**−79.5**	0.94	1.69	78	95
3042	Lena–Tabaga	457,000	81	56.7	**104.4**	43.3	**−12.2**	0.96	1.62	56	94
3821	Lena–Kusur	1,533,000	97	33.2	52.0	32.3	28.1	0.61	1.23	54	63
8013	Angara–Irkutskaya GES	573,000	17	−6.0	**−15.5**	−63.8	**−86.5**	1.91	2.09	34	42
8084	Angara–Boguchany	293,000	29	−20.8	**2.6**	3.8	**−43.9**	1.24	1.92	45	49
9053	Yenisei–Bazaikha	300,000	33	7.4	28.9	79.3	23.5	1.77	1.95	55	58
9803	Yenisei–Igarka	1,274,000	65	77.6	**90.7**	55.7	**64.1**	0.61	1.22	60	65
10006	Ob–Barnaul	169,000	6.2	22.8	**−15.5**	50.7	**−53.6**	0.19	0.70	52	56
10021	Ob–Kolpashevo	317,000	4.8	26.1	41.9	57.2	81.0	0.49	1.03	34	43
10031	Ob–Belogor'e	1,435,000	21	9.2	28.4	142.2	139.3	0.48	1.14	21	21
11048	Irtysh–Omsk	769,000	0	5.2	−0.9	95.2	14.0	0.54	0.67	9	9
11801	Ob–Salekhard	260,000	56	119.4	159.3	131.1	143.3	0.51	1.78	53	59
70801	Sev. Dvina–Ust'-Pinega	348,000	0	−3.5	11.1	42.0	95.1	0.41	0.88	45	40
70850	Pechora–Ust'-Tsilma	248,000	26	90.1	**214.2**	111.7	**74.0**	0.46	1.51	63	82

Source: Shiklomanov and Lammers (2009)
Gray and white lines show the grouping by major river basin.

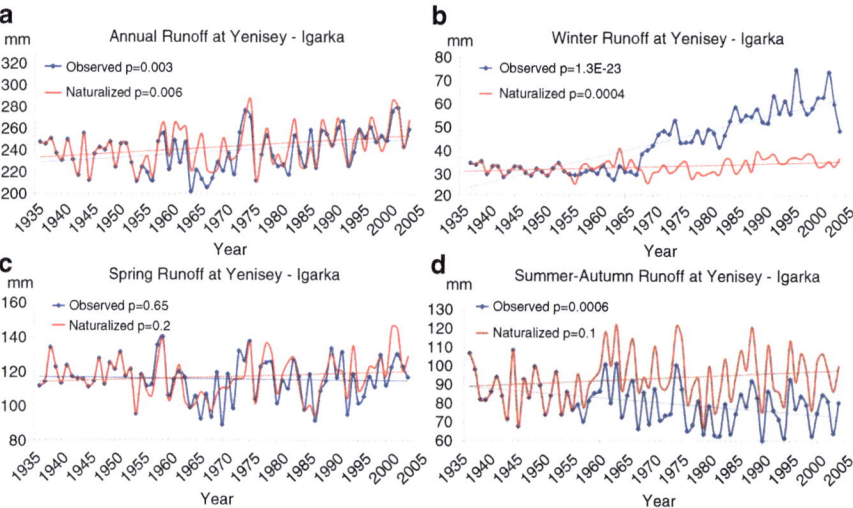

Fig. 4.7 Time series of observed and naturalized (corrected for reservoir impacts) river discharge for the Yenisei River at Igarka over 1936–2004; (**a**) annual, (**b**) winter (November to April), (**c**) spring (May to June), (**d**) summer–autumn (July to October). *Thin dashed lines* show linear trends defined from least squares linear regression analysis, and *p* values represent significance of the trend (Source: Shiklomanov and Lammers 2009)

discharge of the Yenisei River at Igarka (the downstream-most gauge) associated with human impact have negligible influence on the long-term annual trend in river discharge.

4.4   Seasonal Changes in River Discharge

Water management and streamflow regulation play a much larger role in altering seasonal discharge than in annual streamflow variations. In particular, human impacts tend to reduce peak discharges during spring and early summer, while increasing discharges during low-flow periods. This effect on seasonal hydrology is widely documented for the Ob, Yenisei, and Lena Rivers (Shiklomanov 1989; Ye et al. 2003; McClelland et al. 2004; Berezovskaya et al. 2005). To correct for these impacts, which are primarily associated with reservoir regulation, Shiklomanov and Lammers (2009) estimated naturalized streamflows for the Yenisei. The analysis showed that for the Yenisei River, the increase in annual discharge accompanied a significant increase during winter and increasing streamflow during the summer and fall (Fig. 4.7).

Spatial changes in seasonal river discharge of the large rivers flowing to the Arctic Ocean cannot, in general, be assessed using streamflow gauges because so many of them are impacted by large-scale impoundments and other human impacts within the basins. Although at first the reservoirs appear small relative to the annual

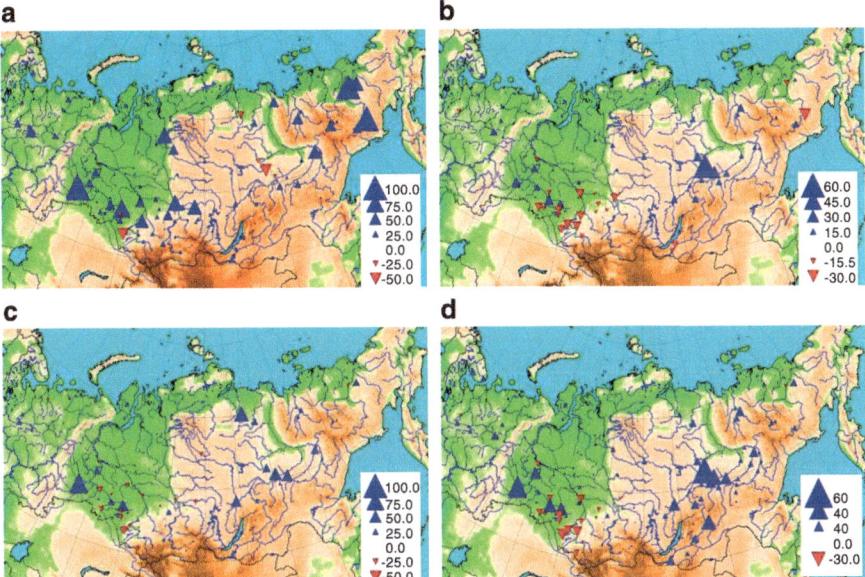

Fig. 4.8 Anomalies of winter (**a**), spring (**b**), summer–fall (**c**), and annual (**d**) runoff across Russian Arctic Ocean drainage basin for 1978–2005 (in % relative to 1946–1977) (Source: Shiklomanov ed. 2008)

discharge of many of the rivers, their impact during low flow (winter) can be important even for gauges located very far from the reservoirs.

Results like those shown in Fig. 4.7 have lead to a reevaluation of previous conclusions that the documented increase in annual discharge from Peterson et al.(2002) is driven mainly due to increases in winter–spring flow (Rawlins et al. 2007; McClelland et al. 2004). The Shiklomanov and Lammers (2009) estimates, in particular, reveal a more significant influence of low-flow changes (summer–fall and winter) on long-term trends in annual discharge. Similar results were obtained by Adam et al. (2007) based on analysis of reconstructed monthly hydrographs simulated with a reservoir-routing model coupled with Variable Infiltration Capacity (VIC) model (Cherkauer et al. 2003) for the Ob, Yenisei, and Lena.

To evaluate seasonal changes in runoff, Shiklomanov ed. (2008) analyzed monthly discharge records longer than 50 years from 75 gauges in the Russian pan-Arctic. The analysis considered only those rivers for which there has been no significant human impact on hydrological regime. The mean seasonal discharge over the period 1978–2005 was compared with mean discharge over the period 1946–1977 (Fig. 4.8). A prevailing pattern of year-round discharge increases across most of Northern and Eastern Siberia, and winter increases across the entire Russian pan-Arctic was found (Fig. 4.8a). Winter runoff has increased by as much as 40–60 % in the Irtysh basin in Southeastern Siberia, and by as much as 15–35 % in

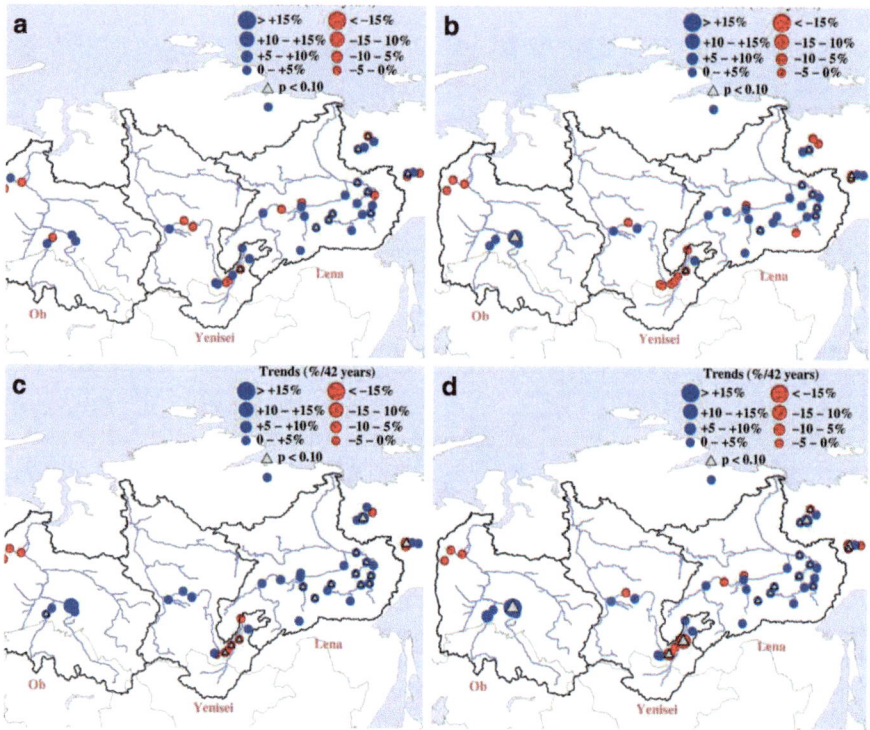

Fig. 4.9 Linear trends in monthly fractional river flow (defined as the ratio of streamflow occurring within a given month or a season to the total streamflow in the water year) for (**a**) December, (**b**) January, and (**c**) February and (**d**) winter season fractional flow (December, January, and February) (From Tan et al. 2011)

Northern Siberia. Additionally, significant increases were found in summer–fall runoff in the Irtysh and Tobol basins (25–40 %) and in Eastern Siberia (10–25 %) (Fig. 4.8c).

Tan et al. (2011) analyzed another subset of discharge data for 45 nonregulated river gauges located in Siberia and also found a significant rise of winter flow over 1958–1999. Increasing winter discharge trends were observed for 41 of the 45 gauges (Fig. 4.9). The trends were statistically significant for 12 of the 41 gauges and were distributed across the entire study domain (Fig. 4.9d) (Tan et al. 2011). It is particularly noteworthy that these trends are similar in character to those observed in larger rivers that have been affected by reservoir construction, suggesting that dams may not be the dominant cause of winter flow trends observed in previous studies (Adam and Lettenmaier 2007; Shiklomanov et al. 2007). Instead, the winter streamflow increases are consistent with the hypothesis of permafrost thawing, expedited by the conductive heat transfer from decreasing snow cover (Stieglitz et al. 2003; Zhang 2005; Yang et al. 2007).

Analysis of data from water balance stations located mainly in non-permafrost areas of European Russia also shows increase in low-flow discharge but for different reasons. It is primarily due to (1) an increase in frequency and duration of winter snowmelt events as the result of rising air temperatures, (2) a reduction in frozen ground extent and thickness, and (3) increases in groundwater levels. For areas underlain by permafrost, the hydrological processes are more complicated. Additional water released from thawing permafrost and improved water draining capacity in winter due to warmer air temperature and thinner river ice may also contribute to river discharge increases (Gurevich 2009).

4.5   Changes in Maximum and Minimum River Discharge

Maximum river discharge is usually associated with floods, which annually cause more damage in Russia than any other natural disaster. Future climate model projections suggest that the frequency and magnitude of extreme hydrological events, including floods, will increase in Russia in response to climate change (IPCC 2007). Shiklomanov et al. (2007) analyzed long-term variability in maximum daily spring discharge for 139 small- and medium-sized Russian rivers (16–50,000 km^2) with minimal human impact. Trend analyses were applied to the magnitude and timing of maximum spring discharge values over the periods 1960–2001, 1950–2001, 1940–2001, and 1930–2001, using both least squares linear regression and the more robust, nonparametric Mann–Kendall test which is widely used in hydrological studies (Helsel and Hirsch 1992). All rivers had clearly pronounced spring peaks in April–June.

Relative changes in spring maximum discharge for different periods are shown in Fig. 4.10 (left column). The number of gauges with significant positive and negative trends is similar for 1960–2001 and 1930–2001, and there are slightly more gauges with significant negative trends for 1950–2001 and 1940–2001. Overall, there were no widespread significant changes in trends of spring maximum discharge across the Russian Arctic drainage basin (Fig. 4.10). Regional patterns, however, are identified. These include significant decreases in spring maximum discharge across the southern parts of the Yenisei and Ob basins in South-Central Siberia. Out of 28 analyzed time series in this region, 8 showed significant negative trends over the period 1950–2001, and none were positive. In the Lena River basin, most gauge records show significant increases in maximum discharge for 1940–2001 ($p < 0.05$) and 1930–2001 and 1960–2001 ($p < 0.1$) (Shiklomanov et al. 2007). There was no uniformity between changes in precipitation over the cold period, and maximum and, in fact, opposing trend directions (insignificant for cold season precipitation) were found over the periods 1950–2001 and 1960–2001. This discrepancy led the authors to suspect that other factors were responsible for the observed discharge changes across the Lena basin, for example, more intense spring snowmelt associated with rising spring air temperatures and increased winter base flow (McDonald et al. 2004; Kilmyaninov 2000; Smith et al. 2007).

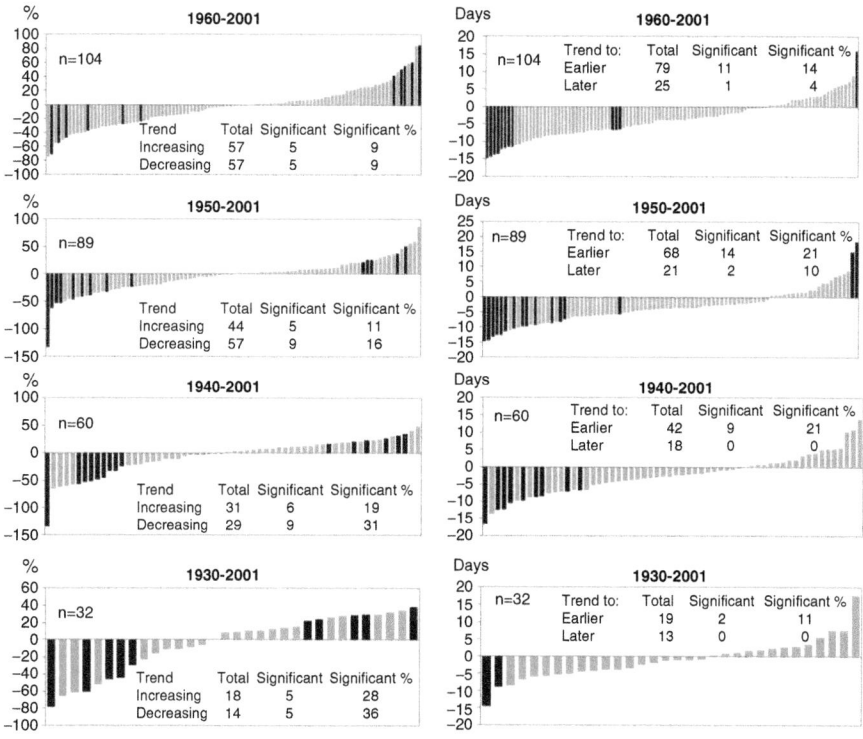

Fig. 4.10 Relative changes in spring daily maximum discharge (%) (*left panels*) and changes in dates of maximum discharge (days) (*right panels*) from linear regressions for different time periods. Changes are ranked from most negative to most positive. *Black bars* represent significant trends ($p < 0.05$) based on Mann–Kendall tests (Source: Shiklomanov et al. 2007)

There has been a distinct shift to earlier dates of spring maximum discharge in drainages across the Russian Arctic (Fig. 4.10, right column). The number of gauges with significant timing changes toward the earlier dates (negative trend) exceeds those with trends toward later peak discharge (positive trend) for all time periods. Furthermore, many gauges with negative changes have significant trends, while statistically significant positive trends are infrequent or nonexistent. Significant trends to earlier dates were also found in time series aggregated across the entire Russian Arctic drainage basin, with average shifts to earlier maximum discharge by 5.0 days for 1960–2001, 4.7 days for 1950–2001, 3.2 days for 1940–2001, and 4.6 days for 1930–2001. In contrast to more recent periods, only two gauging stations had significant shifts over the 1930–2001 period (Fig. 4.10, right column). A likely explanation for this was earlier snowmelt during the 1930s and 1940s, which canceled the more recent trend. Thus, the most significant shift to earlier spring maximum is over the most recent 40 years, a finding consistent with estimates of more intensive air temperature rise in the region (see Fig. 3.4, bottom right panel).

Significant trends to earlier spring discharge peak were consistent over all periods across the eastern part of the Russian Arctic drainage basin including Central and Eastern Siberia and Far Eastern Siberia. There were no significant changes in spring maximum discharge timing in the aggregated time series for Western Siberia, covering most of the Ob River basin. These results are consistent with more recent analysis of discharge timing based on a new data subset for 45 unregulated gauges in the Ob, Yenisei, and Lena basins (Tan et al. 2011). Analysis of time series for the spring pulse onset (SPO), which was defined as the date of the beginning of snowmelt-derived streamflow (Stewart et al. 2005), showed a 0–12-day earlier shift at 27 gauges and 2–4-day later shifts at the remaining 18 stations (Fig. 4.11). However, owing to the considerable interannual variability in the SPO values, only 4 trends of the 27 shifts to an earlier SPO were statistically significant ($p=0.10$), and none of the 18 trends toward a later SPO were significant. Although stations with trends toward earlier SPO were located in all three basins, they tended to be primarily concentrated in the colder Lena and Yenisei basins (Tan et al. 2011).

Smith et al. (2007) analyzed the same Russian time series as Shiklomanov et al. (2007) to examine daily discharge minima, the streamflow quantity most sensitive to groundwater and/or unsaturated zone water inputs. The term "baseflow" is commonly used to describe these inputs and refers to that portion of river discharge produced from water movement through the subsurface into the river channel. In principle, it is possible for minimum daily flows to correspond to true baseflow, typically after a prolonged period with no rainfall. In practice, minimum daily flows are an approximation of, and almost always exceed, true baseflow.

The main results of the Smith et al. (2007) analysis of minimum daily flows are that on balance, minimum daily discharge has risen across most of northern Eurasia since the 1930s, with some of the largest increases on record occurring since ~1985. This overall signal emerges despite a backdrop of intrinsic variability and the presence of decreasing as well as increasing trends. The analysis shows that increases in daily minimum flows have occurred year-round, again contrasting with aforementioned earlier studies describing Eurasian river discharge increases as a wintertime phenomenon. This distinction is a direct result of the separation of minimum flows from mean flows using high-resolution daily discharge records. This approach is necessary during the spring and summer months but is probably unnecessary in winter, when minimum and mean flows converge and therefore provide similar information.

While the broad-scale pattern of minimum-flow trends is clear, regional- and local-scale patterns are not (Fig. 4.12). Even the presence or absence of permafrost, which would seem a primary control on low-flow variability, seems not to matter much. While minimum-flow decreases are more common in summer across South-Central Siberia, suggesting a possible link to agricultural consumption (Yang et al. 2004), some substantial declines also occurred in remote areas of continuous permafrost at the same time (Fig. 4.12). Also, for reasons not fully understood, the minimum-flow increases are frequently found in rivers that have not experienced comparable increases in mean flow. One partial explanation for this is that in terms of absolute magnitude, minimum-flow variations represent a large fraction of overall

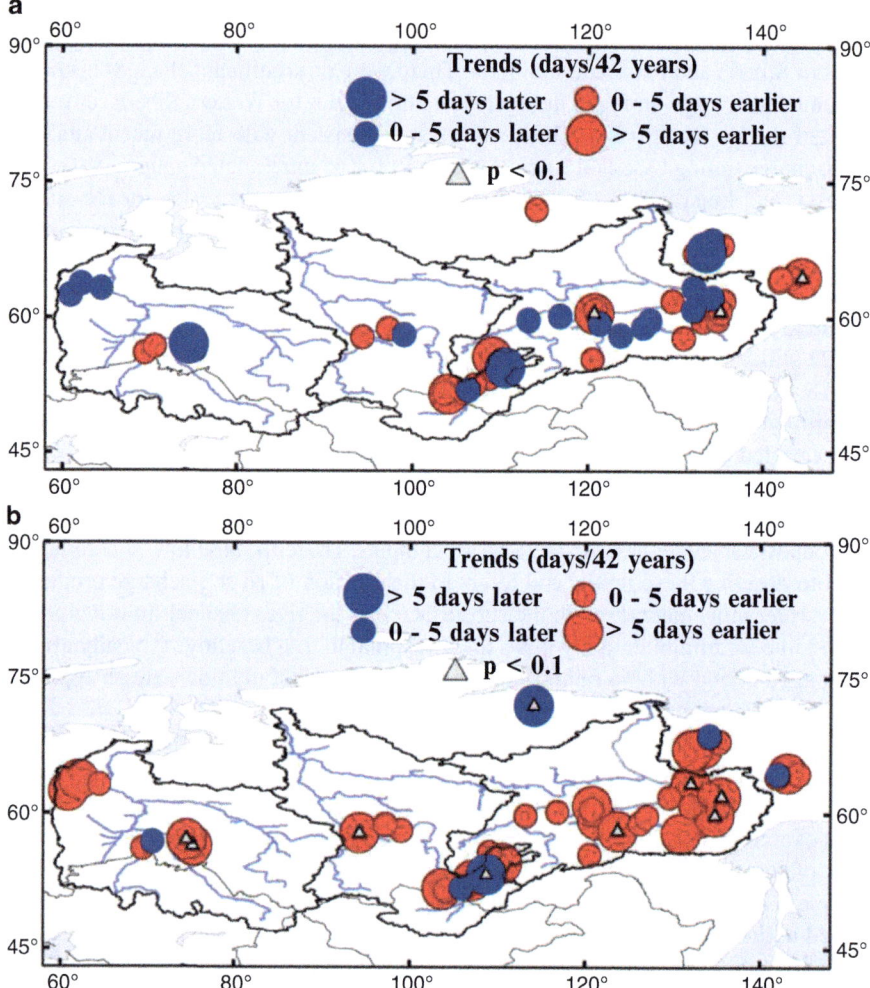

Fig. 4.11 Trends in (**a**) spring pulse onset (SPO; day of year marking the beginning of the snowmelt season) and (**b**) centroid of timing (CT; date on which accumulated flow before and after the date is equal) (Source: Tan et al. 2011)

discharge in winter but a trivial fraction of the annual flow, which is dominated by late spring and summer flows. This results in minimum-flow changes being most noticeable (and most statistically detectable) during winter, despite the presence of correspondingly large increases in summer. From a statistical standpoint, a +2-mm minimum-flow increase during winter will typically be more significant than a +2-mm (or larger) increase in summer, even though in terms of physical process (inferred groundwater contribution) the summer increase is just as meaningful (Smith et al. 2007).

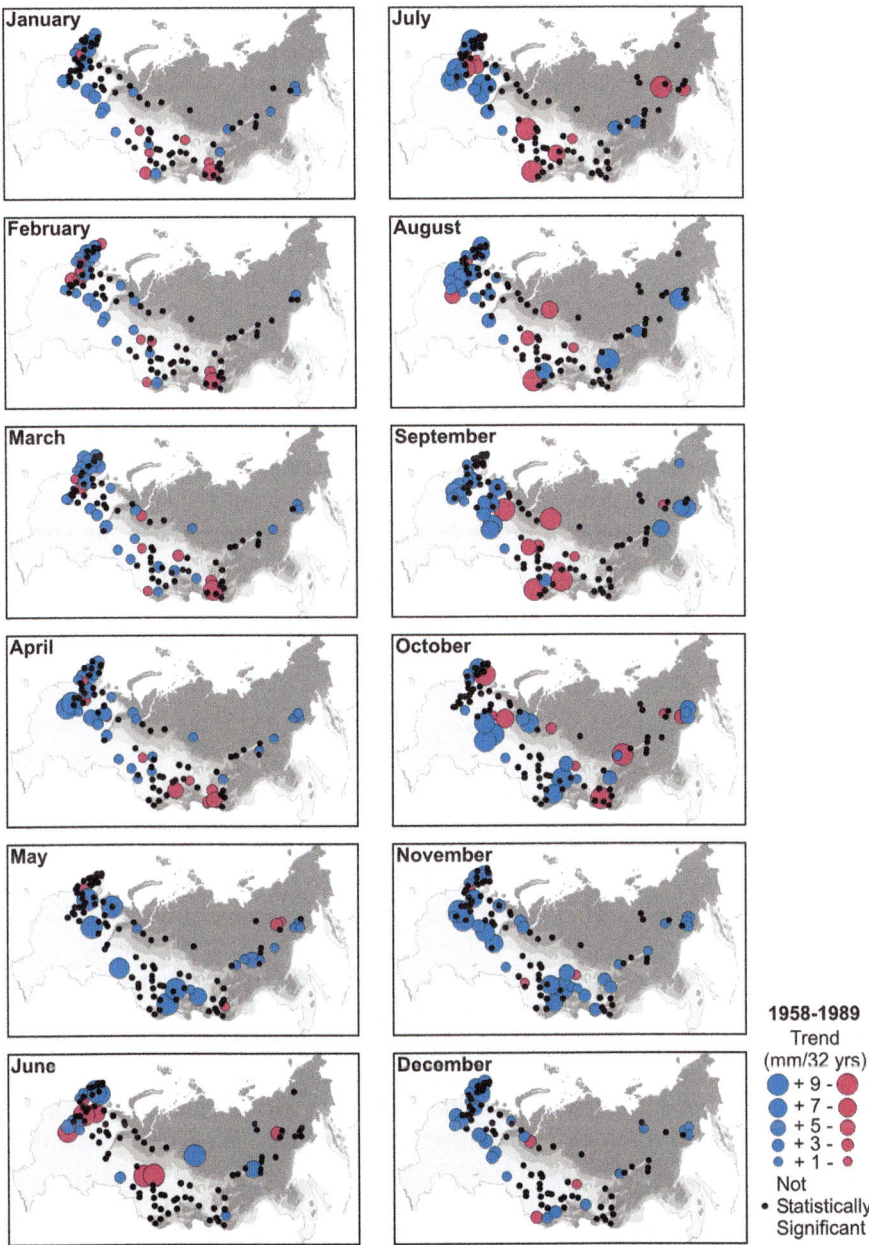

Fig. 4.12 Spatial distribution of statistically significant trends in minimum flow from 1958 to 1989. Increases in minimum flow were particularly evident in May and November, perhaps reflecting seasonal shifts toward earlier spring melt and later autumn freeze-up, respectively. No strongly coherent spatial contrast is apparent between permafrost and permafrost-free areas, and summer trends generally equaled or exceeded winter trends (Source: Smith et al. 2007)

With regard to physical mechanism(s), if minimum flows are presumed to approximate, or at least correlate, with soil- and groundwater inputs to rivers, the results summarized above indicate a broad-scale mobilization of subsurface water activity during the twentieth century. This is not a new idea: the possibility that thawing permafrost and associated melting of ground ice may be releasing stored water to streams has been an important hypothesis in the debate over the terrestrial runoff increases (McClelland et al. 2004). Anecdotal evidence of shrinking or draining lakes in Alaska and Siberia does suggest that thawing of transitional permafrost may promote water infiltration to the subsurface (Yoshikawa and Hinzman 2003; Smith et al. 2005). However, the results suggest little if any unique role for permafrost in the observed minimum-flow increases, because some of the greatest increases have occurred in non-permafrost basins. Nonetheless, the signal is consistent with a "thaw-like" process, one that would promote increased soil infiltration, subsurface water movement, and connections to stream networks. Smith et al. (2007) go on to speculate that decreased winter freezing of soils, caused by warmer winters and/or deeper snowpack, might generate such activity in both permafrost and non-permafrost environments alike. Climate records certainly show warmer winter and spring temperatures over central- and west-northern Eurasia since at least 1979 (Rigor et al. 2000). An argument can be made that a more deeply thawed, or more frequently thawed, landscape would not only accept more infiltration from the surface but also would shift water storage from the surface/near-surface to greater depths, reducing evapotranspiration water losses to the atmosphere. In terms of the regional water balance, this reduced loss would have equivalent impact to a precipitation increase.

4.6   Change in the Thermal and Ice Regimes

Most river temperature studies to date have been conducted on relatively small local or regional spatial scales, and only a limited subset of those has been in the Arctic or cold regions. Yang et al. (2005) analyzed river temperatures for five gauges distributed throughout large tributaries of the Lena basin over the period 1950–1992 and concluded that there has been a consistent warming of stream temperature across the entire basin during the early open water season (early to mid-June), coupled with an increase in peak discharge for the same period. River temperatures for the remainder of the open water season exhibited mixed results, with warming occurring in some subbasins, and cooling in others.

A regionally comprehensive analysis of Russian Arctic river water temperatures and heat fluxes was carried out by Lammers et al. (2007). Annual means of ~10-day water temperatures for the Siberian gauges were in the range 6.8 °C (Anabar) to 10.1 °C (Taz), and maximum 10-day values ranged from 14.3 °C (Norilka) up to 19.3 °C (Taz). The timing of maximum temperature value had a narrow range covering July (Kolyma at Srednekolymsk) to early August (Norilka). The heat flux from all the Russian basins was 3,500 PJ year^{-1} (0.20 W m^{-2}), with the Lena having the largest total heat flux for any basin (15,000 PJ year^{-1}) and the Norilka had the largest area-normalized heat flux (0.62 W m^{-2}). These two basins also had the largest maximum 10-day heat flux for all basins (2,500 PJ year^{-1} for the Lena and 1.58 W m^{-2} for the Norilka). The timing of maximum heat flux ranged from early June (Nadym) to the middle of July (Norilka). Most gauges had similar mean annual heat flux, and the Norilka at Valek appeared to be an outlier.

In terms of long-term water temperature trends, regressions for both least squares and Mann–Kendall analysis (Lammers et al. 2007) showed that most of the significant changes occurred with the maximum 10-day temperature and mean 10-day heat flux (Table 4.2). Temperature changes were observed for the Anabar (decreasing mean and maximum), the Yana (increasing 10-day maximum value), and the Kolyma at Kolymskoye (later timing of the maximum temperature, Mann–Kendall only). Only one Siberian station (on the Olenek River) showed significant change in river discharge with an earlier timing of the maximum spring discharge. Significant increases in mean annual heat flux were seen in the Norilka and Yana while the Yenisei and Kolyma at Srednekolymsk had decreasing trends. The Yana also showed increasing total heat flux, for both total and 10-day maximum values. The Yana and Kolyma (at Srednekolymsk) also displayed decreasing and increasing trends, respectively, in maximum heat flux. The two Kolyma gauges were the only stations showing significant trends with the Mann–Kendall test and not the least squares approach.

Trend line slopes for both water temperature and heat flux at these stations are presented in Table 4.2, together with annual air temperature trends (both local grid cell and aggregate upstream basin average). Using local grid cell air temperature data, three stations showed significant increases: Taz, Norilka, and Kolyma at Kolymskoye. In analyzing upstream air temperature trends, these three stations again displayed significant warming, as well as the Nadym, Yenisei, Anabar, and Kolyma at Srednekolymsk.

In terms of total heat flux, Lammers et al. (2007) compute a mean annual Russian pan-Arctic value (including north-flowing rivers of European Russia), based on 16 representative gauges of 0.19 W m^{-2} but no statistically significant trend. However, the Ob, Yenisei, and Lena subset of the three largest basins (Fig. 4.13) had a mean heat flux of 0.17 W m^{-2} and a significantly decreasing slope (cooling) of −0.00053 W m^{-2} year^{-1} representing a total decrease of 0.029 W m^{-2} from 1938 to 1992 – a 17 % decline relative to the mean for the entire 1938–1992 period. This decreased heat flux is inconsistent with the rising air temperatures found throughout the region. Lammers et al. (2007) postulated this discontinuity could be the result of the introduction of cooler waters into the rivers during the observational periods. Increased groundwater flow (e.g., Smith et al. 2007), melting permafrost, and the release of cooler waters from reservoirs could all contribute to lowering river temperatures. It is also possible that temperature increases may not be observed over larger basins. While there is ample evidence of smaller watersheds displaying increases in river temperature (e.g., Caissie 2006) or significant trends in upstream basins being reduced to insignificance at the basin outlet (Yang et al. 2005), there are none that we know of for the large basins. Perhaps the largest drainage basins simply have too much spatial variability to reflect any consistent trends.

Table 4.2 Least squares regression results for river temperature, heat, and air temperature for each monitoring station

Short-hand	ID	Years	River temperature (T_1) Slope of trend line — Decadal mean °C a⁻¹	10-day maximum °C a⁻¹	Timing of maximum 10-day a⁻¹	Years	Heat flux (using T_2) Slope of trend line — Mean value PJ a⁻¹	10-day maximum PJ a⁻¹	Timing of maximum 10-day a⁻¹	Air temperature Slope of trend line — Local grid cell °C a⁻¹	Upstream average °C a⁻¹
Oneg	18	47	0.0098	**0.0506**	0.0038	40	1.30	−0.10	−0.0273	0.004	0.010
Sevr	16	52	0.0068	**0.0397**	−0.0030	48	−5.47	−1.73	−0.0117	0.001	0.009
Mezn	19	22	0.0014	**0.0578**	−0.0166	20	−0.99	−0.31	−0.0413	0.008	0.009
PchU	20	53	**0.0190**	**0.0423**	−0.0140	51	**26.74**	2.55	−0.0051	0.002	−0.003
PchO	17	28	0.0101	**0.0668**	−0.0127	1			—	0.001	0.002
Ob	12	45	−0.0005	0.0078	−0.0087	45	−14.04	0.67	−0.0194	−0.007	**0.020**
Nadm	13	35	0.0182	0.0267	−0.0085	23	0.00	0.84	0.0404	0.005	0.007
PurU	11	29	−0.0026	−0.0013	0.0021	9	3.88	1.71	0.0077	0.016	0.019
PurS	14	45	0.0090	0.0340	0.0019	41	−0.28	0.04	0.0005	−0.006	0.001
Taz	15	30	−0.0108	−0.0199	−0.0072	20	5.10	2.14	0.0115	**0.036**	**0.037**
Yen	10	59	−0.0107	−0.0066	0.0041	57	**−51.67**	−1.21	−0.0072	−0.002	**0.017**
Nor	9	31	0.0060	0.0352	−0.0079	29	**2.74**	0.42	−0.0242	**0.045**	**0.044**
Khat	4	32	0.0219	−0.0403	−0.0312	3			—	0.020	0.025
Anab	3	34	**−0.0309**	**−0.0736**	0.0047	34	0.29	−0.08	−0.0246	0.016	**0.036**
Olnk	5	25	−0.0236	0.0147	−0.0281	25	−0.93	−1.35	−0.0290	0.017	0.021
Lena	6	50	0.0016	0.0075	−0.0106	50	7.53	−1.42	−0.0088	−0.006	0.011
Yana	7	37	0.0090	**0.0717**	0.0035	17	**34.04**	**9.17**	0.0228	0.000	0.011
Indk	8	51	−0.0024	0.0102	−0.0024	50	2.38	0.55	0.0072	0.001	0.011
KolS	1	55	−0.0086	0.0066	0.0109	52	−9.66	−3.32	0.0038	0.007	**0.011**
KolK	2	37	−0.0076	0.0110	0.0330	24	−32.30	−5.30	0.0265	**0.050**	**0.037**

Source: Lammers et al. (2007)

Significant linear regression slopes ($\alpha=0.05$) shown in **bold**. Positive slope in timing of maximum indicates later in year. Least squares regression results were not calculated for heat flux for basin IDs 17 (PchO) and 4 (Khat) due to insufficient number of years of data (1 and 3 years, respectively)

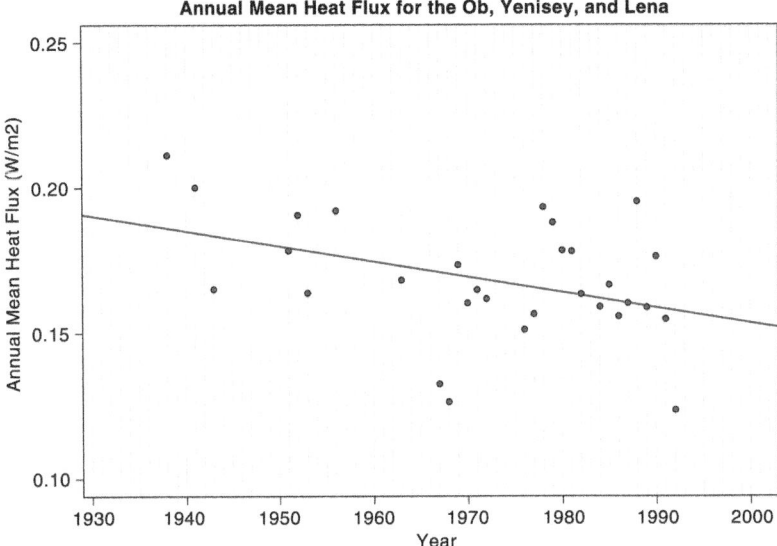

Fig. 4.13 Aggregate annual mean heat flux for the three largest Russian drainage basins, Ob, Yenisei, and Lena. Only years in which all 3 basins had values were used. *Trend line* shows a significant decrease in annual heat flux of −0.00053 W m⁻² year⁻¹ (Source: Lammers et al. 2007)

4.6.1 Ice Regime

Changes in ice regime are especially important in Siberia, where many rivers and lakes are intensively used for navigation in summer–fall and as natural roads in winter. Therefore, ice characteristics as dates of freeze-up and breakup, duration of sustainable ice cover, and ice thickness are very important for regional human activity. Dates and duration of ice cover define length of navigation period, and ice cover thickness is a determining factor for the ice-bearing capacity and length of ice-routes operation on rivers and lakes. Over the last two to three decades, ice regimes in the rivers of Russia have been subject to significant changes mainly due to a stable positive trend in winter air temperatures.

A handful of investigations have utilized the timing of Eurasian surface water freeze-up or breakup to infer information about climate (e.g., Palecki and Barry 1986; Ginzburg et al. 1992; Soldatova 1992, 1993; Ginzburg and Soldatova 1996; Smith 2000). The occurrence of lake ice is primarily a function of air temperature, while river ice formation and breakup are also influenced by water discharge, temperature, and local hydraulics. Despite this complexity, good correlation is found between Eurasian air temperatures and river ice cover (Ginzburg et al. 1992; Soldatova 1993; Ginzburg and Soldatova 1996). It is encouraging that this correlation exists even for regulated rivers like the Volga (Soldatova 1992).

Using historical records, Smith (2000) analyzed long-term river ice trends (1917–1994) for the Varzuga, Onega, Mezen, Pechora, Ob, Yenisei, Olenek, Indigirka,

Table 4.3 Changes in mean ice cover duration and maximum ice cover thickness in large rivers of Russia in 1980–2000 relative to 1950–1979 from Vuglinsky (2002)

River	Changes in mean ice cover duration, days	Changes in mean maximum ice cover thickness, cm
Yenisei (upper stream)	−5 −7	−7 −9
Yenisei (middle stream)	−4 −7	−8 −10
Yenisei (low stream)	−3 −5	−4 −6
Lena (middle stream)	−4 −6	−8 −12
Lena (low stream)	0 −2	−4 −6

and Kolyma Rivers. Twenty of these records were found to display statistically significant trend ($p = 0.1$) using the Mann–Kendall test. Of eight measurement variables (start dates of first winter ice, ice cover formation, melt onset, ice drift, dates of maximum ice drift and ice disappearance, total annual duration of continuous ice cover, and total duration of all ice events between fall and spring of each hydrologic year), the timing of melt onset displayed the highest incidence of change, with statistically significant negative (earlier) shifts of ~1–3 weeks found for the Pechora, Ob, Olenek, Indigirka, and Kolyma Rivers. An observed general pattern of decreased ice cover in the 1950s, increased ice cover in the 1980s, and subsequent decrease during the 1990s was generally consistent with regional temperature trends. However, both interannual and regional was found, including earlier occurrence of autumn freezing for the Onega, Varzuga, Mezen, and Yenisei Rivers, leading to an increase in the duration of winter ice cover for all but the Mezen. An opposite trend was found for the Indigirka and Kolyma Rivers of Far Eastern Siberia. A subsequent MODIS remote sensing study of river breakup processes along the Lena, Ob, and Yenisei Rivers confirmed that the degree of similarity between interannual trends in breakup date at various locations along a river is generally high, thus supporting the use of point-scale historical records to infer regional climatic trends (Pavelsky and Smith 2004).

Magnuson et al. (2000) analyzed ice regimes for 39 rivers and lakes in the Northern Hemisphere for 1846–1995. They found consistent evidence of later freezing and earlier breakup throughout the Northern Hemisphere from 1846 to 1995. Over these 150 years, changes in freeze dates averaged 5.8 days per 100 years later, and changes in breakup dates averaged 6.5 days per 100 years earlier; these translate to increasing air temperatures of about 1.2 °C per 100 years. Interannual variability in both freeze and breakup dates has increased since 1950. Specific features of ice events for 16 Arctic Russian rivers were investigated by Vuglinsky (2002). He found statistically significant negative trends over the last three decades in ice cover duration and maximum ice thickness for the largest Siberian rivers Yenisei and Lena (Table 4.3). Figure 4.14 shows changes in monthly ice cover thickness for the downstream-most gauge on the Lena River based on historical and near real-time data supplied at the University of New Hampshire (http://rims.unh.edu) (Arctic-HYDRA 2010).

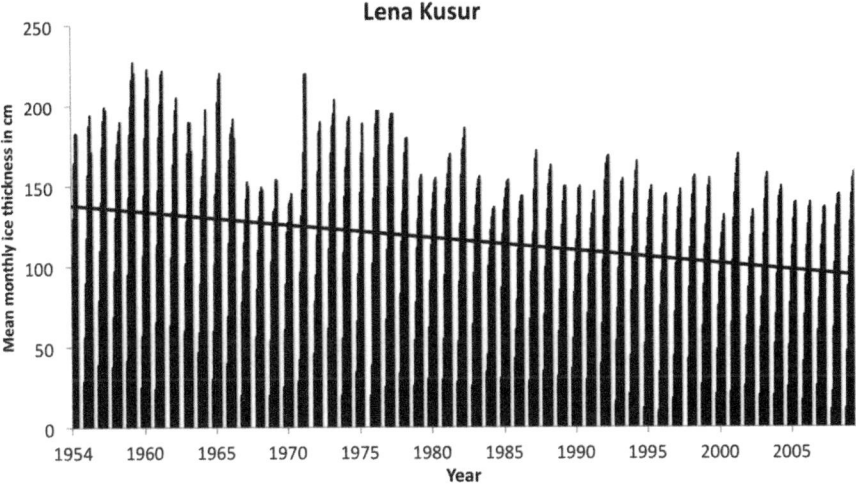

Fig. 4.14 Changes in the mean monthly ice cover thickness in the Lena River, 1955–2009 (Arctic-HYDRA 2010)

4.7 Changes in Groundwater Hydrology

4.7.1 Groundwater Level

Ligotin (2008, 2010), Savichev (2008, 2010), Ligotin and Savichev (2008) and Ligotin et al (2010) studied the groundwater regime of the West Siberian artesian pool in the Ob River basin. All observation wells were free of human impacts. Some were characterized by spring–fall groundwater recharge via infiltration of snowmelt and rain water and terrace type of regime. Others were artesian with spring–fall recharge. These aquifers are characteristic of the taiga zone of Siberia, and the results are probably applicable to significant parts of the Ob and Yenisei River basins as well.

Mean annual groundwater levels in the taiga zone of the West Siberian artesian basins rose on the order of ~10–20+ cm over the period 1965–2005, especially in fall and winter (Fig. 4.15). The largest increasing trends have been observed in the taiga zone of the Ob River basin (Fig. 4.16).

Fluctuations of groundwater levels in wells located in the forest-steppe and steppe zones in the same hydrogeological regime are not consistent with each other. Both increasing and decreasing tendencies were observed from 1980 to 2000 across the region (Figs. 4.16 and 4.17). According to the Ligotin (2006, 2008), changes in groundwater levels (both negative and positive) in some wells exceeded 1 m over the last 40 years.

Changes in groundwater levels are complicated by aftershocks from earthquakes in the Sayan–Altai hydrogeological folded area (upper parts of Ob and Yenisei basins).

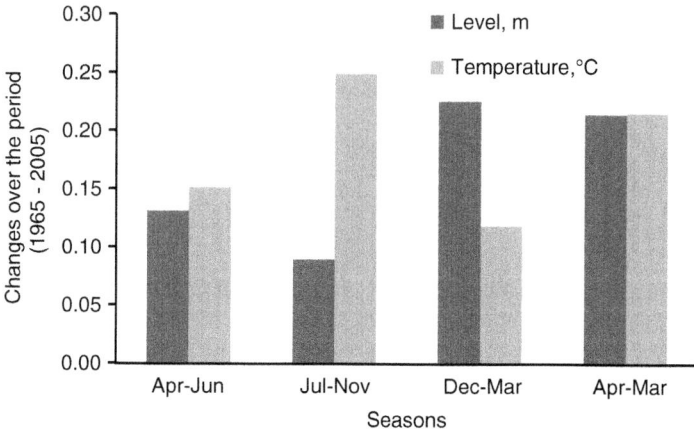

Fig. 4.15 Long-term seasonal changes in groundwater levels and groundwater temperature over the period 1965–2005 in Ob River basin taiga zone

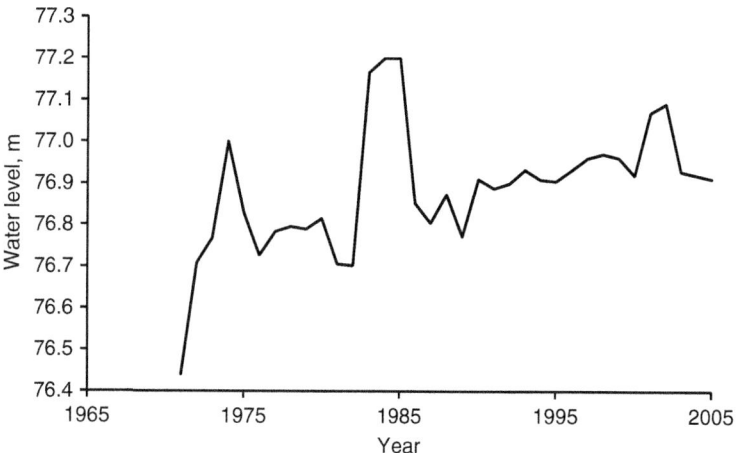

Fig. 4.16 Long-term variability of mean annual groundwater levels (m) in wells at Napas village. Wells draw from the artesian Paleocene aquifer in the taiga forest zone of the Ob River basin, with spring–fall recharge

Regular observations of groundwater levels in other regions of Siberia generally have short records and large gaps and are not suitable for multidecadal trend analysis. Based on a qualitative analysis, Savichev (2010) concluded that groundwater levels in alluvial sediments of Jurassic and Paleozoic formations of the Angara–Lens artesian basin are stable, whereas groundwater levels across the Baikal–Aldan folded system demonstrate variable long-term tendencies.

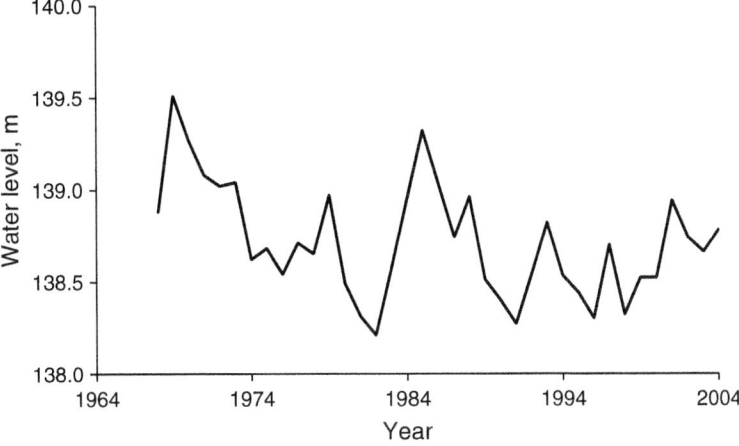

Fig. 4.17 Long-term variability of annual groundwater levels at Talmenka (river in Altay mountains, upper part of Ob basin, lat 53°48′; long 83°34′) in the near-river regime of the forest-steppe zone of the Ob River basin

The general tendency across Siberia is that groundwater levels have increased in the wetter (humid) zones and decreased in dryer (semiarid) areas. These changes have shown a mosaic spatial pattern across the region and are mainly due to increasing regional moisture, seasonal redistribution of atmospheric precipitation, change in the snowmelt timing, and, hence, conditions for infiltration of snowmelt waters (Ligotin and Savichev 2008, 2010).

4.7.2 Groundwater–Surface Water Interactions

Few observational data exist to study interactions between groundwater and surface water in Siberia. However, it is possible to interpret groundwater contributions to streamflow using two assumptions: (1) winter streamflows for rivers with pristine hydrological regimes are supplied mainly by groundwater contributions, and (2) relationships between groundwater level and groundwater contributions, defined for the winter period, hold over the entire year. Taking into account these assumptions and simplifying the equations for infiltration in an unpressurized aquifer, we estimated regional dependencies for the taiga area of the Ob River basin between groundwater runoff, groundwater level, and river water stage. Applying two different empirical methods to separate river hydrographs for surface and groundwater contributions to streamflow (Savichev 2010), we evaluated groundwater contributions for several Siberian rivers with drainage areas ranging from 3,400 to 38,400 km² and sustained groundwater inputs during the cold period

Table 4.4 Long-term mean values of specific discharge in the taiga zone of the Ob River basin and its estimated groundwater component

River – Gauge (drainage area)	Type of runoff	Specific discharge ($1 \text{ s}^{-1} \text{ km}^{-2}$)
The Ket River – MaximkinYar village (38,400 km²)	Total	6.32
	Ground[a]	2.11
	Ground[b]	2.15
The Tym River – Napas village (24,500 km²)	Total	7.92
	Ground[a]	4.45
	Ground[b]	2.88
The Vasugan River – Sredny Vasugan village (31,700 km²)	Total	5.05
	Ground[a]	1.30
	Ground[b]	1.19
The Chuzik River – Pudino village (3,400 km²)	Total	4.10
	Ground[a]	0.74
	Ground[b]	0.72
The Chaya River – Podgornoe village (25,000 km²)	Total	3.24
	Ground[a]	0.96
	Ground[b]	0.99

[a]Computed using simplified approach
[b]Computed using sophisticated approach

(Table 4.4). Both methods provide consistent results for the test basins (Table 4.4), and therefore, we applied this simpler approach to estimate long-term variability of the ground component of streamflow for a number of midsize Siberian rivers (Savichev 2010).

The analysis of annual time series reveals increasing mean annual groundwater runoff across the taiga zone of the Ob, Yenisei, and Lena River basins over the years 1980–2000 (Table 4.5). Mean values of specified groundwater runoff (A) for most river basins are much higher during last 20 years than for observational periods before 1979. In forest-steppe, steppe zones, and mountainous areas, similar tendencies are also found, but they are less consistent across the area. This increase in groundwater contributions to streamflow is very consistent with results obtained from analysis of winter streamflow for small- and medium-sized Russian Arctic rivers discussed in Sects. 4.4 and 4.5. However, observed increases in winter surface runoff across Siberia (Smith et al. 2007; Tan et al. 2011) is much more significant than changes found in ground flow. Unfortunately, there are no reliable experimental data in the region about interactions between ground and surface runoff, and the indirect methods for evaluation of ground flow based on simplified assumptions, discussed above, cannot provide reliable estimates of significant changes in the regional climate system. As a whole, the increasing tendencies in groundwater contributions to most rivers (Table 4.5) are not statistically significant, a conclusion also supported by Simonov and Christoforov (2005) and Kovalevsky (2007).

Table 4.5 Statistical characteristics of annual time series of specified groundwater runoff evaluated for selected Siberian rivers

River – gauge	Drainage area (km^2)	Period	A ($1\ s^{-1}\ km^{-2}$)	σ	Sk/Sk$_a$	Fk/Fk$_a$	Pk/Pk$_a$
Chaya –	25,000	1953–2005	0.98	0.19	1.01	0.57	1.93
Podgornoe		1953–1979	0.89	0.16	–	–	0.14
Village		1980–2005	1.07	0.18	–	–	1.22
Parabel –		1958–2005	1.27	0.21	0.23	0.90	0.34
Novikovo		1958–1979	1.25	0.24	–	–	-0.37
Village		1980–2005	1.30	0.17	–	–	0.60
Vasugan –	31,700	1953–2005	1.42	0.57	0.53	2.00	1.22
Sredny Vasugan		1953–1979	1.28	0.34	–	–	−0.19
Village		1980–2005	1.58	0.71	–	–	1.19
Om –	47,800	1941–1998	0.19	0.15	0.23	0.48	−0.19
Kalachinsk		1941–1979	0.18	0.15	–	–	−0.95
Town		1980–1998	0.21	0.16	–	–	0.27
Vagay – Novo-	9,740	1955–1996	0.16	0.04	1.64	0.38	2.68
vyigryshnaya		1955–1979	0.14	0.03	–	–	0.66
Village		1980–1996	0.20	0.03	–	–	0.43
Konda –	65,400	1936–1996	1.88	0.97	0.59	0.43	1.14
Bolchary		1936–1979	1.77	0.95	–	–	0.77
Village		1980–1996	2.19	0.98	–	–	0.86
Polui –	15,100	1954–1996	2.70	0.35	0.88	1.32	1.46
Polui		1954–1979	2.59	0.39	–	–	0.81
Village		1980–1996	2.87	0.21	–	–	−0.02
Pyakupur –	31,400	1954–1996	3.49	0.52	0.02	0.54	0.21
Tarko-Sale		1954–1979	3.49	0.56	–	–	0.42
Village		1980–1996	3.48	0.47	–	–	0.06
Chadobets –	13,300	1957–1998	0.58	0.20	0.77	0.55	0.75
Yarkino		1957–1979	0.50	0.20	–	–	−0.86
Village		1980–1998	0.67	0.17	–	–	0.32
Big Patom –	27,600	1937–1998	2.38	0.51	0.36	0.56	0.03
Patoma		1937–1979	2.33	0.53	–	–	−0.45
Village		1980–1998	2.48	0.46	–	–	−0.54
Amga –	23,900	1937–1998	0.71	0.22	1.45	0.59	2.09
Buyaga		1937–1979	0.63	0.17		–	0.36
Village		1980–1998	0.90	0.21	–	–	0.53

A average value over the period, σ standard deviation, Sk and Sk_a actual and critical values of Student criterion, Fk and Fk_a actual and critical values of Fisher criterion, k and k_a actual and critical values of Pitman closeness criterion

4.8 Changes in Thermokarst Lakes of West Siberia

Lakes and wetlands dominate the landscape in many areas of the arctic plains and the West Siberian lowlands (Fig. 4.18). These lakes and wetlands play important roles in the regional hydrological and carbon cycles. Climate warming and permafrost

Fig. 4.18 Thermokarst lakes in central part of Nadym–Pur watershed (Photo by S. Kirpotin 1999)

degradation may significantly change the drainage, causing disappearance of many lakes and wetlands with potential effects on the global carbon cycle. Smith et al. (2005) observed a decrease in lake extent between 1973 and 1997 in West Siberia underlain by discontinuous permafrost, based on analysis of Landsat TM imagery. During the same time period, lake extent decreased in the portion of West Siberia underlain by discontinuous permafrost (Smith et al. 2005). In this section, new results are presented on the dynamics of thermokarst lakes across the West Siberian lowlands obtained from detailed analysis of satellite imagery for 30 test sites across the region.

The multiyear archive of satellite optical images from Landsat for the study area was analyzed (Fig. 4.19). The technique of thermokarst dynamics study requires measurement of lake areas by means of the nonsimultaneous space images, and accuracy of results depends on magnitude of intraseasonal changes of lake area. Intraseasonal dynamics of thermokarst lake areas were studied. Because of the small number of cloudless days in these northern regions, only radar space images could be used. Complete methodology and results of this research using radar satellite images ERS-2 are given in Bryksina and Polishchuk (2009). This research has shown that intraseasonal changes of thermokarst lakes areas on different test sites are on average less than 1–2 %. This permits the collection of Landsat satellite images to be augmented by choosing cloudless optical images obtained in different months during the warm season. As a result, the collection formed included 106 cloudless space images received during the period 1973–2009:

- Landsat-1 (scanner MSS, spatial resolution 80 m), 1973;
- Landsat-2, 4, 5 (scanners MSS and TM, spatial resolution 30 m), 1981–2009;
- Landsat-7 (scanner ETM+, spatial resolution 30 m), 1999–2003.

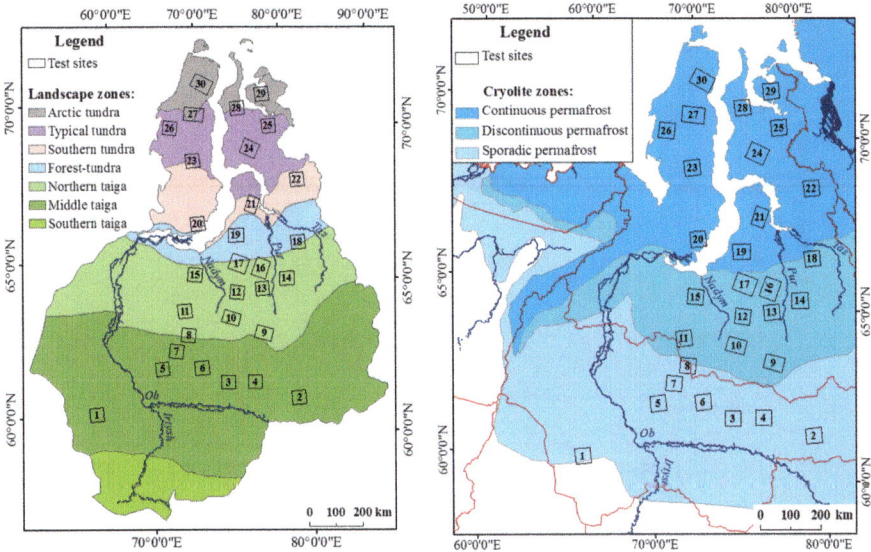

Fig. 4.19 Geocryological map and landscape map of West Siberian lowlands. Distribution of sporadic, discontinuous, and continuous permafrost across the West Siberian lowlands and locations of test sites (*right panel*). Types of regional landscapes and test sites shown in the *left panel*. The locations of test sites were chosen to evenly represent different landscapes and permafrost types

The area of thermokarst lakes on all test sites were measured based on these satellite images, and the total number of lakes in each test site ranged from several hundred to several thousand lakes. Figure 4.18 shows a typical field of thermokarst lakes in the central West Siberian permafrost. The array of measured data has been analyzed statistically for understanding of changes in thermokarst lakes.

Total area of thermokarst lakes was determined at each test site and for each year of observation by latitude and landscape type. Relative changes in total areas (R) were based on

$$R = \frac{S_f - S_i}{S_i},$$

where S_i and S_f were the total area of lakes in the initial and final year of the research period.

The relative change in the total area of lakes (R_m = mean of R across test sites located in each landscape zone) shows that increased total lake areas are found only in two zones of the tundra – the Arctic tundra and typical tundra (Fig. 4.20). In zones of the southern tundra, forest–tundra, northern and middle taiga, reduced total area of the lakes was observed. It should be noted that in areas of the forest–tundra and southern tundra, the tendency toward reduced total area of lakes is large with relative changes in the total area of the lakes in these two zones of −11.1 and −9 %, respectively.

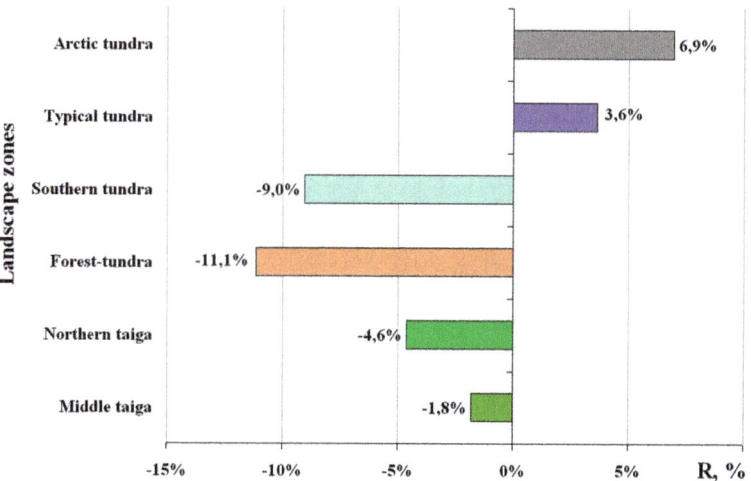

Fig. 4.20 The average value of relative change of total lake area in different landscape zones

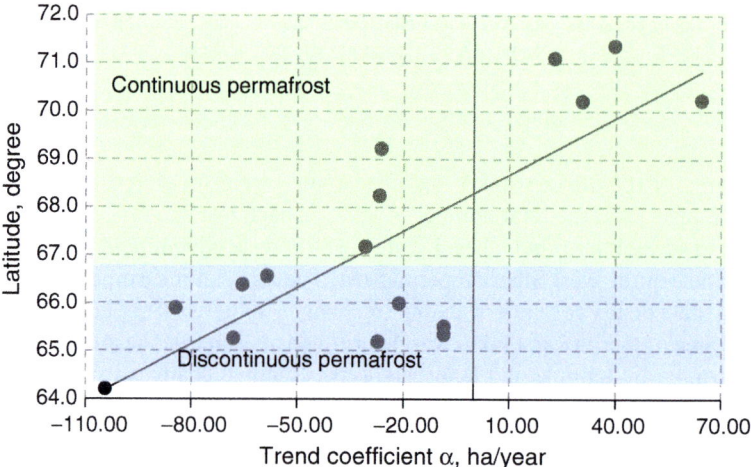

Fig. 4.21 Linear trend coefficients for total area of lakes depending on latitude. *Positive values* of the coefficient indicate increasing lake area, and *negative values* indicate an average reduction in lake area

It is interesting to consider the thermokarst dynamics depending on the latitude. This study used time series of the total area of lakes in the test sites in different years of the study. Linear trends of time series showed that lakes increasing in total area were located in the northern part of the continuous permafrost zone (Fig. 4.21; test sites 27–30 in Fig. 4.19 located above 70°N). The other group of test sites located

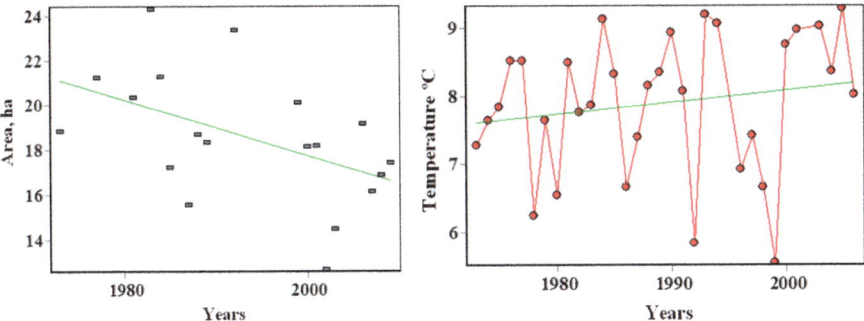

Fig. 4.22 Average thermokarst lake area (*left*) and midsummer temperature (*right*) over the time

south of the 70°N latitude in the southern part of the continuous permafrost zone and in all sites in the discontinuous zone revealed the tendency toward decreasing total lake area.

Thermokarst lake areas were compared to trends of average annual temperature for the continuous and discontinuous permafrost zones in West Siberia (Fig. 4.22). Midsummer temperatures were obtained by averaging data of all weather stations and average areas of lakes in each test site.

It was found that increasing midsummer temperatures are accompanied by decreasing areas of thermokarst lakes that may be explained by the impact of global warming. The mechanism of water draining from small lakes to large lakes, connected with the acceleration of soil drainage with increasing temperature, as set out in Kirpotin et al. (2008), explains the general shrinking of lake area under global warming. If this hypothesis is correct and average temperatures continue to increase, then lake extent through large parts of Siberia will be expected to decline. Such changes will influence the hydrological regime both locally in the immediate vicinity of the lakes and regionally in the seasonal timing of downstream water flows and, in some cases, when critical melt thresholds are reached through the sudden and catastrophic release of these lake waters. In most cases, large releases of water will not directly influence large population centers; however, smaller settlements of humans are certainly located in some of these downstream locales.

4.9   Future Changes in Regional Hydrology

River flow from the Eurasian pan-Arctic, especially the largest river basins (Lena, Yenisei, Ob, Pechora, Sev. Dvina, and Kolyma), has a strong effect on the freshwater balance of the Arctic basin and regional climate and influences the chemical composition of water, the sea ice formation, and the circulation in the Arctic Ocean and North Atlantic.

Table 4.6 Projected changes in river discharge of large Siberian rivers under the impact of climate change from different sources

Source	Climate scenario	River basin	Annual discharge change (%)
Arora (2001)	CGCM, by 2050	Yenisei/Lena/Ob	18/19/–12
Mokhov and Khon (2002)	HadCM3 ECHAM4	Yenisei/Lena/Ob	8/22–24/3–4
Georgievsky et al. (2003)	HadCM3	Lena	12
Manabe et al. (2004)	IS92a by 2035–2065	Yenisei/Lena/Ob	13/12/21
Arnell (2004)	6-model ensemble to 2080; A2 & B2 emissions	Total flow to Arctic Ocean for A2/B2	24/18
Nohara et al. (2006)	19-model ensemble 2081–2100; A1b emissions	Yenisei/Lena/Ob	16/24/10

According to future projections of anthropogenic climate change based on global climate models (GCMs), the largest changes in air temperature and precipitation will take place in northern Eurasia, particularly in Eastern Siberia (IPCC 2007). Regional warming is expected to lead to substantial changes in regional hydrology as a result of soil temperature increases (Varlamov et al. 2002) and consequent permafrost thawing (Anisimov et al. 1997; Pavlov et al. 2005). Changes in the magnitude and seasonality of precipitation may also alter river runoff characteristics and their intra-annual distribution with possible impacts on freshwater and heat flux to the Arctic Ocean, which have potentially important implications for the ocean circulation and climate outside the region.

As noted earlier in this chapter, river discharge to the Arctic Ocean has been increasing, and these changes are primarily from the Siberian part of the domain (Peterson et al. 2002; Berezovskaya et al. 2005; Georgiadi et al. 2008; Shiklomanov and Lammers 2009). Projections of future changes in river discharge and total river flux to the Arctic Ocean due to global warming have been made by many investigators over the last two decades. Results of some of these investigations during the 1990s were summarized by Shiklomanov and Shiklomanov (2003). At that time, most researchers used climate scenarios with GCMs for doubled atmospheric CO_2 concentrations, along with some intermediate concentrations. These climate scenarios were used as inputs to hydrological models of varying complexity, adapted to individual river basins to estimate the effects of climate change on regional hydrological regime (Shiklomanov and Linz 1991).

Table 4.6 summarizes more recent results for expected changes in the discharge of large Siberian rivers due to climate change. These studies are based on different emissions scenarios and future climate simulations using various GCMs archived for the 2001 and 2007 IPCC reports (IPCC 2001, 2007). Both individual and multi-model average climate projections have been used by different authors to evaluate expected changes in Siberian hydrology (see Sect. 3.3 in this book). Despite the diversity of models, scenarios, and approaches, the results of these studies for the

ΔR, %

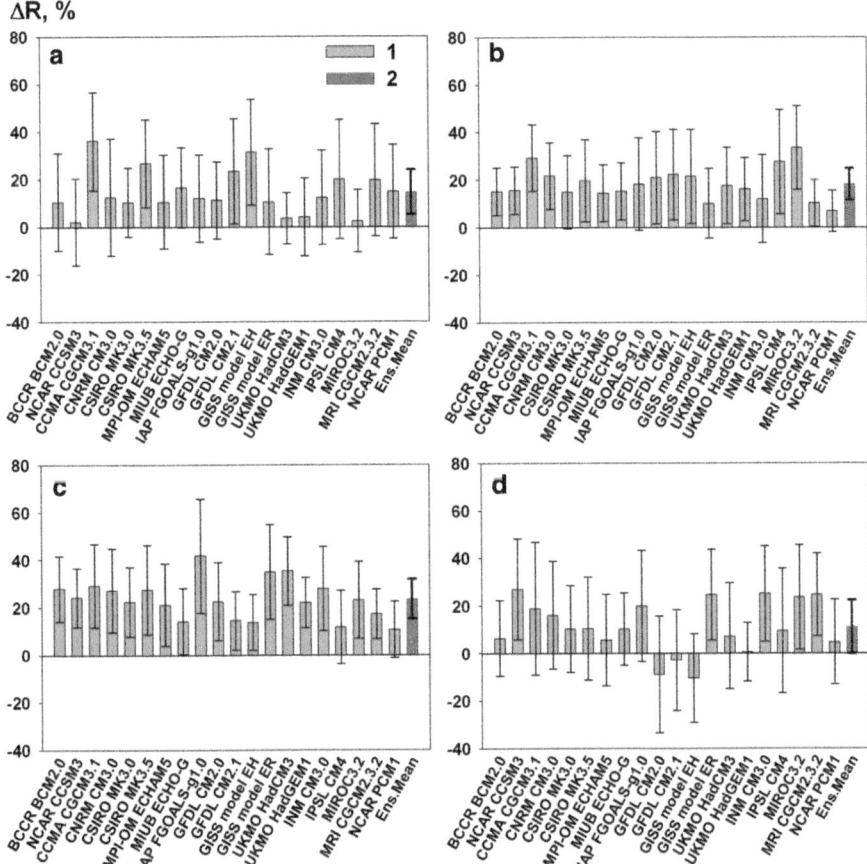

Fig. 4.23 Mean annual runoff changes (%) and its standard deviation for Ob (**a**), Yenisei (**b**), Lena (**c**), and Amur (**d**) Rivers to the end of the twenty-first century according to different models (*1*) and to model ensemble mean (*2*)

large Siberian rivers are qualitatively consistent. According to GCM simulations, further increases of Siberian runoff are expected in the twenty-first century (Mokhov and Khon 2002; Mokhov et al. 2003; Shiklomanov and Shiklomanov 2003; Meleshko et al. 2004; Nohara et al. 2006; Kattsov et al. 2007; Meleshko et al. 2008; Khon and Mokhov 2012) along with a significant shift to earlier dates in spring flood peak flow, associated with rising spring air temperatures and earlier snowmelt. Special analysis of runoff outputs from contemporary GCMs (IPCC 2007; Sect. 3.3 in this book) also demonstrates the consistent increases in mean annual runoff: up to ~15 % for Ob, ~20 % for Yenisei, and ~25 % for Lena Rivers (Fig. 4.23). GCM-derived changes for the Amur River runoff are much lower than for other rivers. Some GCMs project even a small decrease in the Amur River runoff; however, these changes are not statistically significant.

Table 4.7 IPCC model data	AO GCM	Location
in NEESPI-RIMS	BCCR BCM2.0	Bergen, Norway
	CCCMA CGCM3.1	Victoria, Canada
	GFDL CM2.1	Princeton, USA
	INM CM3.0	Moscow, Russia
	CCSR MIROC3.2 Medres	Tokyo, Japan
	MPI ECHAM5	Hamburg, Germany
	NCAR CCCSM3.0	Boulder, USA
	UKMO HADLEYCM3	Exeter, UK

In general, GCMs can adequately reproduce the annual variability of precipitation and runoff in large Siberian river basins (Khon and Mokhov 2012); however, they have large uncertainties in the simulation of intra-annual hydrological regime due to simplified representation of land surface processes, lack of runoff routing computation, and neglected local human impacts (e.g., reservoirs). To eliminate these shortcomings, we analyzed future runoff/discharge changes using simulations from a recently modified version of the UNH water balance model (WBMplus; Wisser et al. 2008). The modifications include a reservoir-routing module that simulates the operation of large reservoirs and their impact on river discharge and an irrigation module that models the interactions of irrigated areas with components of the hydrological cycle. WBMplus is an integrated, physically based, grid cell-based hydrological model that simulates variations of hydrological cycle components. The model has a flexible structure including several methods to estimate evapotranspiration subject to available information. Taking into account the large uncertainty of simulated climate precipitation and other climate characteristics in GCMs, the projected evapotranspiration was defined using the Hamon temperature index method. Discharge was routed through the river network using the Muskingum–Cunge method (Henderson 1966). Gridded temperature and precipitation fields covering contemporary (1959–1999) and future scenarios (SRES A1b, A2, B1; 2001–2099) was generated from eight AO GCMs (Table 4.7 and Sect. 3.3) and are available for analysis through our website: http://neespi.sr.unh.edu/maps/.

WBMplus was run for the IPCC SRES A1B emissions scenario for all AO GCMs listed in Table 4.7. Results indicate a tendency toward increases in river runoff across Siberia although the projected changes are not spatially uniform. Figure 4.24 shows changes in annual river runoff across Siberia by 2040–2060 from the long-term average (1959–1999) for UKMO HadleyCM3, MPI ECHAM5, and NCAR CCSM3 AO GCMs, as well as an average simulation result for all 8 GCMs. Increasing river runoff for the entire Siberian region indicated on the maps is more clearly seen in the frequency diagrams (Fig. 4.24). More significant increases in runoff across Siberia by 2040–2060 are expected based on output from MPI ECHAM5 and NCAR CCSM3. Runoff changes evaluated for UKMO HadleyCM3 and for the ensemble of 8 AO GCMs have a less noticeable increasing tendency across Siberia with some regions showing a decline in river runoff. Wetter conditions are expected in northern

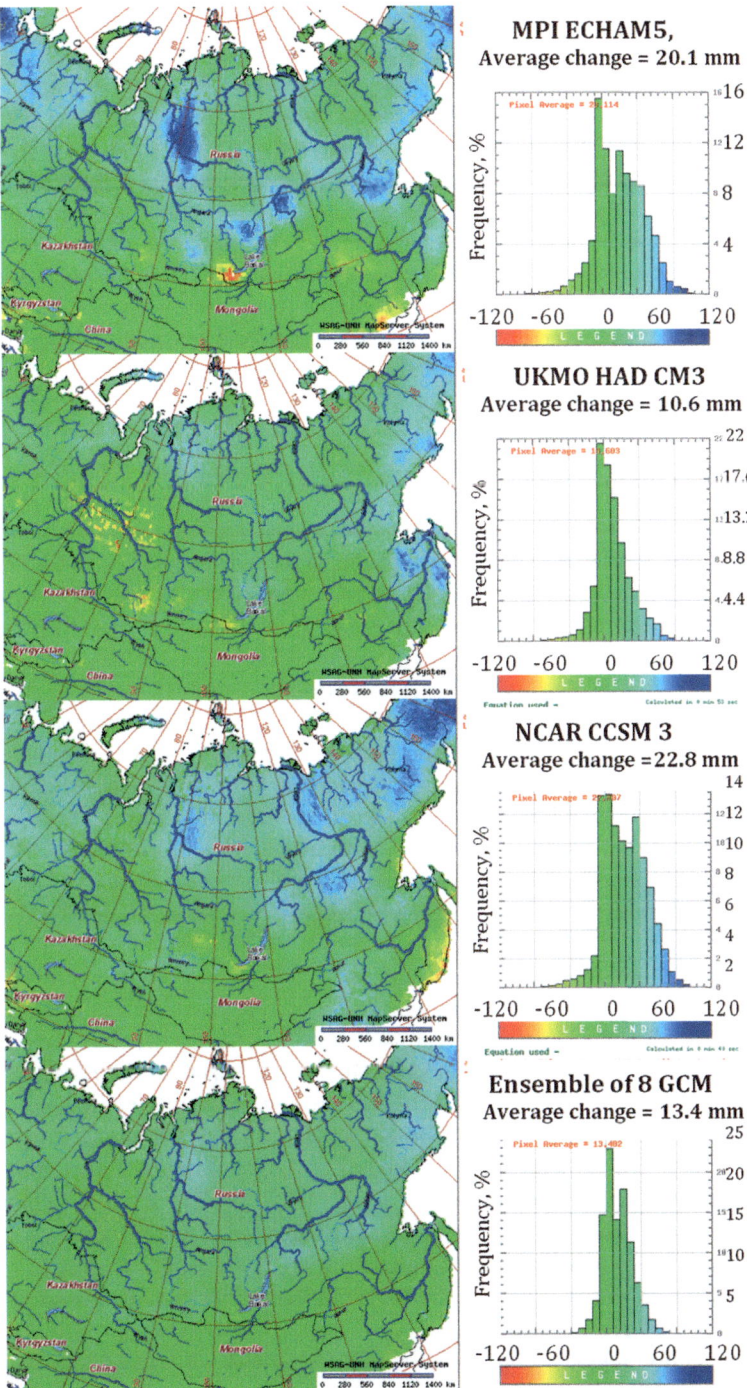

Fig. 4.24 Deviation of simulated mean annual river runoff over 2040–2060 relative to the long-term observed mean (1959–1999)

and eastern regions of Siberia in the Yenisei, Lena, and Kolyma basins. At the same time, runoff across Western and Southern Siberia is projected to change less significantly with both positive and negative alterations.

More significant changes are expected in monthly river discharge of the large Siberian rivers. Figure 4.25 shows projected changes in monthly discharge distribution for the four largest Siberian rivers by 2040–2060 based on the SRES A1B emissions scenario and different GCM simulations. The greatest increases in river discharge are projected during winter and spring months, due to higher winter base flow and earlier spring melt. Some decrease in summer runoff is projected in the Ob and Yenisei basins. It should be noted that similar tendencies of changes in seasonal runoff have been observed from analysis of historical discharge records for Siberia (Shiklomanov et al. 2007; Rawlins et al. 2006b; Smith et al. 2007).

4.10 Conclusion

We discussed changes in different components of the hydrological regime in Siberia based on long-term observational records and modeled future river runoff projections. This concluding section highlights some of the more important findings and gaps derived from regional studies.

From analysis of historical records, it is clear that the hydrologic cycle of Siberia has undergone dramatic changes over the last few decades. Although these changes have not been uniform, some general patterns have emerged. The increase of annual river discharge to the Arctic Ocean is very consistent with the rise of regional air temperature. The discharge increase is strongly associated with significant increases in low flow, often (but not exclusively) in regions underlain by permafrost. Warmer Siberian air temperatures have also led to thinning river ice, shortening of ice cover duration, deepening permafrost active-layer thickness, and earlier snowmelt and spring floods. Unfortunately, these physical mechanisms driving these phenomena are only partially understood. There is also a severe lack of regional experimental investigation on interactions between groundwater and surface water, especially in the permafrost zone. Such studies would help in better understanding changes in intra-annual runoff and would improve hydrological simulations in numerous land surface models.

Analysis of future projections of the hydrological regime based on various scenarios and GCMs and applying more sophisticated hydrological simulations reveals several general tendencies for Siberia: (1) annual runoff will significantly increase in basins of large rivers flowing to the Arctic Ocean but will not change in the drier areas of Southern Siberia; (2) the most significant runoff increases will be observed during winter and spring periods; and (3) a significant shift to earlier snowmelt and earlier spring flood across the entire region. However, these results should be considered as preliminary. We believe that a more detailed approach applying higher-resolution regional climate models (RCM) along with land surface models accounting for changes in permafrost and vegetation is required for more reliable regional-scale projections of future hydrology.

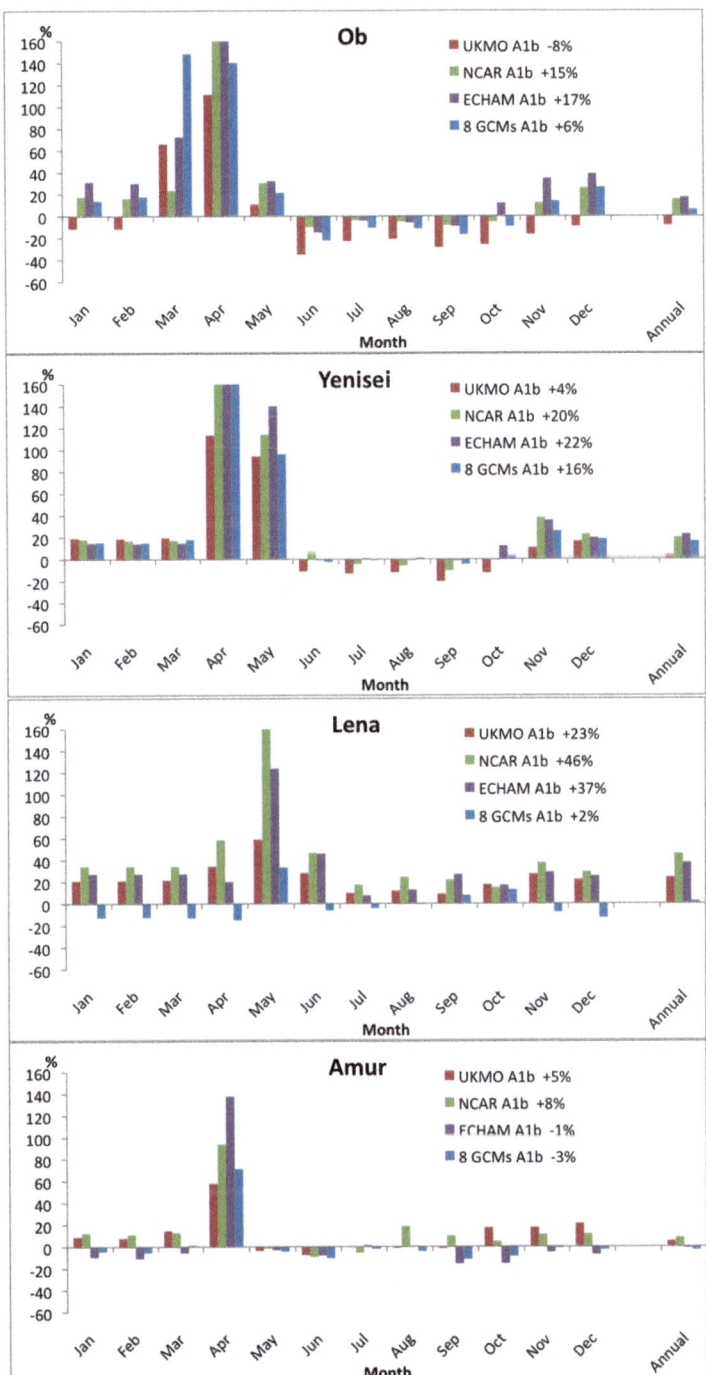

Fig. 4.25 Mean monthly discharge hydrographs calculated with WBMplus for the largest Siberian rivers over 2040–2060 (IPCC A1B emissions scenario) and 1959–1999 (for 20c3m scenario) from different GCMs. Percent change in annual discharge by 2040–2060 versus 1959–1999 for each model is given in the legend

Siberia is an extreme environment with many challenges for detailed hydrological studies due to severe climate, specific conditions of runoff formation associated with permafrost, sparse hydrological networks, and poor logistical support. However, this region is experiencing a most impressive observed air temperature increase (see Chap. 3) and is critically important for understanding responses of northern hydrological systems to climate change. Furthermore, Siberia's vast boreal forests, lakes, and wetlands play a crucial role in global carbon cycle and correspondingly in global climate (Shiklomanov et al. 2011). Therefore, regional hydrological studies should be prioritized despite the many difficulties of accessing this area.

References

ACIA (2005) Arctic climate impact assessment scientific report. Cambridge University Press, New York, 1042 pp

Adam JC, Lettenmaier DP (2007) Application of new precipitation and reconstructed streamflow products to streamflow trend attribution in northern Eurasia. J Clim 21(8):1807–1828

Adam JC, Haddeland I, Su F, Lettenmaier DP (2007) Simulation of reservoir influences on annual and seasonal streamflow changes for the Lena, Yenisei, and Ob' rivers. J Geophys Res 112:D24114. doi:10.1029/2007JD008525

Anisimov OA, Shiklomanov NI, Nelson FE (1997) Effects of global warming on permafrost and active-layer thickness: results from transient general circulation models. Glob Planet Change 15(2):61–77

Arctic and Antarctic Research Institute (AARI) (1985) Arctic Atlas (in Russian), Moscow, 204 pp

Arctic-HYDRA (2010) The Arctic hydrological cycle monitoring, modeling and assessment programme. Science and implementation plan. Available at: http://arctichydra.arcticportal.org/images/stories/Arctic-HYDRA.pdf

Armstrong RL, Brodzik MJ (2001) Recent Northern Hemisphere snow extent: a comparison of data derived from visible and microwave sensors. Geophys Res Lett 28(19):3673–3676

Arnell NW (2004) Climate change and global water resources: SRES emissions and socio economic scenarios. Glob Environ Change 14:31–52

Arora VK (2001) Streamflow simulations for continental scale river basins in a global atmospheric general circulation model. Adv Water Resour 24(7):775–791

Berezovskaya S, Yang D, Kane DL (2004) Compatibility analysis of precipitation and runoff trends over the large Siberian watersheds. Geophys Res Lett 31:L21502. doi:10.1029/2004GL021277

Berezovskaya S, Yang DQ, Hinzman L (2005) Long-term annual water balance analysis of the Lena River. Glob Planet Change 48(1–3):84–95

Bryksina NA, Polishchuk YM (2009) Analysis of seasonal changes of thermokarst lakes areas in permafrost zone of Western Siberia using satellite images ERS-2. Res Earth Space 29(3):90–93 (In Russian)

Caissie D (2006) The thermal regime of rivers: a review. Freshw Biol 51:1389–1406

Cherkauer KA, Bowling LC, Lettenmaier DP (2003) Variable Infiltration Capacity (VIC) cold land process model updates. Glob Planet Change 38(1–2):151–159

Fyodorov-Davydov DG, Kholodov AL, Ostroumov VE, Kraev GN, Sorokovikov VA, Davudov SP, Merekalova AA (2008) Seasonal thaw of soils in the North Yakutian ecosystems. In: Proceedings of the 9th international conference on permafrost, vol 1. Institute of Northern Engineering, University of Alaska, Fairbanks, pp 481–486

Georgiadi AG, Milyukova IP, Kashutina EA (2008) Recent and projected river runoff changes in permafrost regions of Eastern Siberia (Lena River Basin). In: Proceedings of the international conference on permafrost (ICOP), Fairbanks, Alaska, 29 June–3 July 2008

Georgievsky VYu, Shiklomanov IA, Shalygin AL (2003) Long-term variations in the runoff over the Russian territory. Report of the State Hydrological Institute, St. Petersburg, Russia, 85 pp

Ginzburg BM, Soldatova II (1996) Long-term oscillations of river freezing and breakup dates in different geographical zones. Russ Meteorol Hydrol 6:80–85

Ginzburg BM, Polyakova KN, Soldatova II (1992) Secular changes in dates of ice formation on rivers and their relationship with climate change. Sov Meteorol Hydrol 12:57–64

Gurevich E (2009) Influence of air temperature on the river runoff in winter (the Aldan river catchment case study). Meteorol Hydrol 34(9):628–633. doi:10.3103/S1068373909090088

Gutman G, Reissell A (eds) (2011) Eurasian Arctic land cover and land use in a changing climate, vol XXIII, 1st edn. Springer, Dordrecht, 306 p

Hamon WR (1963) Computation of direct runoff amounts from storm rainfall. Int Assoc Sci Hydrol Publ 63:52–62

Helsel DR, Hirsch RM (1992) Studies in environmental science 49 – Statistical methods in water resources. Elsevier, Amsterdam 522 p

Henderson FM (1966) Open channel flow. Macmillan, New York

IPCC (2001) Climate change 2001: The scientific Basis. In: Houghton JT, Dung Y, Griggs DJ et al (eds) Contribution of Working Group I to the 3rd assessment report of the Intergovernmental Panel on Climate Change. Cambridge University Press, Cambridge

IPCC (2007) In: Solomon S, Qin D, Manning M, Chen Z, Marquis M, Averyt KB, Tignor M, Miller HL (eds) Climate change 2007: Working Group I report "The Physical Science Basis", contribution of Working Group I to the fourth assessment report of the Intergovernmental Panel on Climate Change. Cambridge University Press, Cambridge/New York, 996 pp

Kattsov VM, Walsh JE, Chapman WL, Govorkova VA, Pavlova T, Zhang X (2007) Simulation and projection of Arctic freshwater budget components by the IPCC AR4 global climate models. J Hydrometeorol 8:571–589

Khon VC, Mokhov II (2012) The hydrological regime of large river basins in Northern Eurasia in the XX–XXI centuries. Water Resour 39:1–10

Kilmyaninov VV (2000) Conditions for flood formations under ice jams on the middle stream of Lena River in 1998 and 1999. Meteorol Hydrol 10:93–98

Kirpotin S, Polishchuk Y, Zakharova E, Shirokova L, Pokrovsky O, Kolmakova M, Dupre B (2008) One of the possible mechanisms of thermokarst lakes drainage in West-Siberian North. Int J Environ Stud 65(5):631–635

Kovalevsky VS (2007) The Influence of climate change on ground waters. Water Resour 34(2): 158–170, In Russian

Lammers RB, Shiklomanov AI, Vorosmarty CJ, Fekete BM, Peterson BJ (2001) Assessment of contemporary Arctic river runoff based on observational discharge records. J Geophys Res Atmos 106(D4):3321–3334

Lammers RB, Pundsack JW, Shiklomanov AI (2007) Variability in river temperature, discharge, and energy flux from the Russian pan-Arctic landmass. J Geophys Res Biogeosci 112:G04S59. doi:10.1029/2006JG000370

Ligotin VA (ed) (2006) State of the geological environment of the Siberian federal district per 2005, Issue 2. Company "Tomskgeomonitoring", Tomsk, 166 pp (In Russian)

Ligotin VA (ed) (2008) State of the geological environment of the Siberian federal district per 2007, Issue 4. Company "Tomskgeomonitoring", Tomsk, 194 pp (In Russian)

Ligotin VA, Savichev OG (2008) Perennial alterations of hydrogeological and hydrological conditions in the Ob river basin, Western Siberia, Rossia. Global Groundwater Resources and Management. In: Paliwal BS (ed) Selected papers from the 33rd international geological congress, general symposium: hydrogeology, Oslo, Norway, 6–14 Aug 2008. Scientific Publishers (India), Jodhpur, pp 159–173

Ligotin VA, Savichev OG, Makushin JuV (2010) The long-term changes of seasonal and annual levels and temperature of ground waters of the top hydrodynamical zone in Tomsk area. Geoecology 13(1):23–29 (In Russian)

Magnuson JJ, Robertson DM, Benson BJ, Wynne RH, Livingstone DM, Arai T, Assel RA, Barry RG, Card V, Kuusisto E, Granin NG, Prowse TD, Stewart KM, Vuglinsky VS (2000) Historical trends in lake and river ice cover in the Northern Hemisphere. Science 289:1743–1746

Manabe S, Stouffer RJ (1994) Multiple-century response of a coupled ocean-atmosphere model to an increase of atmospheric carbon dioxide. J Clim 7:5–23

Manabe S, Milly PCD, Wetherald R (2004) Simulated long-term changes in river discharge and soil moisture due to global warming. Hydrolog Sci 49(4):625–642

Marchenko S, Gorbunov A, Romanovsky V (2007) Permafrost warming in the Tien Shan Mountains, Central Asia. Glob Planet Change 56:311–327

McClelland JW, Holmes RM, Peterson BJ, Stieglitz M (2004) Increasing river discharge in the Eurasian Arctic: consideration of dams, permafrost thaw, and fires as potential agents of change. J Geophys Res 109:D18102. doi:10.1029/2004JD004583

McDonald KC, Kimball JS, Njoku E, Zimmermann R, Zhao M (2004) Variability in springtime thaw in the terrestrial high latitudes: monitoring a major control on the biospheric assimilation of atmospheric CO_2 with spaceborne microwave remote sensing. Earth Interact 8(20):1–23

Meleshko VP, Golitsyn GS, Govorkova VA, Demchenko PF, Eliseev AV, Katsov VM, Malevsky-Malevich SP, Mokhov II, Nadezhdina ED, Semenov VA, Sporyshev PV, Khon V Ch (2004) Anthropogenic climate changes in Russia in the 21st century: an ensemble of climate model projections. Russ Meteorol Hydrol 29(4):38

Meleshko VP, Kattsov VM, Govorkova VA, Sporyshev PV, Shkol'nik IM, Shneerov BE (2008) Climate of Russia in the 21st century. Part 3. Future climate changes calculated with an ensemble of coupled atmosphere-ocean general circulation CMIP3 models. Russ Meteorol Hydrol 33(9):541–552

Mernild SH, Liston GE, Hiemstra CA, Steffen K (2009) Record 2007 Greenland ice sheet surface melt extent and runoff. EOS Trans Am Geophys Union 90(2):13, 14 (13 January 2009)

Mitchell TD, Carter TR, Jones PD, Hulme M, New M (2004) A comprehensive set of high-resolution grids of monthly climate for Europe and the globe: the observed record (1901–2000) and 16 scenarios (2001–2100), technical report. Tyndall Centre for Climate Change, Norwich, UK

Mokhov II, Khon VCh (2002) Hydrological regime in Siberian river basins: model assessments of changes in 21st century. Meteorol Hydrol 8:77–90

Mokhov II, Semenov VA, Khon VCh (2003) Estimates of possible regional hydrologic regime changes in the 21st century based on global climate models. Izv Atmos Oceanic Phys 39:130–144

Nohara D, Kiton A, Hosaka M, Oki T (2006) Impact of climate change on river runoff. J Hydrometeorol N7:1076–1089

Palecki MA, Barry RG (1986) Freeze-up and break-up of lakes as an index of temperature changes during the transition seasons: a case study in Finland. J Clim Appl Meteorol 25:893–902

Pavelsky TM, Smith LC (2004) Spatial and temporal patterns in Arctic river ice breakup observed with MODIS and AVHRR time series. Remote Sens Environ 99(3):328–338

Pavelsky TM, Smith LC (2006) Intercomparison of four global precipitation datasets and their correlation with increased Eurasian river discharge to the Arctic Ocean. J Geophys Res 111:D21112. doi:10.1029/2006JD007230

Pavlov AV, Skachkov YB, Kakunov NB (2005) Relationship between long-term active-layer changes and atmospheric factors. Earth Cryosph 8(4):3–11

Peterson BJ, Holmes RM, McClelland JW, Vorosmarty CJ, Lammers RB, Shiklomanov AI, Shiklomanov IA, Rahmstorf S (2002) Increasing river discharge to the Arctic Ocean. Science 298:2171–2173, 13 Dec 2002

Rahmstorf S (2002) Ocean circulation and climate during the past 120,000 years. Nature 419:207–214

Rawlins MA, Lammers RB, Frolking S, Fekete BM, Vorosmarty CJ (2003) Simulating pan-Arctic runoff with a macro-scale terrestrial water balance model. Hydrolog Process 17(13):2521–2539

Rawlins MA, Lammers RB, Frolking S, Vorosmarty CJ (2006a) Simulated runoff and evapotranspiration across Alaska: model sensitivity to climate and land cover drivers. Earth Interact 10. doi:10.1175/EI182.1

Rawlins MA, Willmott CJ, Shiklomanov A, Linder E, Frolking S, Lammers RB, Vorosmarty CJ (2006b) Evaluation of trends in derived snowfall and rainfall across Eurasia and linkages with discharge to the Arctic Ocean. Geophy Res Lett 33:L07403. doi:10.1029/2005GL025231

Rawlins MA, Fahnestock M, Frolking S, Vorosmarty CJ (2007) On the evaluation of snow water equivalent estimates over the terrestrial Arctic drainage basin. Hydrol Process 21(12):1616–1623. doi:10.1002/hyp. 6724. 1

Rawlins MA, Schroeder R, Zhang X, McDonald KC (2009) Anomalous atmospheric circulation, moisture flux, and winter snow accumulation in relation to record combined discharge from large Eurasian rivers in 2007. Environ Res Lett 4:045011. doi:10.1088/1748-9326/4/4/045011

Rawlins MA, Steele M, Holland MM, Adam JC, Cherry JE, Francis JA, Ya P, Groisman LD, Hinzman TG, Huntington DL, Kane JS, Kimball R, Kwok RB, Lammers CM, Lee DP, Lettenmaier KC, McDonald E, Podest JW, Pundsack B, Rudels MC, Serreze A, Shiklomanov Y, Skagseth TJ, Troy CJ, Vorosmarty M, Wensnahan EF, Wood R, Woodgate D, Yang KZ, Zhang T (2010) Analysis of the Arctic system for freshwater cycle intensification: observations and expectations. J Clim 23(21):5715–5737. doi:10.1175/2010JCLI3421.1

Rigor IG, Colony RL, Martin S (2000) Variations in surface air temperature observations in the Arctic, 1979–1997. J Clim 13(5):896–914

Savichev OG (2010) Water resources of Tomsk area. Publishing House of the Tomsk Polytechnic University, Tomsk, 248 pp (In Russian)

Serreze MC, Barrett AP, Slater AG, Woodgate RA, Aagaard K, Lammers RB, Steele M, Moritz R, Meredith M, Lee CM (2006) The large-scale freshwater cycle of the *Arctic*. J Geophys Res 111:C11010. doi:10.1029/2005JC003424

Shiklomanov IA (1989) Man's impact on river runoff. L. Hydrometeoizdat, 300 pp (in Russian)

Shiklomanov IA (ed) (2008) Water resources of Russia and their use. Nauka, St. Petersburg, 598 pp (In Russian)

Shiklomanov AI (2010) River discharge, in Chapter 5, Arctic, state of the climate in 2009. Bull Am Meteorol Soc 91(7):116–117

Shiklomanov AI, Lammers RB (2009) Record Russian river discharge in 2007 and the limits of analysis. Environ Res Lett 4, 045015, 9 pp. doi:10.1088/1748–9326/4/4/045015

Shiklomanov IA, Linz G (1991) Effect of climate changes on hydrology and water management. Meteorol Gidrol 1991(4):51–65

Shiklomanov IA, Markova OA (1987) Problems of water availability and river flow transfer in the world. Leningrad, Hydrometeoizdat, 294 pp (in Russian)

Shiklomanov IA, Shiklomanov AI (2003) Climatic change and dynamics of river discharge into the Arctic Ocean. Water Resour 30(6):593–601, November 2003

Shiklomanov IA, Shiklomanov AI, Lammers RB, Peterson BJ, Vorosmarty CJ (2000) The dynamics of river water inflow to the Arctic Ocean. In: Lewis EL et al (eds) The freshwater budget of the Arctic Ocean. Kluwer Academic Publishers, Dordrecht

Shiklomanov AI, Lammers RB, Rawlins MA, Smith LC, Pavelsky TM (2007) Temporal and spatial variations in maximum river discharge from a new Russian data set. J Geophys Res Biogeosci 112:G04S53. doi:10.1029/2006JG000352

Shiklomanov AI, Bohn TJ, Lettenmaier DP, Lammers RB, Romanov P, Rawlins MA, Adam JC (2011) Chapter 7: Interactions between land cover/use change and hydrology. In: Gutman G, Reissell A (eds) Eurasian Arctic land cover and land use in a changing climate. Springer, Dordrecht, pp 137–177

Simonov JA, Christoforov AV (2005) The analysis of long-term fluctuations of a runoff of the rivers of Arctic Ocean basin. Water Resour 32(6):645–652, In Russian

Smith LC (2000) Time-trends in Russian Arctic river ice formation and breakup: 1917–1994. Phys Geogr 21(1)

Smith LC, Sheng Y, Macdonald GM, Dhinzman L (2005) Disappearing Arctic Lakes. Science 308:1429

Smith LC, Pavelsky TM, MacDonald GM, Shiklomanov AI, Lammers RB (2007) Rising minimum daily flows in northern Eurasian rivers suggest a growing influence of groundwater in the high-latitude water cycle. J Geophys Res Biogeosci 112:G04S47. doi:10.1029/2006JG000327

Soldatova II (1992) Causes of variability of ice appearance dates in the lower reaches of the Volga. Sov Meteorol Hydrol 2:62–66

Soldatova II (1993) Secular variations in river breakup dates and their relation to climate changes. Russ Meteorol Hydrol 9:70–76

Stanilovskaya J, Ukhova J, Sergeev D, Utkina I (2008) Thermal state of permafrost in Northern Transbaykalia, Eastern Siberia. In: Proceedings of the ninth international conference on permafrost, vol 1, 29 June–3 July 2008. University of Alaska Fairbanks, pp 1695–1700

Steffen K, Nghiem SV, Huff R, Neumann G (2004) The melt anomaly of 2002 on the Greenland Ice Sheet from active and passive microwave satellite observations. Geophys Res Lett 31:L20402. doi:10.1029/2004GL020444

Stewart IT, Cayan DR, Dettinger MD (2005) Changes toward earlier streamflow timing across western North America. J Clim 18:1136–1155

Stieglitz M, Déry SJ, Romanovsky VE, Osterkamp TE (2003) The role of snow cover in the warming of arctic permafrost. Geophys Res Lett 30(13):1721. doi:10.1029/2003GL017337

Tan A, Adam JC, Lettenmaier DP (2011) Change in spring snowmelt timing in Eurasian Arctic rivers. J Geophys Res 116:D03101. doi:10.1029/2010JD014337

Varlamov SP, Skachkov, Yu B, Skryabin PN (2002) Ground temperature regime of Central Yakutia permafrost landscapes. Yakutsk, SB RAS, 218p (In Russian)

Vörösmarty CJ, Fekete BM, Meybeck M, Lammers R (2000) Global system of rivers: its role in organizing continental land mass and defining land-to-ocean linkages. Global Biogeochem Cycles 14:599–621

Vuglinsky VS (2002) Peculiarities of ice events in Russian Arctic rivers. Hydrol Process 16:905–913

Wisser D, Frolking S, Douglas EM, Fekete BM, Vörösmarty CJ, Schumann AH (2008) Global irrigation water demand: variability and uncertainties arising from agricultural and climate data sets. Geophys Res Lett 35:L24408. doi:10.1029/2008GL035296

Willmott CJ, Matsuura K (1995) Smart interpolation of annually averaged air temperature in the United States. J Appl Meteorol 34(12):2577–2586

Willmott CJ, Matsuura K (2001) Arctic terrestrial air temperature and precipitation: monthly and annual time series (1930–2000) version 1. http://climate.geog.udel.edu/~climate/, 2001 prepared at the University of Delaware

Yang D, Kane D, Hinzman L, Zhang X, Zhang T, Ye H (2002) Siberian Lena river hydrologic regime and recent change. J Geophys Res Atmos 107(D23):4694. doi:10.1029/2002JD002542

Yang D, Ye B, Shiklomanov A (2004) Discharge characteristics and changes over the Ob River watershed in Siberia. J Hydrometeorol 5(4):595–610

Yang D, Liu B, Ye B (2005) Stream temperature changes over Lena River Basin in Siberia. Geophys Res Lett 32:LO5401

Yang D, Zhao Y, Armstrong R, Robinson D, Brodzik M-J (2007) Streamflow response to seasonal snow cover mass changes over large Siberian watersheds. J Geophys Res 112:F02S22. doi:10.1029/2006JF000518

Ye B, Yang D, Kane D (2003) Changes in Lena river streamflow hydrology: human impacts vs. natural variations. Water Resour Res 39(7):1200–1213

Yoshikawa K, Hinzman LD (2003) Shrinking thermokarst ponds and groundwater dynamics in discontinuous permafrost near Council, Alaska. Permafr Periglac Process 14(2):151–160

Zhang T (2005) Influence of the seasonal snow cover on the ground thermal regime: an overview. Rev Geophys 43(RG4002). doi:10.1029/2004RG000157

Chapter 5
Effect of Climate Change on Siberian Infrastructure

Nikolay I. Shiklomanov and Dmitriy A. Streletskiy

Abstract This chapter examines effects of climate change on human infrastructure in permafrost regions of Siberia. The presence and dynamic nature of ice-rich permafrost constitute a distinctive engineering environment. Many engineering problems in Siberia are associated with (1) changes in the temperature of the upper permafrost, (2) increased depth of seasonal thaw penetration, and (3) progressive thawing and disappearance of permafrost. These changes can lead to loss of soil bearing strength, increased soil permeability, and increased potential for development of such cryogenic processes as differential thaw settlement and heave, and development of thermokarst terrain. Each of these phenomena has the capacity for severe negative consequences on human infrastructure in the high latitudes. Results to date indicate that major permafrost-related impacts have already been detected in many Siberian regions, including changes in the temperature and distribution of permafrost, thickening of the seasonally thawed layer (the active layer), and changes in the distribution and quantity of ice in the ground. A quantitative geographic assessment of the ability of frozen ground to support engineering structures under rapidly changing climatic conditions in a variety of settings is provided in this chapter. Results show substantial decreases of permafrost bearing capacity over the last 40 years in some regions of Northern Siberia. Although a substantial proportion of reported deformations of structures and buildings on permafrost can be attributed to climatic warming, other technogenic factors have to be considered. The socioeconomic crisis resulted in reduced infrastructure monitoring and maintenance in many cities on permafrost during the early 1990s which have greatly contributed to the decrease in infrastructure stability.

N.I. Shiklomanov (✉) • D.A. Streletskiy
Department of Geography, George Washington University,
Washington, DC, USA
e-mail: shiklom@gwu.edu

P.Ya. Groisman and G. Gutman (eds.), *Regional Environmental Changes in Siberia and Their Global Consequences*, Springer Environmental Science and Engineering, DOI 10.1007/978-94-007-4569-8_5, © Springer Science+Business Media Dordrecht 2013

5.1 Introduction

Global climate change has been a keystone issue in world climatology over the last several decades. Recent scientific studies have improved our ability to understand and predict the impacts climate change may have on environmental and human systems. Many of the changes have potential to impact the natural environment, sectors of the economy, and socioeconomic conditions adversely. Ongoing and anticipated changes in the climate system can directly expose the environment to risk and cause environmental changes that may threaten human activities.

Although the prospect of climate change presents numerous challenges to human and natural systems throughout the world, there are few regions facing problems of the extent and severity of those affecting the high latitudes. Observational evidence indicates that impacts related to climate warming are well under way in the polar regions (SWIPA 2011). Many studies have provided discussion about the high vulnerability of northern environments to global climate change and expressed concern that anthropogenic warming may have serious impacts on natural systems in the Arctic (ACIA 2005; Gutman and Reissell 2011; SWIPA 2011).

Many of the potential environmental and socioeconomic impacts of global warming in the high northern latitudes are associated with *permafrost*, or perennially frozen ground, which occupy about 22.8×10^6 km^2 (24 %) of the land area in the Northern Hemisphere. Although often perceived in the popular imagination as a vast, unpopulated, and undeveloped wilderness, large tracts of the Arctic have already been affected by intense developmental pressures. According to Kryuchkov (1994, cited by Vilchek et al. 1996), approximately 31,000 km^2 of the Russian Arctic has been subjected to severe environmental disturbance. Large reservoirs of fossil fuels and ore deposits occur in the permafrost regions, and very substantial economic investments have been made in the Arctic. Clusters of economic development and infrastructure, many involving expensive state-of-the-art technology and complex decision-making and administrative processes, are associated with these activities (e.g., Flanders et al. 1998). Examples include the Prudhoe Bay development in the central part of Alaska's North Slope (Walker et al. 1980; Williams 1986) and the oil and gas fields of West Siberia (Andre'eva et al. 1995; Kryukov and Shmat 1995; Seligman 2000).

With a few exceptions (Fairbanks, Barrow, Whitehorse, Yellowknife), human settlements in the permafrost regions of North America are relatively small and are scattered widely. Engineered works are limited to air- and oil fields, transportation corridors, mining facilities, and pipelines. The impact of these facilities is substantial, however, and large areas that were pristine 50 years ago are now traversed by seismic trails and scarred by abandoned petroleum exploration facilities (e.g., Lawson and Brown 1978; Walker et al. 1987). In northern Scandinavia, several cities with populations of more than ten thousand are located in the contemporary zone of sporadic permafrost.

The developments described above are dwarfed by those in the north of Siberia. Mining camps, military bases, power grids, and roads have been constructed to support resource operations and provide quarters for remote populations, which

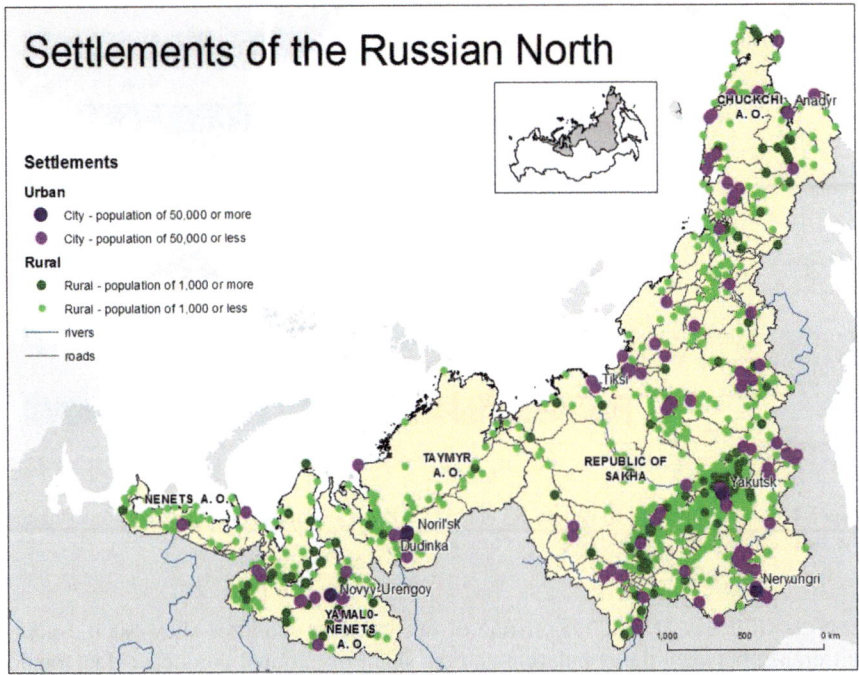

Fig. 5.1 Distribution of population settlements in the nine Russian administrative regions bordering the Arctic Ocean

collectively amount to more than four million people. Figure 5.1 shows distribution of settlements in nine Russian administrative regions bordering the Arctic Ocean. The majority of these northern settlements contain relatively small populations, although several cities in the Siberian Arctic have more than 100,000 inhabitants (Anisimov et al. 2010). Several large river ports also occupy permafrost terrain in the lower Ob' (Salekhard), Yenisei (Igarka and Dudinka), and Lena (Tiksi) River valleys, each with a population of more than 10,000. Smaller settlements are located around terminals on the Arctic coast along the northern sea route from the Barents Sea on the west through the Kara, Laptev, and East Siberian Seas, to the Bering Sea in the east. According to the Russian 2002 census, the majority of Russian Arctic population is considered urban (Fig. 5.2). Several pipelines traverse the discontinuous and sporadic permafrost zones from the West Siberia oil and gas fields to the central parts of Russia. Conventional thermal and hydropower plants are located in several parts of the Siberian Arctic, and a nuclear power station operates in permafrost terrain at Bilibino in the Russian Far East. The network of rudimentary trails and airfields in Siberia is extensive.

In Russia, 5 % of population living in the Northern Siberian regions provides about 11 % of the country's GDP mainly due to extraction of mineral resources. In Russia, 93 % of natural gas and 75 % of oil are produced in permafrost-affected areas. Overall, Russian permafrost regions contribute up to 70 % of total Russian

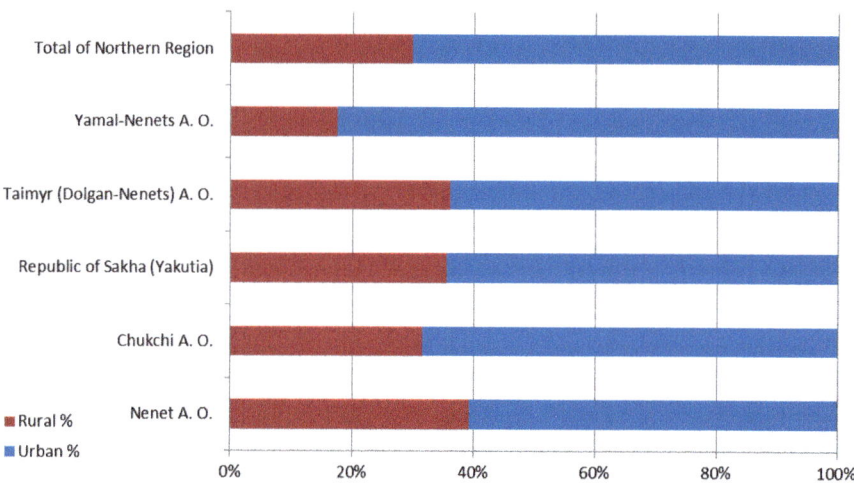

Fig. 5.2 Rural vs. urban population (%) in the nine Russian administrative regions bordering the Arctic Ocean (Based on 2002 Russian census)

exports (Il'echev et al. 2003). None of other Arctic countries show such a great difference between the population and the share of national products (MacDonald et al. 2007). Although extraction of mineral resources accounts for approximately 1/3 of Siberian GDP, the remaining two-thirds is due to other sectors of economy, primarily service, education, construction, transportation, and trade.

5.2 Permafrost Change and Infrastructure Stability

The presence and dynamic nature of ice-rich permafrost constitute a distinctive engineering environment. Many engineering problems in Siberia are associated with (1) changes in the temperature of the upper permafrost, (2) increased depth of seasonal thaw penetration, and (3) progressive thawing and disappearance of permafrost. These changes can lead to loss of soil bearing strength, increased soil permeability, and increased potential for development of such cryogenic processes as differential thaw settlement and heave, destructive mass movements, and development of thermokarst terrain. Each of these phenomena has the capacity for severe negative consequences on human infrastructure in the high latitudes. Several factors are responsible for controlling the stability of ice-rich permafrost. These are primarily temperature at the top of the permafrost and the depth of seasonal thaw penetration. The rate and magnitude of thaw settlement are largely controlled by the volume of ground ice accumulated throughout local permafrost history. In general, the operational life expectancy of infrastructure in permafrost regions is significantly shorter than that in temperate regions. Table 5.1 presents estimates of typical life expectancy of different types of infrastructure in permafrost regions.

Table 5.1 Typical estimated life expectancy of different types of infrastructure in permafrost regions (Adopted from Anisimov et al. 2010)

Paved roads	15–20 years
Pipelines	30 years
Housing (mid to high)	30–50 years
Railroads	50 years
Bridges and tunnels	75–100 years

Results to date indicate that major permafrost-related impacts have already been detected in many Siberian regions, including changes in the soil temperature and distribution of permafrost (Romanovsky et al. 2010a), thickening of the seasonally thawed layer (the active layer), and changes in the distribution and quantity of ice in the ground (Romanovsky et al. 2010b). Such changes in natural systems affect the human environment and have direct and immediate implications for land use, the economy, and human life in Siberia and make geocryological hazards a serious threat to the normal functioning of communities and to economic development. Incorporation of climate change projections for risk assessments of infrastructure on permafrost has become an increasingly important task over the last decade (Nelson et al. 2002; Instanes and Anisimov 2008; Khrustalev and Davidova 2007; Clarke et al. 2008; Nishimura et al. 2009; Shmelev 2010; Streletskiy et al. 2012a, b). Recently, attempts have been made to include economic considerations into analyses of climate change impacts (e.g., Larsen et al. 2008).

However, climate change may already have been taking its toll through deformation of engineered structures in Northern Siberian regions. A survey of infrastructure in industrially developed parts of the Russian Arctic (Kronik 2001) indicates that 10 % of the buildings in Noril'sk, 22 % in Tiksi, 55 % in Dudinka, 35 % in Dicson, 50 % in Pevek and Amderma, 60 % in Chita, and 80 % in Vorkuta are in potentially dangerous states. Analysis of related accidents indicates that in the last decade they increased by 42 % in the city of Noril'sk, 61 % in Yakutsk, and 90 % in Amderma.

A potentially dangerous situation has also been observed with respect to transportation routes and facilities. The long lateral extent of this type of infrastructure makes it difficult to choose an optimum route and apply economically sound strategies for controlling cryogenic processes (Garagulia 1997). According to 1998 data, 46 % of the roadbed under the Baikal–Amur railroad has been deformed by thawing of frozen ground, a 20 % increase over the early 1990s (Kronik 2001). The long-term (1970–2001) monitoring data on Seyda–Vorkuta railroad indicate that the annual ground subsidence has increased from 10 to 15 cm in the mid-1970s to 50 cm in the mid-1990s. Correspondingly, during the same period, the mean permafrost temperature along the railroad has increased by 3–4 °C (from between −6 °C and −7 °C to −3 °C). Runways in Noril'sk, Yakutsk, Magadan, and other major Siberian cities are presently in a state of emergency.

Serious situations have been observed in gas and oil pipelines traversing the Russian north (Seligman 1999, 2000). In 2001, for example, 16 breaks were reported on the single Messoyakha–Noril'sk pipeline, causing significant economic and environmental damage (Kronik 2001). Over the entire West Siberian oil and gas

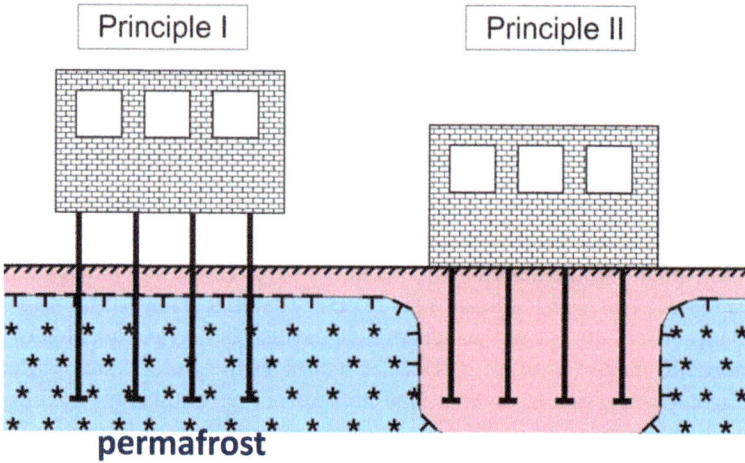

Fig. 5.3 Two types of construction in permafrost regions

region, around 35,000 pipeline accidents of various severity are reported annually. About 21 % of these are thought to be attributable to mechanical deformations related to ground instability (Anisimov and Belolutskaya 2002). For example, 1.5 m of vertical pipeline deformation was reported in the vicinity of Novy Urengoi. According to Oberman (2007) a 1994 pipeline accident in Komi Republic was a result of differential settlement of the ground surface. The results were six bursts of pipe and spillage of more than 160,000 tons of oil. This accident promoted extensive survey and rebuilding of several pipelines underlined by permafrost throughout Siberia. The cost of maintenance, repair, and prevention of the pipeline deformations associated with changes in permafrost conditions is estimated at 55 billion rubles annually (Anisimov et al. 2010).

It should be noted that in many cases it is difficult to differentiate between the effects of climate changes and other factors that may affect a structure on permafrost, such as age, lack of maintenance, or design/construction flaws. However, while other technogenic and environmental factors may or may not have contributed locally, climate change appears to be responsible for the broad patterns of these changes.

There are two common principles of construction design on permafrost (Andersland and Ladanyi 1994; CNR 1990; Grebenets and Rogov 2000; Khrustalev 2005; Shur and Goering 2008). According to "Principle I" in Russia (the "passive method" in North America), permafrost is used as the base for foundations and is protected from thawing during the construction and maintenance of the structure. According to "Principle II" (the "active method" in North America), permafrost is thawed before or during the construction; the ground is protected from aggradation of permafrost during maintenance of the structure. The diagram in Fig. 5.3 demonstrates the conceptual difference between two types of permafrost construction.

It is recommended that only one of the above principles is used inside any one construction area, such as a village, industrial plant, or city district. More than 75 % of the buildings on permafrost in Russia were constructed using Principle I. Foundations are reinforced as they are incorporated in the permafrost (Grebenets and Rogov 2000). Although the exact percentage of buildings constructed on permafrost using the passive principle (II) in the North America is not known, it is economically inefficient to thaw permafrost below −3 °C, limiting use of the active method in areas of cold permafrost (Shur and Goering 2009). Structures built according to the passive method are most susceptible to deformation.

The passive method of permafrost constructions relies on "freezing strength" or bearing capacity of the frozen ground to support structures. Bearing capacity depends on the type of construction and is defined as the maximum stress that can be applied to the foundation without shear failure or catastrophic settlement (Tsytovich 1975). The most common methods use piles to arch structures in permafrost. The bearing capacity of a single post or pile depends on the contact between the side and base areas in contact with permafrost, as well as the temperature of the surrounding medium. The area of the side contact with frozen ground depends on the thickness of the active layer. For a pile of given length, the thicker the active layer, the smaller the area in contact with permafrost and the smaller the load that the pile can support. Increases in near-surface permafrost temperature and thickening of the active layer are likely to result in a decrease of the ability of foundations to support structures to a degree not anticipated at the time of construction. If the decreases are beyond the values of safety coefficients, deformation of foundations may result in severe damage to or even collapse of buildings and structures.

5.3   Infrastructure and Climate Change

Since permafrost temperature and the thickness of the active layer depend on climate, engineering standards and designs have historically utilized the climatic "normals" available prior to construction. For instance, Soviet construction regulations recommended use of decadal climatic averages (CNR 1990). The possible climatic variability and change are accounted for in engineering procedures through a series of "safety factors" that decrease the uncertainty on describing natural environment during construction. While safety coefficients in North America range from 2.5 to 3, in Soviet Russia they rarely exceeded 1.56, making many foundations in Russia especially vulnerable to climate change (Shur and Goering 2009). The rapid change in climatic conditions raises questions about the stability of structures whose design is based on climatic normals from past decades and utilizes the relatively low safety coefficients used at the time of construction. Khrustalev (2000) analyzed the safety coefficient of foundations built using the "passive method" in Russia, as outlined in CNR (1990), and found that the safety coefficient varies from 1.05 to 1.56. Based on these values, if the bearing capacity of foundations under climatic, environmental, or technogenic factors decreases by 5–26 %, a foundation will deform and the building may be subject to collapse. Using an arbitrarily chosen warming value of 1.5 °C, Khrustalev (2000) calculated bearing capacity for common foundation types in Yakutsk, concluding that such a relatively small increase in mean annual air temperature could be enough to trigger deformation of all foundations constructed in the city of Yakutsk. Projected climate warming in Siberia has potential to cause widespread deformation and damage to structures built in permafrost terrain. This could have severe socioeconomic consequences because most existing infra-structure will require expensive engineering solutions to stabilize foundations (ACIA 2005). This is especially true for the urban and industrial centers of Northern Siberia which were extensively developed in the late 1960s to early 1970s.

Streletskiy et al. (2012a, b) have provided geographic assessments of changes in engineering properties of frozen ground due to observed climatic change. The analysis is based on Russian methodology which utilizes the bearing capacity for "standard foundation pile" imbedded in permafrost as a primary variable for engineering assessment of permafrost-affected territory. A set of parameterizations was developed to estimate the bearing capacity of frozen soils as function of TTOP and ALT according to Russian Construction Rules and Regulations 2.02.04–88. The effect of climate on TTOP and ALT was estimated by equilibrium permafrost model based on Kudriavtsev formulation (Anisimov et al. 1997; Shiklomanov and Nelson 1999; Sazonova and Romanovsky 2003).

Changes in bearing capacity under observed climatic changes were analyzed in major industrial centers and settlements. One of the largest towns in Northwest Siberia is Nadym, with a population of about 50,000 people. The town was built at the site of a previous small settlement at the beginning of the 1970s, after discovery of the Medvejie gas-condensate field. It is suspected that climatic averages of the pre-1970 decade were used to estimate the bearing capacity of piling foundations prior to the construction of the town. Comparison of mean annual air temperature from the 1960s and 1990s shows that the latter was about 1.5 °C warmer (Fig. 5.4). Such pronounced warming is likely to cause a substantial decrease in bearing capacity. Changes in bearing capacity associated with air temperature changes were estimated based on meteorological data from the Nadym weather station. It was assumed that heat exchange under buildings is a direct function of changes in air temperature, as ground covers (primarily vegetation and snow) are absent under the buildings. The soil profile was assumed to be a homogeneous sandy loam with a gravimetric soil moisture content of 30 %. Calculated bearing capacity is shown in percent relative to average bearing capacity for the 1963–1973 period. Substantial year-to-year variation in bearing capacity occurs through climatic variability, but there is a general declining trend. Over the 1970s, bearing capacity decreased by less than 4 % compared to the 1960s but by about 22 % in the 1980s. Climatic conditions in the 1990s did not change relative to the 1980s, so bearing capacity did not decrease during that decade. However, a warming trend in the 2000s resulted in a dramatic decrease in bearing capacity – up to 35 % compared to the 1960s. This agrees well with another study conducted in Vorkuta (Oberman and Shesler 2009), according to which catastrophic deformations of buildings were confined to 1980s, the period with the greatest increase of temperatures of permafrost composed of mineral Quaternary deposits and peat.

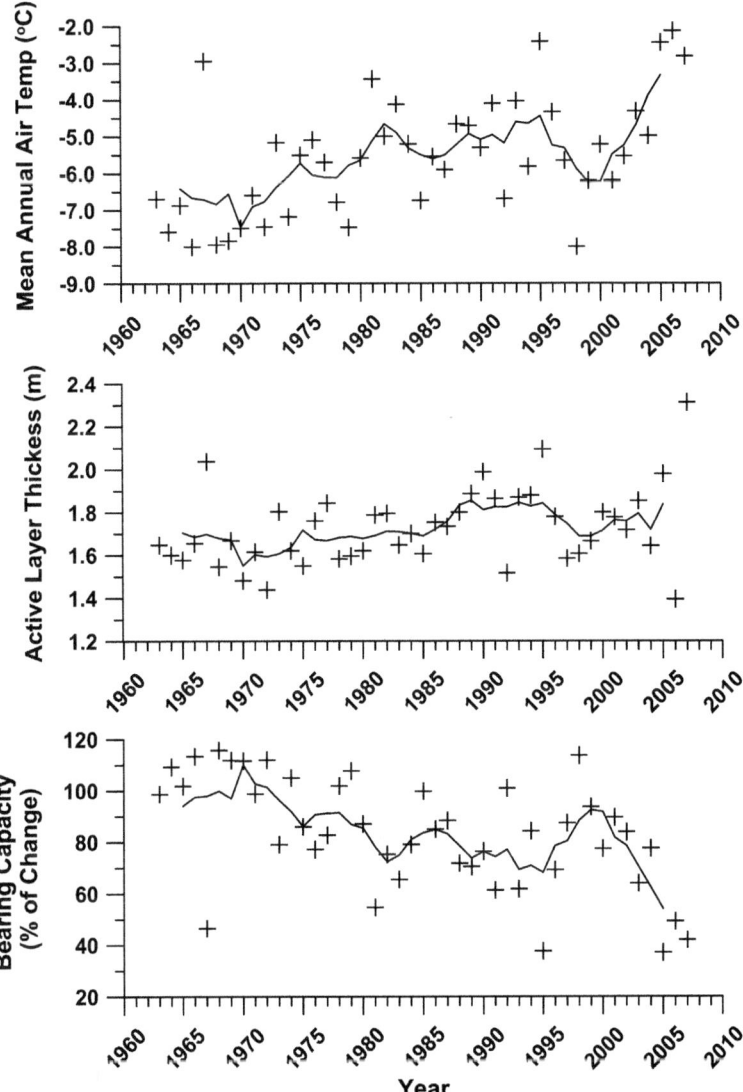

Fig. 5.4 Observed changes of mean annual surface air temperature (*MAAT*) and associated computed changes of *ALT* and bearing capacity (*Fu*) for the Nadym region. *Black crosses* represent estimated annual values. *Solid line* shows 5-year running average

Changes in bearing capacity were calculated for some of the largest settlements on permafrost, representing different parts of the Russian Arctic. Locations were chosen based on the assumption that if modeled temperature at the top of permafrost was less than −3 °C during the 1960s, the foundations were built based on the first (passive) principle. Monthly air temperature from weather stations was used as

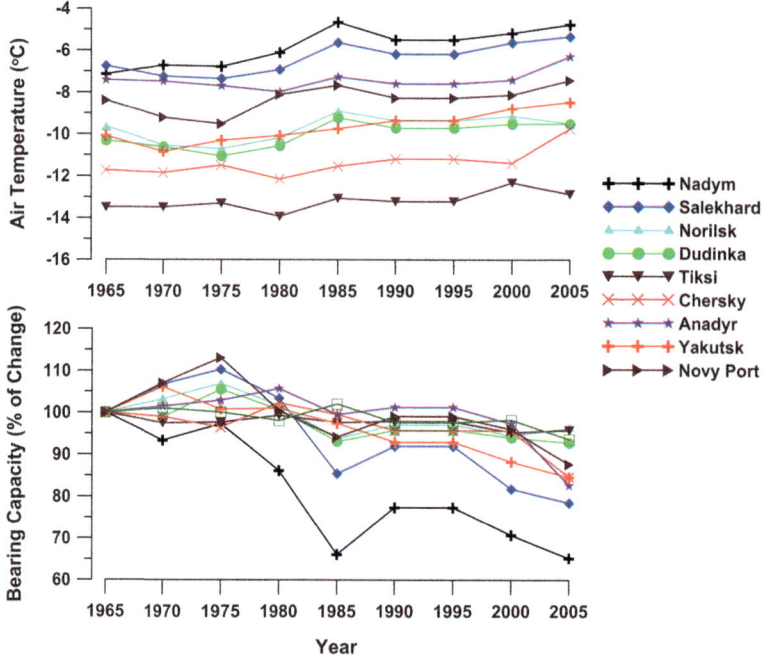

Fig. 5.5 Observed changes in mean annual temperature and corresponding change in modeled bearing capacity for settlements at least partially built based on the first (passive) principle

input. It was assumed that with adequate maintenance of the building, annual temperature at the ground surface approaches that of the air. No snow cover was assumed to be present beneath structures, and no engineering solutions were implemented to control ground temperature (e.g., thermosiphons).

Ground properties can be quite variable inside each settlement. For each location, the ground was assumed to consist of sand, sandy loam, or clay. In each case, high and low estimates of ice content were applied. A characteristic range of bearing capacity was therefore produced for each point. Annual values were averaged to produce 5-year means. Average bearing capacity for 1960–1970 was chosen as a reference point representing 100 %. The percent change was calculated relative to this reference for the series of years, represented in Fig. 5.5. The range of decadal changes in bearing capacity is presented in Table 5.2.

Foundation bearing capacity is quite variable, even at short temporal scales. The end of the 1970s and the beginning of 1980s were colder than the 1960s, resulting in an increase in bearing capacity. Following the general warming observed at all sites, a decreasing trend in bearing capacity occurred. Similar warming results in a more substantial decrease in bearing capacity in the southern permafrost zone, where ground temperatures are warmer than in northern locations. Nadym and Salekhard, for example, are subject to similar climatic conditions. However, the slightly higher warming trend observed in Nadym resulted in a much greater decrease in bearing capacity than at Salekhard (65 and 74 %, respectively). At the same time, similar

Table 5.2   Decadal changes in foundations bearing capacity due to observed climate change

Region	Settlement	Bearing capacity of foundations, %				
		1960s	1970s	1980s	1990s	2000s
West Siberia	Salekhard	100	91–103	72–86	81–82	68–70
	Nadym	100	96–101	77–91	78–100	64–95
	Novy Urengoy	NA	NA	100	97–116	91–96
	Novy Port	100	105–114	86–93	87–92	63–76
Central Siberia	Noril'sk	100	102–105	88–93	84–92	85–94
	Dudinka	100	103–110	93–94	90–94	74–82
East Siberia	Yakutsk	100	91–98	80–92	59–84	54–80
	Bilibino	NA	100	90–98	97–100	80–91
	Tiksi	100	99–100	96–98	95–97	93–96
	Anadyr	100	101–104	92–100	75–94	52–84
	Chersky	100	100–101	97–98	96	76–84

trends were observed in Yakutsk but resulted in significantly smaller changes in bearing capacity (85 %). This geographic pattern shows that areas located in the southern permafrost zone (TTOP higher than −5 °C) are likely to experience more pronounced decreases in bearing capacity under projections of climate warming.

The analysis described above was conducted to evaluate the spatial distribution of changes in bearing capacity over nine Russian administrative regions bordering the Arctic Ocean. Figure 5.6 shows the reduction in bearing capacity between decades of 1960s and 2000s for nine Russian administrative regions bordering the Arctic Ocean. Results demonstrate a diverse pattern of bearing capacity loss over the Russian Arctic. The catastrophic (>20 %) and severe (15–20 %) loss in bearing capacity is found in Nenets and Yamalo-Nenets Autonomous okrugs, southern regions of Yakutia, and entire Chukotka. The Central Siberia (Taymur) and central and northern portions of Yakutia are characterized by relatively stable permafrost (<5 % loss in bearing capacity).

5.4   Regional Assessment of Geocryologic Hazards

Evaluation of near-surface permafrost at regional scales is one of the most difficult tasks of modern geocryology. Unlike modeling at local scales, where input data for landscape parameters are available or modeling at the continental scale where variability of near-surface permafrost can be represented effectively by climatic input, modeling at regional scales requires solid knowledge of both. While the quality of output modeling fields depends directly on that of the model parameterization, differences in climatic input and land-cover characteristics can significantly influence the results.

To provide a more detailed spatial assessment of changes in bearing capacity, the model was applied to the Northwest Siberian region. Widespread deformation of buildings in the region has been attributed to the situation in the late 1980s and early 1990s, a time of great economic and social stress that resulted in poor maintenance

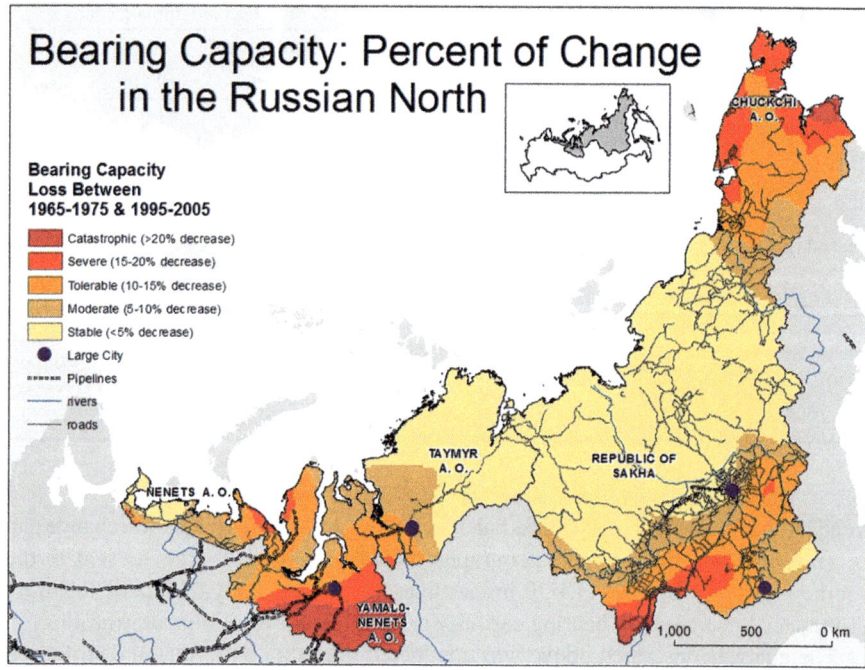

Fig. 5.6 The loss of bearing capacity of permafrost between the decade of 1960s and the decade of 2000s

of many structures. To answer the question whether climate change alone can be responsible for a widespread decrease of foundation bearing capacity or whether socioeconomic factors played a major role, it was hypothesized that climatic normals for the 1960s were used at the time of construction.

The region has relatively good coverage of meteorological stations compared to other Arctic regions. The absence of large topographic barriers in the region creates a primarily zonal distribution of air temperature. Owing to the position of the Ural Mountains, precipitation increases from NW to SE. Comparison of commonly used climatic gridded datasets shows good agreement in the region, so the choice between climate datasets used for the temporal analysis is not as important as in other regions (Streletskiy 2010). For this study, the University of Delaware terrestrial air temperature: 1900–2008 gridded monthly time series version 2.01 was used as climatic input to compare the decades of the 1960s and 2000s.

Environmental variables required for model parameterization were derived from "The Landscape Map of Northwest Siberia" which was downscaled to 1-km resolution (Streletskiy 2010). This resolution was chosen as adequate to represent the geographic variability of near-surface permafrost parameters at the landscape scale (Melnikov et al. 1983). Additional information was obtained from Trofimov (1987) and Melnikov et al. (1983). Soil texture, where missing, was assigned as sandy loam (1,400 kg/m³).

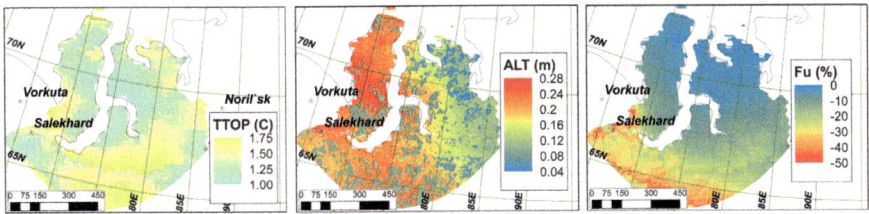

Fig. 5.7 Regional changes in mean annual ground temperature at a permafrost top (*TTOP*), active-layer thickness (*ALT*), and bearing capacity (*Fu*) from 1960s to 2000s

The resulting maps show regional changes between the two decades for mean annual air temperature, mean annual ground temperature, ALT, and foundation bearing capacity (Fig. 5.7). Analysis of the MAAT difference field shows that the eastern part of Northwest Siberia experienced more pronounced warming than did the western part. Climatic averages constructed from the four gridded datasets indicate a temperature increase in the Yamal and Tazovskiy Peninsulas of 0.8–1.1 °C, while in the Gydan Peninsula the increase is 1.1–1.3 °C. Corresponding changes in TTOP are smaller because of the insulative effects of above-ground cover and attenuation of the climatic signal with depth. The increase of MAGT in Yamal and Tazovskiy is 0.6–0.8 °C, while it is 0.7–0.1° in Gydan. The estimated increase in TTOP in the continental part of Northwest Siberia is from 0.1 °C in the west to 0.8 °C in the east. ALT generally increased throughout the region, with the exception of small northern islands such as Oleniy and Sibiryakova. In general, ALT changes between 1990s and 1960s are rather small and do not exceed 10 cm, with a maximum increase in the southern tundra/forest tundra.

Bearing capacity in the same period decreased by up to 45 %, with a regional mean of 17 % (Fig. 5.7d). A small decrease is characteristic of the northern part of the Yamal and Gydan Peninsulas but increases gradually towards the south. Bearing capacity decreases in the Arctic tundra zone are about 5–11 %, in northern tundra 9–13 %, and in southern tundra 12–25 %. Increasing complexity of environmental conditions south of the forest line creates quite variable conditions in bearing capacity fields, so bearing capacity loss generally ranges from 12 to 45 %, with some areas in the west showing increases of up to 12 %. Locations near Salekhard, Nadym, and Novy Urengoy experienced a 15–25 % decrease in foundation bearing capacity. A better situation is expected in settlements along Yenisei River, such as Igarka and Dudinka (10–15 %).

5.5 Concluding Words

This work demonstrates the significant reduction in bearing capacity of foundations due to climatically driven permafrost warming. However, other technogenic factors have to be considered. The socioeconomic crisis that occurred after the collapse of the Soviet Union resulted in reduced monitoring of construction in many cities on permafrost during the early 1990s. In some cases, undetected leaks in sewage and water pipes resulted in rapid warming and chemical contamination of permafrost below the foundations. Increases in permafrost temperature and ground salinity and the resulting decrease in the soil's ability to support foundations resulted in serious deformation of many structures (Grebenets and Kerimov 2001). In some cases the deformation was catastrophic, leading to collapse of entire structures. Permafrost monitoring in Noril'sk indicates that the number of leaks in building basements increased by a factor of 15–20 in the 1990s, compared to the 1970s. The major cause of building deformations (more than 60 % of cases) became the loss of bearing capacity of frozen ground due to unpredictable rates of warming and thawing caused by prolonged leaks of pipes in basements (Khrustalev 2000). A recent study conducted in Yakutsk (Alekseeva et al. 2007) concluded that the main reasons for decreasing foundation strength in Yakutsk were errors in planning and construction and not increased air temperatures. Together with waterlogging and chemical contamination of ground water, these errors resulted in catastrophic situations in residential districts. As a result, the full evaluation of infrastructure stability at local scales requires a comprehensive engendering assessment of every structure. The broad geographic assessments described above can serve as a basis for developing regional *adaptation, mitigation, and risk management options, including changes to construction codes* designed to mitigate the adverse impacts that climate change may have on infrastructure and socioeconomic life.

References

ACIA (2005) Arctic climate impact assessment. Cambridge University Press, New York, 1042 pp

Alekseeva OI, Balobaev VT, Grigoriev MN, Makarov VN, Zhang RV, Shatz MM, Shepelev VV (2007) Urban development problems in permafrost areas (by the example of Yakutsk). Earth Cryosph XI(2):76–83

Andersland OB, Ladanyi B (1994) An introduction to frozen ground engineering. Chapman & Hall, New York

Andre'eva Y, Larichev OI, Flanders NE, Brown RV (1995) Complexity and uncertainty in Arctic resource decisions. Polar Geogr Geol 19:22–35

Anisimov OA, Shiklomanov NI, Nelson FE (1997) Effects of global warming on permafrost and active-layer thickness: results from transient general circulation models. Glob Planet Change 15:61–77

Anisimov OA, Belolutskaya MA (2002) Assessment of climate change impacts on degradation of permafrost and infrastructure in Northern Russia, Meteorol Hydrol 6:15–25 (In Russian)

Anisimov OA, Belolutskaya MA, Grigoriev MN, Instanes A, Kokorev VA, Oberman NG, Reneva SA, Strelchenko YG, Streletskiy D, Shiklomanov NI (2010) Major natural and social-economic consequences of climate change in the permafrost region: predictions based on observations and modeling. Greenpeace, Moscow, Russia, 44 p (in Russian)

Clarke J, Fenton C, Gens A, Jardine R, Martin C, Nethercot D, Nishimura S, Olivella S, Reifen C, Rutter P, Strasser F, Toumi R (2008) In: Kane DL, Hinkel KM (eds) Ninth international conference on permafrost, extended abstracts. Institute of Northern Engineering, University of Alaska, Fairbanks, pp 279–284

CNR (1990) Stoitelnie Normi i Pravila (Construction norms and regulations). Foundations on Permafrost 2.02.04–88. State Engineering Committee of the USSR, 56 pp (in Russian)

Flanders NE, Brown RV, Andre'eva Y, Larichev OL (1998) Justifying public decisions in arctic oil and gas development: American and Russian approaches. Arctic 51:262–279

Garagulia LS (1997) Prognostic estimations of the anthropogenic changes of geocryologic conditions. Izdatel'stvo MGU, Moscow, p 223

Grebenets VI, Kerimov AG (2001) Evolution of natural and man-made complexes in the Noril'sk Region. In: Scientific-practical seminar "geocryolithic and geoecologic construction problems in regions of the extreme north", Norilsk, pp 130–135 (in Russian)

Grebenets VI, Rogov VV (2000) Permafrost engineering. Moscow State University Press, Moscow, 96 pp (in Russian)

Gutman G, Reissell A (eds) (2011) Eurasian Arctic land cover and land use in a changing climate, 1st edn. Springer, Dordrecht/Heidelberg/London/New York, 306 p. ISBN 978-90-481-9117-8

Il'echev VA, Vladimirov VV, Sadovskyi AV, Zamaraev AV, Grebenetz VI, Kuvitskaya NB (2003) Problems of modern developments of northern cities. Russian Academy of Architecture, Moscow, 152 pp (in Russian)

Instanes A, Anisimov OA (2008) Climate change and Arctic infrastructure. In: Proceedings of the 9th international conference on permafrost, Institute of Northern Engineering, vol 1. University of Alaska, Fairbanks, pp 779–784

Khrustalev LN (2000) Allowance for climate change in designing foundations on permafrost grounds. In: Proceedings of the international workshop on permafrost engineering, Longyearbyen, Svalbard, Norway, 18–21 June 2000. Tapir Publishers, Trondheim, pp 25–36

Khrustalev LN (2005) Geotechnical fundamentals for permafrost regions. Moscow State University Press, Moscow, 544 pp (in Russian)

Khrustalev LN, Davidova IV (2007) Forecast of climate warming and account of it at estimation of foundation reliability for buildings in permafrost zone. Earth Cryosph XI(2):68–75

Kronik YA (2001) Accident rate and safety of natural—anthropogenic systems in the permafrost zone. In: Proceedings of the second conference of Russian geocryologists, vol 4. Moscow State University, Moscow, pp 138–146

Kryuchkov VV (1994) Environmental degradation in the Arctic. Narodnoye khozyaystvo Respubliki Komi 3:44–53 (in Russian)

Kryukov V, Shmat V (1995) West Siberian oil and the northern sea route: current situation and future potential. Polar Geogr 19:219–235

Larsen PH, Goldsmith S, Smith O, Wilson ML, Strzepek K (2008) Estimating future costs for Alaska public infrastructure at risk from climate change. Glob Environ Change 18:442–457

Lawson DE, Brown J (1978) Human-induced thermokarst at old drill sites in northern Alaska. North Eng 10:16–23

MacDonald GM, Kremenetski KV, Beilman DW (2007) Climate change and the northern Russian treeline zone. Philos Trans Royal Soc B 363:2285–2299. doi:10.1098/rstb.2007.2200

Melnikov FS, Veysman LI, Moskalenko NG (1983) Landscapes on permafrost in West-Siberian oil-gas province. Novosibirsk, Nauka, 163 pp (in Russian)

Nelson FE, Anisimov OA, Shiklomanov NI (2002) Climate change and hazard zonation in the Circum-Arctic permafrost regions. Nat Hazards 26:203–225

Nishimura S, Martin CJ, Jardine RJ, Fenton CH (2009) A new approach for assessing geothermal response to climate change in permafrost regions. Geotechnique 59(3):213–227

Oberman NG (2007) Global warming and permafrost changes in Pechoro-Ural region of Russia. Prospect Prot Miner Resour 4:63–68 (In Russian)

Oberman NG, Shesler IG (2009) Observed and projected changes in permafrost conditions within the European North-East of the Russian Federation. Problemy Severa i Arctiki Rossiiskoy Federacii (Problems and Challenges of the North and the Arctic of the Russian Federation) 9:96–106 (in Russian)

Romanovsky V, Oberman N, Drozdov D, Malkova G, Kholodov A, Marchenko S (2010a) Permafrost, [in "State of the Climate in 2009"]. Bull Am Meteorol Soc 91(6):S92

Romanovsky VE, Drozdov DS, Oberman NG, Malkova GV, Kholodov AL, Marchenko SS, Moskalenko NG, Sergeev DO, Ukraintseva NG, Abramov AA, Gilichinsky DA, Vasiliev AA (2010b) Thermal state of permafrost in Russia. Permafr Periglac Process 21:136–155

Sazonova TS, Romanovsky VE (2003) A model for regional-scale estimation of temporal and spatial variability of active layer thickness and mean annual ground temperatures. Permafr Periglac Process 14:125–139

Seligman BJ (1999) Reliability of gas pipelines in northern Russia. Pet Econ 66:26–29

Seligman BJ (2000) Long-term variability of pipeline-permafrost interactions in north-west Siberia. Permafr Periglac Process 11:5–22

Shiklomanov NI, Nelson FE (1999) Analytic representation of the active layer thickness field, Kuparuk River basin, Alaska. Ecol Model 123:105–125

Shmelev DG (2010) Forecast of changing of main engineering and geocryological parameters in Russian arctic to 2030 and 2050. Abstracts of third European conference on permafrost, 13–17 June, Svalbard, Norway, p 21

Shur Y, Goering D (2008) Climate change and foundations of buildings in permafrost regions. Soil Biol 16:251–260

Shur YL, Goering DJ (2009) Climate change and foundations of buildings in permafrost Regions. In: Margesin R (ed) Permafrost Soils. Springer, Berlin/Heidelberg, pp 251–260

Streletskiy DA (2010) Spatial and temporal variability of the active-layer thickness at regional and global scales. PhD thesis, University of Delaware, 243 pp

Streletskiy DA, Shiklomanov NI, Grebenetz VA, (2012a) Climate warming-induced changed in bearing capacity of permafrost in the North of West Siberia. Earth Cryosphere XVI(1):22–32 (In Russian)

Streletskiy DA, Shiklomanov NI, Nelson FE (2012b) Permafrost infrastructure and climate change: a GIS-based landscape approach. Arctic Antarctic Alpine Res 44(3):95–116. doi: 10.1657/ 1938-4246-44.3

SWIPA (2011) Snow, Water, Ice and Permafrost in the Arctic (SWIPA): climate change and the cryosphere. Arctic Monitoring and Assessment Programme (AMAP), Oslo, p xii + 538 pp

Trofimov VT (1987) Geocryological regionalization of West Siberia platform. Moscow "Nauka" Press, Moscow, 224 pp (in Russian)

Tsytovich NA (1975) The mechanics of frozen ground. McGraw-Hill, New York, 426 pp

Vilchek GE, Krasovskaya TM, Tsyban AV, Chelyukanov VV (1996) The environment in the Russian Arctic: status report. Polar Geogr 20:20–43

Walker DA, Everett KR, Webber PJ, Brown J (1980) Geobotanical atlas of the Prudhoe Bay region, Alaska. U.S. Army Cold Regions Research and Engineering Laboratory, Hanover, NH, 69 pp

Walker DA, Webber PJ, Binnian EF, Everett KR, Lederer ND, Nordstrand EA, Walker MD (1987) Cumulative impacts of oil fields on northern Alaskan landscapes. Science 238:757–761

Williams PJ (1986) Pipelines and permafrost: science in a cold climate. Carleton University Press, Don Mills, 129 pp

Chapter 6
Terrestrial Ecosystems and Their Change

Anatoly Z. Shvidenko, Eric Gustafson, A. David McGuire,
Vjacheslav I. Kharuk, Dmitry G. Schepaschenko, Herman H. Shugart,
Nadezhda M. Tchebakova, Natalia N. Vygodskaya, Alexander A. Onuchin,
Daniel J. Hayes, Ian McCallum, Shamil Maksyutov, Ludmila V. Mukhortova,
Amber J. Soja, Luca Belelli-Marchesini, Julia A. Kurbatova,
Alexander V. Oltchev, Elena I. Parfenova, and Jacquelyn K. Shuman

Abstract This chapter considers the current state of Siberian terrestrial ecosystems, their spatial distribution, and major biometric characteristics. Ongoing climate change and the dramatic increase of accompanying anthropogenic pressure provide different but mostly negative impacts on Siberian ecosystems. Future climates of the region may lead to substantial drying on large territories, acceleration of disturbance regimes, deterioration of ecosystems, and positive feedback to global warming. The region requires urgent development and implementation of strategies of adaptation to, and mitigation of, negative consequences of climate change.

A.Z. Shvidenko (✉)
Ecosystems Services and Management Program, International Institute for Applied
Systems Analysis, Laxenburg, Austria

Forestry Institute, Siberian Branch, Russian Academy of Science,
Krasnoyarsk, Russian Federation
e-mail: shvidenk@iiasa.ac.at

E. Gustafson
Institute for Applied Ecosystems Studies, USDA Forest Service,
Northern Research Station, K, Rhinelander, WI, USA

A.D. McGuire
Alaska Cooperative Fish and Wildlife Research Unit,
U.S. Geological Survey, Fairbanks, AK, USA

Department of Biology and Wildlife, Institute of Arctic Biology,
University of Alaska-Fairbanks, Fairbanks, AK, USA

V.I. Kharuk • N.M. Tchebakova • A.A. Onuchin • L.V. Mukhortova • E.I. Parfenova
V.N. Sukachev Institute of Forest, Siberian Branch of Russian Academy of Sciences,
Krasnoyarsk, Akademgorodok, Russian Federation

D.G. Schepaschenko
Ecosystems Services and Management Program, International Institute for Applied Systems
Analysis, Laxenburg, Austria

Moscow State Forest University, Mytischi, Moscow Region, Russian Federation

P.Ya. Groisman and G. Gutman (eds.), *Regional Environmental Changes in Siberia*   171
and Their Global Consequences, Springer Environmental Science and Engineering,
DOI 10.1007/978-94-007-4569-8_6, © Springer Science+Business Media Dordrecht 2013

6.1   Introduction

Siberian vegetation is mostly represented by unique ecosystems which evolutionary developed under stable cold climate (Khotinsky 1984). While the region is represented by a vast diversity of ecosystem types, amount of dominant plant species in all land classes is relatively limited (e.g., Lavrenko and Sochava 1956; Zhukov 1969; Scherbakov 1975; Popov 1982; Polykarpov et al. 1986). Plasticity of major forest-forming species is high, and they occupy wide ecological niches. These provided a rather high resilience of boreal forests which coped with stable disturbance regimes during last centuries (Bonan and Shugart 1989, 1990).

The Intergovernmental Panel on Climate Change (IPCC) (IPCC 2007a, b) concluded that for increases in global average temperature exceeding 1.5–2.5 °C, forests globally face the risk of significant, basically undesirable transformations. With this respect, boreal forests of Siberia could be especially affected due to (1) dramatic changes of heat and hydrological balances of northern landscapes over huge territories caused by permafrost thawing, (2) sensitivity of boreal forest ecosystems to warming and the high rates of expected warming in northern high latitudes, (3) different consequences of climate change impacts over the southern and northern ecotones of the forest zone – death and impoverishment of forests in the south and a very limited potential to move northward in the north, and (4) dramatic acceleration of disturbance regimes, particularly fires, outbreaks of insects and diseases, coupled with the tough anthropogenic impacts.

The expected warming generates dangerous risks and challenges for the ecosystems. If the global increase of annual temperature would exceed +3 °C by end of this

H.H. Shugart • J.K. Shuman
Department of Environmental Sciences, University of Virginia,
Charlottesville, VA, USA

N.N. Vygodskaya • J.A. Kurbatova • A.V. Oltchev
A.N. Severtsov Institute of Ecology and Evolution, Russian
Academy of Sciences, Moscow, Russian Federation

D.J. Hayes
Environmental Sciences Division, Oak Ridge National Laboratory, Oak Ridge,
TN 37831-6301, USA

I. McCallum
Ecosystems Services and Management Program, International Institute for Applied
Systems Analysis, Laxenburg, Austria

S. Maksyutov
National Institute for Environmental Studies, Tsukuba, Japan

A.J. Soja
National Institute of Aerospace, Hampton, VA, USA

L. Belelli-Marchesini
Forest Ecology Lab, University of Tuscia, Viterbo, Italy

century, the warming in vast territories of high latitudes of the region is expected to be in range of +7 to +10 °C. Based on an ensemble of 16 coupled atmosphere-ocean GCMs, the increase of temperature by 2080–2099 for entire Russia was predicted at +7.2 ± 1.7 °C and in summer +4.2 ± 1.3 °C (Meleshko et al. 2008). Such a level of warming has never been experienced before. The most likely scenario for the boreal forest, as a major ecosystem type in the region, is a nonlinear response to warming, resulting in the creation of previously nonexistent ecosystems and the extinction of species with limited capacity to adapt. The critical limit for large-scale forest dieback may be a rise of 3–5 °C (Lenton et al. 2008). Experts considered boreal forest dieback as one of nine possible global "tipping elements" (a sudden and dramatic response as global warming exceeds a certain threshold value, presumably 5–6 °C for boreal forests), supposing that the nonlinear response to global warming can happen during this century. The supposed mechanisms behind boreal forest dieback are increased water stress and higher peak summer heat causing increased mortality directly and also indirectly through higher vulnerability to disease and alteration of fire regimes. Once the critical threshold is passed, the process may be rather fast; the transformation of the boreal forest ecosystems may happen over a period of about 50 years (Lenton et al. 2008).

High vulnerability of Siberian ecosystems, particularly forests in south of the boreal zone and on permafrost becomes evident even under ongoing climate. It defines a high importance of understanding current and future behavior of Siberian ecosystems as a background of transition to adaptive ecosystem management.

This chapter has two main objectives: (1) to present a brief overview of the current state of Siberian ecosystems and (2) to identify major ongoing and future dynamic processes in ecosystems of a changing world. Taking into account that part of the available information on Siberian land and ecosystems is either biased or obsolete (e.g., official data on land cover, forest inventory data), the analyses rely on different information sources, particularly recent results of Earth observation from space (e.g., Schepaschenko et al. 2010; Santoro et al. 2011). All areas and biometric indicators in this chapter are defined based on the hybrid land cover 2009 (Fig. 1.1; Schepaschenko et al. 2010).

6.2 Resource and Ecological Services of Siberian Ecosystems

6.2.1 Brief Characteristics of Siberian Ecosystems

The vast expanse of Siberia (about 3,500 km in latitudinal and more than 6,000 km in longitudinal directions) ensures a large diversity of natural conditions, climate, soils, ecosystems, and landscapes. More than 50 % of the territory is mountainous, and about 85 % of the territory is covered by different types of permafrost. While there are vast territories which have been transformed into anthropogenic deserts, about half of the region's area is covered by practically untouched ecosystems.

Diversity of natural conditions of the region requires division of the vast Siberian territory in more homogeneous parts based on different criteria, with many attempts resulting in development of natural, climatic, soil, vegetation, agricultural, etc., divisions (regionalizations). One of the most common was ecological regionalization (Shvidenko et al. 2000), where major spatial classification units – ecoregions – were delineated based on the following major requirements: (1) homogeneity of the natural environment including similar climate, typical soils, and landforms at the level of bioclimatic (vegetation) subzones and large landscape units that led to separation of mountain and plain territories, permafrost regions, etc.; (2) similarity of character and intensity of anthropogenic pressure on ecosystems and landscapes that includes negative impacts (e.g., pollution) and homogeneity of land management (e.g., forest management); (3) similarity in levels of transformation of natural vegetation; and (4) comparable input of each ecoregion in major biogeochemical cycles (i.e., carbon, hydrological, and nitrogen). Ecoregions are combined in bioclimatic subzones and zones. A map and list of Siberian ecoregions is available at http://www.iiasa.ac.at/Research/Forestry/Downloaddata/. A quantitative description of ecoregions of Central Siberia is presented in Farber (2000) and West Siberia in Sedykh (2009). Here we present the most important aggregated characteristics of Siberian ecosystems by bioclimatic zones.

While almost all bioclimatic zones and major vegetation types of the Northern hemisphere are represented in Siberia, this region is usually associated with taiga – coniferous evergreen and deciduous forests of cold climates. Three subzones of taiga cover about 77 % of the total land of the region. Tundra (16.3 %) and steppe (4.7 %) are two other widely distributed zones. The remaining three zones (arctic desert, temperate forests, or sub-taiga, and southern semideserts and deserts) cover only ~2 % (Table 6.1).

Diverse climate, vegetation, and relief of the vast territory of Siberia resulted in the development of various soil types. The majority of the area is covered by typical forest soils: either sandy (*Podzols*) or loamy soils (*Albeluvisols, Cambisols*). Wetlands are also spread widely with *Histosols, Gleysols*, and some *Phaeozems*. Almost every soil type has permafrost signs, especially *Cryosols*. Agricultural area is generally represented by *Chernozems* in Southern Siberia. Weakly developed mountain soils (*Leptosols*) and river valley soils (*Fluvisols*) complete the range of soils of the region. Areas in Table 6.2 were calculated based on a digitized soil map (Fridland 1989, modified to scale 1:1M). More detailed data on Siberian soils are available at the above-mentioned IIASA FOR website.

Siberia is largely a forest territory – forest land (i.e., area covered by forest vegetation or destined for growth of forest) comprises 61.6 % of the total area of which about 20 % are sparse forests (woodlands) – generally in high latitudes, and disturbed forests. Forests (or forested areas, defined by the forest inventory manual of the Russian Federal Forest Service, RFFS 1995) are relatively simple in terms of species composition. They are dominated by seven tree genera – light coniferous larch (*Larix* spp.) 40.3 %, pine (basically *Pinus silvestrys*) 13.0 %, dark coniferous cedar (*Pinus sibirica* and *P. korajensis*) 6.8 %, spruce (*Picea* spp.) and fir (*Abies* spp.) 7.3 %. Softwood deciduous, mostly birch (*Betula* spp.) covers 12.0 % and

Table 6.1 Distribution of area by land classes and bioclimatic zones

Zone	Forest	Agricultural land	Burned forests	Wetland	Grassland	Open woodland	Unproductive	Water	Total
Arctic					0.3		3.3	0.3	3.9
Tundra	11.8	0.1	1.5	30.2	106.4	6.2	19.5	6.5	182.1
FT, NT & SpT[a]	103.3	0.1	2.9	24.4	22.6	24.2	0.3	2.2	179.9
Middle taiga	381.4	10.3	15.4	36.3	74.3	35.5	4.7	4.8	562.6
Southern taiga	89.5	11.2	0.5	12.5	3.4	0.7	0.0	0.9	118.8
Temperate forest	5.5	4.8	0.0	0.6	0.6	0.1	0.0	0.1	11.8
Steppe	9.7	36.4	0.1	2.8	2.5	0.1	0.2	0.9	52.6
SD & D[b]	1.1	4.2	0.0	0.6	0.6	0.0	0.3	0.4	7.2
Total	602.2	67.0	20.5	107.4	210.7	66.8	28.3	16.0	1,119.0

Area, ×10⁶ ha, by land classes

Notes: (1) Administratively, the accounted for territory is represented by the Asian part of Russia without several Far Eastern regions of the monsoon climate – Primorsky and Kamchatka *Krays*, Chukchi Autonomous *Okrug*, and the Sakhalin *Oblast* and Jewish Autonomous *Oblast*. (2) Areas by land classes are estimated based on the hybrid land cover of Russia (see Chap. 1). (3) Abbreviations:

[a]FT, NT and SpT – forest tundra, northern and sparse taiga zones

[b]SD and D – semidesert and desert zones

Table 6.2 Soil distribution by types and bioclimatic zones

Area by bioclimatic zones, ×10⁶ ha

Soil type by WRB[a]	Arctic	Tundra	FT, NT and SpT	Middle taiga	Southern taiga	Temperate forest	Steppe	Semidesert	Total
Albeluvisols			2.3	43.6	49.2	1.9	4.8	0.1	101.9
Cambisols		2.2	10.4	147.0	18.1	3.0	2.6	0.5	183.8
Chernozems				3.7	7.5	6.4	33.3	2.6	53.5
Cryosols	0.7	31.5	0.7						32.9
Fluvisols		12.8	15.8	22.1	7.6	1.2	2.4		61.9
Gleysols		71.0	66.6	53.8	6.1	0.5	0.7	0.3	199.0
Histosols		2.7	24.7	28.1	19.9	0.6	0.1		76.1
Leptosols		3.3	0.0	11.9	0.0	0.0	0.1	0.5	15.8
Phaeozems		7.6	25.5	53.2	4.6	0.5	0.5	0.7	92.6
Podzols		41.7	33.5	186.3	4.7	0.0	0.3	0.5	267.0
Solonchaks		0.4			0.1		1.0	0.3	1.8
Solonetz					0.4	0.4	5.1	1.0	6.9
Total	0.7	173.2	179.5	549.7	118.2	14.5	50.9	6.5	1,093.2

[a]*WRB* world reference base (FAO 2006)

aspen (*Populus tremula*), cover 2.4 %. Shrubs (accounted for as forests in regions where "high" forests cannot grow due to severe natural conditions) cover significant areas (12.4 % of the total forested area) in high latitudes. The rest (5.6 %) of the forested area is presented by hardwood deciduous species, mostly by stone birch (*Betula ermani*) in oceanic territories of Far East and oak (*Quercus mongolica*), ash (*Fraxinus mandshurica*), lime (*Tilia mandshurica* and *T. amurensis*), etc., in broadleaved forests of the south of the Far East. Larch forests (presented mostly by *L. sibirica, L. gmelini,* and *L. kajanderi*) are represented by unique ecosystems occurring on permafrost over huge territories in high latitudes, forming the northern climatic tree line and the most northern forests on Earth (reaching 72°30′N).

The largest global bog-forest systems extend over large areas in West Siberia. Wetlands (defined as forestless territories with a peat layer >30 cm) comprise 107.4×10^6 ha (or 9.6 %). These territories are situated almost exclusively in northern bioclimatic zones. Approximately an area of the same magnitude of relatively shallow peatland is covered by forest. Grasslands (18.8 %) are represented by tundra and meadow vegetation, mostly located in high latitudes. Only 3.4 % of natural grassland occurs toward the south of the middle taiga zone.

Dynamics of major classes of land cover, particularly forests, are documented since 1961, when results of the first complete inventory of Russian forests were published. By January 1, 1961, official statistics reported the total forested area of the region at 505.4×10^6 ha with growing stock of 56.5×10^9 m^3 (FFSR 1962). Official data by 2009 reported the area at 594.7×10^6 with growing stock at 58.0×10^9 m^3. However, it has been shown that methods of forest inventory which were used during this period reported biased estimates of the growing stock volume. An attempt to eliminate the bias, update obsolete inventory data and clarify the area by remote sensing data resulted in the area of 602.2×10^6 ha and growing stock at 60.8×10^9 m^3 for 2009 versus the official 594.7×10^6 ha and 60.1×10^9 m^3, respectively. Taking into account that the unbiased estimate of growing stock in 1961 was 52.2×10^9 m^3 (Shvidenko and Nilsson 2002), we conclude that during the last five decades the forested area in the region increased by 96.8×10^6 ha and growing stock by 8.6×10^9 m^3. Such a large increase in area could be explained by several reasons: (1) improved accuracy and completeness of inventory data, (2) substantial decrease of unforested areas (at $\sim 50 \times 10^6$ ha for Asian Russia), and (3) encroaching forest vegetation in previously nonforest land, particularly in the abandoned agricultural lands. Based on official statistics, the area of cultivated agricultural land in the region decreased by 8.7×10^6 ha during 1990–2009.

However, the dynamics of forested area in most populated regions with intensive forest harvest are substantially different. Typically an analysis of the dynamics of forests of Central Siberia (i.e., Krasnoyarsk Kray including territories of Republic Khakassia, Taimir and Evenkia Autonomous okrugs) generates the following results for 1961–2007 (Vtyurina and Sokolov 2009): (1) forested areas decreased by 5.1×10^6 ha (or 5 %), (2) area of mature and overmature coniferous and deciduous forests decreased by 17.2×10^6 ha (25 %) and 1.7×10^6 ha (17 %), respectively, (3) total growing stock decreased by 3.2×10^9 m^3, and (4) total growing stock in forests available for harvest decreased by 3.7×10^9 m^3 (35 %) in coniferous forests and increased in deciduous forests by 1 %. Similar features of forest industrial development have been reported for the Irkutsk Oblast (Vaschuk and Shvidenko 2006) and Far East (Sheingauz 2008). The process of qualitative degradation of most productive forests in regions with developed infrastructure was observed during the last 50 years.

Overall, man-made changes to forests in Siberia could be briefly summarized as follows:

- Decreasing quality of forests across the southern part of Siberia expressed by worsening species composition, reduction of areas of forests of high productivity, and deconcentration of forests available for industrial exploitation. Major reasons for this stem from the long-period practice of harvest of the most productive coniferous forests in the form of concentrated clear-cuts, nonecological technologies and machinery of logging, ineffective use of harvested wood, particularly of soft deciduous species (birch, aspen).
- Shifting forest harvest toward northern regions with undeveloped infrastructure, particularly roads, that causes a scattered distribution of logged areas. Harvest of basically coniferous species predestines substantial problems for transition to sustainable forest management in future.
- Siberian forests are not managed in a sustainable way. They suffer from large forest fires and outbreaks of dangerous insects (see Sect. 6.3) that mostly damages valuable coniferous forests. Forest fire protection is not sufficient, and each cubic meter of harvested wood is accompanied by losses of about 2.5–3 m^3 killed by disturbances.
- Periodical reforms of Russian forest management (of which those between 2000 and 2007 have been most destructive) substantially decreased governance of Siberian forests. Illegal harvest is widely distributed and reaches from 20 to 40 % of officially reported data, increasing in boundary regions (Vaschuk and Shvidenko 2006; Sheingauz 2001, 2008).

During the last 60 years, about 6×10^9 m^3 of commercial wood, basically represented by valuable coniferous wood, have been harvested in Siberian forests. Besides their resource function, forests provide different environmental, ecological, social, biospheric, cultural, etc., services (e.g., Lebedev et al. 1974). Water-protection services of Siberian forests such as provision of high-quality water, transformation of surface runoff into interflow, and regulation of catchment runoff are among the most important and widely recognized. However, some aspects of interactions of forests with hydrological regimes of landscapes (impacts of forests on total water content in landscapes, river discharge, and evapotranspiration) remain questionable (Voronkov 1988; Protopopov et al. 1991; Geographical... 2007; Onuchin et al. 2006; Onuchin 2009; Liu 2010).

One of the key problems of such debates was the question about increasing precipitation over forest areas caused by roughness of tree canopies that leads to ascending air flows under movements of air masses over forests. However, this theoretically attractive idea on the possible ability of a forest to increase the amount of precipitation has not been unambiguously supported by experimental data in boreal territories and cannot be assumed true over Northern Eurasia (Onuchin 2009; Liu 2010). Assessments of the impacts of forests on redistribution of the water balance between cumulative runoff and evapotranspiration is defined by vegetation, climatic, and geographical conditions (Fedorov 1977; Kabat et al. 2004; Hamilton 2008; Onuchin and Burenina 2008; Onuchin 2009; Liu 2010).

Typically, reduction in biomass and deforestation will increase net runoff, though the characteristics of the timing of such runoff will vary given the condition of soil water and infiltration potential which serves as a trigger mechanism for ground water recharge or surface water runoff (Kabat et al. 2004). At drainage basin scales, the change in the land cover mosaic and secondary vegetation regrowth creates a complex hydrological response as various age classes and successional species have a different effect on major water balance characteristics (Shiklomanov and Krestovsky 1988).

Forest management operations on catchments are recognized as an important factor transforming the major water balance characteristics. According to the type of anthropogenic impacts, geophysical conditions, and restoration dynamics, evapotranspiration and river runoff could increase or decrease (Onuchin et al. 2009). Large-scale logging and forest fires could become a crucial factor that defines conditions of runoff formation in small- and medium-size river basins.

While climate plays a leading role in forming yearly river runoff, non/climatic factors (amount and distribution of forest vegetation over landscapes, soils, size, and geological peculiarities of catchments) are also very important. Both groups of factors are interconnected leading to synergetic effects which could be used for regulation of the runoff by forest management operations taking into account climate and weather specifics (if they are properly understood). Currently, this understanding is insufficient and requires coupled biospheric-hydrological modeling (see Chap. 4, this volume).

Boreal forests with steady snow cover during cold periods play a specific hydrological role because of substantial differences of the balance of snow water between forests and open areas. Depending upon climatic, landscape, and vegetation conditions, forests usually have greater snow storage (cf., Bulygina et al. 2011) and less frozen ground providing more stable feeding of streams throughout the year. However, increased evapotranspiration rate in forests will lead to decrease in annual runoff (Shiklomanov and Krestovsky 1988). For regions where liquid precipitation dominates, the differences in water balance are mostly defined by productivity and morphological structure of ecosystems (but not by the structure of vegetation cover of the landscapes – forest versus treeless areas). More productive ecosystems, regardless of whether they are forests or agricultural lands, have higher evapotranspiration and lower river runoff.

6.2.2 Major Carbon Pools and Productivity of Ecosystems

The term *ecosystem productivity of forests* does not have a monosemantic definition. Usually two groups of indicators are used to quantify productivity which describe either the amount of accumulated organic matter (e.g., live biomass, dead biomass or detritus) or rates of production of organic matter (e.g., net primary production, NPP; net ecosystem production, NEP). Such indicators like growing stock volume, net and gross growth are used in forestry to quantify a resource service of forests.

6.2.2.1 Carbon Pools

Major organic carbon pools include live biomass, dead wood, and soil. Within this study, live biomass (LB) of forests was estimated based on modified forest inventory data and ecoregionally distributed multidimensional models that combine biometric characteristics of forests (species composition, age, site index, and relative stocking) and seven components of live biomass [stem wood, crown wood, bark, foliage, root, understory (shrubs and undergrowth) and green forest floors] (Shvidenko et al. 2007a). LB of agricultural land was assessed based on reported data of crops (FFSSR 2008) and regressions, which connect crop, byproducts, and residuals. LB of the rest of vegetation classes was calculated based on measurements in situ during 1960–2010 [the updated database by N. Bazilevich (1993) was used]. For all nonforest ecosystems, live biomass was assessed by three components – green parts, above ground wood, and below ground biomass.

The most recent (2009) total biomass estimate of terrestrial ecosystems in Siberia is 29.55 Pg C ($1 Pg = 1$ billion $t = 10^{15}$ g), of which 86.5 % are in forests (Tables 6.3 and 6.4, Fig. 6.1). In general, all land classes have a clear zonal gradient of LB distribution with some deviations caused by the characteristics of land classes (e.g., age structure of forests). It corresponds to the fact that the carbon pool of LB of forest ecosystems along the climatic gradient of Central Siberia is positively correlated with mean annual temperature and duration of growing period (Pleshikov 2002).

The structure of live biomass depends on a number of factors – vegetation type, geographical location, level of productivity of ecosystems, and others. In forests, 55.9 % of LB is allocated in stem wood, 9.9 % in branches (both-over bark), 3.5 % in foliage, 23.0 % in roots, 2.2 % in understory, and 5.5 % in green forest floor. These average values vary greatly. For instance, the three- to fivefold increasing of LB of stems toward the south is accompanied by a decrease of the share of ground vegetation LB (such as moss, lichens, grass, and shrubs). In northern low productive ecosystems, the ground vegetation can contribute about half of the total LB storage in the ecosystem and roots – up to 30–35 % and more (Knorre 2003; Shvidenko et al. 2007a).

Other estimates of LB of the entire vegetation cover for the study's region in the above boundaries have never been reported. A number of the assessments that have been done for forest LB either for the Asian part of Russia or for individual administrative regions based on forest inventory of different years present rather consistent results (Alexeyev and Birdsey 1998; Zavarzin 2007; Usoltsev et al. 2011).

Dead wood is an important indicator of the condition and functioning of forest ecosystems. The amount of dead wood (snags, logs, stumps, dry branches of living trees) was estimated in all natural ecosystems based on an extended database. Of the

Table 6.3 Live biomass by major land classes and bioclimatic zones

| | Live biomass, Tg C by land classes | | | | | | |
Zone	Forest	Agro	Burned	Wetland	Grassland	Open woodland	Total
Arctic					0.4		0.4
Tundra	404.2	0.2	1.9	308.4	850.5	50.6	1,615.8
FT, NT & SpT	2,841.7	0.3	5.4	179.1	291.9	187.8	3,506.2
Middle taiga	16,692.5	34.3	58.2	320.6	918.9	296.6	18,321.0
Southern taiga	4,870.7	40.9	2.4	152.9	35.6	11.2	5,113.7
Temperate forest	222.9	18.0	0.2	7.5	6.3	1.0	255.8
Steppe	467.9	126.5	0.4	33.6	22.7	1.8	652.9
D & SD	60.4	14.0	0.1	6.9	3.4	0.1	84.9
Total	25,560.3	234.2	68.5	1,008.9	2,129.7	549.1	29,550.7

Table 6.4 Average (by area unit) live biomass by land classes and bioclimatic zones

| | Live biomass, kg C m^{-2}, by land classes | | | | | | |
Zone	Forest	Agro	Burned	Wetland	Grassland	Open woodland	Total
Arctic					0.13		0.13
Tundra	3.43	0.30	0.13	1.02	0.80	0.81	1.03
FT, NT & SpT	2.75	0.30	0.18	0.73	1.29	0.78	1.98
Middle taiga	4.38	0.33	0.38	0.88	1.24	0.84	3.31
Southern taiga	5.44	0.36	0.44	1.22	1.05	1.51	4.34
Temperate forest	4.02	0.38	0.45	1.25	1.02	1.48	2.20
Steppe	4.84	0.35	0.41	1.21	0.90	1.68	1.27
D & SD	5.47	0.34	0.41	1.10	0.59	1.85	1.31
Average	4.24	0.35	0.33	0.94	1.01	0.82	2.75

total amount of dead wood carbon in all natural ecosystems (Table 6.5) at 6.27 Pg C, about 92 % are in forests. The amount of dead wood in Siberian forests is extremely high (on average 8.33 t C ha^{-1}, in volume units 38.8 m^3 ha^{-1}, almost 40 % of average growing stock volume). The major reasons for that are: (1) an absolute majority of Siberian forests are unmanaged; (2) dominance of mature and overmature, often uneven-aged forests; (3) wide distribution of natural disturbances like fire and insects outbreaks which provide the substantial partial postdisturbance dieback; and (4) slow decomposition rate of dead wood over major part of Siberia.

We calculated carbon content of the above-ground organic layer (with content of carbon >15 %) and the top 1-m layer for mineral soils below; carbon content for organic (peat) soil was estimated for 1 m from the surface (Tables 6.6 and 6.7). Soil organic carbon generates the biggest carbon pools in Siberian ecosystems at 219.5 Pg C or 19.7 kg C m^{-2}. Of this amount, 5.0 % of the carbon lies in the on-ground organic layer.

Evidently, soil carbon content substantially varies by land classes ranging from 2.12 kg C m^{-2} for unproductive land to 44.13 kg C m^{-2} across wetlands. The zonal gradient is not clearly expressed because of availability of different soil types in the

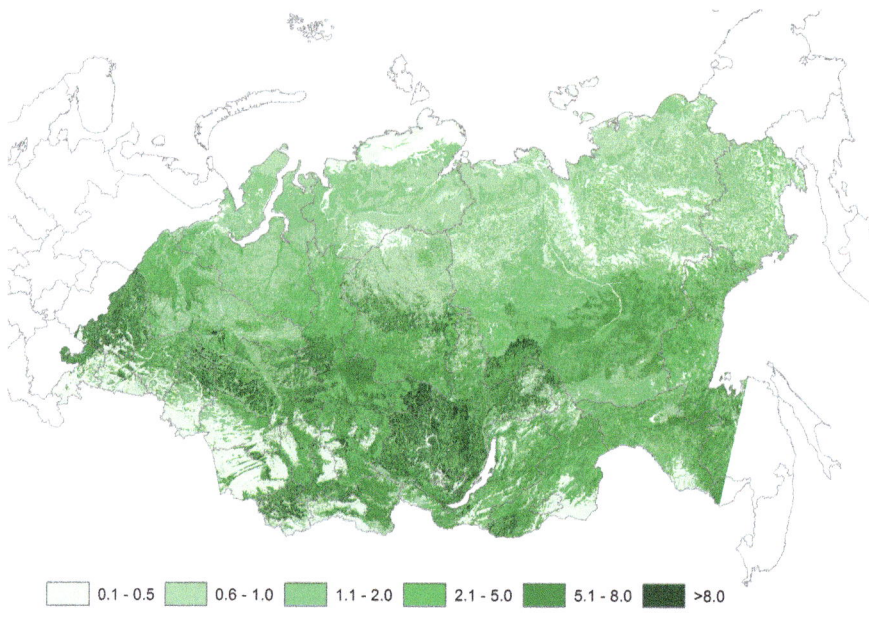

Fig. 6.1 Live biomass of Siberian ecosystems, kg C m^{-2}

Table 6.5 Dead wood by land classes and bioclimatic zones

| | Dead wood by land classes | | | | | | | |
| | Stock, Tg C | | | | Average, kg C m^{-2} | | | |
Zone	Forest land	Wetland	Grass land	Total	Forest land	Wetland	Grassland	Total
Tundra	109.3	4.7	59.5	173.5	0.560	0.016	0.054	0.109
FT, NT & SpT	625.6	20.7	34.3	680.6	0.480	0.085	0.144	0.381
Middle taiga	4,018.1	46.7	293.0	4,357.9	0.926	0.129	0.382	0.797
Southern taiga	890.8	23.0	5.6	919.5	0.978	0.184	0.137	0.854
Temperate forest	46.2	1.0	2.5	49.7	0.713	0.152	0.227	0.605
Steppe	73.7	3.5	0.5	77.7	0.677	0.117	0.015	0.460
D & SD	10.6	0.6	0.2	11.3	0.871	0.082	0.031	0.447
Total	5,774.5	100.2	395.5	6,270.2	0.833	0.093	0.180	0.615

same bioclimatic zones. The average carbon ratio of soil to all types of vegetation (included LB and dead wood) is 6.12 and for forest −3.62.

Previous assessments of organic soil carbon and particularly of the on-ground organic layer vary substantially. For total organic soil carbon, they differ from our estimate in limits of ±10 % (Orlov et al. 1996; Rozhkov et al. 1996). Differences for individual land classes are higher. For example, estimates for forest soils are in the range from −40 (Alexeyev and Birdsey 1998) to +15 % (Rozhkov et al. 1996).

Table 6.6 Soil organic carbon pool (on-ground organic layer + 1 m deep of soil)

Zone	Soil carbon, Pg C by land category									
	UN	FOR	OW	BA	AR	HF & PS	FL & AA	WL	GL	Total
Arctic	0.01								0.03	0.04
Tundra	0.42	1.64	1.15	0.18	0.00	0.12	0.00	11.32	14.71	29.55
FT, NT & SpT	0.01	22.86	6.61	0.31	0.00	0.00	0.00	10.92	3.38	44.11
MT	0.13	61.32	5.27	2.35	0.05	0.93	0.44	16.40	11.00	97.90
ST	0.00	21.48	0.22	0.15	0.69	0.79	0.53	7.91	1.04	32.81
TF	0.00	1.13	0.03	0.01	0.38	0.32	0.37	0.24	0.20	2.68
Steppe	0.01	2.11	0.03	0.02	3.29	4.13	1.07	0.51	0.55	11.73
SD & D	0.00	0.11	0.00	0.00	0.01	0.39	0.03	0.09	0.05	0.68
Total	0.60	110.66	13.32	3.03	4.42	6.68	2.42	47.38	30.97	219.50

Land category: *UN* unproductive area, *For* forest, *OW* open woodland, *BA* burned area, *AR* arable land, *HF & PS* hayfield and pasture, *FL & AA* fallow and abandoned arable, *WL* wetland, *GL* grassland and shrubland. Zone: *SF & NT* sparse forest and northern taiga, *MT* middle taiga, *ST* southern taiga, *TF* temperate forest, *SD & D* semi desert and desert

Table 6.7 Average soil organic carbon density (on-ground organic layer + 1 m deep of soil)

Zone	Soil carbon, kg C m^{-2} by land category									
	UN	FOR	OW	BA	AR	HF & PS	FL & AA	WL	GL	Average
Arctic	0.41								9.40	1.13
Tundra	2.16	13.76	18.56	12.10	0.00	12.16	7.06	37.51	13.30	16.32
FT, NT & SpT	5.27	22.13	27.34	10.70	10.00	6.44	10.35	44.69	14.16	24.63
MT	2.85	16.07	14.87	15.30	14.25	12.59	16.10	45.23	14.31	17.46
ST	4.97	24.02	29.62	27.78	18.61	16.74	18.65	63.16	25.55	27.66
TF	4.82	19.68	47.19	35.44	23.41	18.97	22.05	40.64	29.73	22.19
Steppe	5.35	21.85	27.83	23.57	25.11	21.30	25.50	18.41	20.49	22.44
SD & D	0.99	17.66	7.74	11.43	19.86	16.26	19.99	14.41	10.48	15.04
Average	2.12	18.38	19.94	14.82	23.45	18.24	21.00	44.13	14.10	19.74

A number of reasons contribute to this: use of different land cover in the assessments, incompleteness of information used, lack of planning of the experiment, etc. Soil carbon remains the most uncertain carbon pool of Siberian ecosystems.

6.2.2.2 Net Primary Production of Ecosystems

Historically, numerous empirical data on NPP of Russian forests were biased due to two reasons: (1) they are based on destructive in situ measurements with intervals of weeks and months that do not account for a substantial part of NPP (e.g., root exudates or volatile organic compounds, VOC) which could reach 15–20 % and more in boreal forests (Vogt et al. 1986) and (2) the typical incompleteness of NPP measurements of some components of ecosystems, like fine roots and below-ground parts of shrubs and the green forest floor (Schulze et al. 1999; Vasiliev et al. 2001). In order to eliminate bias, a new "semi-empirical" method for evaluation of NPP of Northern Eurasian forest ecosystems was developed (Shvidenko et al. 2007a, 2008). The method is based on simulation of biological production of forest ecosystems using aggregation of regional empirical regularities of growth of forest stands and dynamic models of live biomass of forest ecosystems. The method does not have any recognized biases in contrast to existing empirical methods of assessment of forest NPP on large areas. This method has been used for assessing forest NPP.

NPP for other classes of natural vegetation has been assessed based on empirical in situ measurements. In general, such measurements also have shortcomings similar to the above mentioned for forests (e.g., Vasiliev et al. 2001); however, we were not able to avoid these. NPP of agricultural land was defined similarly to live biomass of this land class.

The total annual NPP of Siberian ecosystems is estimated at 2.76 Pg C year^{-1}, or 257 gC m^{-2} year^{-1}. By land classes, agricultural lands have the highest NPP (402 gC m^{-2} year^{-1}) and open woodlands – the smallest (84 gC m^{-2} year^{-1}); 97.5 % of open woodlands are situated in northern zones of Siberia, mostly in the forest-tundra ecotone – this is the reason for such low productivity. NPP of forests is estimated at 283 gC m^{-2} year^{-1}. NPP of all land classes has clear zonal gradients (Table 6.9).

To our knowledge, there were no previous estimates of NPP for the entire region. The estimates for forests of individual regions reported by Usoltsev et al. (2011) are consistent with our results. Only one recent estimate of forest NPP for the Asian part of Russia has been reported – at 277 gC year^{-1} for 2003 (Shvidenko et al. 2008). Note that previous estimates of NPP for all terrestrial ecosystems over the entire country are very diverse – from 2.75 Pg C year^{-1} (Filipchuk and Moiseev 2003) for the ~2000s to annual averages of 4.35 Pg C year^{-1} for 1988–1992 (Nilsson et al. 2003) and 4.73 Pg C year^{-1} for 1996–2002 (Zavarzin 2007) and up to 5.1±0.36 Pg C year^{-1} for 2003–2008 (Shvidenko et al. 2010b). The latter paper reports NPP for Asian Russian forests that is very close to the above estimate of forest NPP (Table 6.8) – at 1.68 Pg C year^{-1} against 1.70 Pg C year^{-1}.

An important question is how certain are the above NPP estimates of Siberian ecosystems. Relatively comprehensive analysis of uncertainties could be done for forests (the approach is considered in Shvidenko et al. 2010b) that resulted in ±6–8 % (CI 0.9). This conclusion is supported by comparisons with independent assessment by different methods. Yearly NPP defined by 17 DGVMs (Cramer et al. 1999), which included changing environment, on average differed from our assessment only by +6 %. The NPP product from Terra-MODIS gives the average result almost identical to data of Table 6.8, but shows a substantial bias of different sign for both low and high productive forests of the country.

Siberia faces a significant change of environment, which presumably substantially impacts productivity of ecosystems, i.e., climate change, increasing atmospheric concentration of CO_2, and nitrogen deposition, and along with this a substantial acceleration of anthropogenic impacts. While the two last decades have been the warmest for the entire period of records, still the observed temperature trends and precipitation are rather far from limits of sustainable functioning

Table 6.8 Net primary production of vegetation by land classes and bioclimatic zones

Zone	Net primary production, Tg C year^{-1} by land classes						
	Forest	Agro	Burned	Wetland	Grassland	Open woodland	Total
Arctic					0.2		0.2
Tundra	28.7	0.3	1.4	34.8	126.6	5.4	197.1
Sparse taiga	220.8	0.3	2.9	48.8	38.6	21.8	333.2
Middle taiga	1,065.1	39.5	22.4	94.7	180.9	27.3	1,429.8
Southern taiga	330.0	47.0	1.0	49.9	25.8	1.2	454.8
Temperate forest	19.5	20.7	0.1	3.7	5.4	0.1	49.5
Steppe	35.0	145.4	0.2	56.2	23.6	0.2	260.6
Semidesert and desert	3.8	16.1	0.0	8.4	3.2	0.0	31.6
Total	**1,702.9**	**269.3**	**28.1**	**296.4**	**404.2**	**55.9**	**2,756.8**

Table 6.9 Average net primary production by unit area by land classes and bioclimatic zones

Zone	Net primary production, g C m^{-2} year^{-1} by land classes						
	Forest	Agro	Burned	Wetland	Grassland	Open woodland	Total
Arctic					60		60
Tundra	244	350	95	115	119	86	126
Sparse taiga	214	346	100	200	171	90	188
Middle taiga	279	381	146	261	243	77	258
Southern taiga	369	419	185	399	758	156	386
Temperate forest	351	433	209	623	876	159	425
Steppe	363	400	195	2,021	936	166	506
Semidesert & desert	348	389	162	1,349	552	185	488
Overall average	**283**	**402**	**137**	**276**	**192**	**84**	**257**

of regional ecosystems, particularly forests. However, intensive regional weather anomalies become more frequent and sometimes disastrous.

The average global concentration of CO_2 in the atmosphere continues to grow: in 1995–2005 this indicator on average increased at 1.9 ppm year^{-1} (Forster and Ramaswamy 2007). The impact of increasing concentration of CO_2 on NPP are usually described by "Killing's formulae" NPP (C_0) = NPP$_0$ [1 + β ln $(C_a/C_{a=0})$] where C_a – actual concentration of CO_2, β – growth factor; $C_{a=0}$ is a base concentration of CO_2, e.g., in the preindustrial period; NPP$_0$ = NPP$(C_{a=0})$ (Bacastov and Keeling 1973). As a whole, an invigorative impact of CO_2 on ecosystem productivity is supported by a number of research findings; however, there are different results and opinions about the base parameters of this process. Increasing CO_2 concentration supports earlier maturing of forest trees (under smaller size in age of maturity) and unproportional allocation of carbon in generative organs (de Graaff et al. 2006). Shortage of nitrogen decreases the fertilization effect of elevated concentrations of CO_2 on live biomass increment (Raich and Nadelhoffer 1989; Schneider et al. 2004). Net C accumulation in boreal and temperate forests correlates with regional levels of net deposition of nitrogen (e.g., Magnani et al. 2007). In comparative research of the fertilization impacts of elevated CO_2 concentration on forests of different biomes, Ciais et al. (2005) reported the significant impact of this factor on boreal forests.

There are a number of empirical and modeling studies in support of increasing productivity along the circumpolar boreal and polar belts. It points to increasing of biomass, density, and height of vegetation. For instance, in Canadian tundra the biomass of mosses has increased by 74 % and evergreen shrubs by 60 % during the last two decades (Hudson and Henry 2009). Some models predict the increase of productivity of spruce forests in European Russia (Oltchev et al. 2002; Olchev et al. 2008, 2009) in the future if projected increase of atmospheric CO_2 will be balanced by an increase of available soil nutrients (nitrogen, phosphorus, potassium). A rather consistent conclusion of a number of studies, climatic trends, and environmental change support an increase of productivity of ecosystems in Northern Eurasia. For forest ecosystems, annual average increase of net growth of Russian forests is esti-mated at 0.4–0.6 % per year between 1960 and 2000 (Alexeyev and Markov 2003; Shvidenko et al. 2007b), or about 20–25 % over this period. However, the possibility to distribute these data over Siberia should be used with caution due to the wide distribution of disturbances in the region. A decrease of growing stock in Alaska's forests during 1957–2007 was reported by Rautiainen et al. (2011).

Change of live biomass' stock and NPP integrates all natural processes in the biosphere as well as the human impacts. It is suspected to lead to a reorganization of many components of the production process: reallocation of carbon in above- and below-ground parts of ecosystems, changing rate of decomposition of dead vegeta-tion matter, changing the ratio of major components of carbon cycling (e.g., for NPP and heterotrophic respiration), and others.

It is hypothesized that warming is one of the major drivers of increasing produc-tivity of vegetation ecosystems of Northern Eurasia, particularly forests although the usual climatic norm of temperature of growth periods is lower than the tempera-ture optimum for indigenous tree species of temperate (25–30 °C) and boreal (15–20 °C) latitudes. This, however, requires a corresponding increase in precipita-tion which was not observed over vast continental regions of Northern Eurasia dur-ing recent decades (Lapenis et al. 2005). There are examples from high latitudes when temperature increases lead to a decline of radial increment of trees due to a shortage of available water driven by elevated temperature (Barber et al. 2000). Increasing growth period could enter into the conflict with plant's photoperiodism, whereas insufficient chilling negatively impacts NPP. Besides short-term impacts of elevated temperatures on ecosystems, there are numerous ecosystem indirect responses and feedbacks caused by water regime change, change of succession, change of nutrient availability, species composition, and outbreaks of insects and pathogens (e.g. Lloyd and Bunn 2007).

Impact of climate also yields a change of allometric relationships in boreal forests. Based on more than 3,000 sample plots, Lapenis et al. (2005) showed a pronounced increase in the share of green parts (leaves and needles) and a decrease in the share of above-ground wood in Russian forests between 1960 and 2000. However, there is a large geographical variation of this process. The shift has been largest within European Russia, where both summer temperatures and precipitation have increased.

On the contrary, in the northern and middle taiga of Siberia, where the climate has become warmer but drier, the fraction of the green parts has decreased while the fractions of aboveground wood and roots have increased. These changes are consistent with experiments and mathematical models that predict a shift of carbon allocation to transpiring foliage with increasing temperature and lower allocation with increasing soil drought. This result might be considered as a possible demonstration of the acclimation of trees to ongoing warming and changes in the surface water balance which had a moderate trend during the considered period.

However, there are many unresolved questions in understanding the interaction of a changing environment with boreal vegetation. There is still no clear understanding how tree species and ecosystems function under dynamic conditions of multiple limitations for life resources. While there are no clear answers a number of important questions arise: (1) How stable is direct forcing of photosynthesis and NPP by the direct change of environment? (2) How much do limitations for resources (water, nutrients) restrict CO_2 fertilization effect and how long does the impact of such limitations remain substantial? (3) Could nitrogen depositions alleviate a shortage of nitrogen that is usual for high latitudes? (4) How will cryogenic destruction of permafrost impact landscapes of high latitudes? Future experimental and modeling efforts are needed to answer these questions of fundamental importance.

6.3 Disturbance and Succession Dynamics of Forests

6.3.1 Disturbance Regime and Succession Dynamics

Natural- and human-induced disturbance is an integral feature of the boreal world. Large-scale disturbance shapes the vegetation mosaics that define boreal landscapes and defines the structure of Siberian forest ecosystems by setting the beginning and end of successional dynamics (Heinselman 1978; Shugart et al. 1992; Antonovski et al. 1992). Disturbance is an inherent geographical, landscape, ecosystem, and site-specific phenomenon that interacts as relatively stable processes or combinations of specific or novel disturbance that defines the "disturbance regime" of that ecosystem or biogeocenosis (Chudnikov 1931; Tumel 1939; Shvidenko and Nilsson 2000b). Major types of disturbance (fire, insect/diseases outbreaks, harvest, snow- and windbreaks, pollution and industrial transformation) impact, on average, $10–25 \times 10^6$ ha of Siberian forest land annually.

Successional dynamics form the foundation for the beginning, development and destruction of forests, natural or planted (Sukachev 1972; Kolesnikov 1956; Sedykh 2009). Knowledge of successional regularities serves as the background for understanding the spatial and temporal dynamics of forest ecosystems. Specifics of forest forming successional processes are revealed in the number of stems, individual or species cohorts, spatial extent, health and duration of successional phases (i.e., time periods of a specific quantitative and qualitative morphological forest ecosystem state, which is characterized by homogenous structure and regularities of age dynamics) and stages (successional age development within the phases, e.g., a period with dark coniferous understory succeeding in secondary birch and aspen overstory).

Based on classical works of V. Sukachev and B. Kolesnikov, a comprehensive classification of successions in Siberian forests that has been developed and parametrized by the European Commission project *Siberia* (D. Efremov, A. Isaev, P. Khomentovsky, G. Korovin, V. Rozhkov, V. Sedykh, A. Sheingauz, A. Shvidenko) includes six classes – climatic-morphogenic, biogenic, cenogenic or age-related changes, pyrogenic or postfire, anthropogenic, and climax sparse forests growing in extreme conditions (e.g., on bogs or subalpine sites) which are divided into subclasses, genera, and kinds (Shvidenko et al. 2001; Farber 2000; Sedykh 2009). Of this classification, only two classes of succession – cenogenic (including three major subclasses – succession with change of species, without changes of species and point-dispersion) and climax sparse forests represent dynamics driven by endogenous factors of ecosystem development. The rest of the classes (>90 % of the forest land in Siberia) are driven be exogenous impacts, i.e., by disturbance. Expanding on this thought, the beginning and end of cenogenic phases are also set by disturbance, and disturbance on climax sparse forest lands could alter the system, potentially for centuries. If there is a landscape-scale balance of endogenous and exogenous factors, this maintains stability in landscapes at the regional scale. Thus, disturbance regimes define the specifics of the temporal and spatial dynamics of Siberian forests, and changes in disturbance regimes could alter the mosaic structure of regional-scale landscapes. According to estimates of the Project *Siberia* (Schmullius and Santoro 2005), phases of pyrogenic succession include from 40 to 96 % of the total forest land by individual ecological regions of Siberia.

6.3.2 Fire

Fire is a major natural disturbance in Siberian ecosystems, in particular, in forests, because of: (1) dominance of natural ecosystems in Siberia; (2) almost all Siberian forests are boreal forests, and 71 % of them are dominated by coniferous stands of high fire hazard; (3) a major part of the forested territory is practically unmanaged and unprotected – large fires (>200 ha) play an important role in this region; (4) natural ecosystems accumulate large amounts of dead organic matter due to slow decomposition of plant residuals; and (5) a substantial part of natural ecosystems (particularly in Central and Eastern Siberia) are situated in regions with limited amounts of precipitation during the fire season and/or frequent occurrences of long drought periods during the fire season that often generates fires of high severity. The double-faceted role of forest fires – destructive and creative – is well recognized in high latitudes. In the central and southern part of the boreal zone, forest fires are one of the most dangerous environmental phenomena, which cause significant economic losses and have a strong negative ecological impact on forest ecosystems and environment. In unmanaged and unused forests of northern and sparse taiga, specifically in permafrost regions, on-ground fire with a "normal" fire cycle (80–100 years) is a natural mechanism for preventing the decreased productivity of forests, paludification of forest lands and finally, hinders the distribution of desertification (Sedykh 1990). However, even in high latitudes, frequent human-induced recurrent non-stand replacing fires can significantly decrease the actual productivity and resilience of forests (up to 40–50 %).

Fire activity is basically driven by four major factors – weather/climate, amount and condition of fuel, ignition agents, and human activities. All these together form landscape specific fire (pyrological) regimes – stable combinations of individual types of vegetation (forest) fires and their characteristics, as well as specific interactions between ecosystem, fire, and other disturbance types. Fire regimes are quantified by a number of indicators, which are explicitly defined in space and time (extent of fire; frequency, e.g., in form of return fire interval; intensity of burning, measured, e.g., by radiative energy; indicators of postfire forest transformation, etc.).

On average, for basic forest upland types and geographical localities, the fire-return interval in Siberian taiga forests including all types of fire is 25–70 years. However, the variation of fire frequency is very large: upper limits are 250–300 years for wet sites and dark coniferous forests (and up to 500–700 years for wetlands), lower limits are 7–15 years, and even less, usually observed in dry pine and larch forests in densely populated areas. In a historical perspective, areas in which no fires occurred during a single life cycle of coniferous taiga forests (200–300 years) are negligibly small in drainage sites of the taiga zone (Furyaev 1996). For major forest types of the taiga zone, fire is physically possible if the amount of on-ground fuel exceeds 0.5 kg dm m^{-2}. Particularly severe fires occur in forests disturbed by insects, so-called *shelkoprjadniki*.

The majority of the fire events in Siberia are of human origin and about 20–30 % in some regions (measured by fire extent) can be attributed to natural factors alone (mostly lightning). The majority of fires (~75–85 %) are ground fires, either superficial or steady. Crown fires comprise about 20 %; however, in extremely severe fire years this share could be doubled (Shvidenko and Nilsson 2000a; Achard et al. 2008). Crown and peat fires are stand-replacing fires, while on-ground fires cause a partial dieback which is accounted for from 5–7 % (by initial growing stock) to 70–90 % after steady ground fires, particularly in usually wet sites and on permafrost. Indicators of fire regime depend on many factors: weather specifics during the vegetation period, fuel characteristics of forests and adjoining vegetation, type of forest formation, spatial structure of landscapes, their ecological regimes, inter-annual climate variability (recurrence of extreme droughts), density of population, accessibility of forests, level of forest fire protection, and others.

Severity and extent of fire define the level of transformation of indigenous vegetation that could be expressed by different indicators of disturbance rate (e.g., loss of productivity). Publications report that about 35 % of Siberian forests could be classified as disturbed forests. Irreversible change in forest ecosystems becomes particularly evident after so-called *mega-*or *catastrophic fires*. Catastrophic wildfires are fires covering an area of more than 10,000 ha, resulting in the total destruction of vegetation and organgenic horizons of soils, or the simultaneous occurrence of several fires of the same total area and intensity over a total area of 1,000 km^2

(Sheshukov 1967). Sukhinin (2009) defines catastrophic fires as those which envelop substantial part of a landscape (>20,000 ha) under conditions of a long-period anticyclone and the highest class of drought; they have an extremely high intensity of burning and postfire dieback >50 % of growing stock. During years of high fire danger, a "spring" fire regime is transformed into a "late-summer" one that reflects a substantial increase in fire severity and a following increase of carbon and nitrogen emissions.

Long-term consequences of catastrophic fires are the irreversible transformation of the forest environment, which is obvious beyond the restoration period of indigenous forest ecosystems, i.e., exceeds the lifetime of major forest forming species (that ranges from 150 to 400 years for major forest forming species of Siberia). They reveal the following aspects (Yefremov and Shvidenko 2004):

- A significant (up to several times) decrease of the biological productivity of forest lands due to the destruction of the indigenous ecotopes and replacement of indigenous vegetation formations
- Irreversible changes of the cryogenic regime of soils and rocks
- Change of long-term amplitude of hydrothermal indicators beyond natural fluctuation
- Changes of multi-year average hydrothermal and biochemical indicators of aquatic and sediment runoff, as well as of hydrological regimes and channel processes of water streams
- Accumulative impacts on atmospheric processes resulting in global climate change
- Acceleration of large-scale outbreaks of insects and disease
- Irreversible loss of biodiversity including rare and threatened flora and fauna species
- Transboundary water and air transfer of pyrogenic products
- Change of historical migration routes for migratory birds, ground and water animals

In particular, the correlation coefficient between the share of unforested areas in forest landscapes and the forest fire occurrence rate was estimated at 0.49 (CI 0.95) (Sheingauz 2001). At the level of forest enterprises, a 1 % increase in a forest fire occurrence rate on average causes an 8.4 % decrease in the percentage of forest cover.

By estimates, forest fires increased the total area of deforested lands in the Asian part of Russia by up to 20 million ha over the last 50 years (Yefremov and Shvidenko 2004). Generally, single or repeated catastrophic forest fires transform about 30 % of highly productive forest land (with a total stock of live biomass of up to 200–300 Mg dry matter per ha) to barren land areas for which forest regeneration is postponed for an indefinitely long period of time (process of "green desertification"). On average for the taiga zone, such lands comprise up to 70 % of bogs, 15 % of grass-small shrub and shrub lands, 10 % of open woodlands, and up to 5 % of stone fields and stone outcrops. These territories can only be rehabilitated through targeted and labor-consuming meliorations. The natural restoration of forests in these areas requires several hundred years.

Recent years reveal substantial impacts of catastrophic fires on regional weather. Observations of atmospheric patterns over burning and smoking forests in Eastern Siberia and the Far East in 1998 recognized the presence of anticyclones above huge areas of Northern Asia, from the Yenisei River to the Okhotsk Sea (Sokolova and Teteryatnikova 2002). These territories, where enormous amounts of forest fuels were accumulated and which had a large-scale smoke blanket of a size comparable to the extent of the baric systems (i.e., an area of more than 350–400,000 km^2), had a long period of high atmospheric pressure. It forces the cyclones to take a southern bypass. This caused intensified drought episodes over the fire-affected areas that extenuate the forest fire situation. The presence of anticyclones in temperate latitudes of Eastern Asia both in winter (that is common) and summer time (that is unusual) was due to the increased air density (through the cooling down of near-surface layers caused by the smoke aerosol), and summer anticyclones were duplicating the mechanism of winter ones. Alternatively, such a meteorological situation can generate long periods of rainfall and catastrophic floods as in the basin of Yangtze River in the summer of 1998.

Analysis of meteorological processes based on pressure charts identified an important specific feature. In all years, when in early summer the usual tropospheric ridges at a baric height of AT-500 arose in the smoke affected atmosphere, the anticyclones (associated with a drought) persisted in this area over the entire summer. This smoke-affected anticyclone remains over the entire warm period, and such a phenomenon was observed only in territories affected by the smoke. During the summer, the continental tropospheric ridge is being supported by powerful heat fluxes from fire and hot smoke. Only the decreased temperature in late summer eliminates the influence of smoke atmospheric aerosol that leads to the gradual destruction of the continental tropospheric ridge.

Meteorological conditions which initiated catastrophic forest fire occurred in 1954, 1968, 1976, 1988, and 1998 in the Amur River Region, in 1979, 1985, 1998, and 2003 – in Eastern Siberia (from Krasnoyarsk Kray to Burjatia and Chita regions), in 1996 – in Amur Oblast and in the Republic of Sakha, and in 2002 – in the Republic of Sakha (Sokolova and Teteryatnikova 2002). In spite of incomplete understanding of the mechanism of the above-mentioned regularities, one could assume that catastrophic forest fires have a substantial influence on the formation and alteration of the regional climates.

Substantial negative impacts of catastrophic fires in south taiga on biodiversity were reported (Kulikov 1998) and are particularly dangerous at ecotone boundaries of natural habitats of animal and plants. These fires decrease the amount of fodder, lead to fragmentation of habitats, and eventually substantially decrease populations of animals, reptiles, and birds. Also migrating birds and ungulates now use routes that differ from their traditional ones. However, forest fires of moderate intensity in high latitudes could protect and promote biodiversity at both ecosystem and species levels (Sedykh 2009).

Relatively reliable documented records of areas of vegetation fire in Siberia exist since 1998 when several remote sensing centers in Russia and abroad started systematic reporting of areas enveloped by fire and burned areas. Improving knowledge and availability of new satellite-based fire products substantially improved understanding of fire regimes and emission estimates from wildland fire. Three major remote sensing products that were used for detection of vegetation fire in Siberia – Spot Vegetation (VGT), Terra-MODIS, and Advanced Very High Resolution Radiometer on board of NOAA satellites (AVHRR) – have similar temporal and nadir spatial resolutions. Products from VGT and MODIS have better geometric fidelity, radiometric calibration, multi-spectral registration, multi-temporal registration, and absolute geolocation but are not based on thermal channels. The latter is important for fire detection due to dominance of ground fire in Siberia. Different sensors provide substantial diversity of seasonal variation of the number of fires and the area burned. However, detection of large fires (>200 ha), which comprise about 90 % of burned areas in Siberia are of the same magnitude. This results in relatively consistent estimates of burned areas by the above sensors (e.g., Bartalev et al. 2007; Van Der Werf et al. 2006) – the differences between the total annual areas that are enveloped by fire are in the range of ±20–25 %. However, some publications present a substantially higher variability. For instance, the average area of vegetation fire over the entire country (of which >90 % occurred in Siberia) is reported to be 19.6 million ha from 2003 to 2007 included 6.9 million ha in forest (Vivchar 2010), while the estimate of burned area in forest (Ershov et al. 2009) is 3.9 million ha for the same period. Both of these assessments are based on the same remote sensing products (Terra-MODIS). Reliability of these estimates remains questionable, particularly for large areas of grassland, croplands, woodland, and wooded grassland. Usually coarse resolution remote sensing, particularly those based on optical and infrared bands, underestimate the burned areas (Zhang et al. 2003). Official Russian fire statistics which represented fire in so-called "actively protected forests" of Siberia underestimated actual burned areas on forest land by an order of magnitude (Shvidenko and Goldammer 2001).

In this study we used data presented by the Institute of Forest (Krasnoyarsk) (available also at IIASA FOR website). Wildfire data were acquired based on AVHRR imagery (hot spots) with sampling control of burned area by LANDSAT (Sukhinin 2008). The approach and algorithm used are described in details in Soja et al. (2004). Taking into account that AVHRR (bands 2, 3, 4, and 5) overestimate burned areas, particularly for small fires, the estimates have been corrected by using regression with burned-area estimates from LANDSAT ETM+ (resolution 30 m) for a number of regions. Likely, the AVHRR/LANDSAT ETM+ method currently provides the most reliable estimates of areas of vegetation fires in Russia.

The average areas affected by fire between 1998 and 2010 in Siberia is estimated at 7.4×10^6 ha, of which 50–70 % are on forest land (Fig. 6.2) with the variation from 3.6 (2004) to 16.1×10^6 ha (2003). The change of burned areas for the last 13 years do not give any clear answer about the short-term dynamics of fire danger in Siberia.

However, there is evidence that occurrence of fire danger years have been increasing during recent decades. Specific weather during the two recent decades had a rapid and pronounced effect on fire activity and caused new tendencies of fire regimes. The latter include increase of the extent, frequency, and severity of fires;

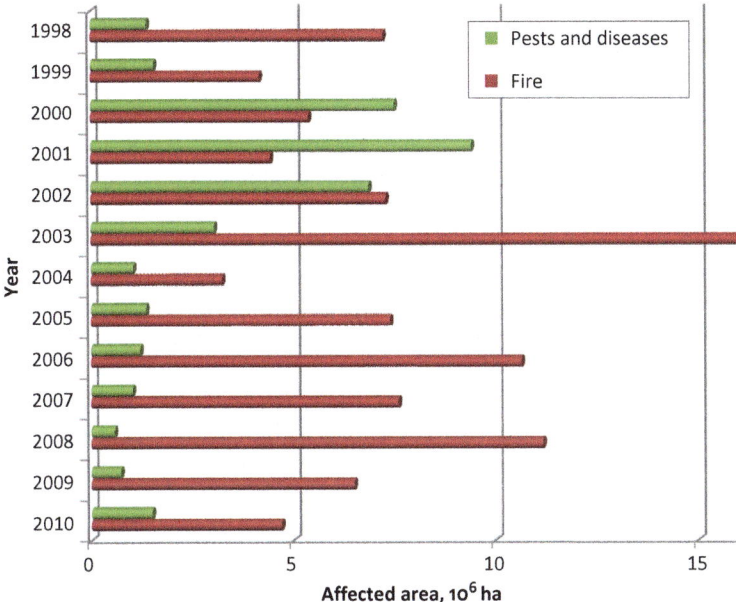

Fig. 6.2 Area affected by fire and pests in Siberia in 1998–2010. Average burned area 8.23×10^6 ha year^{-1} (From 4.2×10^6 ha in 1999 to 17.3×10^6 ha in 2003). Average area of pest and diseases breakout is 2.9×10^6 ha year^{-1} (From 0.6 to 9.4×10^6 ha)

the similar picture is observed during the last several decades across the entire circumpolar belt (Kasischke and Turetsky 2006; Yefremov and Shvidenko 2004; Ivanova 1996, 1998–1999). It has been shown that the observed increases of burned areas during recent decades are the result of climate change (e.g., Gillett et al. 2004). Large fires increased areas of usually nonburned land classes, particularly wetlands, which became more vulnerable to fire. It substantially increased the total amount of greenhouse gas emissions, as well as the amount of such gas components as CO, CH_4, CH_3Br, and CH_3Cl due to increasing smoldering consumption (Manö and Andreae 1994).

Vaganov et al. (1996) based on dendrochronological analysis for the last 380 years reconstructed fire frequency in pine forests of two regions in Krasnoyarsk Kray, and showed a statistically significant correlation between climatic indicators and occurrence of forest fire. The length of the mean fire return interval by the regions (of 35 and 18 years) was clearly impacted by anthropogenic impacts by end of the studied period.

For the zone of northern taiga (River NizhnyayaTunguska basin around Tura in East Siberia, about 65°N), the region of dominance of larch forests, Kharuk et al. (2008) showed that during the nineteenth century the fire return interval (FRI) decreased from 101 to 65 years in the twentieth century. For the more southern territories (~60°N) of the larch-mixed zone ecotone, the decrease was from 97 to 50 years.

Efremov and Shvidenko (2004) reported a trend of increasing annually burned area in Khabarovsk Kray during recent decades.

Evidently dynamics of fire frequency depend on many reasons, both climatic and anthropogenic. An important factor is industrial development of previously untouched territories. Ivanova (1996) reported the following sequence of extremely dangerous fire years for Evenkia: 2 in seventeenth century, 3 in eighteenth century, 9 in nineteenth century, and 29 in twentieth century.

Permafrost and wetlands interacts with fire activity, emissions of carbon, methane, and other greenhouse gases, and postfire mortality. Very likely higher temperatures and permafrost thaw will increase the fire activity and vulnerability of terrestrial ecosystems to burning due to (1) changing hydrological regimes over vast territories due to decreasing water table and increasing evapotranspiration, (2) wider distribution of deep burning, and (3) more intensive postfire dieback.

Interactions between fire regimes land use and climate may become increasingly important for carbon storage and fluxes in ecosystems. One could suppose that climate change and other pressures (such as industrial development and expanding populated areas) will provide a further profound effect on the boreal ecosystem. If current climatic predictions are realized, future fire regimes will be characterized by (1) increasing length of fire season; (2) accelerated fire activities – number of ignitions, large areas with extreme fire hazards, etc.; (3) increased area burned, severity of burning, and amount of consumed fuel; (4) increased occurrence of disastrous fires; and (5) increased postfire impacts on ecosystems and landscapes.

The ability of Siberian forest management to cope with these changes is very limited. Current forest fire protection operates within a narrow margin between success and failure, and disastrous fires of recent decades in Siberia clearly illustrated the scale of problems that arise. It requires substantial changes to forest management policies including setting of priorities, prevention programs, fire monitoring systems, initial attack capabilities, and modification of some legislative and institutional aspects of forest management like access restriction policies.

Research indicates the inability of current fire protection systems to successfully withstand the increased fire pressure even in countries with highly organized fire protection (like Canada and USA) and could lead to catastrophic consequences in countries with fire management in disarray (like Russia). The fire season-2010 in Russia, when large fires occurred in densely populated territories of European Russia, has brought economic losses of about USD ten billion.

Based on GCMs and different vegetation models, a substantial increase of fire hazard is predicted over the entire circumpolar belt mostly due to warming, particularly in southern and continental regions (see also le Goff et al. 2009; Balshi et al. 2009). There is a general tendency of increasing risk, extent, and severity of forest fire, as well as length of fire season. For Siberia, fire danger may increase by a factor of 3–4, with a peak in the middle of August. By end of the twenty-first century, the risk may be decreasing due to increase of precipitation according the contemporary GCM predictions (e.g., Mokhov et al. 2006, 2009, cf., Chapter 3). Intensification of fire activities will be accompanied by substantial shift of bioclimatic zones northward up to 600–1,000 km. New climate and accelerated fire regimes may halve areas of Siberian forests by 2080 (Tchebakova et al. 2009). For Canada, an average increase in the Seasonal Severity Rating and the resulting area burned is expected at

~50 % by mid-twenty-first century ($2 \times CO_2$ scenario, Flannigan and Van Wagner 1991). An earlier start of fire season and significant increases of area large fire in both Canada and Siberia has been predicted by Stocks et al. (1998). Many other studies predict increasing the area burned, particularly due to increasing lightning-caused fires. Altered fire regimes, succession dynamics, and permafrost behavior could result in a positive feedback intensifying rates of climate change (Kurz et al. 1995; Lyons et al. 1998; Soja et al. 2007).

National programs of adaptation to, and mitigation of negative consequences of climate change in the boreal zone, should include fire protection as a cornerstone of current and future sustainable forest management. Development of long-term strategies of adaptation by boreal landscapes to future climates is an urgent problem today. The inherent uncertainty of forecasts is a specific problem of such developments, and hence the need for use of win-win strategies which would be robust to a diversity of possible scenarios. International cooperation in further development of boreal fire protection is an issue of the highest priority.

6.3.3 Biotic and Other Disturbances

Among numerous biotic disturbances, insect and disease outbreaks are most important. Official data on biotic factor impacts remain only a source of information because remote sensing is still not able to identify the partial damage which is usually caused by pests and pathogens. Very likely, official statistics underestimate the real impacts of biotic disturbances, although this underestimation is probably less than for fire because major areas of insect and disease infestations are located in southern regions, where forest pathological monitoring is provided. According to rather consistent opinions of Russian experts, the damage caused by insect and diseases is of the same magnitude of that generated by fire (Isaev 1997; Baranchikov et al. 2001; Isaev et al. 2001). In areas of insect and disease outbreaks, the trees are killed completely or partially, productivity and vitality of forests are substantially decreased, and large amounts of dead wood are accumulated. The latter increases flammability of damaged forests and frequency and severity of fires that slows down the natural regeneration.

Phyllophagous are the most harmful group of insects, particularly, in territories of concurrent interactions between different types of vegetation (e.g., the ecotone forest-steppe, or taiga-subtaiga forests). In zonal taiga vegetation types, insect outbreaks are initiated by special weather and landscape conditions and have a clear cycling character (e.g., Baranchikov 2011). As a rule, warmer and drier weather provokes large-scale outbreaks. For instance, the most dangerous Northern Eurasia forest insect – Siberian silk moth (*Dendrolimus superans sibiricus* Btl.) has favorable conditions for outbreaks in dark coniferous and pine forest, if the sum of active temperature exceeds 1,600 °C and radiation dryness index (RDI) by Budyko is >1.5; for larch forests the weather should be drier (RDI >2.0). Outbreaks of other dangerous insects (*Bupalus piniarius, Dendrolimus pini, Lymantria dispar, L. monacha*) are usual when the amount of heat >1,800–2,000 °C and RDI is 1.5–2.0 (Rozhkov 1965; Pleshanov 1982; Yurchenko et al. 2003). Under such conditions, outbreaks take on an eruptive (pulsating) character of the dynamics of insect populations, occupy significant areas and cause significant economic damage, lead to deep ecological transformation, and result in changes of composition and structure of forest cover (Epova and Pleshanov 1995).

The history of insect outbreaks in Asian Russia is impressive. In the southern taiga of Tomsk Oblast and Krasnoyarsky Kray, seven outbreaks of Siberian silk moth occurred during the period from 1878 to 1970, and the areas of dark coniferous forest that perished totaled more than 8×10^6 ha (Isaev 1997). The area of forests killed in Siberia and the Far East by this dangerous insect between 1880 and 1969 is estimated to be 13×10^6 ha with destroyed growing stock of two billion m^3. Six outbreaks were observed in territories of Irkutsk Oblast from 1870 to 1963, and *Pinus sibirica* forests were killed over an area of 1.06×10^6 ha with losses of 160×10^6 m^3 of wood (Rozhkov 1965). The outbreak of Siberian silk moss in East Siberia (1993–1997) impacted about one million ha and resulted in timber loss of $\sim 50 \times 10^6$ m^3 in highly productive dark coniferous forests (Isaev 1997). Much warmer than usual weather conditions during the last 3 years of the twentieth century provoked an outbreak of Siberian silk moth over a total area of almost ten million ha (2001), mostly in larch forests in the far north, where this pest was not typically observed.

Official data for the last decades shows an increase of the areas affected by biogenic factors. According to official statistics, the total area of outbreaks of insect and diseases (defined as forested area in which pathological die-back exceeds the natural mortality more than two times) in Russian forest under state forest management during the period from 1973 to 1987 ranged between 1.5×10^6 ha and 3.8×10^6 ha, with an average 2.73×10^6 ha (Isaev 1991, 1997). During the decade 1988–1997, the average annual area reported was at 1.55×10^6 ha with the seasonal variability of individual years from 1.43 to 2.28×10^6 ha. The average for 1998–2010 was 5.48×10^6 ha with the peak of 10.4×10^6 ha in 2001 (FAFMRF 2010). The yearly average area for Siberia is estimated at 2.9×10^6, with substantial variation – from 0.6×10^6 to 10.4×10^6 (Fig. 6.2).

Pine forests are damaged by pine silk moth (*Dendrolimus pini*), usually in southern regions of West and Central Siberia, and up to Mongolia, where it produces severe defoliation of pine stands in dry sites with poor sandy soils. Outbreaks occur after several drought years, and could last up to 7–8 years. Curtain dieback of trees begins after 2–3 years of damage, and complete dieback after a longer period. The biggest harm is observed in young planted pine forests. *Bupalus piniaris* significantly damages pine forests in critical site conditions (e.g., in forest steppe of Upper Priobie). Unpaired silkworm (*Lymantria dispar*) damages practically all deciduous and larch forests; often its outbreaks co-occur with other dangerous insects (e.g., *Denrolimus superans, Lymantria monacha*) (Kondakov 1987). Although under one-two defoliations, deciduous and larch forests restored their assimilation organs, decrease of current increment can reach 50 %. *Lymantria monacha* is a very heat-loving species, and its outbreaks happen only under the heat availability of GDD at about 2,200 °C, usually in the forest steppe of West Siberia, where single severe defoliations are usually observed.

Large-scale outbreaks of dangerous insects during recent decades were also reported for the north of the American continent (e.g., Kurz et al. 2008b; Volney and Fleming 2007). As a rule, similar to Siberia, the massive outbreaks were facilitated by the impacts of extremely warm and dry summers of several years in a row and milder than usual winters that together supported shortening the life cycle and fertility of insects. It is expected that projected climate change in the boreal zone will cause the increase of frequency and intensity of outbreaks in both direct (through direct impact of climate) and indirect (through decreasing resilience of ecosystems and disruption of community interactions between different components of biogeocenosis) ways (e.g., Stireman et al. 2005).

Other large-scale disturbance processes, like drying of large areas of forests, caused by a complicated combination of biotic and abiotic factors, are reported for different regions of the country. This is a rather typical process in Far Eastern spruce-fir forests. During recent decades, several waves of increasing dieback of trees were observed here. During the large wave of the second half of the 1960s, the areas of drying forests were estimated to be 5.5×10^6 ha in territories of the two administrative regions of the Russian Far East – Khabarovsk and Primorsk Krays (44 % of the total spruce-fir forests there) with the storage of dead wood of more than 360×10^6 m^3 (24 % of the total living and dead growing stock by the moment), with average storage of dead wood at about 100 m^3 ha^{-1} (Ageenko 1969). The next wave occurred between 1970 and 1980 in Sikhote-Alin where only in seven forest enterprises, 165,000 ha of forest died with a growing stock of 14 million m^3 (data of forest pathological survey of 1987–1988). The last wave of dryness was observed during the period from 1989 to 1993. Many hypotheses have been expressed attempting to explain this phenomenon, but there is no generally accepted answer (Manko and Gladkova 2001).

Air pollution, industrial destruction of sites and unfavorable weather conditions, wind and snowbreak also damage forests over large territories. As a rule, these data are not reported in any regular way. A special survey on impacts of air pollution (1991) indicated 321×10^3 ha of dead forests with 465×10^3 ha strongly disturbed due to this impact. The survey was substantially incomplete. Surveys in the area around Norilsk (the northern part of Krasnoyarsk Kray) in 1992 estimated 2×10^6 ha of dead forest-tundra landscapes, of which 650×10^3 ha of forests has been killed; another 1.0×10^6 ha forests were found with evident serious negative impact (Isaev 1991, 1997). Kharuk et al. (2006) estimated areas of forests declined in Siberia by pollution at $3–3.5 \times 10^6$ ha, – three to four times more than the officially reported data for all the country.

6.4 Carbon Budget of Siberian Ecosystems

There is a substantial set of publications considering different aspects of the carbon budget of ecosystems of Northern Eurasia for individual regions, different vegetation types and countries as a whole (e.g., for Russia). These publications used diverse methods, had different information background, and reported a wide range of estimates. In this section, we present the carbon budget for the territory delineated for this study (for the first time), as well as important results obtained by major methods of carbon accounting for parts of the region and sometimes – for comparison – for the entire country.

Current philosophy of ecosystem carbon accounting is based on two prerequisites: (1) a need to follow the major requirements of applied systems analysis, and (2) that a proper understanding of the role which terrestrial ecosystems play in the global carbon cycling requires a *verified terrestrial ecosystems full carbon account* (FCA). The FCA means (1) assessing all fluxes of major carbon-containing greenhouse gases, aerosols, and lateral carbon fluxes in the hydrosphere and lithosphere should include all ecosystems and processes in a spatially and temporally explicit way; (2) uncertainties are assessed comprehensively and transparently at all stages and for all modules of the account; and (3) the methodology used presents information how to optimally manage the uncertainties (Shvidenko et al. 2010b).

The FCA is considered a complicated, stochastic dynamic underspecified (fuzzy) system that cannot be directly verified due to obvious cost and resource limitations. Fuzziness of the FCA defines that any approach of carbon accounting, individually implemented, is unable to recognize structural uncertainties, and the usually reported uncertainties in essence represent only a part (i.e., "within model") of the real uncertainties. Thus, the FCA requires a systematic combination of the relevant methods (landscape-ecosystem method which covers bottom-up semi-empirical assessments of C pools and fluxes, direct measurements of C exchange with the atmosphere by eddy covariance, dynamic vegetation models (DVMs), and top-down inverse modeling) and harmonization and multiple constraints of independent results.

6.4.1 Carbon Account by Landscape-Ecosystem Approach

The landscape-ecosystem approach (LEA) is an attempt to implement major systems requirements to the FCA based on as comprehensive as possible description of land and ecosystems and the use of all related empirical data and models. This allows for consideration of the LEA as a basis for designing carbon accounting schemes and as a background for comparative analysis. The LEA includes a relevant combination of pool-based and flux-based approaches. The pool-based method estimates the change of ecosystems' carbon pools for a definite period of time. The flux-based approach is used as a chain calculation:

$$NECB = NPP - HR - D - L$$

where NECB denotes net ecosystem carbon balance (or net biome production, NBP), NPP – net primary production, HR – ecosystem heterotrophic respiration, D – flux due to disturbances (including consumption of plant products), L – lateral fluxes to the hydrosphere and lithosphere. HR includes heterotrophic soil respiration and the flux due to decomposition of dead wood. Sign of NECB in this study is used according to the micrometeorological convention: "minus" means carbon sink (uptake from the atmosphere) and "plus" means carbon source (emission to the atmosphere).

The pool-based method has substantial practical limitations due to high uncertainties of the size and dynamics of some pools, particularly soils. An Integrated Land Information System that includes (1) a hybrid land cover (HLC) and (2) numerous attributive datasets of available measurements serves as an information background of the LEA. The HLC for the Russian territory was developed based on the system integration and harmonization of multi-sensor remote sensing products (GLC-2000, MODIS VCF, AVHRR, LANDSAT TM, ENVISAT ASAR, others), available on-ground data (e.g., State Land Account, State Forest Account), and other appropriate information. The downscaling and parameterization of the HLC have been done for each 1-km pixel using a special optimization algorithm. Details of the approach are considered in Schepaschenko et al. (2010).

Sets of different empirical models were used to assess major components of the FCA (NPP, HR, D, and L) by land classes of the HLC. Taking into account that some initial data, both official statistics (e.g., areas of forest fire, illegal harvests) and results of measurements (e.g., NPP for forests) are biased or obsolete, corresponding corrections have been done based on independent sources of information. NPP for forests has been assessed based on a new, presumably unbiased approach (see Sect. 6.2).

In order to asses HR, a special model has been developed (Schepaschenko et al. 2011). Based on measurements of HR in situ, the model provides corrections by zonal specifics of soil types, land uses, and vegetation classes. While the fluxes due to disturbances were defined by a fairly common approach (as a product of the disturbed area and amount of consumed [transformed] carbon), the calculation schemes accounted for the specifics of each individual type of disturbance (e.g., Kajii et al. 2002). The consumption of plant products (agriculture, forest) was calculated based on official statistical data including imports and exports.

The total HR is estimated at 2.09 Pg C year^{-1}, or ~76 % of NPP (Tables 6.8–6.11). HR of two land classes (disturbed forests and open woodlands) exceeds NPP. Consequently, all these ecosystems are unlikely to be in equilibrium. Litter in these semi-natural ecosystems decomposes slowly, e.g., Mukhortova and Evgrafova (2005), so that fire and other disturbances are a major regulator of the amount of dead organic material and substantially affects soil respiration. Additional imbalance is caused by removal of harvested products in forests and agriculture. Independent estimates in the northern taiga of Central Siberia by Klimchenko (2007) estimated heterotrophic soil respiration consistent to data reported by Quegan et al. (2011) – between 73 and 59 % of NPP. Similarly, Wirth et al. (2002) estimated 48 % for postfire chronosequences in pine forests of Siberian middle taiga, although reported values for NEP (or NEE) for taiga forests of Central Siberia vary greatly – from 50–60 to 250–270 gC m^{-2} year^{-1} (e.g., Röser et al. 2002; Shibistova et al. 2002; Schulze 2002). Substantial differences between NPP and soil respiration can occur, particularly for crops, since a major part of the NPP is removed as yield, e.g., the sink of agro-ecosystems of the taiga zone of the region was estimated to be

Table 6.10 Heterotrophic soil respiration by bioclimatic zone and land classes

Zone	Heterotrophic respiration, Tg C year⁻¹ by land classes						
	Forest	Agro	Burned	Wetland	Grassland	Open woodland	Total
Arctic					0.2		0.2
Tundra	14.2	0.1	1.4	28.3	108.9	5.7	158.6
FT, N & SpT	133.0	0.2	3.1	41.5	36.9	27.6	242.4
Middle taiga	700.2	31.9	26.7	93.5	202.4	43.3	1,097.9
Southern taiga	227.1	42.3	2.2	49.6	19.6	2.1	342.9
Temperate forest	18.1	16.2	0.1	2.6	2.7	0.2	39.9
Steppe	24.9	126.2	0.4	17.6	10.6	0.3	180.0
Semidesert and Desert	2.5	8.2	0.0	2.3	10.7	0.0	23.8
Total	1,120.0	225.2	33.9	235.4	391.9	79.2	2,085.6

Table 6.11 Average heterotrophic soil respiration by bioclimatic zone and land classes by area unit

Zone	Heterotrophic respiration, g C m⁻² year⁻¹ by land classes						
	Forest	Agro	Burned	Wetland	Grassland	Open woodland	All land classes
Arctic					59		59
Tundra	121	166	91	94	99	91	99
Sparse taiga	129	215	107	170	155	114	136
Middle taiga	184	308	174	258	263	122	198
Southern taiga	254	377	399	396	483	275	289
Temperate forest	326	339	325	440	394	307	341
Steppe	257	347	375	634	397	271	348
Semidesert & Desert	229	198	249	373	259	223	238
All zones	186	336	166	219	176	119	192

111 gC m⁻² year⁻¹, or 42 % of NPP (Vedrova 2002). Such effects usually are not represented in DGVMs.

Temperature and precipitation are considered as the best predictors of the annual and seasonal dynamics of the soil respiration rate (Raich et al. 2002). The high positive correlation between CO_2 emission and soil temperatures was found in natural and agricultural ecosystems of the Russian taiga zone (Kudeyarov and Kurganova 1998). Soil temperature and soil moisture are considered the crucial environmental factors controlling soil surface carbon dioxide exchange rate. These factors interact to affect the productivity of terrestrial ecosystems and the decomposition rate of soil organic matter (Feiziene et al. 2010).

On the global scale, Raich and Schlesinger (1992) have found that soil respiration rates are positively correlated with mean annual air temperatures and mean annual precipitation. A close correlation exists between mean annual net primary productivity (NPP) of different vegetation biomes and their mean annual soil respiration rates.

Evidently, extrapolation of decomposition rates into a future warmer world based on observations of current apparent temperature or moisture sensitivities may not be adequate. Rather, it is necessary to understand how substrate availability will change and how a changing set of environmental constraints to decomposition in a future climate will determine the future apparent temperature and climate sensitivity of decomposition (Davidson and Janssens 2006).

Hence, it seems likely that soil organic C will decrease with increasing temperature due to climate change (e.g., Shaver et al. 1992). This knowledge has been incorporated by many authors (Schimel et al. 1990; Jenkinson et al. 1991; Thornley et al. 1991; Kirschbaum 1993) who used soil organic matter models to show how a future temperature increase could lead to the release of large quantities of C from the world's soils. Gifford (1994) on the other hand, conducted a similar analysis, but concluded that there should be no loss of C with increasing temperature. However, the most critical factor in all these analyses is the relativity between the climate response functions of net primary productivity and soil organic matter decomposition. There is a general expectation that increasing temperature leads to increases in both net primary productivity which provides the input to soil organic C, and the rate of soil organic matter decomposition which determines the loss of soil organic C. The critical question is whether NPP or organic matter decomposition rate is stimulated more by increasing temperature (Kirschbaum 1995).

Fire carbon emissions (for 2009) have been defined based on data of V.N. Sukachev Institute of Forest (see Sect. 6.3) by the methods described in Kajii et al. (2002) and Shvidenko et al. (2011), using also an emissions ratio by biomes and type of vegetation collected by Andrea (2010, personal communication) (Table 6.12). By the area, this year was relatively favorable by fire danger. Of the total 6.57×10^6 ha, about 50 % of burned area was in forest. The clear "spring – early summer" fire regime of this year defined the relatively small fire C emissions – 91.2 Tg C, of which 55 % were in forest and 22 % on wetlands. Almost 80 % of the total C emissions were in the form of CO_2. Interannual variation of fire carbon emissions is large – from ~40 to 220 Tg C year^{-1} (Shvidenko et al. 2011).

Table 6.13 contains a net ecosystem carbon balance of Siberian ecosystems. Overall, Siberian terrestrial ecosystems serve as a carbon sink of 349 Tg C year^{-1}, or 32.5 g C m^{-2} year^{-1}. If consumption of plant products is included, the sink decreases to 245 Tg C year^{-1}. Practically all of the sink is provided by forests. Disturbed forests are a clear source. On average, grasslands are neutral. However, tundra grasslands become a small source. Figure 6.3 presents spatial distribution of NECB across the region.

NECB(NBP):NPP ratio for all vegetation classes is ~0.11. The overall C sink is basically provided by forests (-71 ± 25 g C m^{-2} year^{-1}) which is very close to the long-term carbon sink of EU-25 forests – 75 ± 20 g C m^{-2} year^{-1} (Luyssaert et al. 2010), although major drivers of that are different. Within agricultural land, cultivated arable land served as a C source (about +40 g C m^{-2} year^{-1}). This flux is almost completely compensated by increasing carbon sequestration on abandoned agricultural land. Large areas on permafrost were also estimated as a weak source, generally considered as neutral. Uncertainties ("within an approach"

Table 6.12 Carbon emissions 2009 in Siberia by products of burning

Vegetation	Area, 10^3 ha	Emission, 10^3 t C	Including the main emission species, 10^3 t C							
			CO_2	CO	CH_4	NMHC	OC	BC	$PM_{2.5}$	TPM
Forest	3,390.7	50,961.9	42,770.6	4,278.7	519.1	367.8	514.7	51.0	553.1	805.7
Arable	235.0	337.0	292.3	25.5	3.6	3.2	1.7	0.3	2.7	4.4
Hayfield	322.9	1,050.1	903.9	83.7	12.5	10.4	5.3	1.1	8.7	13.9
Pasture	517.4	1,583.4	1,379.0	116.5	16.4	14.3	7.9	1.6	12.2	20.1
Fallow	92.7	273.1	239.0	19.4	2.6	2.4	1.4	0.3	2.1	3.4
Abandoned arable	232.4	801.5	699.5	58.0	8.0	7.2	4.0	0.8	6.1	10.1
Wetland	666.8	20,193.5	17,567.0	1,456.9	260.3	117.2	119.2	20.2	137.4	218.2
Open woodland	245.7	5,683.8	4,698.4	530.9	71.1	39.4	56.4	5.7	66.0	94.4
Disturbed forest	99.9	1,395.2	1,180.2	108.2	12.1	10.6	14.8	1.4	14.7	21.7
Grassland	769.8	8,914.7	7,862.0	568.7	96.6	48.9	51.2	8.9	57.9	93.5
Total	6,573.2	91,194.1	77,592.0	7,246.5	1,002.4	621.4	776.6	91.2	860.7	1,285.4

Table 6.13 Net ecosystem carbon balance of Siberia in 2009

Indicators	Total, Tg C year⁻¹	Fluxes by land classes, Tg C year⁻¹					
		For	Agro	Burned	Wetland	Grassland	Open woodland
NPP	2,756.9	1,702.9	269.3	28.1	296.4	404.2	55.9
HR	2,085.6	1,119.9	225.2	33.9	235.4	391.9	79.2
Fire	91.2	51.0	4.0	1.4	20.2	8.9	5.7
Other disturbances	53.9	53.9					
Decomposition	133.9	106.0		12.4	2.3	11.4	1.8
Lateral flux	43.2	24.2	2.7	0.8	4.3	8.5	2.7
Subtotal	−349.1	−347.9	−37.4	+20.4	−34.2	+16.5	+33.5
Consumption	104.6	12.0	85.1		4.0	3.5	
Grand total	−244.5	−335.9	+47.7	±20.4	−30.2	+20.0	+33.5

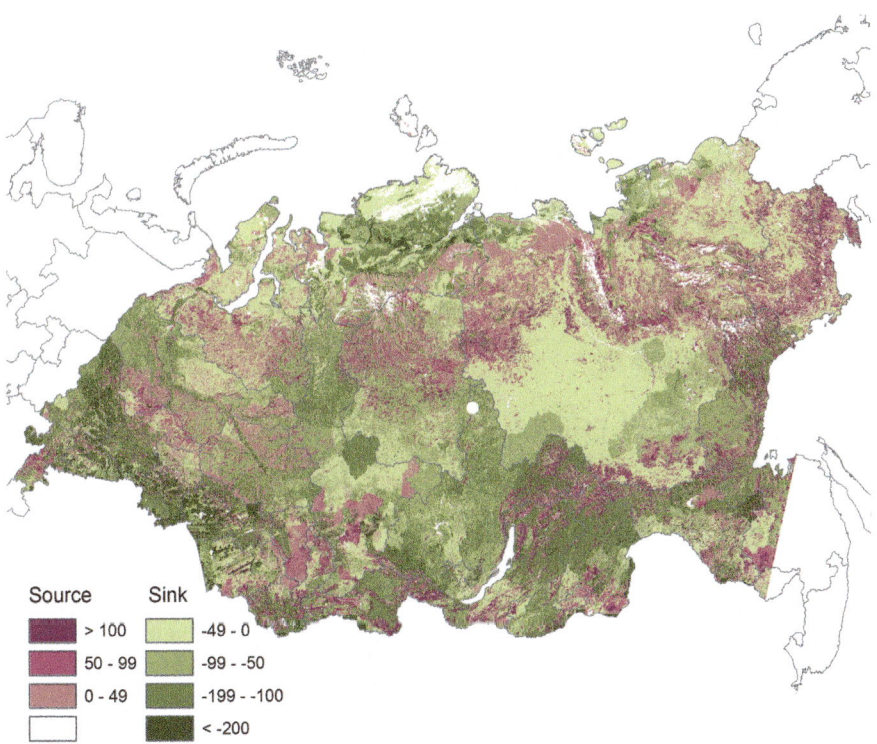

Fig. 6.3 Net ecosystem carbon balance for 2009, g C m⁻² year⁻¹

estimate, the confidential interval, CI, is 0.9) of major fluxes, such as NPP and HR, are estimated at ±7–12 %.

Recently, Pan et al. (2011) within the global estimate reported change of carbon stock in forests of Asian Russia based on the pool-based approach. They estimated forests of Asian Russia as the net carbon sink at 260 Tg C year^{-1} on average for 1990–2007. This estimate is consistent with data of Table 6.13 (−244.5 Tg C year^{-1}).

Overall a similar character of the NECB has been reported for Central Siberia (Quegan et al. 2011). This study presented five independent estimates of the full carbon account of a large region of Central Siberia (about 307 million ha in boundaries of Krasnoyarsk Kray and Irkutsk Oblast) using three different methodologies: a landscape-ecosystem approach, two DVMs – Sheffield DGVM (Woodward and Lomas 2004) and LPJ (Sitch et al. 2003; Gerten et al. 2004), along with two atmospheric inversions. Apart from one of the DVMs, all methods produce estimates of the NECB that are consistent both among themselves and with a range of other estimates. They indicate the region to be a carbon sink with a NECB of 27.1 ± 7.4 gC m^{-2} year^{-1}, which is equivalent to 347 ± 95 TgC year^{-1} if considered representative for boreal Asia. This is comparable with fossil fuel emissions for the Russian Federation, currently estimated as 427 TgC year^{-1}, but implies that boreal Asia does not play the major role in the northern hemisphere land sink, typically estimated to be of magnitude 1.5–2.9 Pg C year^{-1}. LEA and DVM produce substantially different partitioning of NBP into its component fluxes. The DVMs find NPP to be nearly balanced by soil respiration, disturbance being a relatively small term pushing the system closer to equilibrium. Due to the LEA, soil respiration is significantly less than NPP, and disturbance plays a much larger role in the overall carbon balance. This study concluded that the use in the LEA of observationally based estimates of soil respiration and a more complete description of disturbance fluxes suggests that the partitioning derived by the LEA is more credible, and that improved process descriptions and constraints by data are needed in the DVMs.

6.4.2 Analysis of Carbon Balance of Boreal Asia Based on Terrestrial Ecosystem Model

Based on the process-based Terrestrial Ecosystem Model (TEM, Hayes et al. 2011), McGuire et al. (2010a, b) analyzed the C balance of boreal Asia for 1997–2006. The boreal Asia region is defined in boundaries settled by the TransCom 3 model experiments (Gurney et al. 2002). The TEM considers the effects of a number of factors on its simulations of C dynamics including changes in atmospheric CO_2 concentration, tropospheric ozone pollution, nitrogen deposition, climate variability and change, and disturbance/land use including fire, forest harvest, and agricultural establishment and abandonment. TEM also calculates pyrogenic emissions of CO_2, CH_4, and CO from the combustion of vegetation and soil carbon in wildfires. The dissolved organic carbon (DOC) leaching dynamics of TEM are a function of soil C decomposition rate, soil DOC concentration, and water flux through the soil. The methane dynamics module (MDM-TEM) was used to estimate the exchange of CH_4 with atmosphere of both wetlands, which generally emit CH_4 to the atmosphere, and uplands, which generally consume CH_4 from the atmosphere. The MDM-TEM assesses the effects of a number of factors on its simulations of CH_4 dynamics including the area of wetlands, fluctuations in the water table of wetlands, temperature, and labile carbon inputs into the soil solution derived from the NPP estimates of TEM.

The results of these simulations were compared with estimates of CO_2 and CH_4 exchange from atmospheric inversion models and with observations of terrestrial C export from Arctic watersheds. The simulated transfer of land-based C to the Arctic Ocean was compared against estimates based on a sampling of DOC export from major Arctic rivers (McClelland et al. 2008). The land-atmosphere CO_2 exchange estimate was compared with results from the TransCom 3 atmospheric inversion model intercomparison project (Gurney et al. 2008), and CH_4 to results from atmospheric inversion-estimated surface emissions (Chen and Prinn 2006). To compare the "bottom-up" results from our model simulations with the "top-down" estimates from these inversion studies, we summarize our estimates of surface-atmosphere CO_2 and CH_4 exchange for the land area matching the boreal Asia region defined in the TransCom 3 model experiments (Gurney et al. 2002).

The TEM estimates a net uptake of 42.7 Tg C year^{-1} as CO_2 from the atmosphere to the land area of boreal Asia (limited by basins of large Siberian rivers) over the 1997–2006 time period (Fig. 6.4). This terrestrial sink is the result of a positive uptake of 302 Tg C year^{-1} by the terrestrial component as net ecosystem production (NEP), the balance between NPP (3,260 Tg C year^{-1}) in vegetation and the decomposition of soil organic matter through heterotrophic respiration (HR 2,958 Tg C year^{-1}). However, NEP during this time period was substantially offset by the release of C as CO_2 from fires across the region (255 Tg C year^{-1}) and the decomposition of agricultural and forestry products (4.3 Tg C year^{-1}). Nearly 70 % of the pyrogenic CO_2 emissions is attributed to the combustion of soil C (178 Tg C year^{-1}), which when added to the 75 Tg C year^{-1} gain to the soil from the difference between litter inputs from vegetation and HR and to other losses from net biogenic CH_4 emissions and DOC export, results in a net loss of 202 Tg C year^{-1} from the Arctic Basin soil C pool between 1997 and 2006. The loss of soil C is estimated to be greater than the net gain in vegetation C simulated in this study (118 Tg C year^{-1}), resulting in an estimated overall loss of 84 Tg C year^{-1} in ecosystem C stores in the land area of boreal Asia over this time period.

The estimates of land-atmosphere CO_2 exchange in this study are generally consistent with the range of uncertainty in estimates from the atmospheric perspective, although the TEM estimates differ overall in the direction of flux compared to the model means of the TransCom 3 results. For boreal Asia, TransCom model means suggest a net sink ranging from 30 to 470 Tg C year^{-1} (±33 to 48 Tg C year^{-1}), while TEM estimates of Net Ecosystem Exchange (NEE) range from −90 to 0 Tg C year^{-1}, depending on the time period. Monthly NEE estimates from TEM match the seasonal patterns in the TransCom estimates and are within the range of uncertainty for most months during the 1997–2006 time period. The exceptions occur primarily during the peak drawdown of the summer months, where forward process-based models typically predict less uptake relative to inverse models (Pacala et al. 2009). The differences in peak uptake between the two estimates are greatest in large fire years when TEM estimates large fire emissions, e.g., 2002–2003 in boreal Asia,

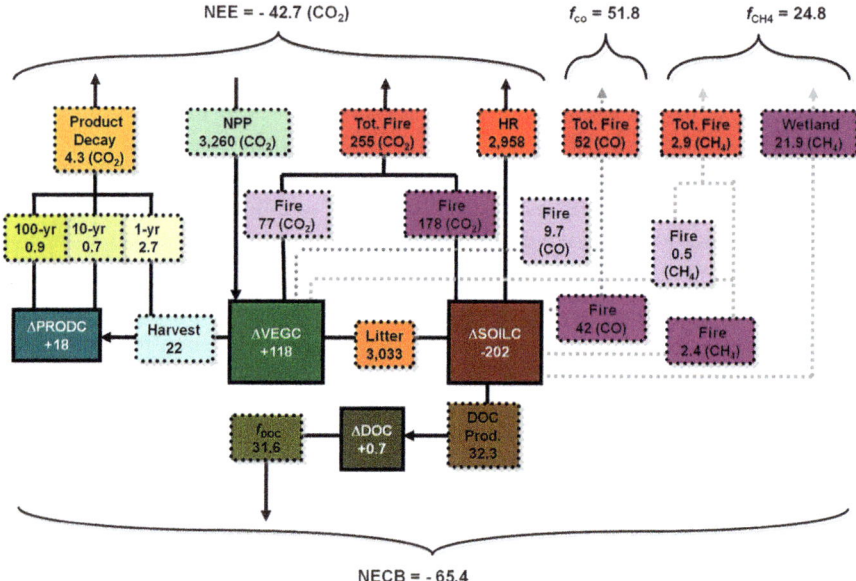

Fig. 6.4 Simulated carbon stock changes and fluxes (Tg C year⁻¹) of the terrestrial ecosystems of boreal Asia, 1997–2006 (Source: McGuire et al. 2010a, b)

although the TransCom estimates do show some negative effect on land uptake during those years.

Although the area of this simulation by TEM is somewhat different from that used in the LEA assessment of major components of carbon cycle above (Tables 6.8 and 6.10), comparison of these two results points out some specific features of applications of DGVMs to polar and boreal domains. On average, estimates of NPP by TEM, like other DGVMs, seem unbiased and relatively closed to empirical data while modeled HR for these regions is higher. The reason for that is likely a balance between NPP and HR which in one way or another is hypothesized in major DGVMs, but such a balance does not exist in northern ecosystems with a slow decomposition rate and regulating role of fire.

The combination of the biogenic MDM-TEM and the pyrogenic estimates of CH_4 fluxes indicate that the terrestrial areas of boreal Asia annually released approximately 41.5 Tg CH_4 to the atmosphere between 1997 and 2006, with most of the emissions (38.0 Tg CH_4 year⁻¹) from biogenic sources. There was substantial interannual variation of emissions with the peak emissions in 1998 and lowest emissions in 1997. TEM simulations indicate that emissions in boreal Asia increased ($R=0.66$; slope $=0.6$ Tg CH_4year⁻¹; $n=11$; $p=0.03$) during the analysis period. Over the decade of analysis, there were very weak trends for increases in air temperature (0.6 °C; $R=0.33$; $p=0.3$) and simulated soil temperature (0.14 °C; $R=0.33$; $p=0.3$) in boreal Asia that could have stimulated methanogenesis. While interannual variability in emissions in boreal Asia was not significantly correlated with changes in soil temperature ($R=0.25$; $n=11$; $P=0.45$), it was marginally correlated with precipitation variability ($R=0.52$; $n=11$; $p=0.09$). Although precipitation seems to control interannual variability in CH_4 emission on boreal Asia, the increase in simulated water table depth ($R=0.14$; slope$=0.10$ mm year^{-1}; $p=0.67$) is weaker than the increases in soil temperature.

The monthly emissions for 1996–2001 estimated by Chen and Prinn (2006) using a 3D model inverse method for the North American and Eurasian Arctic sectors with the MDM-TEM estimates were compared (McGuire et al. 2010a, b). The inversions show well-defined seasonal cycles peaking in July–August. Significant inter-annual variations in these peaks typically ±4 TgC year^{-1} in boreal Asia occur. The annual average emissions and timing and average amplitudes of the seasonal cycles were in reasonable agreement with the process-based estimates of CH_4 exchange. The monthly emissions agreed well between inversion and process-based modeling estimates in boreal Asia ($R=0.90$; slope$=0.65$; $n=60$; $p<0.001$).

The TEM estimated the DOC delivery to the ocean from boreal Asia to be 31.6 Tg C year^{-1} between 1997 and 2006, which is consistent with empirically based estimates of DOC delivery to the ocean. About a third of the total DOC exported to the Arctic Ocean each year is contributed by rivers emptying into the Kara Sea. Boreal needleleaf deciduous forests and forested wetlands had the largest loss rate of DOC, averaging about 6 gC m^{-2} year^{-1}, and the analysis suggested that these biomes in the Kara Sea and Laptev Sea watersheds are responsible for over 61 % of the terrestrial DOC (22.2 Tg C year^{-1}) delivered to the Arctic Ocean each year. Between 1997 and 2006, this analysis did not indicate any trend in DOC export.

This analysis indicated that from 1997 to 2006, the terrestrial areas of boreal Asia annually gained 42.7 Tg C as CO_2 from the atmosphere, lost 31.6 Tg C as DOC to the ocean, lost 24.8 Tg C as CH_4, and lost 51.8 Tg C as CO to the atmosphere for a total negative carbon balance of 65.4 Tg C year^{-1} (Fig. 6.4). While boreal Asia is estimated to be a sink for C, it can generally be considered as a source for greenhouse gas forcing based on the conversion of net CH_4 emissions to CO_2 equivalents. Between 1997 and 2006, these simulations estimated that the region was a net source of 33.1 Tg CH_4 year^{-1}, a magnitude that is comparable with the atmospheric inversion of CH_4 that was presented in this study and within the uncertainty of other CH_4 inversions (McGuire et al. 2009). This magnitude of CH_4 emissions is equivalent to 761.3 Tg CO_2 year^{-1} calculated on a 100-year time horizon for which 1 g of CH_4 is equivalent to 23 g of CO_2 in terms of global warming potential (IPCC 2001). This is over 600 Tg CO_2 greater than the net CO_2 sink of 156.6 Tg CO_2 year^{-1} (42.7 Tg C–CO_2 year^{-1}) estimated by our simulations. A key issue is whether the global warming potential of the Arctic Basin is evolving to be a greater sink or source of greenhouse gases.

The net uptake of atmospheric CO_2 simulated for the terrestrial ecosystems of the boreal Asia between 1997 and 2006 is a result of the net uptake in vegetation C through NPP being greater than the release of vegetation C to the atmosphere through fire and harvest and the release of soil C through decomposition and pyrogenic CO_2 emissions (Fig. 6.4). The net increase in vegetation C is primarily associated with the positive effects of CO_2 fertilization, N deposition, and climate (e.g., effects of longer and warmer growing seasons on NPP). The net release of soil C to the atmosphere is a function of warming on decomposition and the combustion of C in more frequent fires. Although not part of NEE, any effect of climate variability that increases biogenic emissions of CH_4 and the leaching of DOC can also contribute to a decrease in the soil C pool. To understand how the effects of these various controlling factors interact to produce the contemporary (1997–2006) dynamics of land-atmosphere CO_2 exchange, it is necessary to compare the importance of the various factors influencing terrestrial CO_2 exchange between the analysis period of this study to previous decades when various studies have suggested that high-latitude terrestrial ecosystems acted as a stronger CO_2 sink (McGuire et al. 2009).

The historical simulations of the C balance of boreal Asia using TEM indicate a strengthening atmospheric CO_2 sink in tundra ecosystems (Fig. 6.5a), and a strengthening source from boreal forest ecosystems (Fig. 6.5b). The results suggest that CO_2 fertilization effects on NPP are playing the primary role in the C uptake component for both tundra and boreal forest ecosystems, and that these effects are increasing along with the rise in atmospheric $[CO_2]$ over the past several decades. However, the effect of CO_2 fertilization on NEE has been dampened in tundra ecosystems with a weakening sink/increasing source effect from changes in climate since the 1970s. The source effect of climate variability in tundra ecosystems suggests that increases decomposition have been outpacing any warming-driven increases in NPP. The increases in decomposition in the simulations are a function of greater microbial activity in response to warmer temperatures, as well as a result of more soil C available for decomposition from increasing active layer depths due to permafrost degradation. Other studies suggest that the microbial decomposition of previously frozen organic matter can overcome uptake from increased vegetation productivity to alter the C balance in tundra ecosystems over decadal time-scales (Schuur et al. 2009). The simulations show an increasing net C sink in the boreal forest ecosystems of boreal Asia prior to the 1970s that has been trending toward a weakening sink/increasing source since the 1970s as a result of climate and disturbance effects. In the last decade of analysis, it is the large impact on NEE (the increased source) from fire that is driving the net C source from boreal forests.

Similar to previous findings that the regional net CH_4 emissions have increased during the twentieth century (Zhuang et al. 2004), this simulation indicates that emissions continued to increase between 1996 and 2006. While precipitation seems to control interannual variability in CH_4 emission on boreal Asia, the increase in simulated water table depth is weaker than the increases in soil temperature over the decade. Estimates of CH_4 emissions by the MDM-TEM are very sensitive to increases in soil temperature, which increase methanogenesis, and changes in water table position, which affect the amount of soil carbon subject to methanogenesis – an anaerobic process (Zhuang et al. 2007). Thus, the slight regional increase of CH_4 emissions appears mainly due to the increasing soil temperature.

A previous analysis with MDM-TEM suggests that climate change has the potential to substantially increase biogenic CH_4 emissions throughout northern high latitudes during the twenty-first century (Zhuang et al. 2006). Another analysis focused

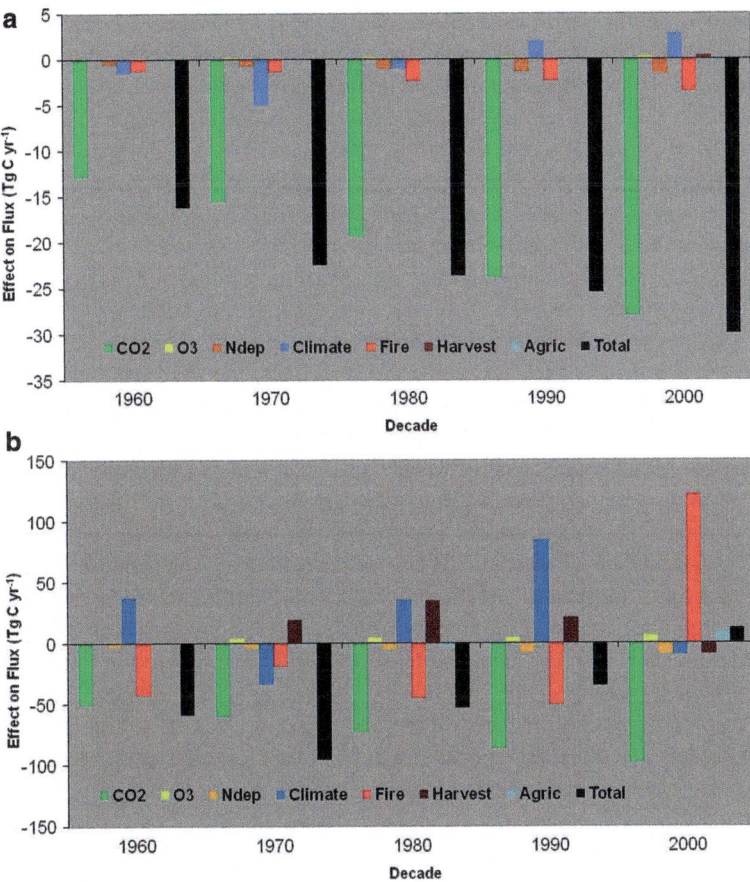

Fig. 6.5 Total and individual average annual effects (Tg C year^{-1}) of temporal variability in atmospheric [CO_2], tropospheric O_3 levels, N deposition rates, climate, fire, forest harvest and agricultural establishment and abandonment on NEE for each decade since the 1960s across (**a**) tundra and (**b**) boreal forest ecosystems in boreal Asia (Source: McGuire et al. 2010a, b)

on Alaska indicates that methane emissions can potentially double by year 2100, and that the temperature sensitivity of methanogenesis dominates over the water table sensitivity (Zhuang et al. 2007). It is also expected pyrogenic CH_4 emissions to increase with climate change if fire becomes more frequent in northern high latitudes as projected by several studies (Flannigan et al. 2009; Balshi et al. 2009).

Although no trends in DOC export occur during the 1997–2006 study period, the simulations indicate that DOC delivery to the ocean has been increasing by 0.047 Tg C year^{-1} over the twentieth century throughout northern high latitudes. Based on a factorial simulation experiment, most of this increase (0.039 Tg C year^{-1}) has been the result of climate variability and change. The response of DOC export to climate variability and change is primarily a result of changes in decomposition rates over this time period whereas the response to land-use change and fire is primarily caused by changes in the availability of the soil organic carbon substrate. As described earlier, warming increases decomposition to enhance the production of DOC from soil organic matter and the amount of DOC available for leaching to the neighboring river networks. Warmer temperatures also increase the active layer depth due to permafrost degradation to influence DOC export. This exposes more soil organic matter to decomposition and hence produces more DOC. Thus, permafrost degradation accentuates the effects of warming on the DOC concentration of runoff in our simulations. These results suggest that the delivery of DOC to the Arctic Ocean will likely increase as the climate warms.

The study indicates that between 1997 and 2006, the terrestrial ecosystems of boreal Asia were a source of carbon to the atmosphere and to the Arctic Ocean (McGuire et al. 2009, 2010a, b). Because of CH_4 emissions of wetlands, boreal Asia continues to act as source of greenhouse gas forcing to the warming of the Earth. These analyses suppose that CH_4 emissions are increasing, and other studies have indicated that CH_4 emissions are expected to generally increase with continued warming of the Arctic (Zhu et al. 2011). It appears that the sink for atmospheric CO_2 in boreal Asia is diminishing because of more fire disturbance in this decade compared to previous decades. Continued warming of the Arctic is expected to substantially increase fire activity. The results of this study emphasize the importance of analyzing the dynamics of the C balance of boreal Asia as a linked system of CO_2 and CH_4 exchange with the atmosphere and delivery of DOC to the ocean. Earth system models that consider carbon-climate feedbacks should treat the carbon processes of boreal Asia as a linked system.

6.4.3   Eddy Covariance

Systematic long-term annual eddy covariance measurements and accompanying climatology in Siberia became possible in the framework of the EuroSiberian Carbon Flux (1998–2000) and TCOS-Siberia (2000–2002) projects initiated by the Max-Planck Institute of Biogeochemistry, Jena, Germany.

The first results of net ecosystem exchange of CO_2, water, and heat in various ecosystems in Siberia for the 1998 growing season were summarized by Schulze et al. (1999) and later, for 1998–2000 and 2002–2004, were overviewed by Tchebakova et al. (2011). Seasonal and annual dynamics of the carbon exchange in a 200-year pine forest, a *Sphagnum* bog, a true grass steppe and a tussock tundra, and possible feedbacks of the CO_2 exchange to the atmosphere in a changing climate were evaluated.

Net ecosystem exchange of CO_2 (NEE) in various ecosystems is related to sensible heat and thus, the Bowen ratio (β), the ratio between sensible and latent heat (Karelin and Zamolodchikov 2008; Schulze et al. 1999). Maximal fluxes of CO_2 of 10–12 μmol m^{-2} s^{-1} were usually found in the afternoon and in midsummer with the

Table 6.14 Annual carbon balance components of major ecosystems in central Siberia

	Forest[a] (1998–2000)	Bog[b] (1998–2000)	Steppe[c] (2002–2004)	Tundra[d] (2004)
Max. CO_2-flux, μmol m^{-2} s^{-1}, July	10	6	12	8
Net ecosystem exchange of CO_2, gC m^{-2} year^{-1}	−156	−52/−30[e] (200 days)	−122	−38 −6[e]
Ecosystem respiration, gC m^{-2} year^{-1}	372	170 (200 days)	375	201[e]
Ecosystem assimilation, gC m^{-2} year^{-1}	−534	−222 (200 days)	−497	−207[e]

[a]Tchebakova et al. (2002)
[b]Arneth et al. (2002)
[c]Belelli-Marchesini (2007)
[d]Corradi et al. (2005)
[e]Karelin and Zamolodchikov (2008)

maximum Bowen ratios in the forest and steppe where $\beta > 1$ and 6–8 μmol m^{-2} s^{-1} in the bog and tundra with the less intensive energy exchange above water surfaces $\beta < 1$.

Daily NEE depends on meteorological parameters like PAR (photosynthetically active radiation), temperature of both the air and soil, and the air humidity. In winter, early spring, and late fall, NEE is slightly positive. During the growing season, the NEE course is distinct: it is increasing from spring and reaches its maximum in the midsummer and then is decreasing by fall. In mid-summer, NEE increases after sunrise, changes its sign from the positive to negative at 6 a.m., and reaches its maximum around noon, and then decreases and changes its sign from negative to positive about 9 p.m.

Table 6.14 summarizes the main components of the carbon balance of the four ecosystems studied. Net ecosystem exchange (NEE) is close to the net ecosystem production (NEP) in annual and seasonal terms (Schmid et al. 2000). The tussock tundra in East Siberia was a carbon sink of−38 gC m^{-2}year^{-1} while the southern tundra in Taimyr were a minor carbon sink of −10 gC m^{-2}year^{-1} (Karelin and Zamolodchikov 2008; Zamolodchikov et al. 2003). The pine forest in middle taiga was a sink of −156 gC m^{-2}year^{-1} while the Sphagnum bog was a sink of only −29 gC m^{-2}year^{-1} (Arneth et al. 2002). The true steppe in Khakassia was a significant sink at −122 gC m^{-2}year^{-1}. In the annual course, tundra started working as a carbon sink in June, the forest and the bog from late May or early June, and the steppe from the end of April. In the fall, these ecosystems become a source: tundra from the beginning of September, the pine forest and the bog from the beginning of October, steppe from the end of October.

Averaged for 1999–2000, the annual NEP of CO_2 was equal to 156 gC m^{-2} (13 mol m^{-2}), and the annual ecosystem respiration (summed from respiration in the growing season as 312 gC m^{-2} modeled from the air temperature and respiration of 66 gC m^{-2} in winter measured by the eddy covariance method) resulted in a total photosynthetic production of 534 gC m^{-2} (Tchebakova et al. 2002). Lloyd et al. (2002) estimated annual NEP for this pine forest to be 174 gC m^{-2} (14.5 mol m^{-2}) and ecosystem respiration to be 384 gC m^{-2} totaling 558 gC m^{-2} of photosynthetic production.

The annual NEP of the pine forest was 156 gC m^{-2} which fell within 90–210 gC m^{-2} of the NEP of boreal conifer forests in North America and western Europe measured by the eddy covariance method (Goulden et al. 1996; Valentini et al. 2000). Schulze et al. (1999) and Wirth et al. (1999) evaluated the NPP of this particular pine forest from inventory measurements correspondingly 123 and 144 gC m^{-2}. Bazilevich (1993) and Monserud et al. (1996) found the NPP of pine forests in the middle taiga of West Siberia to be 206 gC m^{-2} and to be 161 gC m^{-2} in Central Siberia. Comparative results were received by eddy covariance measurements on two forested sites in European southern taiga (Kurbatova et al. 2002).

As a result of low respiration, gross primary production (GPP) is 1.5–2 times lower than that of conifer forests in Western Europe (Valentini et al. 2000). From 25 various forest ecosystems in western Europe cited in this study, only 3–4 forests from the harsh climates of Iceland and highlands of the Alps in Italy produce about the same NEP as the pine forest in the Siberian middle taiga at the latitude of 60°N. Interestingly, despite the different weather conditions of 1998 and 2000, NEE varied insignificantly – only 4 % of the annual value. Anthoni et al. (1999), for instance, noted that the difference in NEP in years with El Niño and without may reach 90 %.

As the climate warms and gets drier in the future, studied ecosystems may turn from a sink to a significant source. Zamolodchikov et al. (2003) pointed the temperature 14 °C as a limit above which photosynthesis in tundra is inhibited and respiration simultaneously increases, together promoting additional carbon emissions into the atmosphere enhancing climate warming. In a warmed and dried climate, carbon emission from bogs increases: as carbon dioxide from drying bogs (Morishita et al. 2003) or as methane from watered bogs. In current climate methane emissions from the cryolithozone accounts some 6 % (Karelin and Zamolodchikov 2008).

In the pine forest, NEE increases in the course of the growing season. The carbon uptake in the 1999 growing season that was 2 weeks longer than the 1998–2000 average was 18 % greater than C-uptake in other years. Thus, one may suggest that in a longer growing season in a warming climate, the forest production would first increase in Siberia. However, because climate is predicted to be not only warmer but also drier, the forests would die out due to lack of water. Thus, because of the decomposition of dead wood (Knohl et al. 2002; Vygodskaya et al. 2002) and the potential for unprecedented fires (Furyaev et al. 2001), in a warmed climate the forest may become a carbon source for many years.

Boreal bog ecosystems are usually a carbon sink of 30–100 gC m^{-2} during the growing season or year (Arneth et al. 2002; Corradi et al. 2005) and a GPP of some 200 gC m^{-2} for the growing season (Table 6.14). However, during dry summers, the upper bog may turn from a carbon sink into a carbon source of 50 gC m^{-2} (Arneth et al. 2002) suggesting that some bogs may dry out in a warming climate and become a significant carbon source in the atmosphere. Comparing different studies, Corradi et al. (2005) expect, though, that global warming would increase the CO_2–C sink in wetlands and tundras.

Subboreal true steppes may be both a significant carbon sink under different disturbance regimes (fire, pasture), or a slight sink if they work at a "stationary" regime (Titlianova and Tisarzhova 1991) when microbes consume NPP during several years depending on weather conditions. In our study, in an "nonstationary" regime, the true steppe was a sink of 115 gC m^{-2} $year^{-1}$ for three seasons recovering after multiyear pasturing and that of 130 gC m^{-2} $season^{-1}$ recovering after fire (Belelli-Marchesini 2007; Belelli-Marchesini et al. 2007). The true steppe produces as much GPP as the pine forest; GPP of both the steppe and pine forest is 2–2.5 times larger than that of the bog and the tussock tundra (Table 6.14).

Thus, along with increased carbon emissions due to increased fire events and area burned in a warming climate, carbon uptake by plants also increases, smoothing the negative effects of large fires.

6.4.4   Synthesis

Each of the major methods of carbon accounting has its own strengths and weaknesses. Under proper systems designing, the LEA allows to achieve the most reliable assessment for recent past and current, in a spatially and temporarily explicit way. However, this method does not consider mechanisms of the processes and lacks any substantial predictive capacity. DVMs (and, particularly DGVMs) describe processes and explain drivers. However, they are a somewhat rough tool for regional and national consideration due to the impossibility of properly including regional peculiarities in the models and sufficiently represent the transformation (disturbances) of vegetation cover. There is no solid methodology yet for estimating the uncertainty of DGVMs. However, they are practically a single tool for prediction. They also supply estimates of important components of carbon cycling for cross-checking uncertainties. Applications of DGVMs to northern ecosystems show that on average DGVMs estimates NPP rather reliably while overestimate HR. This results in underestimating the NECB (NBP) of terrestrial ecosystems in Siberia by DGVMs.

Eddy covariance methodology supplies important direct estimates of exchange of atmospheric carbon between ecosystems and the atmosphere. These estimates are extremely important for understanding the productivity processes of individual ecosystems and parameterization of models. However, these data have substantial uncertainties and – what is important – there are substantial problems with upscaling of the results of local measurements.

Inverse modeling presents a unique possibility for top-down verification of the above bottom-up estimates. Recent results for Russian natural land received from four different inversion approaches for 2000–2004 gave a mean of carbon sink -0.65 ± 0.12 Pg C $year^{-1}$ (inter-model variability) and the median -0.61 Pg C $year^{-1}$ (P. Ciais, 2010, personal communication). In earlier work, Gurney et al. (2003) calculated the average C flux using 17 different inverse models at -0.58 Pg C $year^{-1}$ for boreal Asia (45 ± 44 gC m^{-2} $year^{-1}$). A smaller sink for this region has been reported by Baker et al. (2006) – 29 ± 19 gC m^{-2} $year^{-1}$ and Patra et al. (2006) – 26 ± 61 gC m^{-2} $year^{-1}$.

Taking into account that European Russia provides about one-third of the entire country's sink and the level of reported uncertainties, all the above results are fairly consistent.

Still, the uncertainty of inverse modeling is high (up to 50–80 %) due to imperfect transport models and rare stations for measuring atmospheric concentration in Russian territories. However, the inverse estimates of the last decade are consistent and all together provide valuable information in understanding the carbon cycle of Russian terrestrial ecosystems. Eddy covariance measurements are important for the parameterization of the models and a perception of how terrestrial ecosystems function but lack proper gradients for upscaling over large territories.

In spite of substantial uncertainties, different methods give rather consistent estimates for Northern Eurasia (Table 6.15).

The above estimates basically refer to the $C-CO_2$ part of carbon cycling. Atmospheric methane provides the second-largest radiative forcing after CO_2. Globally, the contribution of CH_4 to the radiative forcing from preindustrial to present time is estimated at about 20 % of all greenhouse gases (Le Mer and Roger 2001). To our knowledge, there were no estimates for the region considered in this study. Rather detailed studies of methane emissions were provided in West Siberia based on detailed in situ measurement (Glagolev et al. 2010, 2011) and inverse modeling (Kim et al. 2011). These estimates are rather consistent varying in limits from 2.9 to 3.2 Tg CH_4 year^{-1}. In a recent inventory, Shvidenko et al. (2010a) estimated the total biosphere methane flux for entire Russia at 16.2 Tg $C-CH_4$ year^{-1} including 10.5 Tg $C-CH_4$ year^{-1} from wetlands and other wet soils, 1.5 Tg $C-CH_4$ year^{-1} from agriculture, 1.1 Tg $C-CH_4$ year^{-1} from fire, and 3.1 Tg $C-CH_4$ year^{-1} from water reservoirs. This study estimated methane fluxes from West Siberian wetlands at 3.4 Tg $C-CH_4$ year^{-1} that is consistent with the above studies. Taken into account that wetlands of Siberia comprise about 75 % of Russian territories in this land cover class (107.4 million ha of the total of 144.6 million ha), one could suppose that a major part of the country's methane emissions is provided by the study's region. Modeling results that were obtained by TEM are substantially higher – about 38 Tg CH_4 year^{-1} from biogenic sources (McGuire et al. 2010a, b). These estimates show that by the global warming potential, $C-CO_2$ sink in Siberia is almost compensated by the methane emissions, although uncertainties of methane's assessment for Russia are high.

Given the large stores of carbon in northern high latitude regions, the response of the carbon cycle of boreal Asia to changes in climate is a major issue of global concern (McGuire et al. 2006). Analyses to date indicate that the sensitivity of the carbon cycle of Siberia during the twenty-first century is also highly uncertain.

Table 6.15 A comparison between values of NBP (gC m^{-2} year^{-1}) from different studies

Method	Model	NBP (gC m^{-2} year^{-1})	Description	Period	Study Area (Mha)	Citation
LEA	FCA	−32.5 ± 21	All land types	2009	Siberia, 1019.0	This study
	FCA	−33 ± 8	All land types	2003	Central Siberia 297	Quegan et al. (2011)
	FCA	−21 ± 10	All land types	1988–1992	Russia 1709	Nilsson et al. (2003)
	FCA	−30 ± 8	All land types	1998–2002	Russia 1709	Nilsson et al. (2003)
	FCA	−23 ± 10	Forest land (vegetation only)	1961–1998	Russia 882	Shvidenko and Nilsson (2002)
	FCA	−36 ± 23	Forest land	1961–1998	Russia 886	Shvidenko and Nilsson (2003)
Inventory	PCA	−34–57	Forest land	1987–1990	Russia 884	Dixon et al. (1994)
	PCA (TBFRA- 2000)	−48	Forest & other wooded land	1993	Russia 886	Liski et al. (2000)
DVM	LPJ	+6 ± 6	All land types	1995–2003	Central Siberia 297	Quegan et al. (2011)
	SDGVM	−19 ± 14	All land types	1995–2003	Central Siberia 297	Quegan et al. (2011)
	LPJ	−17	Forest land	1981–1999	Russia 774	Beer et al. (2006)
Atmospheric inversion	MPI	−23 ± 13	All land types	1995–2003	Central Siberia 297	Quegan et al. (2011)
	LSCE	−24 ± 34	All land types	1995–2003	Central Siberia 297	Quegan et al. (2011)

(continued)

Table 6.15 (continued)

Method	Model	NBP (gC m^{-2} year^{-1})	Description	Period	Study Area (Mha)	Citation
IM		-49 ± 28	All land types	1992–1996	Boreal Asia 1280	Maksyutov et al. (2003)
	IM	-45 ± 44	All land types	1992–1996	Boreal Asia 1280	Gurney et al. (2003)
	IM	-29 ± 19	All land types	1998–2003	Boreal Asia 1280	Baker et al. (2006)
	IM	-26 ± 61	All land types	1999–2001	Boreal Asia 1280	Patra et al. (2006)
	IM	-38 ± 7	All land	2000–2004	Russia 1709	Ciais et al. (2010)

Modified from Quegan et al. (2011)

Note: The definition of boreal Asia in each of the atmospheric inversion studies varies; we have used a value of 1280 Mha throughout (Kaplun et al. 1994). In the Table, FCA and PCA refer to Full Carbon Account and Partial Carbon Account, respectively and TBFRA-2000 is the Temperate and Boreal Forest Resources Assessment 2000. IM – Inverse models. Sign "minus" denote the carbon sink

6.5 Changes Within Climate-Driven Ecotones in Siberia

The response of trees to climate change is expected to be significant at the climate-driven ecotones, e.g., in the alpine and northern forest-tundra ecotones. There are a number of observations of tree response to warming, including species invasion into the tundra, stand densification, and growth increment increase along the northern and alpine tree-lines during recent decades (e.g., Kullman 2007). These phenomena are also considered for different regions of Asian Russia (e.g., Shiyatov 2003; Shiyatov et al. 2007; Soja et al. 2007; Devi et al. 2008; Kharuk et al. 2004, 2010a, 2010b). These studies were located within the Ural and Altai – Sayan Mountains, the northern tree line around the Yenisei meridian, and the "larch-mixed taiga" ecotone in Central Siberia. The results obtained showed that Siberian pine and larch growing in the alpine forest-tundra ecotone are strongly responding to warming by an increasing increment, stand densification and regeneration density, upward tree line shift, and transformation of *krummholz* to arboreal forms. Similar phenomena were observed for the northern forest-tundra ecotone.

6.5.1 Northern Tree Line

The Northern tree line on the Polar Ural Mountains (the western border of the study's region) and westward from the Yenisei meridian is formed by a combination of elevation and northward temperature gradients. Along the northern edge of boreal forests in the Polar Ural Mountains, Shiyatov et al. (2007) reported a 35-m upward shift of closed larch dominated stands between the years 1910 and 2000. Along with the current shift of tree-lines, dendrochronology analysis describes former tree-line dynamics in the forest-tundra ecotone (Shiyatov 2003). Observations within the most northward forest stand (Ary-Mas, 72°+N) dominated by larch showed regeneration advance into tundra and stand densification (Kharuk et al. 2004, 2006). For the period 1970–2000, stand density was increased (at ~1.6 times); the advance of regeneration was approximately 3 m year^{-1}.

6.5.2 Tree-Line Evolution in South Siberian Mountains

Historically, climate-induced waves of upslope and downslope tree migration were reported for the alpine forest-tundra ecotone (Fig. 6.6). Tree mortality was observed during the Little Ice Age and followed the cooling with a lag. Living tree mortality dates showed that the tree-line advance began at the end of the nineteenth century and lagged behind temperature changes. Larch and Siberian pine regeneration now survives at elevations up to 160 m higher in comparison with the maximum observed tree-line recession during the Little Ice Age and surpasses its historical maximum

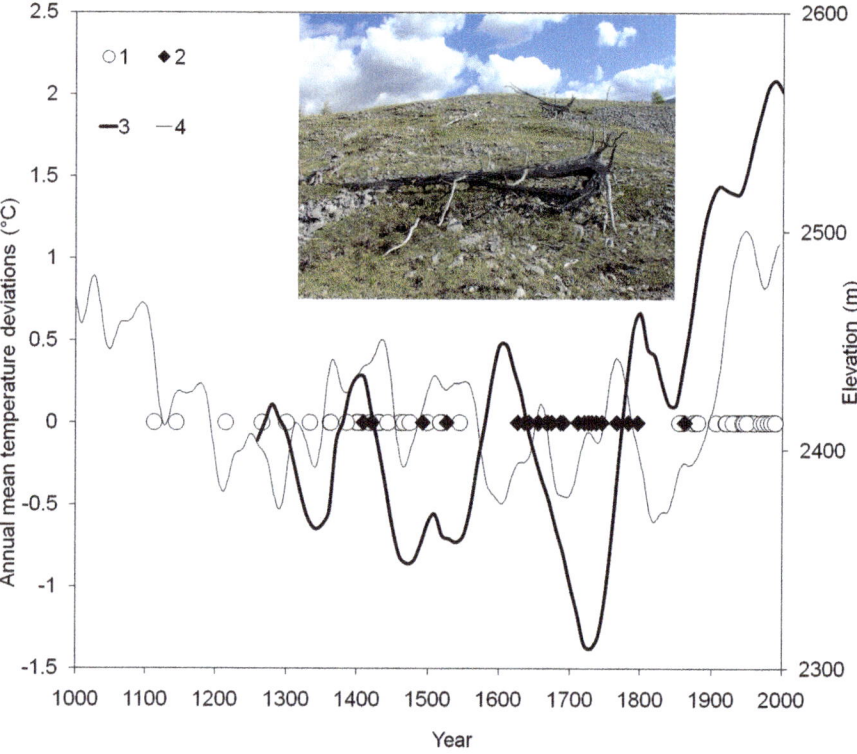

Fig. 6.6 The history of tree establishment and mortality at the Sengilen site and reconstructed air temperatures for northern Siberia and the Northern Hemisphere. 1, 2 are the dates of tree origin and mortality, respectively. 3 and 4 are the reconstructed air temperature deviations for southern Siberia (March–November temperature; Schwikowski et al 2009) and the Northern Hemisphere (annual temperature), respectively. Bars show elevation of treelines: A – "historical," B – "refugee," C – "postcooling" treelines; D – regeneration line. Insert: a view on the "refugee tree line"

during the last millennium by up to 90 m. Temperature change of 1 °C promoted an upward shift in the tree-line of about 80 m. The tree-line advance rate was estimated at 0.90 ± 0.22 m year^{-1} (Kharuk et al. 2010b). Similar values were found for the Polar Ural Mountains (0.4 m year^{-1}; Shiyatov et al. 2007). The mean rate of regeneration advance during the last century was 1.2 ± 0.24 m year^{-1}, increasing to 2.35 ± 1.26 m year^{-1} during the last three decades. The observed warmer winter temperatures were important for regeneration recruitment because of reduced desiccation and snow abrasion, which are the main causes of tree/regeneration damage and mortality (Shiyatov 2003; Kharuk et al. 2006, 2009). Meanwhile at high elevations, seedlings are still in the vulnerable stage and can be killed by cold winters (in synergy with winds), resulting in recession of the tree-line.

6.5.3 Forest Densification and Area Change

Studies within the Altai-Sayan Mountain System sites showed a 150 % increase of dense forest stand area within upper forest belt during the last four decades due to growth of preestablished trees (Kharuk et al. 2010a, 2010c). An increase of growth increment in trees was observed starting in the mid 1980s, which was strongly correlated with mean summer temperatures.

Stand densification was also observed along rivers and streams due to earlier snowmelt, which increases the growing period. At the southern edge of the boreal forests in Siberia an upward shift of the closed forest border was observed to be about 63 ± 37 m between 1960 and 2000. Studies within the Polar Ural Mountains also showed a considerable stand densification over the last four decades (Shiyatov 2003; Devi et al. 2008). Forest densification itself improves the microclimate for tree survival under harsh environmental conditions by reducing desiccation and snow abrasion. Greater stand density also leads to increased snow accumulation, which facilitates the local regeneration potential in a positive feedback loop (Kharuk et al. 2007). During recent decades, in the Polar Ural Mountains, a doubling of winter precipitation was correlated with an upward shift of the tree-line (Devi et al. 2008). Substantial densification in tree-line populations seems to be a common phenomenon in northern and high-elevation environments and occurs more frequently than actual elevational tree-line advance (Kullman 2007; Kharuk et al. 2008).

6.5.4 Forest Response to Climate Variables and Relief Features

Forest response to climate variables at high elevations is nonuniform because tree establishment and survival depend on the availability of sheltered (wind protected) areas. In mountains, the forest spatial distribution is dependent on azimuth, elevation, and slope steepness (Vygodskaya 1981; Polykarpov et al. 1986), and this pattern changed over recent decades. With respect to aspect, the forest area distribution is asymmetric and elevation-dependant. At lower elevations, forest patterns were oriented approximately northward with minimal forest cover observed on the southern slopes where trees experience water stress. As the elevation increased, the forest area distribution orientation changed in a clockwise direction, becoming oriented eastward at the highest elevation. The observed changes can be attributed to wind impact, which increases as elevation increases. The typical upper boundary is a mosaic as tree and regeneration survival depends on the availability of sheltered relief which is provided by rocks or local depressions. It was found that at any given elevation, the majority of forests occupied slopes with greater than mean slope values, and at higher elevations forests shifted to steeper slopes (Kharuk et al. 2010c).

Thus, forest response to warming was dependant on topographic relief features and this response significantly modified the spatial patterns of high elevation forests in southern Siberia during the last four decades.

6.5.5   Transformation of Krummholz into Vertical Forms

Milder climate also promotes changes in tree morphology, i.e., transformation of mat and prostrate *krummholz* into the vertical form (Kharuk et al. 2006, 2010b; Shiyatov et al. 2007; Devi et al. 2008). Recent decades of warming caused a widespread transformation of larch and Siberian pine mat and *krummholz* to a vertical form, which began mainly in the late 1980s. This date coincides approximately with the period when winter temperatures surpassed the mean value during the twentieth century. Larch was much less likely than Siberian pine to be found in *krummholz* forms. Larch surpasses Siberian pine in frost and wind resistance and grew in arboreal forms where Siberian pine was still prostrate. Shiyatov et al. (2007) and Devi et al. (2008) reported similar transformation of larch *krummholz* into vertical forms. Meanwhile, periodic shoot and needle desiccation, caused by synergy of low temperatures during cold winters and wind impact, decrease vertical growth.

6.5.6   Ecotone "Zone of Larch Dominance – Mixed Taiga"

For Central Siberia it was shown (Kharuk et al. 2007) that dark coniferous species (Siberian pine, spruce and fir) are expanding into habitat of larch. The invasion of dark coniferous species into historical larch habitat was quantified as an increase of the proportion of those species both in the overstory and regeneration. Siberian pine and spruce have high K_i [propagation coefficient: $K_i = (n_i - N_i)/(n_i + N_i)$, where N_i and n_i – the proportion of a given species in the overstory and regeneration, respectively] values both along the margin and in the center of zones of absolute larch dominance, where their presence in the overstory is <1 %. The age structure of the regeneration (with mortality control) showed that it was formed mainly during the last two to three decades. The results obtained indicate climate-driven migration of Siberian pine, spruce, and fir into traditional larch habitat. Very likely, this process is substantially driven by changes in permafrost regimes. On the western and southern margins of the larch-dominance zone, regeneration of dark coniferous species formed a second layer in the forest canopies, which could eventually replace larch in the overstory. With stand densification, Siberian pine received an additional advantage since larch is a shade intolerant species (Kharuk et al. 2007).

In a warming climate, Siberian pine should enjoy a competitive advantage due to its higher temperature response and shade tolerance. Thus, current climate change should lead to the shift of Siberian pine into larch forests. Substitution of deciduous larch by evergreen conifers, decreases albedo and provides a positive feedback for even greater warming. The other expected consequence is an increase of biodiversity since Siberian pine-dominated communities provide a better food base for animals and birds. Larch will continue to maintain its advantage in drier areas and in zones of temperature extremes, particularly on permafrost.

6.6 Future Trajectories of Forest Ecosystems in Siberia

6.6.1 Individual-Based Models of Forest Dynamics

Climate change has been identified as a driver of structural compositional change in the Russian forest. Purves and Pacala (2008) reviewed predictive models of forest dynamics and structure and suggested that the world's relatively low-diversity boreal forests are prime areas for the examination of ecological models because enough inventory data has been collected to make the species parameter estimations needed by the models. We concur with the spirit of their assertion. The model discussed here is the product of two decades of an ongoing synthesis of the silvics of boreal forest trees for species parameter estimations for boreal forests (Shugart et al. 1992). In the past 20 years, individual-based models (IBMs) have expanded to provide increasingly accurate predictions and simulations of forest change over time (Mladenoff 2004). These models have been applied to investigate forest disturbance and succession, to manage forest stands, and to predict how forests will respond to alterations in environmental scenarios.

The examples considered here use one of a class of forest individual-based models called "gap models" (Shugart and West 1980) to project responses of the Russian boreal forests to changes in climate. Individual-based gap models simulate individual trees, their growth, mortality, and decomposition into litter in a relatively small area, typically the size of a forest gap (Urban and Shugart 1992). These models simulate the vertical dimension but often are incorrectly classified as nonspatial models in that they do not interact in the horizontal. This limitation ended with subgrid-based and horizontally explicit versions (Urban et al. 1991; Weishampel et al. 1992). The models often include ecosystem processes such as nutrient cycling and interactions with the local abiotic environment (Scheller and Mladenoff 2007).

One of the primary results of the importance of the dynamics of structure is illustrated in Fig. 6.7. Successional carbon dynamics is shown as predicted by the FAREAST model (Yan and Shugart 2005) simulating the forest dynamics of Changbai Mountain (intrazonal boreal forests on the North Korean and Chinese border). In this succession, deciduous conifer forest (with *Larix* as the dominant species) converts over time to a mixed Spruce-Fir (*Picea* and *Abies*) forest. The deciduous stands begin to break up after about 100 years of stand development. The forest acts as a strong carbon sink for the first 100 years of forest dynamics after stand initiation on bare ground (in case of regeneration after disturbance the decomposition of legacy material would exceed C accumulation in live biomass for the first 10–20 years, possibly longer). After about 100 years, the mortality and decomposition of the initial early successional forest biomass causes the forest to be a relatively strong carbon source for almost 200 years. Then the carbon standing crop of the forest settles into a quasi-equilibrium where the forest is neutral in its carbon source-sink relation to the atmosphere. The leaf area of the forest is relatively constant from about 50 years onward. The dynamic response of this forest to disturbance from a leaf-canopy model (based solely on leaf gross-primary-production kinetics) would

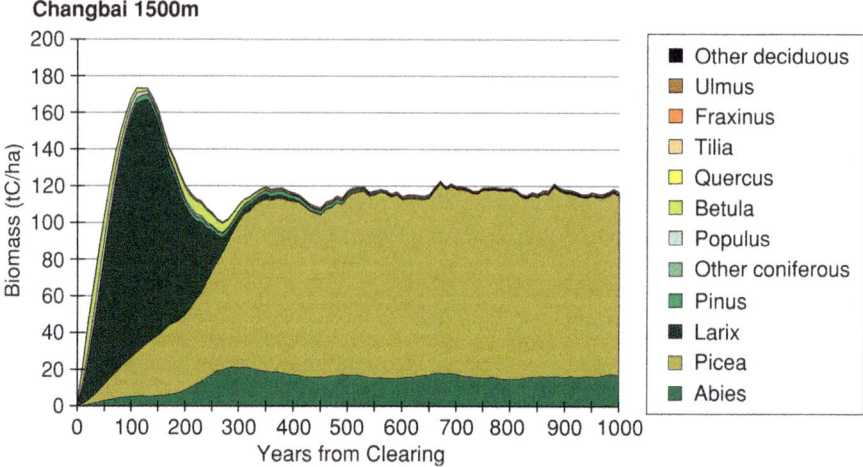

Fig. 6.7 The simulated forest species composition dynamics expressed as cumulative biomass (t C ha^{-1}) for succession from bare ground Changbai Mountain (People's Republic of China/Democratic People's Republic of Korea) at 1,500 m in elevation (From Yan and Shugart 2005)

not capture these significant shifts in the source-sink status of the forest because it ignores several essential factors that control the retention of C accumulated through production processes such as tree size, longevity, and species succession. Furthermore, these factors are critical for simulating tree mortality which are a major pathway in C cycling of forest ecosystems. Yet this process is difficult to capture in most current biophysical models, because of their mathematical structure (ordinary differential equations applied over relatively large areas with an assumption of spatial homogeneity). An important additional consideration is the need to include a woody debris pool which also plays a significant role in C storage and cycling.

These are specific examples of an important general problem. Many of the models that are currently being applied to determine regional and global productivity use mathematical functions that are derived from the physiological and biophysical performance of leaves – even though the models themselves are typically formulated and calibrated for landscapes. The actual heterogeneity of leafy canopies found in both the horizontal and the vertical dimensions of vegetation can create significant problems in such upscaling (Shugart 2000). For example, a leaf area of 4 held by a single large tree is *not* metabolically equivalent to a leaf area of 4 held by a collection of small trees – the plant respiration rates in the two cases could be very different. Essentially, vegetation of different structures could be similar for one ecosystem property (e.g., productivity, biomass, height, etc.) but not for any of the others.

This might be thought of as an interesting theoretical problem, were it not for the fact that highly aggregated leaf-based models (in some cases augmented by remotely sensed variables such as photosynthetically active radiation or PAR absorbed by the vegetation) are the backbone of our prediction of global productivity, carbon uptake and loss, and vegetation cover. These models can characterize the productivity and biophysical properties of large regions and are used for assessing carbon fluxes associated with these regions. Thus, the issue of properly including the effects of vertical and horizontal variability in our understanding of regional ecosystem dynamics has arisen as an important problem to improve our predictive capability.

Forest gap models are well suited to evaluate the effect of climatic fluctuations on forest structure and composition. Following the addition of species range data, the forest gap model, FAREAST, was tested across the broad geographic and climatic variation of Russia using independent forest inventory biomass data (Shuman et al. 2011).

The version of the FAREAST model (Yan and Shugart 2005) applied here simulates the composition of the Russian boreal forest in response to current and changing climate conditions and is suitable for exploring the feedback between climate and forest composition at both the continental and regional scale. Species silvics for the more temperate-forest species of the southern border with China and Russia are already in the FAREAST model; parameters for species in Northern Eurasia were added for a total of 52 tree species. Species range maps for these 52 species are used to include appropriate diversity on a site-specific basis. The model was used to simulate the impact of changes in temperature and precipitation on both total and genera-specific tree biomass across Siberia and the Russian Far East.

For the model runs with temperature or precipitation change, a linear increase in temperature or precipitation or both takes place from years 0 to 200 of the simulation. This is followed by an additional 150 years of simulation during which the climate stabilizes around the conditions attained in year 200. A nonparametric factorial ANOVA was used to assess differences in the biomass (t C ha^{-1}) of *Larix* spp., dark conifers, and the total forest biomass among model runs that employed different climate treatments at 10-year intervals.

Under the influences of a warming climate, the FAREAST model simulates cases in which the patterns of succession change qualitatively. Two examples are shown in Fig. 6.8. In the top panels of Fig. 6.8, the simulated forest dynamics (200 simulated plots averaged annually over 350 years) shows succession of the *Larix*-dominated forest for conditions appropriate to Irkutsk region in southern Siberia. The successional pattern is for an increase in *Larix* biomass during initial stand development that reaches a maximum biomass in simulated year 200. As the initial even-aged stands of *Larix* break up and transition to a mixed-aged mosaic, the biomass over the landscape drops (Fig. 6.8) over the next 150 years (from years 200 to 350). Under climate warming, *Larix* persists as a dominant until the even-aged stands break-up and then there is a transition to a *Betula –Pinus* mosaic landscape. The composition shift is much like a successional response with a species shift as forests mature. This compositional shift is part of the simulated "normal" successional development seen in forests of the Russian Far East (Fig. 6.8) in which an initial pioneering successional stage dominated by *Larix* transitions to *Pinus-Picea* forest. One also can see this classic successional replacement pattern in the simulations of the northern Chinese forests (Yan and Shugart 2005, Fig. 6.7). Under climate change, the forests of the Russian Far East retain the replacement successional

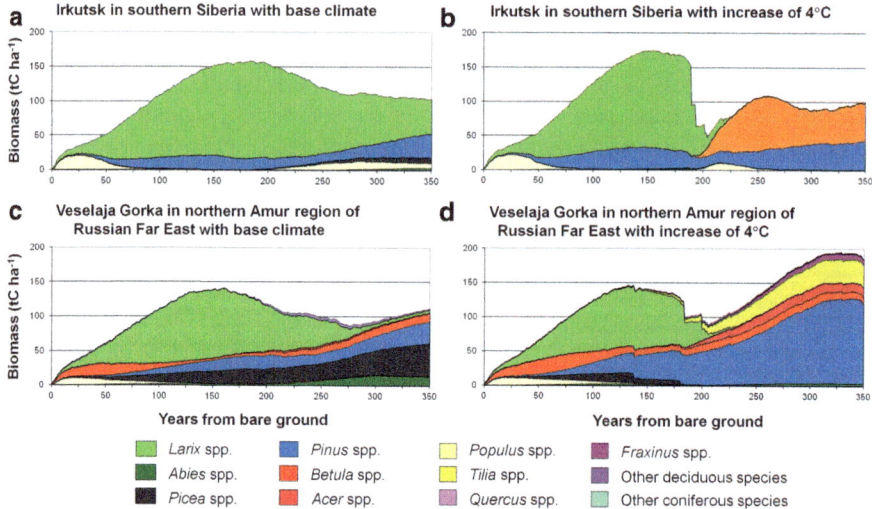

Fig. 6.8 Cumulative biomass (t C ha⁻¹) of species from model simulations for two locations in Russia. Each graph displays composition by the dominant genera. The base cases in each of the pairs of graphs represent the successional dynamics from a bare ground condition in year 0 for 350 years of ecological succession. Also, in each pair of simulations is a climate-change case with a temperature increase ramping up over the first 200 years of succession to an increase in average temperature of 4 °C; this level of change in then continued until year 350. (**a**) and (**b**) Simulated mixed species biomass dynamics (t C ha⁻¹) for Irkutsk in southern Siberia. The effect of temperature increase of 4 °C across 200 years of forest succession causes transition from *Larix* spp. dominance to mixed species forest (**b**) when compared to base scenario (**a**). Legend for (**a** and **b**) shown to the *right* side of (**b**). (**c**) and (**d**) Simulated mixed species biomass dynamics (t C ha⁻¹) for Veselaja Gorka in the Russian Far East. The effect of temperature increase of 4 °C across 200 years causes transition to a mixed-species forest dominated by *Pinus* spp. (**d**) when compared to base scenario (**c**). for (**c** and **d**) shown to the *right* side of (**d**)

pattern (Fig. 6.8) but the composition of the mature-phase forest is different (a mixed *Pinus* with broad-leaved deciduous-tree-species forests). In the Irkutsk forest, landscape climate change induces a new successional pattern of stand replacement; in the Far East forest the same pattern of climate change has the same broad type of successional replacement but with different dominant species.

Distributed gap-models can project forest response to climate change (Fig. 6.9) including forest structural properties over the Siberian boreal forest. These model responses can be compared to those of other more aggregated models (Vygodskaya et al. 2007; Tchebakova et al. 2009) to determine the sensitivity of the predicted responses to forest structure. Using the FAREAST model distributed on points across Siberia, one can inspect the response of the forest to different climate-change scenarios. The case shown in Fig. 6.9 is for a simple sensitivity analysis with an increase of 4 °C across 372 sites. What emerges is an expected pattern of biomass distribution for Siberia and the Russian Far East (Fig. 6.9, left) with high levels of biomass (t C ha⁻¹) in the Russian Far East and in the southern part of Siberia.

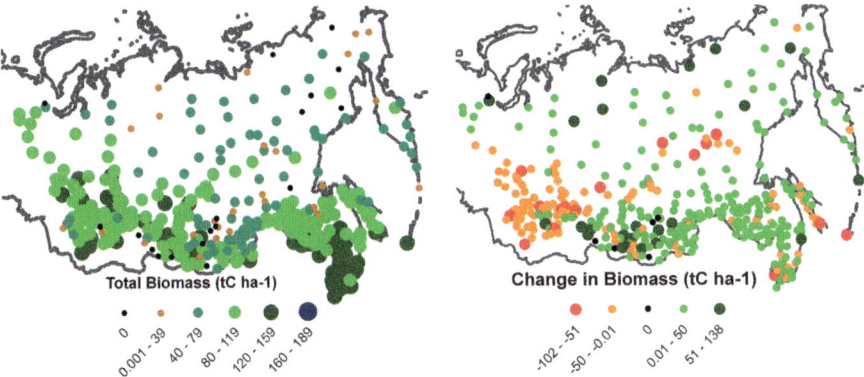

Fig. 6.9 230-year-old forest landscapes at 372 sites across Siberia and Russian Far East simulated by the FAREAST model for Each point is the average of 200 one-twelfth *ha* plots simulated through succession from bare ground and then averaged at Each point. *Left*: Simulated total mixed-species biomass for historical climate data. *Right*: Difference in total biomass (t C ha⁻¹) between a +4 °C climate warming and the historical baseline. Decrease in biomass is in *red* and *orange*; increase in shades of *green*

Under a warming, there is a substantial decrease in biomass through middle Siberian sites, particularly in the forests of the region surrounding Novosibirsk and Tomsk in southwestern Siberia. The diverse high-biomass forests of the Russian Far East maintain their biomass levels and even increase due to a warming, but there is a potential change in composition (Fig. 6.8). The change is complex across all of Siberia and the indication is that wide forest monitoring is necessary to obtain a complete picture of the nature of the change. Please note that the responses do not include the possibilities of increased rates of disturbance (notably fire and insect pests), which could be expected to increase in their effects with warming.

6.6.2   Potential Land Cover Change in Siberia Predicted by Siberian Bioclimatic Model

The changing climate rearranges land cover and impacts biophysical and biochemical processes directly and indirectly through land cover change. To understand these climate-induced vegetation changes, bioclimatic models of either a static (time-independent), dynamic (time-dependent), or mixed nature are used. To simulate zonal vegetation across Siberia (within the window: 60–140°E and 48–72°N, about 12 million km²) during the twenty-first century, a Siberian bioclimatic model, SiBCliM was designed (Tchebakova et al. 2003), a static, envelope-type large-scale bioclimatic model based on the Siberian vegetation classification of Shumilova (1962). SiBCliM uses three bioclimatic indices to characterize a climatic envelope of each vegetation class: (1) growing degree-days above 5 °C, characterizing plant requirements for warmth;

(2) negative degree-days below 0 °C, characterizing plant tolerance to cold; and (3) an annual moisture index (ratio of growing degree days above 5 °C to annual precipitation), characterizing plant resistance to water stress. SiBCliM was updated to include permafrost (the active layer depth), a critical ecosystem determinant in the extremely cold continental climates of Siberia (Tchebakova et al. 2009, 2010). SiBCliM includes 14 vegetation classes, 10 boreal (tundra, forest-tundra, northern, middle, southern dark-coniferous and light-coniferous taiga, forest-steppe, and steppe), and four temperate (conifer-broadleaved, forest-steppe, steppe, and semidesert) vegetation classes that do not exist in the current Siberian climate but are included in SiBCliM because of their potential importance in future climates. The SiBCliM performance was evaluated by comparing modeled and actual vegetation maps using the kappa statistics (Monserud and Leemans 1992). The overall agreement between these two maps was a "fair" match (kappa 0.53).

To simulate vegetation cover and hot spots of vegetation change in Siberia in a changing climate by the end of the twenty-first century, we used two IPCC climate change scenarios that reflect opposite ends of the spectrum, the Hadley Centre HadCM3 A2, the largest temperature increase, and B1, the smallest temperature increase (IPCC 2007a, b). We evaluated then possible feedbacks of vegetation change to the climatic system (albedo and thus net radiation change) that may mitigate/accelerate vegetation shifts at the end of the twenty-first century in response to the harsh HadCM3 A2 scenario (Fig. 6.10 1C). According to this scenario, northern vegetation types (tundra, forest-tundra, and taiga) would decrease from 81.5 to 30 %, with southern (forest-steppe, steppe, and semidesert) vegetation prevailing on 67 % of Siberia (Table 6.16). A moderate change in vegetation is predicted from the B1 scenario (Fig. 6.10 1B). According to this scenario, habitats for northern vegetation classes would decrease from 81.5 to 50 % enabling southern habitats to expand from 18.5 to 50 %.

Fire and the thawing of permafrost are considered to be the principal mechanisms that will promote zonal vegetation shifts across Siberia in a rapidly changing climate.

In a future warm and dry climate in Siberia, zonobiomes could shift northward as far as 600 km (Tchebakova et al. 2009). Trees at the northern tree-line would move by means of migration. Migration rates of boreal tree species as estimated from paleoecology may reach 300–500 m year^{-1} or even less (Udra 1988; King and Herstrom 1997), thus taking a millennium for a vegetation zone of hundreds of kilometers in width to be completely replaced by vegetation of a neighboring zone under warming. Species with broad climatic niches and high migration rates could adjust to a rapidly warming climate while species with a restricted range of suitable habitats and limited dispersal are likely to disappear first (Solomon and Leemans 1990). While extirpation and immigration are the main processes at the margins of the forest distribution, within the forest zone, natural selection and gene flow are the primary processes favoring tree adaptation to climate change (Davis and Shaw 2001; Rehfeldt et al. 2004). Evolutionary processes of adjusting to predicted climate change would take a time at least one order of magnitude greater than the time it takes for the climate itself to change. Estimates for *Pinus sylvestris* in Siberia

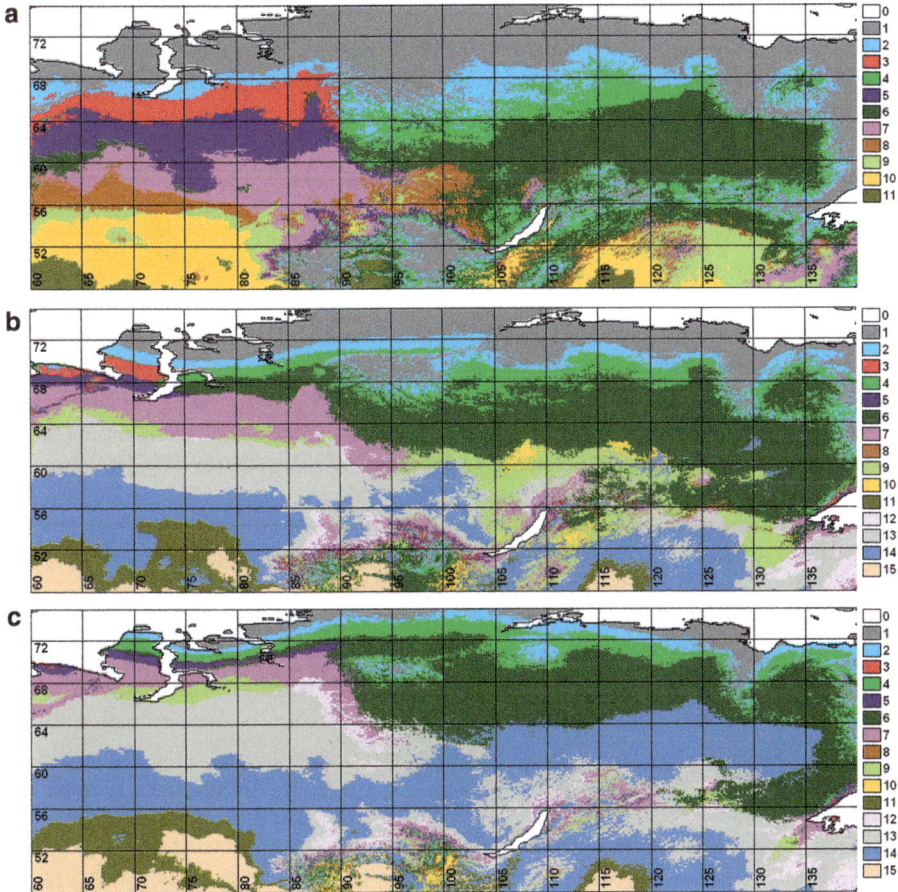

Fig. 6.10 Vegetation distribution in 2080 predicted from current climate (**a**) and the moderate HadCM3 B1 (**b**) and the harsh A2 (**c**) climate change projections. Vegetation class key: 0 - Water; Boreal: 1 - Tundra; 2 - Forest-Tundra; Northern Taiga: 3 - dark, 4 - light; Middle taiga: 5 - dark, 6 - light; Southern Taiga: 7 - dark, 8 - light; 9 - Subtaiga, Forest-Steppe; 10 - Steppe; 11 - Semidesert; Temperate: 12 - Broadleaf; 13 - Forest-Steppe; 14 - Steppe, 15 - Desert

Table 6.16 Potential vegetation change (%) in Siberia by 2080 predicted from two climate change scenarios

Vegetation	Current climate, %	Had CM3 A2, %	Δ, %	Had CM3 B1, %	Δ, %
Tundra	18.3	2.0	−16.3	7.5	−10.8
Forest-tundra	8.5	2.4	−6.1	4.0	−4.5
Taiga					
Light-conifer	36.1	19.9	−16.2	25.7	−10.4
Dark-conifer	18.5	9.0	−9.5	12.1	−6.4
Forest-steppe	6.1	23.8	17.7	22.5	16.4
Steppe	10.0	30.1	20.1	19.1	9.1
Semidesert	2.5	12.9	10.4	9.0	6.5

suggest that 5–10 generations (above 150 years) are required for the evolutionary process to follow a predicted warming. Genotypes would be reorganized within tree distributions, and tree boundaries would follow a changing climate. The forest adjustment to climate change would occur, but it would require a long time due to the large amount of change predicted by the end of the current century (Rehfeldt et al. 2004).

At the southern tree-line, forest fire would promote a replacement of forest vegetation by grasslands. A drier climate would result in increased tree mortality in the southern taiga, thus increasing fire fuel accumulation. When paired, both factors, increased fuel load and fire weather, show that risks of large fires would significantly escalate in dry southern Siberia and in central Yakutia advancing new habitats for steppe and forest-steppe rather than forests. In a warmer and drier climate, postfire forest regeneration may not be possible due to decreased precipitation and increased evapotranspiration. Grasses would replace forest, as suggested in several model scenarios (Rizzo and Wiken 1992; Smith and Shugart 1993), because they are adapted to less precipitation and droughts, and they are able to recover after frequent fires due to a short life cycle.

Future climate is often regarded as analogous to the mid-Holocene climate (Borzenkova and Zubakov 1984). In Siberia, the mid-Holocene climate as reconstructed from mid-Holocene vegetation (Chap. 3) was warmer and wetter than the present (Monserud et al. 1996). The climate of the twenty-first century as predicted from GCMs is warmer and drier. The mid-Holocene and future climates are likely to be dissimilar and impact terrestrial ecosystems differently, so they may not be regarded as analogous.

Altered land cover would generate additional regional forcing and feedback to the climate system resulting in a potential nonlinear response to changes in climate. Predicted significant changes in land cover across Siberia by the end of the century would initiate change in surface albedo and thus energy fluxes between the biosphere and the atmosphere. To evaluate possible effects of feedbacks of vegetation-induced albedo change to net radiation change to accelerating/mitigating vegetation shifts over Siberia, we first evaluate albedo change by 2080. Albedo of both current and the 2080 vegetation was calculated for each pixel as a sum of winter albedo (vegetation covered by snow), summer albedo (snow-free vegetation), and albedo of the winter-to-summer and summer-to-winter transition periods. Albedo values were ascribed to each pixel for each year period from Budyko (1974). The snow period was calculated from the regression relating it to July and January temperatures ($R^2=0.72$, $p<0.0000$); the winter-to-summer and summer-to-winter transition period was calculated from the regression relating this period to the snow period ($R^2=0.51$, $p<0.0000$). Average annual albedo change by 2080 was calculated as the difference between 2080 and current albedo. According to the harsh A2 scenario, albedo would increase over 44 % of the area in the southern and middle latitudes in Siberia due to the forest retreat. Albedo would decrease in 56 % of the territory mainly in the northern latitudes and highlands, where tundra would be replaced by forest with smaller albedo and in southern grasslands where the snow-free period would extend. With consideration of changes in snow cover, tundra characterized by a high seasonal variation in albedo showed increased feedbacks (greater atmospheric heating) compared to forests characterized by less seasonal variation in albedo that showed lesser feedbacks (less atmospheric heating) (Chapin III et al. 2000). Thus, resulting warming due to effects of albedo and snow cover change would be greater at high latitudes and lesser warming or some cooling would occur at middle and low latitudes.

Change in albedo would change energy fluxes: shortwave radiation (Rs) and thus net radiation (R). Net radiation evaluated as $R = (0.6*Rs + \alpha)$ with $R^2 = 0.93$, $p < 0.0000$, and shortwave radiation evaluated as $Rs = Q*(1 - \alpha)$, where Q is total radiation and α – albedo, results then in $\Delta R \sim 0.6* Q* (\Delta\alpha)$, assuming Q and $R = (0.6*Rs + a)$ are constant in the future climate.

The simulations show that under a greater warming (scenario A2), net radiation balance would increase by $2,200 \times 10^{13}$ MJ year^{-1} in half of the area in the north and would decrease by 700×10^{13} MJ year^{-1} in the other half of the area in the south totaling $1,500 \times 10^{13}$ MJ year^{-1} over the entire Siberian window, about 1.2×10^{13} m^2. Compared to the net radiation 1,000–2,000 MJ m^{-2} year^{-1} in current climate, this change is about 10 % in the A2 climate and 6 % in the B1 climate.

To estimate albedo-induced net radiation feedbacks to vegetation, net radiation was transformed to growing degree-days, GDD, the climatic constraint we use in our SiBCliM, with $R^2 = 0.82$, $p < 0.0000$. Corrections for growing degree-days were calculated coupling this regression with the net radiation maps in the A2 and B1 climates at 2080 (Figs. 6.11 and 6.12). SiBCliM was run again for the A2 and B1 climates to understand how vegetation would shift with regard to the predicted albedo-induced feedbacks. Potential albedo feedbacks due to land cover change predicted from IPCC scenarios may result in additional warming in the north promoting the further forest advance into tundra, and some cooling in the forest-steppe ecotone promoting the forest return in the south (cf., Vygodskaya et al. 2007). Due to albedo feedback effects, the forest may gain some more area of tundra and may lose less area in grasslands gaining an additional 6–8.5 % (Fig. 6.13, green).

Thus, global and regional circulation models need to account for feedbacks from terrestrial ecosystems to get more reliable projections.

6.6.3 Prediction at Landscape Level

Reliable prediction of future states of the boreal forest under global change requires understanding the complex interactions among forest regenerative processes (succession), natural disturbances (e.g., fire, wind, and insects), and anthropogenic disturbances (e.g., timber harvest). At a landscape level, long-period predictions have been simulated based on the process-based, spatially explicit dynamic model of forest succession and disturbance LANDIS-II (LANDscape DIsturbance and Succession) (Mladenoff 2004; Scheller et al. 2007). The model independently simulates multiple ecological and anthropogenic processes so that interactions of these processes are an emergent property of the simulations. LANDIS-II consists of a

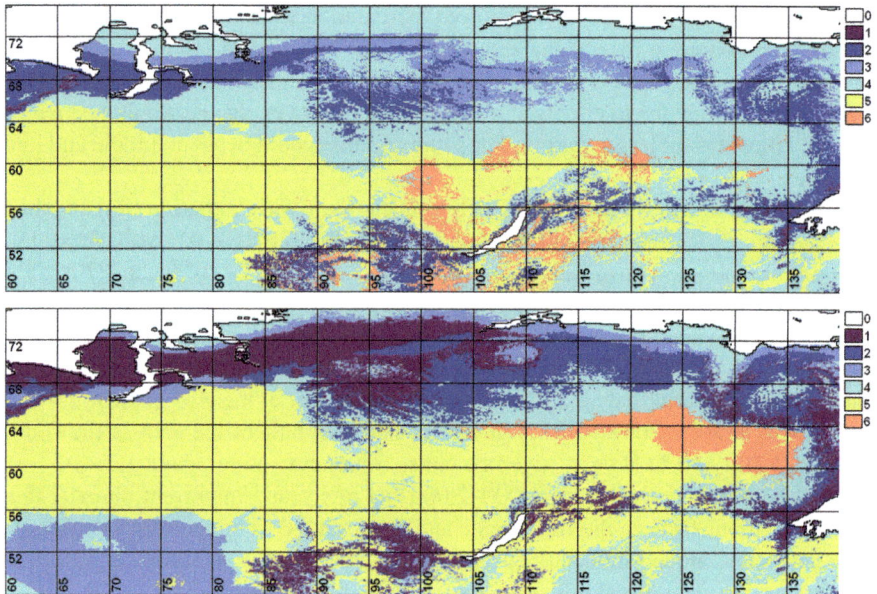

Fig. 6.11 Vegetation-induced albedo change in 2080 predicted from the moderate HadCM3 B1 (*upper*) and the harsh A2 (*lower*) climate change projections projections. Albedo class key: 0. Water; 1. - <-0.3, 2. -0.3-(-0.2), 3.- 0.2 -(-0.1), 4. -0.1-0, 5. 0-0.1, 6. >0.1

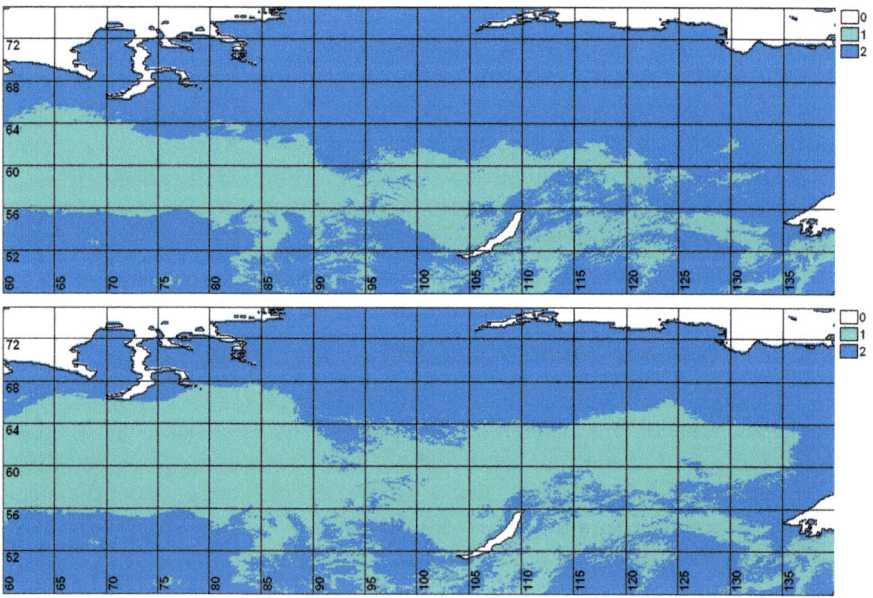

Fig. 6.12 Albedo-induced net radiation change in 2080 predicted from the moderate HadCM3 B1 (*upper*) and the harsh A2 (*lower*) climate change projections. Dark-blue marks net radiation increase (warming), light-blue marks net radiation decrease (cooling)

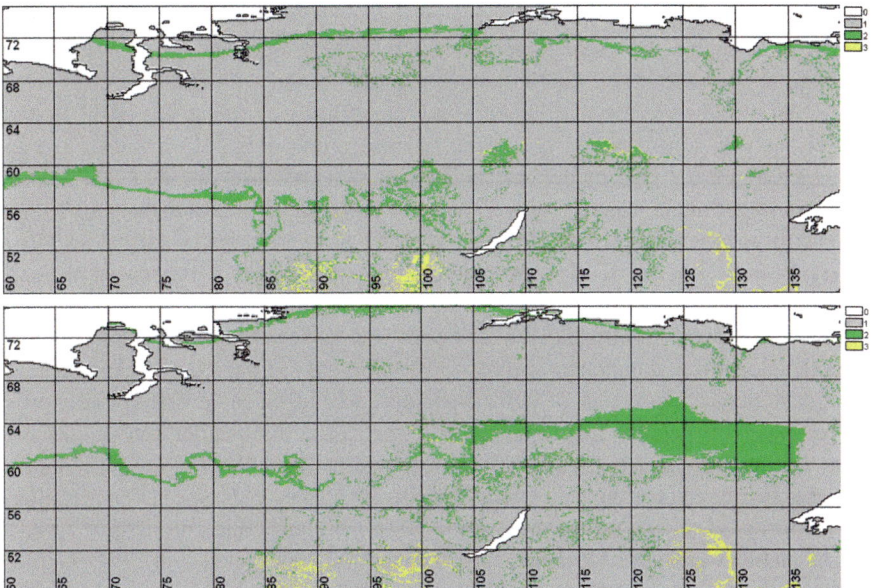

Fig. 6.13 Potential albedo feedback effects on the forest area: *green* – forest area increased due to additional warming (albedo decrease) in the north and some cooling (albedo increase) in the south promoting the forest return; *yellow* – forest area decreased

core collection of libraries and optional extensions that represent the ecological processes of interest (e.g., succession, wind disturbance, biotic disturbance, fire, biomass harvest, etc.). The model can be linked to the outputs of global circulation models (GCMs) to allow climate change to interact with landscape processes in the simulation environment and to process-based physiological models.

LANDIS-II represents the landscape as a grid of interacting cells at an appropriate resolution (e.g., 100 m), and each cell may contain multiple species and each species can be represented by one or many age cohorts. Each cohort will establish and respond to disturbance as a function of its life history attributes. Various disturbance extensions can be turned on or off, and the timber harvest extension allows the examination of alternative management regimes. The primary model outputs are maps of forest conditions, including species, age classes, aboveground biomass (living and dead), disturbance types, and their respective severities.

The model has been used to study the relative effects of climate change, timber harvesting, fire and insect outbreaks on forest composition, biomass (carbon), and landscape pattern in south-central Siberia (Gustafson et al. 2010, 2011a, b). The study area was represented by an almost intact forest region near Ust'-Ilmsk, at the boundary of southern and middle taiga, with a total area of about 315,000 ha. Permafrost is rare in the study area. Future climate parameters were based on the Hadley GCM (HadCM3) A2 scenario (Gordon et al. 2000) for the years 2080–2099. Mean monthly temperatures in the study area were predicted to rise by about 5 °C during the twenty-first century, and annual precipitation was predicted to increase by about 20 %. For the study area, the Hadley GCM projections were intermediate among those of the major GCMs, and precipitation and temperature trends were linear through the twenty-first century and variability was stable. LANDIS-II was coupled with a forest carbon and water balance model PnET-II, version 4.1–1.2c (Aber et al. 1995). This model matched well empirical observations in the region under current climate (e.g., for aboveground NPP within 4 %, Gustafson et al. 2010).

The most interesting result of the study was that major response variables (e.g., forest composition, Fig. 6.14) were more strongly influenced by timber harvest and insect outbreaks than the direct effects of climate change. The effect of the expected future climate was significant, but its effect was minor compared to harvest and insects, except the abundance of Scots pine. Climate did have a modest effect on the fire regime. The total area burned per decade and mean severity of fires was projected to be slightly increased, with higher variability under future climate. However, both the area burned and fire severity were lower by end of the simulation period under the future climate scenario because of changes in the species composition of the forest. The amount of live aboveground biomass and the level of forest fragmentation were related to the amount of disturbance associated with each scenario. Biomass increased during the last 100 years of the simulations under the insect scenario because insects favor tree species with higher growth rate.

The direct effects of climate change in the study area are not as significant as the exploitation of virgin forest by timber harvest and the potential immigration of the Siberian silk moth. Direct climate effects generally increased tree productivity and modified probability of establishment, but the indirect effects on the fire regime often seemed to counteract the direct effects. Harvest and insects significantly produced changes in forest composition, reduced living biomass, and increased forest fragmentation. It is expected that a warming climate will allow a major tree-killing insect, the Siberian silk moth (*Dendrolimus sibiricus superanse*), which currently does not occur in the study area, to exploit resources north of its current distribution by reducing the frequency of extreme cold events that kill overwintering individuals.

Simulation of the historical climate and disturbance regimes (historical regime of natural variability – HRNV) showed variation within the empirical range of proportions for most species and age classes, but the range of variability for some species (and ages) is currently near the extreme. Several reasons may contribute to this. The initial conditions map was partially impacted by human intervention during the recent decade that has generated more perturbation compared to the HRNV. This may shift the regularities of successions which have a "wavy" character and do not reach equilibrium during relatively short (e.g., 200 years) periods. Finally, while the model reproduces the empirical (current) fire regime quite closely, gradation of disturbance parameters (e.g., fire tolerance) may be too coarse to describe subtle differences of the fire tolerance of some species (e.g., spruce and fir).

Of the multiple global changes studied, climate had the least direct effect on species and seral stage composition. However, the arrival of the Siberian silk moth would be an indirect effect of climate warming, and the consequences may be quite

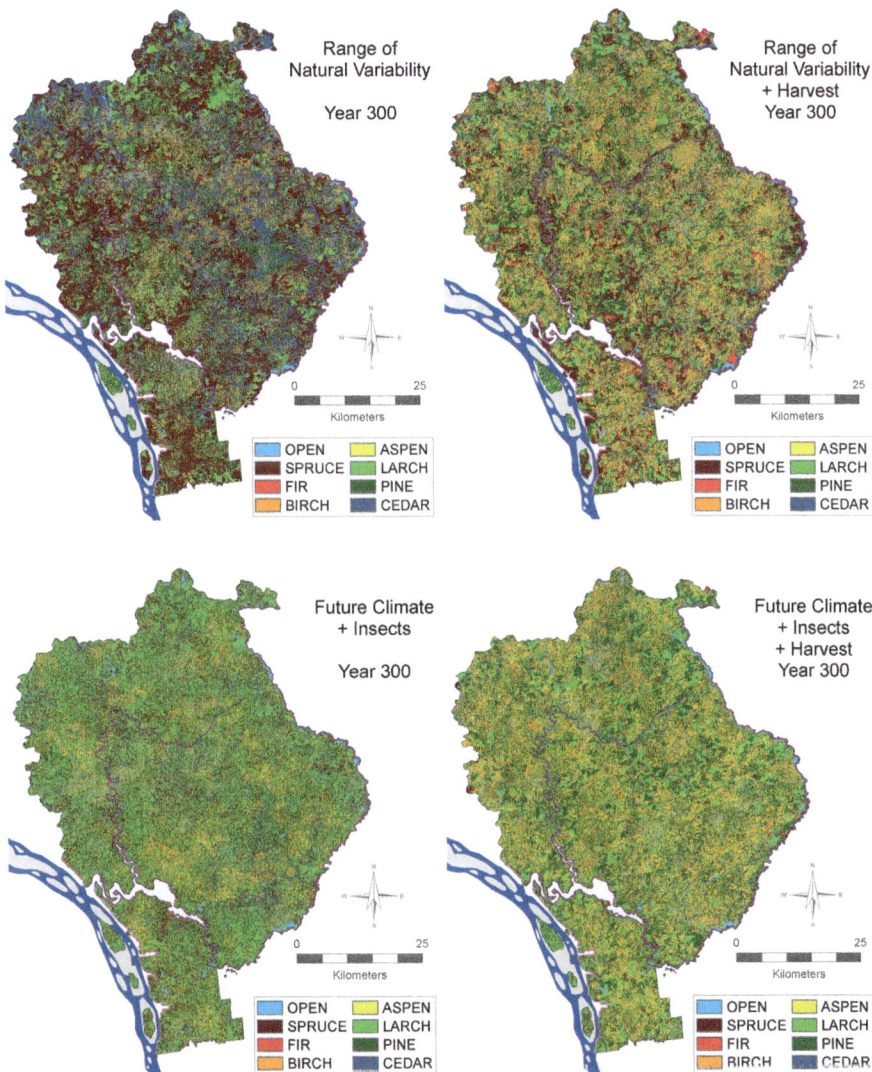

Fig. 6.14 Maps of change of forest composition under four different scenarios. Prediction for intact forests at ecotone of southern and middle taiga (the area at 315×10^3 ha, Ust'-Ilimsk region of Irkutsk Oblast) by LANDIS-II model

devastating. More direct climate effects on composition have a lag time and they induce long-lasting ecological changes that interact with other disturbance processes to fundamentally alter the ecosystem dynamics of these forests. Timber harvest activity produces a sudden and significant change in composition, particularly age class distribution. Climate also did not have a significant effect on biomass, likely because the positive effects of warmer temperatures on net primary production and establishment probability were canceled out by increased losses to fire. This is consistent with projections for boreal forests in Canada (Kang et al. 2006; Kurz et al. 2008a; Girardin and Mudelsee 2008). However, this study did not incorporate CO_2 fertilization effects on net primary production that may significantly interact with other global change effects (Alo and Wang 2008; Peng and Apps 1999).

Another application of LANDIS-II to the same region examined the ability of broad silvicultural strategies to reduce losses to disturbance, maintain the abundance of preferred species, mitigate fragmentation and loss of age class diversity, and sequester carbon under future climate conditions (Gustafson et al. 2011b). A factorial experiment was conducted manipulating three factors representing timber cutting strategies (cutting method, cutting rate and cutblock size). Simultaneously, the model simulated natural disturbances (fire, wind, insect outbreaks) and forest succession under projected future climate conditions as predicted by an ensemble of four global circulation models (CCCma, DFDL-CM2, HadCM3, and MPI-echam5).

As a result, the cutting method and cutting rate treatments had a large effect on species and age class composition, residual living biomass and susceptibility to disturbance. Clear-cutting reduces productivity in the short term, but increases it over the long term (>200 years), an effect that is likely related to changes in species composition. However, some of the more productive species may not be those of greatest commercial value. Cutblock size (within ecologically accepted limits) seems to have little effect except on fragmentation, so it can be used to achieve fragmentation objectives without fear of compromising other objectives. Based on the results, a "recommended" strategy was simulated and compared to the current forest management practice ("business as usual") (Gustafson et al. 2011b). The recommended strategy resulted in higher forest productivity, increased abundance of favored species, and reduced fragmentation, but it did not significantly reduce losses to disturbance. No single strategy appears able to achieve all possible forest management objectives. Forest management has limited power to reduce total losses to disturbance because the species most likely to reduce disturbance rates are not currently economically valuable. Silviculture can have an effect on losses by a single disturbance type, but not all types simultaneously.

Although these results do not point to a clear, universal management solution to sustaining healthy forests in south-central Siberia, they do provide insight into the direction and magnitude of the effects of very general strategies in the face of climate change and interacting disturbances.

The application of LANDIS-II to forests of central Siberian landscapes lead to conclusions that global change is likely to significantly change forest composition of central Siberian landscapes, reduce the ability of Siberian forests to sequester carbon, and may significantly alter ecosystem dynamics and wildlife populations by increasing forest fragmentation. However, the model does not predict irreversible decline and death of forests in the region, as some other models considered above do.

6.7 Brief Summary

In order to understand how Siberian terrestrial ecosystems might respond to ongoing and anticipated climate and environmental change, at least four different important processes should be considered – expected ecosystem change in vitality and productivity, acceleration of natural disturbance, migration and alteration of the land cover (including changes in patterns of species cohorts), and character and intensity of anthropogenic pressure. Global change in both of its major interacting drivers – climate and humans – generates diverse but mostly dangerous challenges for Siberian ecosystem, particularly forests.

Vulnerability of Siberian ecosystems, particularly in a long run, is high. Overall, global change could provide both positive and negative impacts on ecosystems, their spatial distribution, structure, and functioning. Major drivers of increasing ecosystems productivity are: (1) elevated atmospheric CO_2 concentration, (2) increased nitrogen deposition, and (3) longer and warmer growing seasons. However, the combined impacts of these processes on productivity of northern ecosystems (and particularly, forests) and the long-term C balance are not completely clear (see also De Vries et al. 2006; Euskirchen et al. 2006; Juday et al. 2005), particularly the interaction between processes (i.e., dynamics of NPP and HR in the changing environment, intensification of disturbance regimes, and changes in ecosystems). There is much empirical and modeling evidence of changes in ecosystems in boreal Asia, including productivity, both increasing and decreasing, particularly at northern and southern tree-lines and altitudinal transition zones. Overall, terrestrial ecosystems of Siberia still serve a net carbon sink of about 0.3–0.4 Pg C annually. This sink is provided by forests and – to a much smaller extent – by wetlands. Other major land classes serve, on average, as a relatively small carbon source.

Some empirical and modeling results show that future climate and environmental change, together with increasing natural disturbance could have a clearly negative impact on vegetation and change some ecosystems from C sinks to C sources, particularly in disturbed forests and ecosystems on permafrost. These changes also serve as prerequisites to feedbacks to weather and climate, which could accelerate future disturbance regimes in Siberia. The future trajectories of Siberian ecosystems, particularly forests, will strongly depend upon a number of large-scale social and economic decisions which inter alia include transition to integrated land management on the landscape basis, implementation of an ecologically friendly paradigm of industrial development of Siberian territories, increase overall governance of renewable natural resources, and transition to adaptive forest management. These problems are still far from any practical solution in the region.

Risks to terrestrial ecosystems, which include steppe, forest, tundra, and agricultural lands, that are initiated by climate change and anthropogenic pressure are summarized:

- Negative impacts of processes that degrade or destroy permafrost structure including physical destruction of sites, thermokarst, and solifluction
- Dramatic changes of the heat balance of landscapes

- Irreversible changes in hydrological regimes on large territories
- If IPCC climatic predictions were realized, the region would become a large source of carbon, mostly in form of methane
- Loss of soil fertility due to water erosion, soil compaction, desertification, lack of nutrients, salinization, change of water table and other changes in water regimes, and soil contamination
- Decrease of resilience and the decline of biological productivity of boreal forests, particularly at southern edge of the forest domain and in permafrost territories
- Critical increase of mortality of trees in boreal forests
- Lack of water resources in arid areas of the region
- Damage of agriculture lands in river valleys due to an increase in inundation
- An increased or expanded spatial distribution of traditional and new insects and more frequent occurrence of pan-Siberian outbreaks
- Dangerous alteration of forest fire regimes
- Loss of biodiversity due to disturbances and loss of species which are not able to cope with the changing environment
- Impacts of air pollution, soil and water contamination in regions of intensive industrial development

In spite of numerous studies on global change in Siberia, many ecological processes and tendencies are poorly understood. There are a number of fundamental problems that require urgent investigation: setting thresholds of acceptable (nondestructive) impacts on ecosystems taking into account nonlinear and multivariant responses of ecosystems to a long-term accumulation of stress; system (holistic) analysis of a complicated dynamic system to include assessment of ecosystem impacts, responses, and feedbacks in ecosystems that are biophysically unique (i.e., ecologically, socially, economically, and distinctly evolved); theory and practice of decision making in a changing world under a variety of uncertainties; and the development of integrated observing systems. Uncertainty of climatic, social, and economic predictions is high. These uncertainties create special difficulties to the development of future strategies of coevolution of human and nature in high latitudes.

This region requires the urgent development of an anticipatory strategy of adaptation to, and mitigation of, the negative impacts of global change, because this is a large enough region to feedback to regional and global climate systems and this is a region that is predicted to experience temperature changes up to two- to threefold greater than the global mean.

References

Aber JD, Ollinger SV, Federer CA, Reich PB, Goulden ML, Kicklighter DW, Melillo JM, Lathrop RG (1995) Predicting the effects of climate change on water yield and forest production in the northeastern United States. Clim Res 5:207–222

Achard F, Eva HD, Mollicone D, Beuchle R (2008) The effect of climate anomalies and human ignition factor on wildfires in Russian boreal forests. Philos Trans R Soc B Biol Sci 363:2331–2339

Ageenko AS (ed) (1969) Forests of the Russian Far East. Far Eastern Forestry Research Institute, Khabarovsk, 390 pp (in Russian)

Alexeyev VA, Birdsey RA (1998) Carbon storage in forests and peatlands of Russia. General technical report NE-244. USDA, Forest Service, Northeast Research Station, Radnor, PA, 137 pp

Alexeyev VA, Markov MV (2003) Statistical data about forest fund and change of productivity of forests of Russia during the second half of the 20th century. Saint Petersburg Forestry Research Institute, Saint Petersburg, 271 pp (in Russian)

Alo CA, Wang G (2008) Potential future changes of the terrestrial ecosystem based on climate projections by eight general circulation models. J Geophys Res G Biogeosci 113(1):G01004, 16 pp

Anthoni PM, Law BE, Unsworth MH (1999) Carbon and water vapor exchange of an open-canopied ponderosa pine ecosystem. Agric Forest Meteorol 95:151–168

Antonovski MY, Ter-Mikaelian MT, Furyaev VV (1992) A spatial model of long-tem forest fire dynamics and its applications to forests in western Siberia. In: Shugart HH, Leemans R, Bonan GB (eds) A systems analysis of the global boreal forest. Cambridge University Press, New York, pp 373–403

Arneth A, Kurbatova J, Kolle O, Shibistova OB, Lloyd J, Vygodskaya NN, Schulze ED (2002) Comparative ecosystem-atmosphere exchange of energy and mass in a European Russian and a central Siberian bog II. Interseasonal and interannual variability of CO_2 fluxes. Tellus B Chem Phys Meteorol 54:514–530

Bacastov R, Keeling CD (1973) Atmospheric carbon dioxide and radiocarbon in the natural carbon cycle: II. Changes from A.D. 1700–2070 as deducted from a geochemical model. In: Woodwell GM, Pekan EV (eds) Carbon and the biosphere. U.S. Atomic Energy Commission Series 30, Springfield, pp 86–135

Baker DF, Law RM, Gurney KR, Rayner P, Peylin P, Denning AS, Bousquet P, Bruhwiler L, Chen YH, Ciais P, Fung IY, Heimann M, John J, Maki T, Maksyutov S, Masarie K, Prather M, Pak B, Taguchi S, Zhu Z (2006) TransCom 3 inversion intercomparison: impact of transport model errors on the interannual variability of regional CO_2 fluxes, 1988–2003. Glob Biogeochem Cycles 20 (art no. GB1002). doi:10.1029/2004GB002439 17 pp

Balshi MS, McGuire AD, Duffy P, Flannigan M, Walsh J, Melillo J (2009) Assessing the response of area burned to changing climate in western boreal North America using a multivariate adaptive regression splines (MARS) approach. Glob Chang Biol 15:578–600

Baranchikov YuN (2011) Siberian moth – a relentless modifier of taiga forest ecosystems in Northern Asia. In: Boreal forests in a changing world: challenges and needs for actions. IBFRA, Krasnoyarsk, pp 105–107

Baranchikov YN, Kondakov YP, Petrenko ES (2001) Catastrophic outbreaks of Siberian silk moth in forests of Krasnoyarsk Kray. In: Lepeshev AA (ed) Safety of Russia. Regional problems of safety. Krasnoyarsk Kray. Znanie Publ, Moscow, pp 146–147

Barber VA, Juday GP, Finney BP (2000) Reduced growth of Alaskan white spruce in the twentieth century from temperature-induced drought stress. Nature 405:668–673

Bartalev SA, Egorov VA, Loupian EA, Uvarov IA (2007) Multi-year circumpolar assessment of the area burnt on boreal ecosystems using SPOT-VEGETATION. Int J Remote Sens 28:1397–1404

Bazilevich NI (1993) Biological productivity of ecosystems of Northern Eurasia. Nauka, Moscow, 293 pp

Beer C, Lucht W, Schmullius C, Shvidenko A (2006) Small net carbon dioxide uptake by Russian forests during 1981–1999. Geophys Res Lett 33(15):L15403, 4pp

Belelli-Marchesini L (2007) Analysis of the carbon cycle of steppe and old field ecosystems of Central Asia. PhD thesis, Italy University of Tuscia, Viterbo, 227 pp

Belelli-Marchesini L, Papale D, Reichstein M, Vuichard N, Tchebakova N, Valentini R (2007) Carbon balance assessment of a natural steppe of southern Siberia by multiple constraint approach. Biogeosciences 4:581–595

Bonan GB, Shugart HH (1989) Environmental factors and ecological processes in boreal forests. Ann Rev Ecol Syst 20:1–28

Bonan GB, Shugart HH, Urban DL (1990) The sensitivity of some high latitude boreal forests to climatic parameters. Clim Chang 16:9–31

Borzenkova II, Zubakov VA (1984) The climate optimum of the Holocene as a model of the beginning of the 21st century. Russ Meteorol Hydrol 8:69–77

Budyko MI (1974) Climate change. Nauka, Leninggrad, 285 pp (in Russian)

Bulygina ON, Groisman PYa, Razuvaev VN, Korshunova NN (2011) Changes in snow cover characteristics over Northern Eurasia since 1966. Environ Res Lett 6:045204. doi:10.1088/1748-9326/6/4/045204, 10 pp

Chapin Iii FS, McGuire AD, Randerson J, Pielke R Sr, Baldocchi D, Hobbie SE, Roulet N, Eugster W, Kasischke E, Rastetter EB, Zimov SA, Running SW (2000) Arctic and boreal ecosystems of western North America as components of the climate system. Glob Chang Biol 6:211–223

Chen YH, Prinn RG (2006) Estimation of atmospheric methane emissions between 1996 and 2001 using a three-dimensional global chemical transport model. J Geophy Res D Atmos 111(10):D10307, 25 pp

Chudnikov PI (1931) Impact of fire on regeneration of Urals Forests. Selkhozgiz (Agricultural State Publishing House), Moscow, 160 pp (in Russian)

Ciais P, Janssens I, Shvidenko A, Wirth C, Malhi Y, Grace J, Schulze ED, Heimann M, Phillips O, Dolman AJ (2005) The potential for rising CO_2 to account for the observed uptake of carbon by tropical, temperate, and boreal forest biomes. SEB Exp Biol Series 109–149

Ciais P, Canadell JG, Luyssart S et al (2010) Can we reconcile atmospheric estimates of the Northern terrestrial carbon sink with land-based accounting? Curr Options Environ Sustain 2:1–6

Corradi C, Kolle O, Walter K, Zimov SA, Schulze ED (2005) Carbon dioxide and methane exchange of a north-east Siberian tussock tundra. Glob Chang Biol 11:1910–1925

Cramer W, Kicklighter DW, Bondeau A, Moore Iii B, Churkina G, Nemry B, Ruimy A, Schloss AL (1999) Comparing global models of terrestrial net primary productivity (NPP): overview and key results. Glob Chang Biol 5:1–15

Davidson EA, Janssens IA (2006) Temperature sensitivity of soil carbon decomposition and feedbacks to climate change. Nature 440:165–173

Davis MB, Shaw RG (2001) Range shifts and adaptive responses to quaternary climate change. Science 292:673–679

de Graaff MA, van Groenigen KJ, Six J, Hungate B, van Kessel C (2006) Interactions between plant growth and soil nutrient cycling under elevated CO_2: a meta-analysis. Glob Chang Biol 12:2077–2091

De Vries W, Reinds GJ, Gundersen P, Sterba H (2006) The impact of nitrogen deposition on carbon sequestration in European forests and forest soils. Glob Chang Biol 12:1151–1173

Devi N, Hagedorn F, Moiseev P, Bugmann H, Shiyatov S, Mazepa V, Rigling A (2008) Expanding forests and changing growth forms of Siberian larch at the Polar Urals treeline during the 20th century. Glob Chang Biol 14:1581–1591

Dixon RK, Brown S, Houghton RA, Solomon AM, Trexler MC, Wisniewski J (1994) Carbon pools and flux of global forest ecosystems. Science 263:185–190

Efremov D, Shvidenko A (2004) Long-period consequences of catastrophic fires in forests of the Russian Far East and their impacts on global processes. In Furyaev VV (ed) Forest fire management at ecoregional level. World Bank, Moscow, pp 66–73 (in Russian)

Epova VI, Pleshanov AS (1995) Zones of injuriousness of insects-phyllophagous of Asian Russia. Nauka, Novosibirsk, p 46

Ershov DV, Kovganenko KA, Sochilova EN (2009) GIS-tecnology for estimation of fire carbon emissions based on Terra-MODIS data and state forest account. Curr Probl Earth Obs Space 6(II):365–372 (in Russian)

Euskirchen ES, McGuire AD, Kicklighter DW, Zhuang Q, Clein JS, Dargaville RJ, Dye DG, Kimball JS, McDonald KC, Melillo JM, Romanovsky VE, Smith NV (2006) Importance of recent shifts in soil thermal dynamics on growing season length, productivity, and carbon sequestration in terrestrial high-latitude ecosystems. Glob Chang Biol 12:731–750

Fedorov SV (1977) Studying of water balance elements in forest zone of the European area of the USSR. Hydrometeoizdat, Leningrad, 264 pp (in Russian)

FAFMRF (2010) Report on forest pathological monitoring in forests of Russia in 2009. Federal Agency of Forest Management of Russian Federation, Moscow, 86 p

FAO (2006) World reference base for soil resources 2006, 2nd edn. World Soil Resources Reports. IUSS Working Group WRB. FAO, Rome, 145 pp. ISBN: 92–5–105511–4. ftp://ftp.fao.org/agl/agll/docs/wsrr103e.pdf

Farber SK (2000) Formation of stands of East Siberia. Nauka, Novosibirsk, 433 pp (in Russian)

Feiziene D, Feiza V, Vaideliene A, Povilaitis V, Antanaitis S (2010) Soil surface carbon dioxide exchange rate as affected by soil texture, different long-term tillage application and weather. Zemdirbyste 97:25–42

FFSR (1962) Forest fund of RSFSR (state by January 1, 1961). Main Forest Management Department of Council of Ministers of RSFSR. Moscow, 628 p (in Russian)

FFSSR (2008) Regions of Russia. Social and economic indicators. Official edition. Federal Service of State Statistics, Moscow, 1000 pp (in Russian)

Filipchuk AN, Moiseev BN (2003) Assessment of atmospheric carbon uptake by vegetation cover in Russia. World climate conference, Moscow, 29 Sept–3 Oct, 503 p

Flannigan MD, Van Wagner CE (1991) Climate change and wildfire in Canada. Can J For Res 21:66–72

Flannigan M, Stocks B, Turetsky M, Wotton M (2009) Impacts of climate change on fire activity and fire management in the circumboreal forest. Glob Chang Biol 15:549–560

Forster P, Ramaswamy V (2007) Changes in atmospheric constituents and in radiative forcing. In: Solomon S.e.a (ed) Climate change 2007: the physical science basis. Contribution of WG I to the fourth assessment report of the IPCC. Cambridge University Press, Cambridge

Fridland VM (1989) Soil map of the USSR. Committee on Cartography and Geodesy, Moscow, (in Russian)

Furyaev VV (1996) Rope of fire in process of forest development. Nauka, Novosibirsk, 252 pp

Furyaev VV, Vaganov EA, Tchebakova NM (2001) Effects of fire and climate on successions and structural changes in the Siberian boreal forest. Eurasian J For Res 2:1–15

Geographical… (2007) Geographical studies in Siberia. In: Landscape hydrology: fundamental and practical research. Geo Publishing, Novosibirsk, 262 p

Gerten D, Schaphoff S, Haberlandt U, Lucht W, Sitch S (2004) Terrestrial vegetation and water balance – hydrological evaluation of a dynamic global vegetation model. J Hydrol 286:249–270

Gifford RM (1994) The global carbon cycle: a viewpoint on the missing sink. Aust J Plant Physiol 21:1–15

Gillett NP, Weaver AJ, Zwiers FW, Flannigan MD (2004) Detecting the effect of climate change on Canadian forest fires. Geophys Res Lett 31:L18211 1–4

Girardin MP, Mudelsee M (2008) Past and future changes in Canadian boreal wildfire activity. Ecol Appl 18:391–406

Glagolev MV, Sirin AA, Lapshina ED, Filippov IV (2010) Carbonaceous greenhouse gases flux from West Siberian wetland ecosystems. Herald Tomsk State Polytech Univ 3(93):120–127

Glagolev M, Kleptsova I, Filippov I, Maksyutov S, Machida T (2011) Regional methane emission from West Siberia mire landscapes. Environ Res Lett 6:045214

Gordon C, Cooper C, Senior CA, Banks H, Gregory JM, Johns TC, Mitchell JFB, Wood RA (2000) The simulation of SST, sea ice extents and ocean heat transports in a version of the Hadley Centre coupled model without flux adjustments. Clim Dyn 16:147–168

Goulden ML, Munger JW, Fan SM, Daube BC, Wofsy SC (1996) Measurements of carbon sequestration by long-term eddy covariance: methods and a critical evaluation of accuracy. Glob Chang Biol 2:169–182

Gurney KR, Law RM, Denning AS, Rayner PJ, Baker D, Bousquet P, Bruhwiler L, Chen YH, Clals P, Fan S, Fung IY, Gloor M, Heimann M, Higuchi K, John J, Maki T, Maksyutov S, Masarie K, Peylin P, Prather M, Pak BC, Randerson J, Sarmiento J, Taguchi S, Takahashi T, Yuen CW (2002) Towards robust regional estimates of CO_2 sources and sinks using atmospheric transport models. Nature 415:626–630

Gurney KR, Law RM, Denning AS, Rayner PJ, Baker D, Bousquet P, Bruhwiler L, Chen YH, Ciais P, Fan S, Fung IY, Gloor M, Heimann M, Higuchi K, John J, Kowalczyk E, Maki T, Maksyutov S, Peylin P, Prather M, Pak BC, Sarmiento J, Taguchi S, Takahashi T, Yuen CW (2003) TransCom 3 CO_2 inversion intercomparison: 1. Annual mean control results and sensitivity to transport and prior flux information. Tellus B Chem Phys Meteorol 55:555–579

Gurney KR, Baker D, Rayner P, Denning S (2008) Interannual variations in continental-scale net carbon exchange and sensitivity to observing networks estimated from atmospheric CO_2 inversions for the period 1980 to 2005. Global Biogeochem Cycles 22:GB3025

Gustafson EJ, Shvidenko AZ, Sturtevant BR, Scheller RM (2010) Predicting global change effects on forest biomass and composition in south-central Siberia. Ecol Appl 20:700–715

Gustafson EJ, Sturtevant BR, Shvidenko AZ, Scheller RM (2011a) Using landscape disturbance and succession models to support forest management. Chapter 5. In: Li C, Lafortezza R, Chen J (eds) Landscape ecology in forest management and conservation. Challenges and solutions for global change. Higher Education Press and Springer, Beijing/New York, pp 99–118

Gustafson EJ, Shvidenko AZ, Sheller RM (2011b) Effectiveness of forest management strategy to mitigate effects of global change in south-central Siberia. Can J For Res 41:1405–1421

Hamilton LS (2008) Forest and water. Food and Agriculture Organization of the United Nations, Rome, p 78

Hayes DJ, McGuire AD, Kicklighter DW, Gurney KR, Burnside TJ, Melillo JM (2011) Is the northern high latitude land-based CO_2 sink weakening? Global Biogeochem Cycles 25(3):GB3018. doi:10.1029/2010gb003813

Heinselman (1978) Fire intensity and frequency as factors in the disturbance and structure of northern ecosystems. In: Mooney HA, Bonnicksen TM, Christensen NL, et al (eds) Fire regimes and ecosystem properties. General technical report WO-26. USDA, Honolulu, pp 7–57

Hudson JMG, Henry GHR (2009) Increased plant biomass in a high arctic heath community from 1981 to 2008. Ecology 90:2657–2663

IPCC (2001) Climate change 2001: mitigation. Contribution of working group III to the third assessment report of the intergovernmental panel on climate change. Cambridge University Press/IPCC, Cambridge, 881 p

IPCC (2007a) In: Metz B, Davidson OR, Bosch PR, Dave R, Meyer LA (eds) Climate change 2007: mitigation. Contribution of working group III to the fourth assessment report of the intergovernmental panel on climate change. Cambridge University Press, Cambridge/New York

IPCC (2007b) Climate change 2007: impact, adaptation and vulnerability. Chapter 4: Ecosystems, their properties, goods and services. Contribution of working group II to the fourth assessment report. Cambridge University Press, Cambridge, pp 211–272

Isaev AS (1991) Forecast of use and reproduction of forest resources by economic regions of the USSR by 2010. State Committee of USSR on Forest, Moscow, p 508

Isaev AS (ed) (1997) Program of the extraordinary activities on biological struggle with pests in forests of Krasnoyarsk Krai. Federal Forest Service of Russia, Moscow, 157 pp (in Russian)

Isaev AS, Khebopros RG, Nedorezov LV, Kondakov YP, Kiselev VV, Sukhovolsky VG (2001) Population dynamics of forest pests. Nauka, Moscow, 374 pp

Ivanova GA (1996) Extreme fire seasons in Evenkia forests. Siberian Ecol J 1:29–34

Ivanova GA (1998–1999) The history of forest fire in Russia. Dendrochronologia 16–17, 147–161

Jenkinson DS, Adams DE, Wild A (1991) Model estimates of CO_2 emissions from soil in response to global warming. Nature 351:304–306

Juday GP, Barber V, Duffy P, Linderholm H, Rupp S, Sparrow S, Vaganov Eu, Yarie J (2005) Forests, land management, and agriculture. Chapter 14. In: ACIA, Arctic climate impact assessment 2005, Cambridge University Press, pp 782–862

Kabat P, Claussen M, Dirmeyer PA, Gash JHC, Bravo de Guenni L, Meybeck M, Pielke RA Sr, Vörösmarty CJ, Hutjes RWA, Lütkemeier S (eds) (2004) Vegetation, water, humans and the climate: a new perspective on an interactive system. Springer, Berlin, 566 pp

Kajii Y, Kato S, Streets DG, Tsai NY, Shvidenko A, Nilsson S, McCallum I, Minko NP, Abushenko
  N, Altyntsev D, Khodzer TV (2002) Boreal forest fires in Siberia in 1998: estimation of area
  burned and emissions of pollutants by advanced very high resolution radiometer satellite data.
  J Geophys Res D Atmos 107(24):ACH 4-1–ACH 4-8 art. no. 4745
Kang S, Kimball JS, Running SW (2006) Simulating effects of fire disturbance and climate change
  on boreal forest productivity and evapotranspiration. Sci Total Environ 362:85–102
Kaplun F, Brumarova N, Vidyapin V (eds) (1994) New Russia. Information and statistical book.
  International Academy of Informatization, Moscow, 740 pp (in Russian)
Karelin DV, Zamolodchikov DG (2008) Carbon exchange in cryogenic ecosystems. Nauka,
  Moscow, 343 pp
Kasischke ES, Turetsky MR (2006) Recent changes in the fire regime across the North American
  boreal region – spatial and temporal patterns of burning across Canada and Alaska. Geophys
  Res Lett 33:1–5
Kharuk VI, Im ST, Ranson KJ, Naurzbaev MM (2004) Temporal dynamics of larch in the forest-
  tundra ecotone. Dokl Earth Sci 398:1020–1023
Kharuk VI, Ranson KJ, Im ST, Naurzbaev MM (2006) Forest-tundra larch forests and climatic
  tends. Russ J Ecol 37:291–298
Kharuk V, Ranson K, Dvinskaya M (2007) Evidence of evergreen conifer invasion into larch dominated forests during recent decades in central Siberia. Eurasian J For Res 10:163–171
Kharuk VI, Dvinskaya ML, Im ST, Ranson KJ (2008) Tree vegetation of the forest-tundra ecotone
  in the western Sayan mountains and climatic trends. Russ J Ecol 39:8–13
Kharuk VI, Ranson KJ, Im ST (2009) Climate-induced mountain tree line ecotone dynamics in
  southern Siberia. International conference on computational information technologies for environmental sciences CITES-2009, Krasnoyarsk, 5–15 July, p 83
Kharuk VI, Im ST, Dvinskaya ML (2010a) Forest-tundra ecotone response to climate change in the
  western Sayan mountains, Siberia. Scand J For Res 25:224–233
Kharuk VI, Im ST, Dvinskaya ML, Ranson KJ (2010b) Climate-induced mountain tree-line evolu-
  tion in southern Siberia. Scand J For Res 25:446–454
Kharuk VI, Ranson KJ, Im ST, Vdovin AS (2010c) Spatial distribution and temporal dynamics of
  high-elevation forest stands in southern Siberia. Glob Ecol Biogeogr 19:822–830
Khotinsky NA (1984) Holocene vegetation history. In: Velichko AA, Wright HE, Barnosky CW
  (eds) Late quaternary environments of the Soviet Union. University of Minnesota Press,
  Minneapolis, pp 179–200
Kim H-S, Maksyutov S, Glagolev MV, Machida T, Patra PK, Sudo K, Inoue G (2011) Evaluation
  of methane emissions from West Siberian wetlands based on inverse modeling. Environ Res
  Lett 6:035201, 8pp
King GA, Herstrom AA (1997) Holocene tree migration rates objectively determined from fossil
  pollen data. In: Huntley B et al (eds) Past and future rapid environmental changes: the spatial
  and evolutionary responses of terrestrial biota. Springer, New York, pp 91–101
Kirschbaum MUF (1993) A modelling study of the effects of changes in atmospheric CO_2 concen-
  tration, temperature and atmospheric nitrogen input on soil organic carbon storage. Tellus Ser
  B 45 B:321–334
Kirschbaum MU (1995) The temperature dependence of soil organic matter decomposition, and
  the effect of global warming on soil organic C storage. Soil Biol Biochem 27:753–760
Klimchenko AV (2007) Parameters of carbon cycle in restoration-age-specific row of larch for-
  ests with small shrubs and green mosses of northern taiga of Middle Siberia. PhD thesis,
  Institute of Forest of Siberian Branch, Russian Academy of Sciences, Krasnoyarsk, 21 pp (in
  Russian)
Knohl A, Kolle O, Minayeva TY, Milyukova IM, Vygodskaya NN, Foken T, Schulze ED (2002)
  Carbon dioxide exchange of a Russian boreal forest after disturbance by wind throw. Glob
  Chang Biol 8:231–246
Knorre AA (2003) Using of registering structures to estimate annual production of phytocoenoses
  components in forest and swamp ecosystems of Yenisey Siberia. V.N. Sukachev Institute of
  Forest, Abstracts of PhD thesis, Krasnoyarsk, 24 pp (in Russian)

Kolesnikov BP (1956) Cedar forests of the Far East. Transactions of the Far Eastern branch of the Academy of Sciences of the USSR. Ser Bot 2(4), 264 pp (in Russian)

Kondakov YP (1987) Phytocenotic specifics of outbreaks of needle- and leaf-eating insects in forests of Siberia. In: Vladyshevsky DV, Petrenko ES (eds) Ecological assessment of habitats of forest animals. Nauka, Novosibirsk, pp 29–40

Kudeyarov VN, Kurganova IN (1998) Carbon dioxide emissions and net primary production of Russian terrestrial ecosystems. Biol Fertil Soils 27:246–250

Kulikov AN (ed) (1998) Forest fires in Russian Far East. Project's report WWE RU1029, Khabarovsk

Kullman L (2007) Tree line population monitoring of *Pinus sylvestris* in the Swedish Scandes, 1973–2005: Implications for tree line theory and climate change ecology. J Ecol 95:41–52

Kurbatova J, Arneth A, Vygodskaya NN, Kolle O, Varlargin AV, Milyukova IM, Tchebakova NM, Schulze ED, Lloyd J (2002) Comparative ecosystem-atmosphere exchange of energy and mass in a European Russian and a central Siberian bog I. Interseasonal and interannual variability of energy and latent heat fluxes during the snowfree period. Tellus B Chem Phys Meteorol 54:497–513

Kurz WA, Apps MJ, Stocks BJ (1995) Global climate change: disturbance regimes and biospheric feedbacks of temperate and boreal forests. In: Woodwell G (ed) Biotic feedbacks in the global climate system: will the warming speed the warming? Oxford University Press, Oxford, pp 119–1333

Kurz WA, Stinson G, Rampley GJ, Dymond CC, Neilson ET (2008a) Risk of natural disturbances makes future contribution of Canada's forests to the global carbon cycle highly uncertain. Proc Natl Acad Sci U S A 105:1551–1555

Kurz WA, Dymond CC, Stinson G, Rampley GJ, Neilson ET, Caroll AL, Ebata T, Safranyik L (2008b) Mountain pine beetle and forest carbon feedback to climate change. Nature 452:987–990

Lapenis A, Shvidenko A, Shepaschenko D, Nilsson S, Aiyyer A (2005) Acclimation of Russian forests to recent changes in climate. Glob Chang Biol 11:2090–2102

Lavrenko EM, Sochava VB (eds) (1956) Vegetation cover of the USSR. Academy of Sciences of the USSR, Moscow/Leningrad, 460 pp (in Russian)

le Goff H, Flannigan MD, Bergeron Y (2009) Potential changes in monthly fire risk in the eastern Canadian boreal forest under future climate change. Can J For Res 39:2369–2380

Le Mer J, Roger P (2001) Production, oxidation, emission, and consumption of methane by soils: a review. Eur J Soil Biol 37:25–50

Lebedev AV, Gorbatenko VM, Krasnoschekov UN, Reshetkova NB, Protopopov VV (1974) Environmental role of forests of the lake Baikal basin. Nauka, Novosibirsk, 215 pp (In Russian)

Lenton TM, Held H, Kriegler JW, Lucht W, Rahmstorf S, Schellnhuber HJ (2008) Tipping elements in the earth climate system. PNAS 105(6):1786–1793

Liski J, Karjalainen T, Pussinen A, Nabuurs GJ, Kauppi P (2000) Trees as carbon sinks and sources in the European Union. Environ Sci Policy 3:91–97

Liu J (2010) Investigation in forest and water relations with an emphasis on China. IUFRO world congress 2010, Seoul

Lloyd AH, Bunn AG (2007) Responses of the circumpolar boreal forest to 20th century climate variability. Environ Res Lett 4:045013

Lloyd J, Shibistova O, Zolotoukhine D, Kolle O, Arneth A, Wirth C, Styles JM, Tchebakova NM, Schulze ED (2002) Seasonal and annual variations in the photosynthetic productivity and carbon balance of a central Siberian pine forest. Tellus B Chem Phys Meteorol 54:590–610

Luyssaert S, Ciais P, Piao SL, Schulze ED, Jung M, Zaehle S, Schelhaas MJ, Reichstein M, Churkina G, Papale D, Abril G, Beer C, Grace J, Loustau D, Matteucci G, Magnani F, Nabuurs GJ, Verbeeck H, Sulkava M, van der Werf GR, Janssens IA (2010) The European carbon balance. Part 3: Forests. Glob Chang Biol 16:1429–1450

Lyons WA, Nelson TE, Williams ER, Cramer JA, Turner TR (1998) Enhanced positive cloud-to-ground lightning in thunderstorms ingesting smoke from fires. Science 282:77–80

Magnani F, Mencuccini M, Borghetti M, Berbigier P, Berninger F, Delzon S, Grelle A, Hari P, Jarvis PG, Kolari P, Kowalski AS, Lankreijer H, Law BE, Lindroth A, Loustau D, Manca G, Moncrieff JB, Rayment M, Tedeschi V, Valentini R, Grace J (2007) The human footprint in the carbon cycle of temperate and boreal forests. Nature 447:848–850

Maksyutov S, Machida T, Mukai H, Patra PK, Nakazawa T, Inoue G (2003) Effect of recent observations on Asian CO_2 flux estimates by transport model inversions. Tellus B Chem Phys Meteorol 55:522–529

Manko UI, Gladkova GA (2001) Drying of spruce in the context of global degradation of dark coniferous forest. Dalnauka, Vladivostok, 231 pp

Manö S, Andreae MO (1994) Emission of methyl bromide from biomass burning. Science 263:1255–1257

McClelland JW, Holmes RM, Peterson BJ, Amon R, Brabets T, Cooper L, Gibson J, Gordeev VV, Guay C, Milburn D, Staples R, Raymond PA, Shiklomanov I, Striegl R, Zhulidov A, Gurtovaya T, Zimov S (2008) Development of Pan-Arctic database for river chemistry. Eos 89:217–218

McGuire AD, Chapin Iii FS, Walsh JE, Wirth C (2006) Integrated regional changes in arctic climate feedbacks: implications for the global climate system. Annu Rev Environ Resour 31:61–91

McGuire AD, Anderson LG, Christensen TR, Scott D, Laodong G, Hayes DJ, Martin H, Lorenson TD, Macdonald RW, Nigel R (2009) Sensitivity of the carbon cycle in the Arctic to climate change. Ecol Monogr 79(4):523–555

McGuire AD, Hayes DJ, Kicklighter DW, Manizza M, Zhuang Q, Chen M, Follows MJ, Gurney KR, McClelland JW, Melillo JM, Peterson BJ, Prinn R (2010a) An analysis of the carbon balance of the Arctic Basin from 1997 to 2006. Tellus 62B:455–474. doi:10.1111/j.1600-0889.2010.00497.x

McGuire AD, Hayes DJ, Kicklighter DW, Zhuang Q, Chen M, Gurney RK e a (2010b) Analysis of the carbon balance of boreal Asia from 1997 to 2006. In: Proceedings of the international conference on environmental observations, modeling and information system ENVIROMIS, Tomsk, Russia, 5–11 July, 2010, pp 53–58

Meleshko VP, Katsov VM, Govorkova VA (2008) Climate of Russia in the XXI century. 3. Future climate changes obtained from an ensemble of the coupled atmosphere-ocean GCM CMIP3. Meteorol Hydrol 9:5–22

Mladenoff DJ (2004) LANDIS and forest landscape models. Ecol Model 180:7–19

Mokhov I, Chernokulsky AV, Shkolnik IM (2006) Regional model assessment of fire risks under global climate change. Dokl Earth Sci 411A(9):1485–1488

Mokhov II, Chernokul'Skii AV, Akperov MG, Dufresne JL, Le Treut H (2009) Variations in the characteristics of cyclonic activity and cloudiness in the atmosphere of extratropical latitudes of the Northern Hemisphere based from model calculations compared with the data of the reanalysis and satellite data. Dokl Earth Sci 424:147–150

Monserud RA, Leemans R (1992) Comparing global vegetation maps with the Kappa statistic. Ecol Model 62:275–293

Monserud RA, Tchebakova NM, Kolchugina TP, Denissenko OV (1996) Change in Siberian phytomass predicted for global warming. Silva Fenn 30:185–200

Morishita T, Hatano R, Desyatkin RV (2003) CH4 flux in an Alas ecosystem formed by forest disturbance near Yakutsk, Eastern Siberia, Russia. Soil Sci Plant Nutr 49:369–377

Mukhortova LV, Evgrafova SI (2005) Dynamics of organic matter decomposition and microflora composition of forest litter in artificial biogeocenoses. Proc Russ Acad Sci (Biol) 6:731–737

Nilsson S, Vaganov EA, Shvidenko AZ, Stolbovoi V, Rozhkov VA, MacCallum I, Ionas M (2003) Carbon budget of vegetation ecosystems of Russia. Dokl Earth Sci 393:1281–1283

Olchev A, Ibrom A, Ross T, Falk U, Rakkibu G, Radler K, Grote S, Kreilein H, Gravenhorst G (2008) A modelling approach for simulation of water and carbon dioxide exchange between multi-species tropical rain forest and the atmosphere. Ecol Model 212:122–130

Olchev A, Novenko E, Desherevskaya O, Krasnorutskaya K, Kurbatova J (2009) Effects of climatic changes on carbon dioxide and water vapor fluxes in boreal forest ecosystems of European part of Russia. Environ Res Lett 4:045007 (8 pp)

Oltchev A, Cermak J, Nadezhdina N, Tatarinov F, Tishenko A, Ibrom A, Gravenhorst G (2002) Transpiration of a mixed forest stand: field measurements and simulation using SVAT models. Boreal Environ Res 7:389–397

Onuchin AA (2009) Hydrological role of forest ecosystems of boreal zone. In: Proceedings of the international conference on ecological and geographical aspects of forest forming process, 23–25 Sept 2009, Krasnoyarsk, pp 11–14

Onuchin AA, Burenina TA (2008) Hydrological role of forests in Siberia. In: Prescott AP, Barkely TU (eds) Trends in water research. Nova, New York, pp 67–92

Onuchin A, Balzter H, Borisova H, Blyth E (2006) Climatic and geographic patterns of river runoff formation in Northern Eurasia. Adv Water Resour 29:1314–1327

Onuchin A, Burenina T, Gaparov K, Zyrukina N (2009) Land use impacts on river hydrological regimes in Northern Asia. In: Proceedings of the international conference "Hydroinformatics in hydrology, hydrogeology and water resources", Hyderabad, India, 6–12 September, pp 163–170

Orlov DS, Biryukova ON, Sukhanova NI (1996) Organic matter in soils of the Russian Federation. Nauka, Moscow, 254 pp (in Russian)

Pacala S, Birdsey RA, Bridgham SD, Conant RT, Davis K, Hales B, Hougthon RA, Jenkins JC, Johnston M, Marland G, Paustian K (2009) The North American carbon budget past and present. Chapter 3. In: The first state of the carbon cycle report (SOCCR). The North American carbon budget and implications for the global carbon cycle. National Oceanic and Atmospheric Administration, National Climatic Data Center, Asheville, NC pp 29–36

Pan Y, Birdsey R, Fang J, Hougthon R, Kaippi PE, Kurz AA, Phillips OL, Shvidenko A, Lewis SL, Canadell JG, Ciais P, Jackson RB, Pacala SW, McGuire AD, Piao S, Rautianen A, Sitch S, Hayes D (2011) A large and persistent carbon sink in the world's forests. Science 333:988–993

Patra PK, Gurney KR, Denning AS, Maksyutov S, Nakazawa T, Baker D, Bousquet P, Bruhwiler L, Chen YH, Ciais P, Fan S, Fung I, Gloor M, Heimann M, Higuchi K, John J, Law RM, Maki T, Pak BC, Peylin P, Prather M, Rayner PJ, Sarmiento J, Taguchi S, Takahashi T, Yuen CW (2006) Sensitivity of inverse estimation of annual mean CO_2 sources and sinks to ocean-only sites versus all-sites observational networks. Geophys Res Lett 33:L05814

Peng C, Apps MJ (1999) Modelling the response of net primary productivity (NPP) of boreal forest ecosystems to changes in climate and fire disturbance regimes. Ecol Model 122:175–193

Pleshanov AS (1982) Insects-defoliators of larch forests of East Siberia. Nauka, Novosibirsk, 209 pp

Pleshikov FI (ed) (2002) Forest ecosystems of the Yenisey Meridian. Publ. House of SB RAN, Novosibirsk

Polykarpov NP, Tchebakova NP, Nazimova DI (1986) Climate and mountain forests of Southern Siberia. Nauka, Novosbirsk, 225 pp (in Russian)

Popov LV (1982) Southern taiga forests of Middle Siberia. Irkutsk State University, Irkutsk, 330 pp (in Russian)

Protopopov VV, Lebedev AV, Bizyukin VV (1991) Water protective role of forests of the lake Baikal basin. In: Proceedings of the scientific conference on problems of water resources of the Far Eastern economic region and Zabaikalie. Hydrometeoizdat, St. Petersburg, pp 298–310

Purves D, Pacala S (2008) Predictive models of forest dynamics. Science 320:1452–1453

Quegan S, Beer C, Shvidenko A, McCallum I, Handoh IC, Peylin P, Rödenbeck C, Lucht W, Nilsson S, Schmullius C (2011) Estimating the carbon balance of central Siberia using a land-scape-ecosystem approach, atmospheric inversion and dynamic global vegetation models. Glob Chang Biol 17:351–365

Raich JW, Nadelhoffer KJ (1989) Belowground carbon allocation in forest ecosystems: global trends. Ecology 70:1346–1354

Raich JW, Potter CS, Bhagawati D (2002) Interannual variability in global soil respiration, 1980–1994. Glob Chang Biol 8:800–812

Raich JW, Schlesinger WH (1992) The global carbon dioxide flux in soil respiration and its relationship to vegetation and climate. Tellus 44B: 81–99

Rautiainen A, Wernick I, Waggoner PE, Ausubel JH, Kauppi PE (2011) A national and international analysis of changing forest density. PLoS One 6:e19577

Rehfeldt GE, Tchebakova NM, Parfenova EI (2004) Genetic responses to climate and climate-change in conifers of the temperate and boreal forests. Recent Res Dev Genet Breed 1:113–130

Rizzo B, Wiken E (1992) Assessing the sensitivity of Canada's ecosystems to climatic change. Clim Chang 21:37–55

Röser C, Montagnani L, Schulze ED, Mollicone D, Kolle O, Meroni M, Papale D, Marchesini LB, Federici S, Valentini R (2002) Net CO_2 exchange rates in three different successional stages of the "Dark Taiga" of central Siberia. Tellus B Chem Phys Meteorol 54:642–654

Rozhkov AS (1965) Outbreaks of Siberian silk moth and measures of their control. Nauka, Moscow, 180 pp

Rozhkov VA, Wagner VB, Kogut VM (1996) Soil carbon estimates and soil carbon map for Russia. International Institute for Applied Systems Analysis, Laxenburg, 44 pp

Santoro M, Beer C, Cartus O, Schmullius C, Shvidenko A, McCallum I, Wegmüller U, Wiesmann A (2011) Retrieval of growing stock volume in boreal forest using hyper-temporal series of Envisat ASAR ScanSAR backscatter measurements. Remote Sens Environ 115:490–507

Scheller RM, Mladenoff DJ (2007) An ecological classification of forest landscape simulation models: tools and strategies for understanding broad-scale forested ecosystems. Landsc Ecol 22:491–505

Scheller RM, Domingo JB, Sturtevant BR, Williams JS, Rudy A, Gustafson EJ, Mladenoff DJ (2007) Design, development, and application of LANDIS-II, a spatial landscape simulation model with flexible temporal and spatial resolution. Ecol Model 201:409–419

Schepaschenko D, McCallum I, Shvidenko A, Fritz S, Kraxner F, Obersteiner M (2010) A new hybrid land cover dataset for Russia: a methodology for integrating statistics, remote sensing and in-situ information. J Land Use Sci 6(4):245–259. doi:10.1080/1747423X.2010.511681, Published online 22 December 2010

Schepaschenko DG, Shvidenko AZ, Mukhortova LV, Schepaschenko MV (2011) Soil in the estimation of biospheric role of terrestrial ecosystems of Russia. In: Aparin BF (ed) International conference V.V. Dokuchaev 165 anniversary. Resource potential of soils – basis of food and ecological safety of Russia, Saint Petersburg State University, Saint Petersburg, 1–4 March, pp 511–512

Scherbakov IP (1975) Forest cover of North-East of the USSR. Nauka Publisher, Novosibirsk, 344 p (in Russian)

Schimel DS, Parton WJ, Kittel TGF, Ojima DS, Cole CV (1990) Grassland biogeochemistry: links to atmospheric processes. Clim Chang 17:13–25

Schmid HP, Grimmond CSB, Cropley F, Offerle B, Su HB (2000) Measurements of CO_2 and energy fluxes over a mixed hardwood forest in the mid-western United States. Agr Forest Meteorol 103:357–374

Schmullius C, Santoro M (2005) SIBERIA-II. Multi-sensor concepts for greenhouse gas accounting of Northern Eurasia, Contract Number EVG1-CT-2001–0048, EC Deliverable, Final report, 167 pp

Schneider MK, Lüscher A, Richter M, Aeschlimann U, Hartwig UA, Blum H, Frossard E, Nösberger J (2004) Ten years of free-air CO_2 enrichment altered the mobilization of N from soil in Lolium perenne L. swards. Glob Chang Biol 10:1377–1388

Schulze ED (2002) Understanding global change: lessons learnt from the European landscape. J Veg Sci 13:403–412

Schulze ED, Lloyd J, Kelliher FM, Wirth C, Rebmann C, Luhker B, Mund M, Knohl A, Milyukova IM, Schulze W, Ziegler W, Varlagin AB, Sogachev AF, Valentini R, Dore S, Grigoriev S, Kolle O, Panfyorov MI, Tchebakova N, Vygodskaya NN (1999) Productivity of forests in the Eurosiberian boreal region and their potential to act as a carbon sink – a synthesis. Glob Chang Biol 5:703–722

Schuur EAG, Vogel JG, Crummer KG, Lee H, Sickman JO, Osterkamp TE (2009) The effect of permafrost thaw on old carbon release and net carbon exchange from tundra. Nature 459:556–559

Schwikowski M, Eichler A, Kalugin I, Ovtchinnikov D, Papina T (2009) Past climate variability in the Altai. Pages News 17:44–45

Sedykh VN (1990) Thermal effect in the forest forming process of West Siberia. In: Processes of transfer in natural and artificial covers of the Earth. Tjumen, pp 10–25 (in Russian)

Sedykh VN (2009) Forest forming process. Nauka, Novosibirsk, 163 pp (in Russian)

Shaver GR, Billings WD, Chapin FSI, Giblin AE, Nadelhoffer KJ, Oechel WC, Rastetter EB (1992) Global change and the carbon balance of arctic ecosystems. Bioscience 42:433–441

Sheingauz AS (2001) Forest complex of Khabarovsk Kray. RIOTIP, Kabarovsk, 103 pp (in Russian)

Sheingauz AS (2008) Selected works. FEB RAS, Khabarovsk, 654 pp (in Russian)

Sheshukov MA (1967) On classification of forest fire based on extent of burnt areas. Forestry For Manage 1:53–57 (in Russian)

Shibistova O, Lloyd J, Zrazhevskaya G, Arneth A, Kolle O, Knohl A, Astrakhantceva N, Shijneva I, Schmerler J (2002) Annual ecosystem respiration budget for a *Pinus sylvestris* stand in central Siberia. Tellus B Chem Phys Meteorol 54:568–589

Shiklomanov IA, Krestovsky OI (1988) Influence of forests and forest reclamation practice on streamflow and water balance. In: Reynolds ERC, Thompson FB (eds) Forests, climate, and hydrology: regional impacts. United Nations University Press, Tokyo, pp 78–116

Shiyatov SG (2003) Rates of change in the upper tree line ecotone in the Polar Ural Mountains. Pages News 11:8–10

Shiyatov SG, Terent'ev MM, Fomin VV, Zimmermann NE (2007) Altitudinal and horizontal shifts of the upper boundaries of open and closed forests in the Polar Urals in the 20th century. Russ J Ecol 4(38):223–227

Shugart HH (2000) Importance of structure in the longer-term dynamics of landscapes. J Geophys Res D Atmos 105:20065–20075

Shugart HH, West DC (1980) Forest succession models. Bioscience 30:308–313

Shugart HH, Leemans R, Bonan GB (eds) (1992) A systems analysis of the global boreal forest. Cambridge University Press, Cambridge, 542 pp

Shuman JK, Krankina ON, Shugart HH (2011) Validation of the dynamic forest gap model FAREAST across Russia ecological applications (in preparation)

Shumilova LV (1962) Botanical geography of Siberia. Tomsk State University, Tomsk, 440 pp (in Russian)

Shvidenko A, Goldammer JG (2001) Fire situation in Russia. Int Fire News 24:41–59

Shvidenko A, Nilsson S (2000a) Fire and the carbon budget of Russian forests. In: Kasischke ES, Stock BJ (eds) Fire, climate change, and carbon cycling in the boreal forest. Springer, New York, pp 289–311

Shvidenko AZ, Nilsson S (2000b) Extent, distribution, and ecological role of fire in Russian forests. In: Kasischke ES, Stocks BJ (eds) Fire, climate change, and carbon cycling in the boreal forest. Springer, New York, pp 132–150

Shvidenko A, Nilsson S (2002) Dynamics of Russian forests and the carbon budget in 1961–1998: an assessment based on long-term forest inventory data. Clim Chang 55:5–37

Shvidenko A, Nilsson S (2003) A synthesis of the impact of Russian forests on the global carbon budget for 1961–1998. Tellus B Chem Phys Meteorol 55:391–415

Shvidenko AZ, Strakhov VV, Nilsson S, Sedykh VN, Sokolov VA, Efremov DE (2000) Spatial scale of productivity of Russian forests. For Manage Inf 1–2:7–23 (in Russian)

Shvidenko A, Strakhov V, Nilsson S, Sedykh V, Sokolov V. Efremov D (2001) Productivity of forests of Russia. 3. Spatial scale of productivity's assessment. Forest Management Information, Moscow 1–2, pp 7–23 (in Russian)

Shvidenko A, Schepaschenko D, Nilsson S, Bouloui Y (2007a) Semi-empirical models for assessing biological productivity of Northern Eurasian forests. Ecol Model 204:163–179

Shvidenko AZ, Schepaschenko D, Nilsson S (2007b) Materials to knowledge of current productivity of Russian forest ecosystems. In: Basic problems for going over to sustainable forest management of Russia – forest inventory and forest organization. V.N. Sukachev Institute of Forest SB RAS, Krasnoyarsk, pp 7–37

Shvidenko AZ, Schepashchenko DG, Vaganov EA, Nilsson S (2008) Net primary production of forest ecosystems of Russia: a new estimate. Dokl Earth Sci 421A(6):1009–1012

Shvidenko A, Schepaschenko D, McCallum I (2010a) Bottom-up inventory of the carbon fluxes in Northern Eurasia for comparison with GOSAT level 4 products. Unpublished manuscript. International Institute for Applied Systems Analysis, Laxenburg, 210 p

Shvidenko A, Schepaschenko D, McCallum I, Nilsson S (2010b) Can the uncertainty of full carbon accounting of forest ecosystems be made acceptable to policymakers? Clim Chang 103:137–157

Shvidenko A, Schepaschenko D, Vaganov EA, Sukhinin AI, Maksyutov SS, McCallum I, Lakyda IP (2011) Impacts of vegetation fires of 1998–2010 in Russia on ecosystems and global carbon budget. Proc Russ Acad Sci (Doklady RAN) 441(2):544–588

Sitch S, Smith B, Prentice IC, Arneth A, Bondeau A, Cramer W, Kaplan JO, Levis S, Lucht W, Sykes MT, Thonicke K, Venevsky S (2003) Evaluation of ecosystem dynamics, plant geography and terrestrial carbon cycling in the LPJ dynamic global vegetation model. Glob Chang Biol 9:161–185

Smith TM, Shugart HH (1993) The transient response of terrestrial carbon storage to a perturbed climate. Nature 361:523–526

Soja AJ, Cofer WR, Shugart HH, Sukhinin AI, Stackhouse PW Jr, McRae DJ, Conard SG (2004) Estimating fire emissions and disparities in boreal Siberia (1998–2002). J Geophys Res D: Atmos 109:D14S06

Soja AJ, Tchebakova NM, French NHF, Flannigan MD, Shugart HH, Stocks BJ, Sukhinin AI, Parfenova EI, Chapin Iii FS, Stackhouse PW Jr (2007) Climate-induced boreal forest change: predictions versus current observations. Glob Planet Change 56:274–296

Sokolova GA, Teteryatnikova EP (2002) Estimation of dynamics of forests of Russian Far East and Eastern Siberia caused by natural and anthropogenic factors. Research report. Far Eastern Research Forestry Institute, Khabarovsk, 180 p (in Russian)

Solomon AM, Leemans R (1990) Climatic change and landscape ecological response: issues and analysis. In: Boer MM, de Groot RS (eds) Landscape ecological impact of climatic change. Ios Press, Amsterdam, pp 293–316

Stireman JO et al (2005) Climatic unpredictability and parasitism of caterpillars: implications of global warming. PNAS 102(48):17384–17387

Stocks BJ, Fosberg MA, Lynham TJ, Mearns L, Wotton BM, Yang Q, Jin JZ, Lawrence K, Hartley GR, Mason JA, McKenney DW (1998) Climate change and forest fire potential in Russian and Canadian boreal forests. Clim Chang 38:1–13

Sukachev VN (1972) Selected scientific works, vol 1: Foundations of forest typology and biogeocenology. Nauka, Leningrad, 418 pp (in Russian)

Sukhinin AI (2008) Satellite data on wild fire. Available from www.iiasa.ac.at/Research/FOR/forest_cdrom/index.html. Accessed 5 May 2012

Sukhinin AI (2009) Airspace monitoring of catastrophic fires in forests of East Siberia. Unpublished report. Institute of Forest of Siberian Branch of the Russian Academy of Sciences, Krasnoyarsk, 91 p (in Russian)

Tchebakova NM, Kolle O, Zolotuknine DA, Lloyd J, Arneth A, Parfenova EI, Schulze E-D (2002) Annual and seasonal dynamics of energy and mass exchange in a middle taiga pine forest. In: Pleshikov FI (ed) Forest ecosystems of the Yenisei meridian. Siberian Branch of Russian Academy of Sciences Publishing House, Novosibirsk, pp 252–264

Tchebakova NM, Rehfeldt GE, Parfenova EI (2003) Distribution of vegetation zones and populations of *Larix sibirica* and *Pinus sylvestris* in central Siberia under climate warming. Siberian Ecol J 6:677–686 (in Russian)

Tchebakova NM, Parfenova EI, Soja AJ (2009) Effects of climate, permafrost and fire on vegetation change in Siberia in a changing climate. Environ Res Lett 4:045013, 10 pp

Tchebakova NM, Rehfeldt GE, Parfenova EI (2010) Vegetation zones to climate types: effects of climate warming on Siberian ecosystems. In: Osawa A, Zyranova OA, Matsurura Y et al (eds) Permafrost ecosystems: Siberian larch forests. Springer, Berlin, pp 427–447

Tchebakova NM, Vygodskaya NN, Arneth A, Belelli-Marchesini L, Corradi C, Kurbatova JA, Parfenova EI, Valentini R, Vaganov EA, Schulze E-D (2011) Energy and mass exchange and productivity of major ecosystems in central Siberia based on the eddy covariance method. Biol Bull (submitted)

Thornley JHM, Fowler D, Cannell MGR (1991) Terrestrial carbon storage resulting from CO_2 and nitrogen fertilization in temperate grasslands. Plant Cell Environ 14:1007–1011

Titlianova AA, Tisarzhova M (1991) Regimes of the biological turnover. Nauka, Novosibirsk, 150 pp

Tumel VF (1939) Changes in the soils and permafrost regimes following vegetation cover burning. Academy of Sciences of the USSR, Moscow/Leningrad. Reports of the Permafrost Commission, Issue VIII, pp 3–80 (in Russian)

Udra IF (1988) Settling and migration of woody plants in temperate belt of Eurasia. Naukova Dumka, Kiev, 200 pp (in Russian)

Urban DL, Shugart HH (1992) Individual-based models of forest succession. In: Glenn-Lewin DC, Peet RK, Veblen TT (eds) Plant succession: theory and prediction. Chapman & Hall, London, pp 249–293

Urban DL, Bonan GB, Smith TM, Shugart HH (1991) Spatial applications of gap models. For Ecol Manage 42:95–110

Usoltsev VA, Voronov MP, Chasovskikh VP (2011) Net primary production of forests of Ural: methods and results of automized estimation. Ecology 4:334–343 (in Russian)

Vaganov EA, Arbatskaya MK, Shashkin AV (1996) History of climate and fire frequency in central part of Krasnoyarsk Kray. 2. Dendrochronological analysis of interdependence of ibcrement of trees, climate and fire frequency. Siberian Ecol J 1:19–28

Valentini R, Matteucci G, Dolman AJ, Schulze ED, Rebmann C, Moors EJ, Granier A, Gross P, Jensen NO, Pilegaard K, Lindroth A, Grelle A, Bernhofer C, Grünwald T, Aubinet M, Ceulemans R, Kowalski AS, Vesala T, Rannik Ü, Berbigier P, Loustau D, Guomundsson J, Thorgeirsson H, Ibrom A, Morgenstern K, Clement R, Moncrieff J, Montagnani L, Minerbi S, Jarvis PG (2000) Respiration as the main determinant of carbon balance in European forests. Nature 404:861–865

Van Der Werf GR, Randerson JT, Giglio L, Collatz GJ, Kasibhatla PS, Arellano AF Jr (2006) Interannual variability of global biomass burning emissions from 1997 to 2004. Atmos Chem Phys Discuss 6:3175–3226

Vaschuk LN, Shvidenko AZ (2006) Dynamics of forests of Irkutsk region. Federal Agency of Forest Management, Irkutsk, 392 pp (in Russian)

Vasiliev SV, Titlyanova AA, Velichko AA (eds) (2001) West Siberian peatlands and carbon cycle: past and present. In: Proceedings of the international field symposium. Russian Academy of Sciences, Novosibirsk 250 pp

Vedrova EF (2002) Intensity of mineralisation flux in forest ecosystems. In: Pleshikov FI (ed) Forest ecosystems of Enisey meridian. SB RAS, Novosibirsk, pp 248–252

Vivchar A (2010) Wildfires in Russia in 2000–2008: estimates of burnt areas using the satellite MODIS MCD45 data. Remote Sens Lett 2:81–90

Vogt KA, Grier CC, Vogt DJ (1986) Production, turnover, and nutrient dynamics of above- and belowground detritus of world forests. Adv Ecol Res 15:303–377

Volney MJ, Fleming RA (2007) Spruce budworm (*Choristoneura* spp.) biotype reactions to forest and climate characteristics. Glob Chang Biol 13(8):1630–1643

Voronkov NA (1988) Role of forests in water protection. Hydrometeoizdat, Leningrad, 286 pp (in Russian)

Vtyurina OP, Sokolov VA (2009) Dynamics of forests. In Semechkin IV (ed) Management of sustainability in Krasnoyarsk krai. Institute of Forest SB RAS, Krasnoyarsk, pp 103–123

Vygodskaya NN (1981) Solar radiation regime and structure of mountain forest. Gidrometeoizdat, Leningrad, 261 pp (in Russian)

Vygodskaya N, Schulze E-D, Tchebakova N, Karpachevskii L, Kozlov D, Sidorov K, Panfyorov M, Shaposhnikov E, Solnzeva O, Minaeva T, Jeltuchin A, Wirth C, Pugachevskii A (2002) Climatic control of stand thinning in unmanaged spruce forests of the southern taiga in European Russia. Tellus 54B(5):443–461

Vygodskaya NN, Groisman PY, Tchebakova NM, Kurbatova JA, Panfyorov O, Parfenova EI, Sogachev AF (2007) Ecosystems and climate interactions in the boreal zone of northern Eurasia. Environ Res Lett 2:045033, 7 pp

Weishampel JF, Urban DL, Shugart HH, Smith JB (1992) Semivariograms from a forest transect gap model compared with remotely sensed data. J Veg Sci 3:521–526

Wirth C, Schulze ED, Schulze W, von Stunzner-Karbe D, Ziegler W, Miljukova IM, Sogatchev A, Varlagin AB, Panvyorov M, Grigoriev S, Kusnetzova W, Siry M, Hardes G, Zimmermann R, Vygodskaya NN (1999) Above-ground biomass and structure of pristine Siberian Scots pine forests as controlled by competition and fire. Oecologia 121:66–80

Wirth C, Czimczik CI, Schulze ED (2002) Beyond annual budgets: carbon flux at different temporal scales in fire-prone Siberian Scots pine forests. Tellus B Chem Phys Meteorol 54:611–630

Woodward FI, Lomas MR (2004) Vegetation dynamics – simulating responses to climatic change. Biol Rev Camb Philos Soc 79:643–670

Yan H, Shugart HH (2005) FAREAST: a forest gap model to simulate dynamics and patterns of eastern Eurasian forests. J Biogeogr 32:1641–1658

Yefremov DF, Shvidenko AZ (2004) Long-term ecological consequences of catastrophic forest fires in forests of Far East and their contribution to global processes. In: Forest fire management at ecoregional level. World Bank, Publishing house "Alex", Moscow, pp 66–73

Yurchenko GI, Turova GI, Kuzmin EA (2003) Consequences of outbreaks of Siberian silk moth in Far Eastern conifer-broadleaves forests. Science background of use and reproduction of forest resources of Far East. Trans Far-Eastern Res For Inst 36:176–193

Zamolodchikov DG, Karelin DV, Ivaschenko AI, Oechel WC, Hastings SJ (2003) CO_2 flux measurements in Russian Far East tundra using eddy covariance and closed chamber techniques. Tellus B Chem Phys Meteorol 55:879–892

Zavarzin GA (ed) (2007) Carbon pools and fluxes in terrestrial ecosystems of Russia. Nauka, Moscow, 315 pp (in Russian)

Zhang YH, Wooster MJ, Tutubalina O, Perry GLW (2003) Monthly burned area and forest fire carbon emission estimates for the Russian Federation from SPOT VGT. Remote Sens Environ 87:1–15

Zhu X, Zhuang Q, Chen M, Sirin A, Melillo J, Kickligther D, Sokolov A, Song L (2011) Rising methane emissions in response to climate change in the Northern Eurasia during the 21st century. Environ Res Lett 6:045211. doi:10.1088/1748-9326/6/4/045211

Zhuang Q, Melillo JM, Kicklighter DW, Prinn RG, McGuire AD, Steudler PA, Felzer BS, Hu S (2004) Methane fluxes between terrestrial ecosystems and the atmosphere at northern high latitudes during the past century: a retrospective analysis with a process-based biogeochemistry model. Global Biogeochem Cycles 18(GB3010):1–23

Zhuang Q, Melillo JM, Sarofim MC, Kicklighter DW, McGuire AD, Felzer BS, Sokolov A, Prinn RG, Steudler PA, Hu S (2006) CO_2 and CH_4 exchanges between land ecosystems and the atmosphere in northern high latitudes over the 21st century. Geophys Res Lett 33(17):L17403, 5 pp

Zhuang Q, Melillo JM, McGuire AD, Kicklighter DW, Prinn RG, Steudler PA, Felzer BS, Hu S (2007) Net emissions of CH_4 and CO_2 in Alaska: implications for the region's greenhouse gas budget. Ecol Appl 17:203–212

Zhukov AB (ed) (1969) Forests of Ural, Siberia and Far East, vol 4, Forests of the USSR. Nauka Publisher, Moscow, 768 pp (in Russian)

Chapter 7
Human Dimensions of Environmental Change in Siberia

Kathleen M. Bergen, Stephanie K. Hitztaler, Vyacheslav I. Kharuk,
Olga N. Krankina, Tatiana V. Loboda, Tingting Zhao,
Herman H. Shugart, and Goquing Sun

Abstract This chapter provides background on socioeconomic contexts followed by synthesis of remote sensing-based case studies highlighting major human influences on the Siberian landscape during three eras: *Soviet* (1917–1991), *early post-Soviet transformation* (starting after 1991), and *recent/emerging*. During 1975–2001, Landsat-based LCLUC data in East Siberia showed characteristic patterns including: high rates of logging during the Soviet era that declined abruptly and remained low after 1989, a decline in agriculture (and subsequent reforestation) beginning prior to 1991, and a decline in mature conifer and increase in deciduous forest. In the far north, multiple remote sensing data over time demonstrated the degradation and mortality of the larch forests surrounding the Norilsk nickel mining complex. In East Siberia, multiple remote sensing data showed that oil and gas reconnaissance

K.M. Bergen(✉) • S.K. Hitztaler
School of Natural Resources and Environment, University of Michigan,
Ann Arbor, MI, USA
e-mail: kbergen@umich.edu

V.I. Kharuk
V.N. Sukachev Institute of Forest, Siberian Branch of Russian Academy of Sciences,
Krasnoyarsk, Academgorodok, Russia

O.N. Krankina
College of Forestry, Oregon State University, Corvallis, OR, USA

T.V. Loboda
Department of Geographical Sciences, University of Maryland, College Park, MD, USA

T. Zhao
Department of Geography, Florida State University, Tallahassee, FL, USA

H.H. Shugart
Department of Environmental Sciences, University of Virginia, Charlottesville, VA, USA

G. Sun
Department of Geographical Sciences, University of Maryland,
College Park, MD, USA

P.Ya. Groisman and G. Gutman (eds.), *Regional Environmental Changes in Siberia and Their Global Consequences*, Springer Environmental Science and Engineering, DOI 10.1007/978-94-007-4569-8_7, © Springer Science+Business Media Dordrecht 2013

directly disturb the landscape but that their indirect influence on increased fire occurrence is of greater consequence. Along the Siberia-China Amur River border, replanting and the almost complete removal of mature conifer are evident in Landsat data on the Chinese side, whereas fire predominates on the Siberian side. In the recent/emerging era, LCLUC in Siberia is being influenced by greater transnationalism and increased demand for wood from other Asian countries. Oil and gas development is shifting to East Siberia. Pipelines and infrastructure are being built across Siberian lands directly to the Pacific and to China. Remote sensing–based analyses have been integral to increased knowledge of past and emerging trends in human dimensions of environmental change across the vast geographic region of Siberia.

7.1   Introduction

Many important environmental changes in regions of the world are ultimately driven by changes in human activities. Environmental changes have been associated with human drivers ranging from gradual historical settlement and development, major shifts in political and socioeconomic paradigms, changes in resource management institutions, to local human adaptation strategies. In Siberia, all of these drivers and their conse-quences have shaped and continue to influence the environment over time, and over the region's vast and varied geographic extent. The purpose of this chapter is to characterize both human dimensions and environmental changes in Siberia and to provide evidence of their interrelationships through synthesis of completed scientific case studies.

Human dimensions of environmental change could be considered from a number of different fields and perspectives. This chapter (and its case studies in particular) focuses on environmental changes that have been observed directly or indirectly via spaceborne remote sensing. Given that Siberia is a predominantly forested region and that its primary socioeconomic activities center on exploitation of forests, energy, and mineral resources, a representative balance of human dimensions of environmental change related to these resources is emphasized. Synthesis case stud-ies selected for inclusion were largely conducted at the landscape level and where dominant human drivers of change were clearly present.

Because there have been dramatic political and socioeconomic shifts in Siberia over the past century, the fundamental organization of this chapter is designed to investigate and compare human dimensions of environmental change in Siberia by political and socioeconomic eras. Specific chapter sections and organization are as follows: Section 7.1 provides definitions and sets the stage for the investigation of the human dimensions of environmental change in Siberian Russia. It is followed by three sections representing roughly chronological eras delineated as: *Soviet* (1917–1991, Sect. 7.2), *early post-Soviet transformation* (starting after 1991, Sect. 7.3), and *recent/emerging* (Sect. 7.4). The information in Sects. 7.2, 7.3, and 7.4 is structured such that, first, a more detailed context of human dimensions (specifically demographics, institutions, and resource management) that define each particular era is provided. This background is followed in Sects. 7.2, 7.3, and 7.4 by synthesis case studies on human-influenced land-cover and environmental change, emphasizing what has been observed and quantified via landscape-level remote sensing. Section 7.5 provides a brief summary and conclusions. Because the primary focus of this chapter is on human dimensions, treatment of socio-economics *and* environmental change is given equal emphasis.

7.1.1 *Research on Human Dimensions of Environmental Change in Siberia*

Global change came to the fore as a field in the 1990s through increased efforts by the scientific community to understand the sources and sinks of global carbon. Within this field, inquiry into *human dimensions of environmental change* has focused on the human drivers and consequences which influence or alter, either singularly or in combination, the biosphere, atmosphere, and hydrosphere. Some human drivers and consequences of global change may be direct (e.g., the addition of carbon to the atmosphere from internal combustion engines); however, research has shown that many complex human-driven environmental changes (including those that affect CO_2 in the atmosphere) have their basis in land-cover and land-use changes (Houghton et al. 2004; Houghton 1995; Vitousek 1994). Interest in this component we now call land-cover/land-use change is not new (Marsh 1885; Thomas 1956; Turner 1990) but has grown in sophistication of theory, methods, and programs over time (Gutman et al. 2004). Satellite remote sensing data which became available to environmental scientists in the 1970s have been instrumental in more accurate and consistent characterization of land-cover change trends. Today, most land-cover/land-use studies using remote sensing data are undertaken with some combination of field data, socioeconomic data, and modeling methods (Brown et al. 2004; Janetos 2004; Turner et al. 2004; Rindfuss et al. 2004).

Over the past several decades, the US and other national and international organizations have become active in the programmatic investigation of human dimensions of environmental change and land-cover/land-use change *over global regions*. These include the NASA Land-Cover Land-Use Change (LCLUC) Program and the International Geosphere-Biosphere Programme's (IGBP) core projects, such as Land Use and Cover Change (LUCC) and its continuation, the Global Land Project (GLP). The NASA LCLUC research program uses NASA and other primarily spaceborne sensors to observe many global environments, and has focused on both human and natural drivers and consequences of changes in forests, agriculture, natural and managed grasslands, wetlands, lakes, and urban systems (Gutman et al. 2004). In the late 1990s, several NASA LCLUC project teams began working with organizations in Russia to quantify and understand the past, present, and future land-cover and land-use trends within the Russian Federation. Bergen et al. (2003) reviewed objectives and results of these early research projects in this region in four categories: forest and land-cover dynamics, fire and fire behavior, carbon budgets, and new remote sensing analysis methods. Krankina et al. (2004) focused on early results from three intensive case study sites in western Russia, central Siberia, and the Russian Far East (RFE).

Presently, NASA LCLUC research in Russia related to human dimensions of environmental change also continues as part of the broader NEESPI program (Northern Eurasia Earth Science Partnership Initiative; Groisman et al. 2009). NEESPI focuses on research projects in Northern Eurasia and operates in partnership with academia, government agencies from the United States, Russian Federation, Europe, and Japan and with international programs such as IGBP and others (Gutman 2007). Gutman and Reissell (2011) have compiled results of NEESPI-related research in their book focused specifically on land-cover/land-use and climate change in the Eurasian Arctic (generally above 60° latitude), which includes northern Siberia. Also involved in assessing human-driven environmental change in Russia are several Institutes of the Russian Academy of Sciences (including the Sukachev Institute of Forests in Krasnoyarsk), the International Institute for Applied Systems Analysis (IIASA) in Austria (Shvidenko et al. 2007), the IGBP GLP, and the European Commission Joint Research Center (JRC) in Ispra, Italy, among others. Investigations from these research programs form the basis of the case studies syntheses in this chapter.

7.1.2 Historical Human Dimensions: Leading up to the Soviet Era

Prior to the Soviet era (and the era of aerial photography and satellite imagery), humans had been altering the Siberian landscape. The legacy of these activities includes the present-day patterns of settlement and infrastructure (Fig. 7.1), summarized in Chap. 1, as well as their influence on ongoing environmental changes. Given the challenges of Siberian terrain and climate (vast distances, cold temperatures, and permafrost soils), one may well ask how its large and important cities, continent-spanning transportation routes, and expansive industrial and forestry development come to be established. In addition to the geographies of terrain and climate, the other important factor influencing present human population patterns of Siberia is historical expansion. This has consisted of immigration and migration, increasingly associated over the centuries not only with nation-building and military concerns but also with the presence and locations of natural resources and the development of those resources (Table 7.1).

Russia calls itself the nation that "looks both east and west" (symbolized by the east-west facing double eagles on its historic national emblem). Siberia was occupied prior to the thirteenth century by differing groups of indigenous peoples, some of whose ethnic groups remain part of the Siberian population today (e.g., Yakuts, Buryats). The southern fringes of Siberian lands were conquered by the Mongols of the East early in the thirteenth century who subsequently established the Siberian Khanate, which persisted until the sixteenth century when Russia to the west became a growing power. The first major groups of peoples to expand eastward from Russia were the Cossacks who sought Siberia's natural resources for hunting, fishing, and agriculture. The Cossacks were also led by the spirit of exploration and adventure and, along with the Russian army, established military and trading stations (*ostrogs*), some of which remain the sites of important present-day cities.

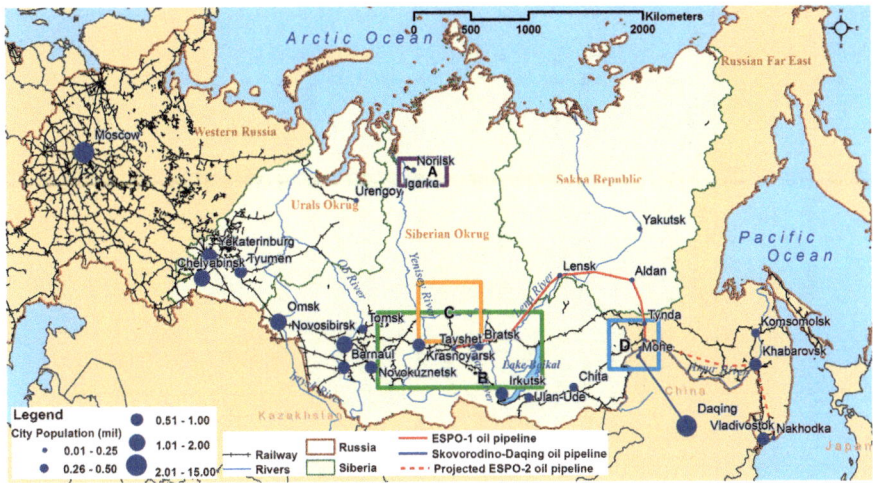

Fig. 7.1 Map of Siberia showing: geography including land, oceans, and major rivers; human dimensions including regions of the Russian Federation, federal provinces of Siberia, cities, rail transportation, and major east-west pipelines; locations of synthesis case study sites: Norilsk mining (*A, purple*), central Siberian LCLUC (*B, green*), central East Siberian oilfields (*C, orange*), and Amur LCLUC (*D, blue*) (University of Michigan, School of Natural Resources and Environment, Environmental Spatial Analysis Laboratory)

Following the Cossacks, peasants from Russia migrated to Siberia in search of land and agricultural livelihoods free from the suppression of feudal landlords (*pomeshchiks*). Siberia also attracted "Old Believers," Russian Orthodox Christians who did not accept the church reforms of the seventeenth century, and were persecuted by the government and the official Church. Still, by the mid-seventeenth century, the population of ethnic Russians in what is present-day Siberia amounted to probably <200,000 persons. Nevertheless, when Peter the Great became Tsar (1682–1725), Russian-controlled lands had been extended to the Pacific and Russia was already a great empire in terms of the geographic territories it controlled, including much of those of present-day Siberia.

Up through these preindustrial eras, human economic activity in the boreal forests regions of northern Eurasia included hunting, fishing, conversion of meadows for growing hay, and clearing of forest patches for shifting and permanent small cultivation. However, Yaroshenko et al. (2001) in their mapping of intact landscapes (in northern European Russia) do not consider such types of preindustrial land uses as "anthropogenic disturbances," but instead as "anthropogenic factors that have formed the ecosystems," along with natural disturbances (i.e., wildfire).

Starting in the nineteenth century, the identification of mineral and energy resources began to spur the establishment and growth of many Siberian cities, industrial districts, and transportation routes. During this period, metallic ores, including iron, bauxite (in West Siberia), and gold were discovered. Wood for local consumption from high-quality forests near access routes (rivers) was the focus of pre-Soviet era logging activity. Largely constructed over the period 1891–1916, the Trans-Siberian

Table 7.1 Major human population centers of Siberia (top ten by population), their 1989 and most recent population estimates, and their socioeconomic activities

City	Federal Okrug	Krai/Oblast	1989 Population estimate	Most recent population estimate (date)	Main socioeconomic activities
Novosibirsk	Siberian	Novosibirsk Oblast	1,436,516	1,391,400 (2009)	Oblast capital, Siberian Federal District capital, air transport, electric power, gas supply, water supply, metallurgy, metal working, mechanical engineering
Yekaterinburg	Urals	Sverdlovsk Oblast	1,364,621	1,293,537 (2002)	Machinery, metal processing, ferrous and nonferrous metallurgy, rail transport, air transport
Omsk	Siberian	Omsk Oblast	1,148,418	1,134,700 (2009)	Oblast capital, rail transport, tourism
Chelyabinsk	Urals	Chelyabinsk Oblast	1,141,777	1,077,174 (2002)	Oblast capital, metallurgy, military machinery, electronics, air transport
Krasnoyarsk	Siberian	Krasnoyarsk Krai	912,629	927,200 (2009)	Krai capital, rail transport, tourism, air transport
Barnaul	Siberian	Altai Krai	601,811	649,600 (2007)	Fuel processing, heavy machinery, tires, furniture, footwear, diamond faceting, restaurants, rail transport, air transport
Tyumen	Urals	Tyumen Oblast	476,869	609,100 (2010)	Oblast capital, gas, oil, rail transport, air transport
Irkutsk	Siberian	Irkutsk Oblast	626,135	575,900 (2009)	Oblast capital, aircraft production, rail transport, air transport, Lake Baikal
Novokuznetsk	Siberian	Kemerovo Oblast	599,947	562,200 (2008)	Coal mining, metal plants, rail transport, air transport
Tomsk	Siberian	Tomsk Oblast	501,963	493,000 (2009)	Oblast capital, rail transport, information technology, air transport

Statistical Sources: Goskomstat Russia (1996), Russian Federal State Statistics Service (2005, 2010), and Europa Publications (2010)

railway linked Siberia to rapidly industrializing western Russia. Between 1801 and 1914, several million settlers migrated from European Russia to Siberia to colonize available agricultural lands. Many convicted criminals and other exiles (Lenin and Stalin among them) were also relocated to Siberia at this time.

Throughout the industrial twentieth century and continuing into the present era, the amount and effect of human-influenced changes in Siberia has increased greatly. As introduced in Chap. 1, forests, energy (oil and gas), and minerals/metals have become the predominant natural resources developed in Siberia. Other resources include biodiversity, agriculture, and water. A more in-depth look at the ecology of these resources and their geographic distributions may be found in Chap. 6. As a consequence of industrialization, dominant environmental changes at the landscape-level include the effects of fire, logging, forest succession, agricultural conversion and abandonment, and urban-industrial pollution (Ranson et al. 2003; Bergen et al. 2008). Sections 7.2, 7.3, and 7.4 first provide in-depth reviews of the human drivers of environmental changes that have occurred over the past century in Siberia in terms of demographics, institutions, and resource management. Synthesis of case studies focused on important types of human-driven land-cover and land-use change are then provided in each of these sections.

7.2 The Soviet Era

7.2.1 Soviet Era Human Dimensions

7.2.1.1 Introduction

The colonization, urbanization, and industrialization of Siberia began in earnest following the 1917 revolution. At that time, building communism in the Soviet Union demanded the rapid industrialization of the economy and the collectivization of agri-culture. In Siberia, where these objectives were fully implemented (Naumov and Collins 2006), they altered land-use patterns. The holdings of large landowners were expropriated and distributed to farmers or consolidated into state collective farms (*sovkhozy*); later, smaller-scale collective farms (*kolkhozy*) were instituted. Millions of acres of new land were also cultivated and the mechanization of agriculture freed labor to work in other industries.

The 1930s saw an even greater burst of socioeconomic change throughout the Soviet Union driven by the state's concentrated effort to catch up with the developed world (Afontsev et al. 2008). Large-scale development of Siberian natural resources during this period was in part fueled by the institution in 1929 of a penal labor system (also known as the GULAG) designated to colonize "the least accessible and most difficult to develop" regions of the Soviet Union (Hill and Gaddy 2003). During this time, development of mining and metallurgical complexes was predominant (instead of forestry), as they served to support the defense industry (Naumov and Collins 2006).

During World War II, Siberia acquired heightened importance as a strategic redoubt: vital industries and economic activities were shifted from the European part of the Soviet Union further eastward in an effort to spare them from invading German forces (Hill and Gaddy 2003). After World War II another socioeconomic boom led to the development of giant industrial complexes in almost every sector, including oil and gas, coal, nonferrous metals, hydropower, diamond production, and increasingly, timber. Along major rivers (e.g., Ob, Yenisei, Vilyui) that drain toward the Arctic Ocean, major hydroelectric projects were constructed. A combination of water transportation and hydropower from Siberia's great rivers enabled the construction of industrial-scale processing plants to exploit mineral and timber resources.

This socioeconomic transformation of Siberia, during which remote and undeveloped lands were transformed into industrialized and urbanized complexes, is cited as one of the Soviet Union's greatest achievements, the epitome of the transformation from an agrarian to a modern industrial state (Hill and Gaddy 2003). At the same time, however, the vision of the Siberian industrial utopia also involved high costs associated with extremely rapid land-use development, the need for transportation across immense areas, and sustaining human populations in challenging climatic conditions. This extensive development took a toll on the environment, creating complex ecological problems (Naumov and Collins 2006). The next sections take a closer look at the Soviet-era human dimensions of environmental change, namely, demographics, institutions, and forest resource management.

7.2.1.2 Soviet Era Demographics

Major shifts in population dynamics surrounded the early rise of communism in Russia. World War I and the Civil War (1914–1922) resulted in an estimated population loss of 16 million people (Afontsev et al. 2008). The revolutionary government disbanded traditional social and economic structures, replacing them with radical measures, such as the collectivization of agriculture and a rapid shift to socialized industrialization (Coale et al. 1979). Consequently, an extensive out-migration from rural to urban areas occurred, and there was a continual increase of women entering the work force. Together, these events also suppressed the birth rate (Coale et al. 1979; Perevedentsev 1999; Hoffmann 2000).

Despite a lowered birth rate in Russia, the population in Siberia expanded owing to extensive in-migration. As early as October 1924, the Soviet government issued a decree emphasizing the need to transfer population to Siberia to develop the country's vast resource base. The state offered incentives to potential settlers including lowered taxes, postponement of military service, and payment of transportation costs (Hill and Gaddy 2003). After World War II, the Soviet state continued to lure people to remote areas in Siberia by introducing favorable wage differentials, living standards, and labor market conditions (Heleniak 2001; Gerber 2006). Beyond economic impetus, people were attracted to Siberia on the ideological grounds of building communism; military draftees and officers were also sent to these regions and many stayed, employed by the growing civil and military-industrial complexes (Davis 2003). Siberia (and the RFE) became paragons of a frontier region where people flocked to extract valuable raw materials, which were shipped to the European part of the Soviet Union for processing, then distributed as finished goods.

The results of settlement efforts were also reflected in population statistics and age structures. Siberia's population expanded by nine million people (from 23.5 to 32.5 million) between the years 1959 and 1989 (Naumov and Collins 2006). Overall, those who migrated to Siberia tended to be educated and of prime working age (25–44). These patterns led to a demographic phenomenon in peripheral regions of Siberia in which population growth was primarily a product of migration instead of natural growth (Voinova et al. 1993), thereby setting the stage for a significant shift in population and resource-use patterns in the post-Soviet era.

7.2.1.3 Soviet Era Institutions

State institutions played a critical role in the development of Siberia. A strong institutional framework was imperative in sustaining newly arrived populations of skilled workers, particularly in oil, gas, mining, and forestry operations (Ryabushkin 1978; Eikeland et al. 2004). The same institutions that regulated and organized natural resource use often assumed a social responsibility for their workers as well. For instance, they equipped working towns and villages with municipal, cultural (community centers), educational (schools), and health services (including day-care centers and nursing homes) (Malmlöf 1998). This phenomenon was particularly evident in the forestry sector due to the occurrence of logging in remote, unsettled regions.

In the early years following World War II, the state enacted new forestry policies. It enforced these through the establishment of local forestry management units (*leskhozy*) (Eikeland et al. 2004), in particular, in the forest surplus zones of the European North and Siberia (Barr and Braden 1988). The *leskhozy* were intended to supervise the planning and the activities of the state-owned forest enterprises (*lespromkhozy*) and to implement new practices aimed at sustainable forestry, including forest regeneration. Beyond forestry, the *leskhozy* were critical in anchoring a permanent labor force of skilled workers in remote areas and, along with the *lespromkhozy*, in providing for the welfare of these workers (Malmlöf 1998; Pallot and Moran 2000, both quoted in Eikeland et al. 2004). The *leskhozy* answered to the Regional Forestry Administration, which reported to the Soviet State Forest Committee (or Soviet Forest Service – *Gosleskhoz* until 1988, and *Goskomles* from 1988 to 1993). Established in 1966, the Committee managed approximately 95 % of the state forests (Barr and Braden 1988), oversaw approximately 2,500 *leskhozy*, and administered around 20 research institutes (Blandon 1983, quoted in Malmlöf 1998). In this capacity it was responsible for forest inventories, regeneration measures, thinning operations and salvage logging, managing selected forests for agricultural purposes, and safeguarding forests from fire, insects, and disease (Kukuev et al. 1997).

7.2.1.4   Soviet Era Forest Management

Three basic stages of timber management and harvesting in the former Soviet Union are delineated (Barr and Braden 1988): (1) the period prior to the mid-1930s during which time harvesting was concentrated in the European areas of the country, much like it was during the Tsarist rule; (2) the late 1930s until 1960 where the timber frontier in the European Soviet Union pushed further to the north and northeast into dense forest surplus zones where impressive increases in harvesting occurred; (3) from 1960 to 1989 during which time declines from the surplus regions in the European Soviet Union were compensated for by increased production from Siberia.

During the first stage, the emerging Soviet state embraced the long-standing Russian system of open access to forests through the Forest Decree of 1918 – the first forest legislation of the Soviet Union – in which all forests were declared common national property (Malmlöf 1998). The advent of centralized planning and industrial development in 1928 overrode pre-existing regional forest codes. Throughout the Sovietperiod, forest management was premised on Five-Year Plans that were the linchpins in the centralized planned economy. The state commenced a geographical expansion of forestry by setting up GULAG forestry camps in which prisoners engaged in "forest mining" (Eikeland and Riabova 2002). This system followed the model of extensive development: as the accessible timber in one area was cut, new infrastructures were built to exploit increasingly remote forests (Shvidenko and Nilsson 1994; Eikeland and Riabova 2002).

This initial management system took its toll on the forests; thus, a hallmark of the second stage was a unified forest management system for the entire Soviet Union established in 1947. A significant outcome of this step was the division of forests into three groups based on their functions and qualities (Krankina and Dixon 1992). Group One forests in riparian zones, urban peripheries, and sensitive tundra and sub-alpine areas were given protected status. Group Two forests were found in densely populated areas and may have been overharvested; thus, their use was restricted in accordance with management needs. Group Three forests constituted the largest group and were suitable for commercial exploitation. Group Three also included surplus forests, resulting in the world's largest area of protected forests, although they received such status by default. In Siberia, vast and remote stretches of mature forests as well as forests in various successional stages following fire were given Group Three status (Kukuev et al. 1997).

In the final three decades of the Soviet era, timber production in Siberia helped meet a rising demand for forest products fueled by considerable industrial growth, both in the Soviet Union and abroad. For instance, in 1968 and again in 1974, the Soviet Union signed a joint agreement with Japan on forest development in Siberia (and the RFE) that granted the Japanese a supply of raw round wood in return for substantial infrastructure investment in the Soviet forestry sector. This investment included payment for the relocation of timber operations to eastern regions (Siberia and RFE) (Mathieson 1979). Overall, the eastward shift helped compensate for diminished forest resources in the European part of the Soviet Union.

Through planning, the Soviet state expressed its commitment to the "correct and versatile utilization of forests," claiming that its logging methods promoted natural reforestation by preserving over 70 % of young, commercially valuable coniferous species (Precoda 1970). The Soviet system also boasted what was arguably the world's largest collection of scientific field forest data acquired through comprehensive inventories that consistently covered forests throughout the country (Kukuev et al. 1997). Yet despite this commitment and wealth of scientific information, the on-ground activities of forest harvesting proceeded on a scale often marked by losses and inefficiencies (Shvidenko and Nilsson 1994). These occurred in part due to geography and the state's insistence on regional development, which together resulted in large distances separating forest markets, industries, and the wood supply from each other.

7.2.2   Soviet Era: Synthesis Case Studies of Environmental Change

7.2.2.1   Synthesis Case Study: Soviet Era Forest- and Land-Cover Change in Central Siberia

During the Soviet era, Krasnoyarsk Krai, Irkutsk Oblast, and Tomsk Oblast were three of the four largest timber-producing federal provinces of Siberia. To quantify and understand differences in land-cover and land-use change over the Soviet and post-Soviet eras in dominantly forested regions, Bergen et al. (2008) used the 1972–2001 Landsat archive along with official Russian statistics. Three case study sites each the size of a Landsat scene footprint (approximately 185 × 185 km) were established: (1) just west of the Yenisei River in the Tomsk Oblast; (2) north of the city of Krasnoyarsk in the Krasnoyarsk Krai; and (3) along the western border of Lake Baikal north of the city of Irkutsk in the Irkutsk Oblast (Fig. 7.1). Landsat MSS, TM, and ETM+ time series included scenes from approximately 1975, 1990, and 2000 (±1 year). The following classes were delineated: conifer, mixed, deciduous, and young forest; cut, burn, and insect disturbance; and wetland, agriculture, bare, urban, and water land covers. Despite the relatively remote locations of the case study sites, analysis showed that over 70 % of the area in the sites classified as forested in the 1975 Landsat MSS scenes was likely disturbed prior to 1974. Mature (primary) conifer forest decreased over the 1975–2000 study period, with the greatest decrease occurring during the Soviet era period of 1975–1990. This trend of decreasing mature conifer in the landscape appears to corroborate state statistics, where for Siberia total mature and overmature coniferous growing stock decreased on average by 2.7 % between 1983 and 1988, and by 5.6 % between 1973 and 1988, in contrast to the total forested area of Siberia which increased over the same interval (Shvidenko and Nilsson 1994). Regional inventory statistics (Stolbovoi and McCallum 2002) and the results from the study sites (Fig. 7.2) also are in agreement that the decrease in mature conifer was lower in West Siberia (Tomsk site) than East Siberia (Krasnoyarsk

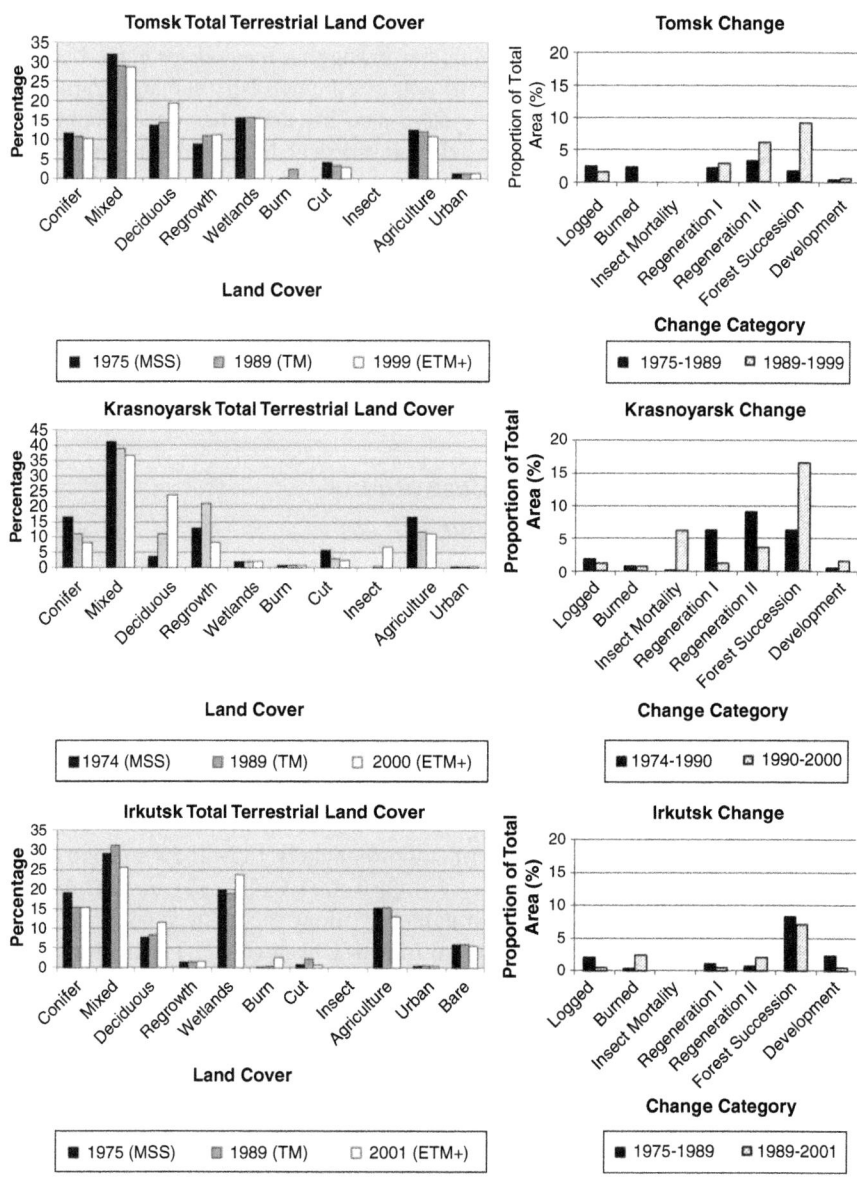

Fig. 7.2 Terrestrial land cover and land-cover change for the central Siberian study sites for three dates (~1975, ~1989, ~2000) derived from Landsat MSS, TM, and ETM+, respectively. Land-cover change is computed between change pairs of ~1975–1989 and ~1989–2000. Regenerated Type I refers to areas of Conifer or Mixed forest classes that were cut or burned at some time between change pair dates and were forest at the earlier date; regenerated Type II refers to areas that were labeled Cut, Burn, Insect, or Agriculture at the first date and were forest at the later date. Forest Succession refers to forests that transitioned to later stages (e.g., class of Young at time 1 and Deciduous at time 2) between mapped dates (University of Michigan, School of Natural Resources and Environment, Environmental Spatial Analysis Laboratory; total terrestrial land cover from Bergen et al. 2008 with permission from ASPRS)

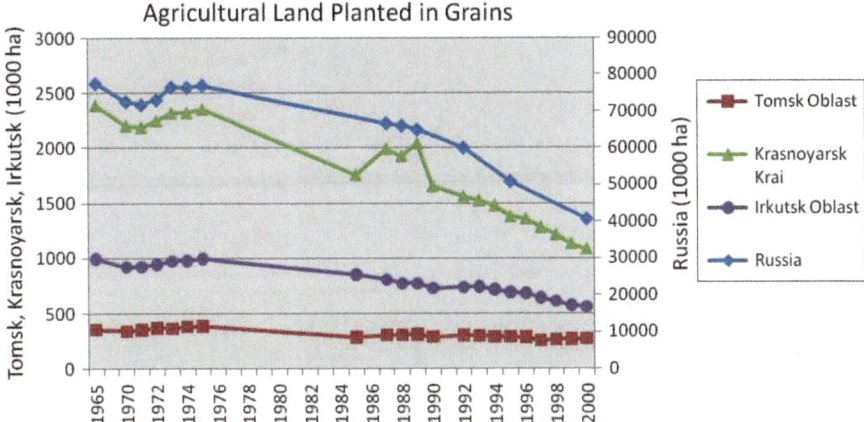

Fig. 7.3 The temporal trends for total grain-sowed agriculture 1965–2000 in Russia and in the three provinces in which the central Siberian sites are located. The trend in all three case sites, Krasnoyarsk in particular, is similar to that of Russia overall (Goskomstat Russia, 1965–1989; Federal State Statistics Service 2010)

and Irkutsk sites) during this time period. Areas classified as young and deciduous forests in study sites increased over the entire study period 1975–2000 and at a generally greater rate during the Soviet era period of 1975–1990. This increase cannot be attributed solely to regeneration from earlier cutting. Agriculture in all three of the case study sites also declined in the Soviet era period 1975–1990, contributing to forest regrowth. This is also in general agreement with Russian statistical data. For example, official statistics showed declines in the total area in sowed grain fields starting prior to dissolution of the Soviet Union (Fig. 7.3). These trends are evident in the land-cover statistics extracted from the images of the study sites (Fig. 7.2) and are generally consistent across the three sites.

In terms of spatial pattern, patches of forest regenerating from logging or fire, or both prior to 1989 were generally large landscape scale patches (although possibly comprised of merged smaller clear-cuts). These patches were irregular in shape and appeared to be regenerating naturally to noncommercial species of birch and aspen (Fig. 7.4). This appears to corroborate Shvidenko and Nilsson's (1994) assertion that nearly 85 % of clear-cuts in Siberia during the Soviet-era were large-scale and created using heavy equipment that inhibits regeneration; also, during this time the most common method of regrowth was natural regeneration. Fires in logged forests likely increased, owing to considerable logging refuse. Specifically, about 40 % of felled biomass remained in the forest after harvest due to inefficient methods (Shvidenko and Nilsson 1994). Thus, fire stemming from human activities is probably also partially responsible for large irregular patches of regenerating forests from the Soviet era, such as those observed in 1975 and 1989 Landsat MSS/TM data at the Tomsk site (Fig. 7.4).

In order to quantify and explain spatial relationships of disturbance with natural and human explanatory features, data for all three study sites were used in logistic

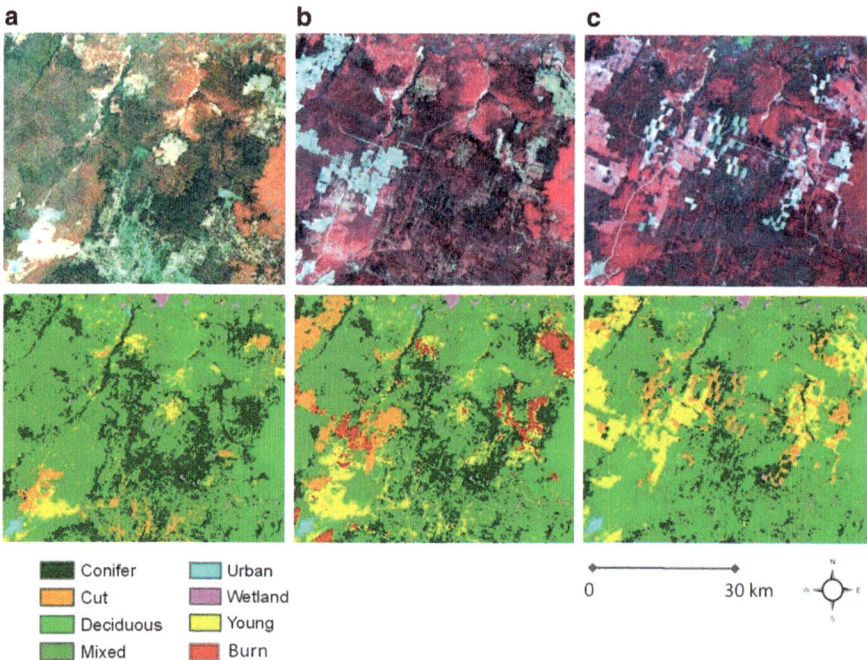

Fig. 7.4 The central Siberian Tomsk study site showing patterns of forest change and harvest over three dates from Landsat: (**a**) 1975 (MSS), (**b**) 1989 (TM), and (**c**) 1999 (ETM+). The *top* tier shows a small subset of the entire Landsat images as false-color infrared composite. The *bottom* tier shows the Landsat-derived land-cover classifications at each date for the same small subsetted area (University of Michigan, School of Natural Resources and Environment, Environmental Spatial Analysis Laboratory)

regression models (Zhao and Bergen 2006; Peterson et al. 2009). Model relationships using data pooled over three sites are shown in Table 7.2 as "+" (positive association) and "−" (negative association), respectively. Those that are neither + or − do not necessarily imply lack of relationship, but did not meet statistical significance tests ($p < 0.05$). There was a significant negative correlation between cuts (very recent logging) in the 1975 images with elevation (i.e., found at lower elevations, typically river valleys); there was also a significant negative correlation between this disturbance type and distance to rivers (i.e., logging took place closer to rivers). The fact that these relationships were statistically significant over all three dispersed sites suggests the importance and use of rivers as access and transportation for logging activities during the earlier Soviet period. This finding is in agreement with historical documentation suggesting that much timber cut during the Soviet-era was floated or rafted on rivers, especially during the 1970s where 70 % of long-distance log transport occurred on rivers (Shvidenko and Nilsson 1994). In the 1975 and 1990 Landsat classifications, data and models pooled over the three sites resulted in relationship between fire and elevation; however, individual models for the more topographically varied Irkutsk site showed a positive relationship between fire and elevation. Pooled data showed a significant relationship between fire and increasing

Table 7.2 Relationships between dependent variables of forest regeneration, fire, logging and agriculture, and several independent environmental and human land-use variables in central Siberian forested landscapes assessed using logistic regression methods

Logistic regression analysis of three-site pooled model

		Independent variable					Previous forest type		
Year	Dependent variable	Elevation	Slope	Distance to rivers	Distance to roads	Distance to urban	Conifer	Mixed	Deciduous
MSS	Regeneration			+	+	−		N/A	
TM						−			
ETM+		+			−	−			
MSS	Burn				+	−			
TM								−	
ETM+					+	+			
MSS	Cut	−		−	+				
TM					−	+	+		
ETM+					−	+		−	−
MSS	Agriculture		−		−	−		N/A	
TM			−		−	−			
ETM+			−	+	−	−			

Zhao and Bergen (2006)

The model is a pooled model of three central Siberian sites (Tomsk, Krasnoyarsk, and Irkutsk) comprised of Landsat path/rows 147/20, 141/20, and 133/23, respectively, and for 3 years (MSS – 1975, TM – 1989/1990, ETM+ – 2000). Significant relationships only are shown ($p < 0.05$). For distance variables, a positive relationship (+) indicates that greater occurrence of the dependent variable is positively associated with greater distance from the independent variable. Previous forest type was investigated only in cases of direct disturbance to the forest by logging or fire. All models were created with a lag variable to account for spatial autocorrelation distance to roads. Logging was positively associated with greater distance to roads in the ~1975 Soviet era pooled dataset. This changeed in the 1990 and 2000 data (pooled and individual) where logging was generally negatively associated with greater distance to roads. There was a significant negative correlation of agriculture with slope, distance to roads, and distance to urban areas in the Soviet era, a condition which did not change in later eras. Although not all dates were significant, where they were significant, there was a positive association between areas classified as logged where conifer was the prior forest type; similarly there were significant negative associations for areas classified as either fire or logged where deciduous was the prior forest type. These same relationships generally held for the non-pooled individual site data.

Landsat-derived land-cover data were used to develop predictive models based on extending LCLUC trends of Soviet vs. post-Soviet eras using Markov and cellular automata methods (Zhao and Bergen 2006; Peterson et al. 2009). For example, for the Irkutsk site, when predictions for 2001 landscapes using 1975–1989 Soviet era transition rates were compared with observed 2001 land cover, the proportion predicted was 74 % higher than the actual proportion classified as logged (Cut category) in 2001. This reflects the higher timber harvest and forest industry production prior to the dissolution of the Soviet Union. In the modeled scenario using Soviet-era transition rates, conifer forest had a 14 % lower proportion compared to the observed. This result suggested that higher rates of logging in conifer forests would have prevailed should Soviet-era trends continued. The proportion of agriculture was also slightly higher based on the 1975–1989 Soviet-era data compared to that observed in the 2000 Landsat TM scene, indicating a trend of collective agriculture abandonment during the Soviet era, but at a lower rate compared to the early post-Soviet era. As would be expected, future projections modeled for the Irkutsk site in 2013 showed a greater decrease in conifer forests and a greater amount of logging on the landscape when Soviet-era (1974–1989) rates of change were projected to 2013 as compared with early post-Soviet rates. The proportion of the logged (Cut) class projected for 2013 was about 80 % lower when based on post-Soviet era probabilities, instead of those for the Soviet era (Peterson et al. 2009).

7.2.2.2 Synthesis Case Study: Effect of Soviet-Era Industrial Mining on Sub-Arctic Forest Landscapes

Norilsk and the area of the Norilsk Nickel plant (69°N, 88°E; Fig. 7.5) lie on the East Siberian Taimyr Peninsula in Krasnoyarsk Krai. This industrial area overlies one of the world's largest deposits of nickel (over 30 % of global reserves), copper (12 %) and cobalt, gold and platinoids (15 %). Also, it has been arguably the most significant pollution site in the boreal zone for the past 60 years (Derome and Lukina 2011; Kozlov and Zvereva 2007; Kharuk 2000; Kharuk et al. 1996a). Here we present a case study synthesis of the influence of the Soviet-era Norilsk Nickel on the forested land-cover in its vicinity based on multiple remote sensing analyses and air quality data.

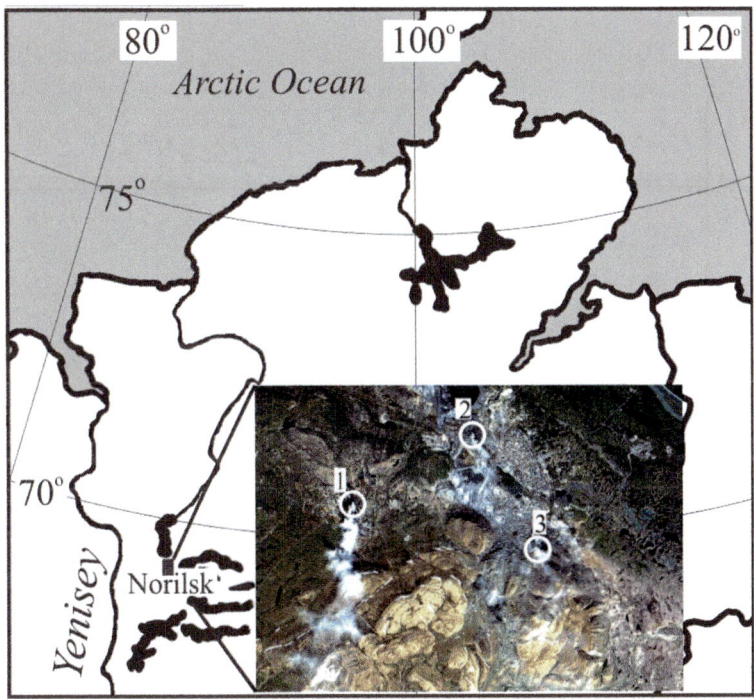

Fig. 7.5 The Norilsk area on the Taimyr Peninsula of East Siberia. Inset: A subset of a Landsat ETM+ (1999) natural color composite over the Norilsk city area with locations of three smelting plants: *1* "Nadezdha" plant, *2* "Copper" plant, *3* "Nickel" plant (Sukachev Institute of Forest)

The presence of copper and nickel deposits in the Taimyr Peninsula has been known since the seventeenth century, with intensive exploration of the deposits beginning in the 1920s. The city of Norilsk was established in 1935 and is today the world's second largest city (population ~105,000 in 2009) north of the Arctic Circle. During the years 1930–1950, prison labor was used to construct much of Norilsk Nickel. The first copper-nickel concentrate was extracted in 1939 and in 1942 the first nickel and copper was produced to meet military needs. By 1953, Norilsk Nickel produced 35 % of the nickel output, 12 % of copper, 30 % of cobalt, and about 90 % of platinoids in the Soviet Union. The main smelter smokestacks in operation today have the following heights: 138 m, 150 m (constructed in the 1940s), 180 m (1950s), and 250 m (the smokestack of the Nadezhda ("Hope") plant constructed in 1980).

The Norilsk industrial region lies in the transition area between tundra and northern taiga forest ecological zones. The soils are very poorly drained with an average depth of soil thaw during summer of 0.4–0.6 m. The forest floor is composed of thick "pillows" of moss and lichen. The growing season is approximately 60 days, the July mean temperature is +12–14 °C, the January mean temperature is −37 °C, and

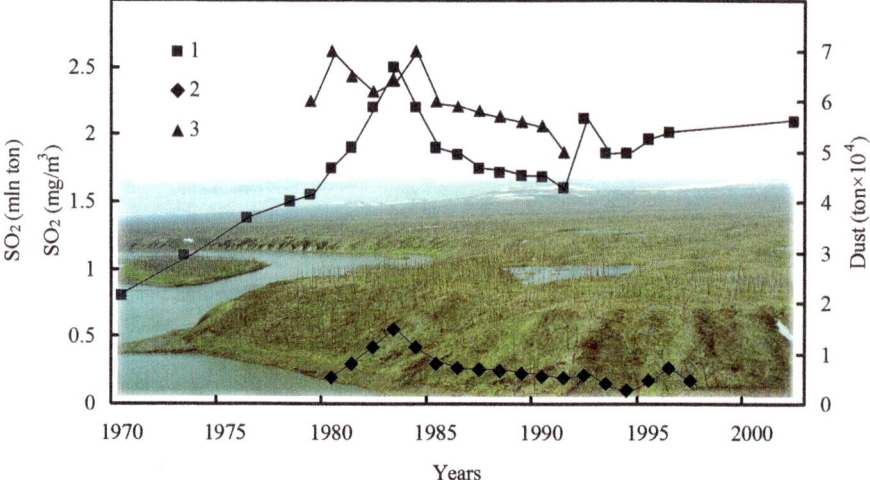

Fig. 7.6 Time series of *1* SO$_2$ total emission (million tons), *2* Norilsk city ambient air SO$_2$ concentrations (mg/m^3), and *3* dust (tons ×10^4) (Annual Report on the Krasnoyarsk Region Environment. State Committee on Nature Protection, Krasnoyarsk, 1996–1999)

the mean annual wind speed is 1.9–9.4 m/s, yet can reach up to ~40 m/s during winter. Precipitation is 400–500 mm annually in the valleys, and snow cover lasts from September to June or July with snow depths of 0.4–0.8 m on the plateau, and exceeding 10 m in the canyons. The forests are concentrated in Norilsk valley and around its rivers and numerous bogs and lakes. The much higher Lontokoy Ridge is on the west and Putorana plateau borders the valley on the east. Forest stands are formed of larch (*Larix sibirica, L. gmelinii*, the dominant species) with spruce (*Picea obovata*), birch (*Betula pendula*), and willow species (*Salix* spp.). The stands are mostly old (200–300 years) and of low density.

Extreme climatic conditions make vegetative communities in this region especially sensitive to pollution impacts. These subarctic forests, the "last frontier" of boreal forests, have been subject to both damage and mortality from the pollutants of Norilsk Nickel since the 1940s. The earliest reliable data on pollution-caused forest decline in the Norilsk area goes back to the late 1960s when the area of dead forest was estimated to be approximately 5,000 ha. Local residents, however, consider the mid-1950s to be the beginning of forest degradation. From the 1990s to today, the annual output of the primary pollutant (~96 % of total), sulfur dioxide (SO$_2$), is about 2.0 million tons, or approximately 20 % of total SO$_2$ emissions in Russia (Savchenko 1998). Maximum SO$_2$ emissions (2.5 million tons/year) were registered in the 1980s (Fig. 7.6), and this periodic high abnormal emission of SO$_2$ due to emission violations during the 1970s–1980s may have been an important factor in forest damage (Nilsson et al. 1998).

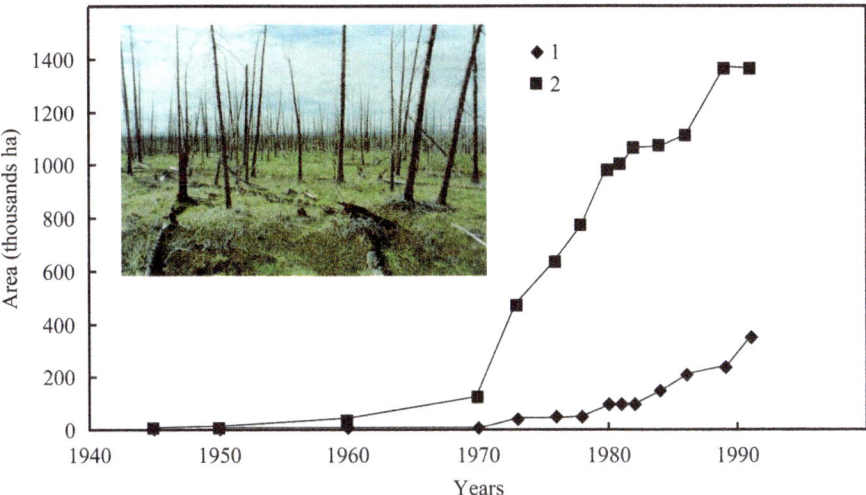

Fig. 7.7 Change in the area of *1* dead and *2* damaged stands 1945–1992 based on Russian Forest Service Forest Inventory maps. *Inset*: Tree mortality along the Ribnaya River (Sukachev Institute of Forest)

Measurement of total emissions and concentrations of SO_2 and related pollutants in the air conveys only a partial impact of the industry. The remainder of this case study presents a synthesis of mapping, remote sensing, and statistical analyses focused on analyzing the effect and spatial distribution of Norilsk Nickel and its SO_2 output on the forests in the region. Work reviewed includes that based on Russian Forest Service forest inventory maps, several types of satellite data (KFA-1000, AVHRR and MODIS imagery), a lichen vigor map (lichen are highly sensitive to SO_2), and ground data from permanent test plots along the pollution gradient running up to 200 km (Kharuk 1992).

Forest inventory maps of forest damage were generated through the year 1992 by the Russian Forest Service using manual interpretation of false-color infrared airborne images (1:15,000 scale) and observations from ancillary test plots (Vlasova and Klein 1992). The inventory maps have four damage categories: (1) very heavily damaged (>75 % of dying and dead trees), (2) heavily damaged (51–75 %), (3) moderately damaged (26–50 %), (4) slightly damaged (<25 % of dying and dead trees), and background. Forest damage and mortality statistics from 1945 to 1992 have been derived from these maps (Fig. 7.7). Statistics show that intensive forest damage and mortality was observed from the 1970s to the beginning of the 1980s. This increased damage coincides with an increase in SO_2 emissions from about 0.7 to more than two million tons/year over the same time period. Statistics from the inventory maps also show that the rate of increase in forest damage and mortality slowed down and reached an approximate saturation at the end of the 1990s–2000s.

Resurs KFA-1000 satellite image scenes were used for mapping and spatial analysis of late Soviet-era trends of the impact Norilsk pollution on forests, and to corroborate

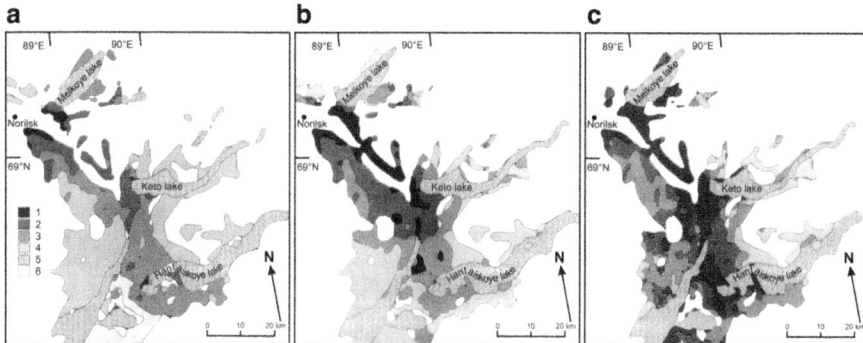

Fig. 7.8 Forest stand vigor over time based on interpretation of Resurs KFA-1000 imagery:
(**a**) 1979, (**b**) 1982, (**c**) 1984; *1* dead stands; *2* severely damaged stands, *3* moderately damaged
stands, *4* slightly damaged and healthy stands, *5* water bodies, *6* background (Modified from
Kharuk et al. 1996a, with permission from Taylor & Francis)

forest inventory mapping. KFA-1000 was a Soviet satellite camera with four spectral
bands in the visible and near infrared and with an on-ground resolution of 5–6 m.
Analysis of these remotely sensed data showed a rapid increase of pollution between
1979 and 1984 (Fig. 7.8; Kharuk et al. 1996a); additionally, it revealed that the
spatial patterns on the ground were related to terrain and prevailing winds. The period
1979–1984 also corresponds to a sharp increase in the SO_2 output (Fig. 7.6; Kharuk
et al. 1996b) caused by construction of the Nadezdha plant (Fig. 7.5, inset). Statistical
analyses showed that forest damage was linearly related ($r^2 = 0.98$) to the annual
SO_2 output. Effects of Norilsk Nickel pollution have continued in the post-Soviet
era and are further discussed in Sect. 7.3.2.2.

7.3 Post-Soviet Era

7.3.1 Post-Soviet Transformation Human Dimensions

7.3.1.1 Introduction

During the Gorbachev era, deep murmurs of structural instability in the Soviet
economy began to surface, owing in part to a disproportionate dependence on raw
natural resources and accompanying environmental pollution. This volatility contributed to a steady rise in economic difficulties and social unrest (Naumov and
Collins 2006). The ultimate dissolution of the Soviet Union in 1991 ushered in sweeping
socioeconomic and political reforms throughout the new Russian Federation (Williams
and Kinard 2003). Also known as "shock therapy," the extremely intricate work of
economic reformation involved the following elements: price, trade, and exchange
liberalization; extreme decreases in state spending; and the quick privatization of

state enterprises (Naumov and Collins 2006; Gerber 2006). Rooted in a system paralyzed by institutional and economic breakdown, these reforms, however, had adverse effects, including recession, hyperinflation, and escalating inequality. They also wreaked havoc in the labor market, causing recurring nonpayment or the delay of wages and salaries (Clarke 1998, quoted in Gerber 2006; Gerber 2006).

This period also marked the wholesale transformation of power structures into an oligarchy where the political and economic spheres were fused. At the regional level, local authorities forged very close ties with the new owners of privatized industries and others who oversaw financial, industrial, and commercial capital (Naumov and Collins 2006). For the majority of people outside of the ruling elite or oligarchy, socioeconomic and political reforms translated into mass impoverishment. Inflation quickly diminished people's wages, leaving them unable to cover the costs of even basic food goods whose prices increased by 100–125 times (Williams 1996). The gravity of this situation was exacerbated by the dearth of new social policies to accompany the economic reforms, leaving a vacuum in place of the former social organization under communism. As a result, along with wage and job loss, people were also deprived of a host of other services that the state had once provided, such as universal health care (Chernina 1996).

Russia's deepening economic and social crises were acutely felt in Siberia where hundreds of enterprises were shut down, slashing industrial production to half (or more, in the case of timber production) of its 1989/1990 level (Naumov and Collins 2006). Siberia's northern rural regions were particularly marginalized during this economic and social transformation (Pallot and Moran 2000; Kasten 2002). This crisis situation forced people to assume greater responsibility for their own livelihood and survival. Semi-subsistence and subsistence economies built on household self-help and livelihood strategies emerged in these regions. One unintended outcome of this crisis was an improvement in Siberia's ecological situation.

7.3.1.2 Post-Soviet Demographics

The dissolution of the Soviet Union greatly exacerbated already troubling demographic patterns, leading to a deep demographic crisis in Russia (Perevedentsev 1999). From 1985 to 1995, the birth rate fell from 16.6 births per 1,000 to 9.6; at the same time, the death rate during this period rose from 11.3 per 1,000 to 15.7 (Williams 1996). Russian Federation population numbers continued to decline between 1995 and 2000, so that by 2000 they were essentially the same as in 1987 (Anderson 2002). The demographic crisis encompassed all of Russia, yet it was particularly acute in eastern Siberia (Granasen et al. 1997).

People also responded to changing socioeconomic conditions in their migration decisions (Zayonchkovskaya 1999). In Siberia, a reversal of migration trends occurred during the initial post-Soviet period. Because Russia was unable to maintain its population redistribution policies, including wage differentials and the heavy subsidization of regional economies, there was a massive outflow of people from the remote areas of northern and eastern Siberia to the densely populated southern and western regions of the Russian heartland (Heleniak 2001; Kumo 2007). The latter offered better economic opportunities and a comparatively higher level of social infrastructure. This large out-migration had a deep impact on Siberia's population, which by 2002 had declined by approximately 1.5 million people from its 1989 level (Naumov and Collins 2006).

Not everyone who desired to leave for European Russia or other places where life was less expensive and easier had the opportunity to do so. Those who remained behind, particularly in rural and remote regions, had to adjust to a life of intermittent formal employment accompanied by a great reduction in available goods and services that came with the phasing out of subsidies (Pallot and Moran 2000). Remote logging communities were increasingly isolated from the outside world (Malmlöf 1998). Despite the dismal economic outlook for many smaller towns and villages in Siberia (and the RFE), legal and illegal immigration of Chinese immigrants into these regions became a significant demographic trend in the 1990s. Some estimates put the number of these immigrants in Siberia (the majority of whom are illegal) in the hundreds of thousands (Naumov and Collins 2006). Some Russians who left their homes in the Central Asian Republics also arrived in Siberia.

7.3.1.3 Post-Soviet Institutions

During the 1990s many of the Soviet-era institutional regulations on natural resource protection were rewritten. And, in May 2000 a cardinal change occurred: a federal decree issued by then-president Vladimir Putin fundamentally transformed the bureaucratic structure of Russia's environmental agencies to make natural resources more readily accessible. Specifically, this decree abolished the Federal Forestry Service, the State Committee on Ecology, and the Ministry of Nature as separate entities; instead, their roles were consolidated into the single Ministry of Natural Resources. This move eliminated the existing system of built-in checks and balances among the three agencies that had allowed them to take legal action against one another for law violations (Mischenko 2002, quoted in Metzo 2003). In effect, the Ministry of Natural Resources assumed responsibility for both the exploitation and protection of resources. Many environmentalists viewed this decree as opening the door to unrestricted natural resource exploitation (Shvidenko and Nilsson 1994). Moreover, this decree lowered the political profile of the Russian forestry sector as the Ministry of Natural Resources was focused more on oil and gas resources (Williams and Kinard 2003).

7.3.1.4 Post-Soviet Forest Management

Prior to the 2000 decree, the dissolution of the Soviet Union and the subsequent transformation of its economy and society profoundly affected the forestry sector. Beginning in 1991, reported harvest volumes significantly declined such that by 1994 the total timber harvest was approximately 175 million cubic meters compared to 439 million cubic meters in 1989 (Obersteiner 1995; Granasen et al. 1997). Net growth did not occur again in the forest industry until 2009. In addition to a lack of state subsidies, other significant factors affecting this dramatic drop included outdated technology and equipment and an inadequate transportation infrastructure (Williams and Kinard 2003). Moreover, there was a steep slide in domestic demand, especially after 1992 (Malmlöf 1998). Finally, inconsistency in expectations for forest management at the federal and local levels was also a major contributor to great declines in timber output.

At the federal level new forest legislation was first enacted in 1993 in the form of a revised Forest Code (followed several years later by the Forest Code of 1997). This organizational reform differentiated between management and production functions. On this basis, the former State Committee for Timber Harvesting and Forestry (*Goskomles*) was divided into two entities: the Russian Federal Forest Service (*Rosleskhoz*) responsible for forest management and the Russian Forest Industry Company (*Roslesprom*) in charge of production (Eikeland et al. 2004). The new Forest Code also provided an overall framework for the monitoring, protection, exploitation, and regeneration of state forests (Metzo 2003). In practice, however, these new management strategies proved difficult to implement due to a lack of specificity, including on the issue of forest property rights (Shvidenko and Nilsson 1994; Malmlöf 1998). The forests were declared the property of the Russian state, giving it the jurisdiction to determine harvest quantities (Malmlöf 1998). At the same time, regional and local authorities gained greater control by choosing who could harvest the forests.

At the local level, *leskhozy* (forestry management units) assumed the unprecedented and daunting challenge of integrating the forest sector into a market environment (Eikeland et al. 2004). This responsibility involved the dual tasks of designating forest sections for logging and regulating private timber enterprises. To this end, a leasing system was established in the 1990s in which timber rights were granted through auctions and other bidding mechanisms (Eikeland and Riabova 2002). In this system, the state as owner of the forests imposed a stumpage fee, or forestry tax, on land leased through auctions. The *leskhozy* received a portion of these fees, which in theory was intended to push them to forge partnerships with as many private companies as possible. Yet, the high stumpage fees resulting from the auctions created a problematic business environment. The *leskhozy* also had a distinct advantage over private enterprises in terms of the production capacity they inherited from the Soviet era; private enterprises typically did not have the same capacity. Both factors kept private enterprises away and ultimately resulted in decreased timber harvests.

Diminished harvests against the backdrop of economic crisis meant that the federal government was unable to channel tax revenue back into the logging communities. The ensuing budget shortfall created a dire situation in remote logging communities, including those in Siberia (and the RFE), that continued to rely on timber resources (Malmlöf 1998; Naumov and Collins 2006). Forced to seek ways to make up for insufficient funding, the *leskhozy* resorted to corruption in the allocation of forest resources, which had the unintended effects of building a barter economy and stalling the institution of a new forest management system (Eikeland and Riabova 2002).

Illegal logging also proliferated, particularly in Siberia where it was spurred by growing demand in China and Japan. Insufficient funding also quelled conservation efforts and forest-fire prevention and fighting measures. This inability of the *leskhozy* to be able to manage resources effectively had an impact on the forests (Eikeland et al. 2004).

7.3.2 Post-Soviet Transformation: Synthesis Case Studies of Environmental Change

7.3.2.1 Synthesis Case Study: Post-Soviet Era Forest- and Land-Cover Change in Central Siberia

As described in the preceding section, the abrupt political and socioeconomic changes near the end of the twentieth century created a distinct turning point in institutional factors influencing natural resources land management (Korovin 1995). To observe the influence of such changes on the Siberian landscape, the Landsat archive was used to map early post-Soviet land-cover, land-cover change, and spatial relationships between LCLUC and environmental variables (Bergen et al. 2008; Peterson et al. 2009) as well as carbon stock (Zhao et al. 2009). This analysis covered the same three central Siberian study sites in Tomsk Oblast, Krasnoyarsk Krai, and Irkutsk Oblast that were evaluated for the Soviet era. Other studies have also investigated early post-Soviet era land change and spatial relationships with environmental factors in Siberia using remotely sensed imagery from newer sensors, including SPOT VEGETATION (Achard et al. 2006) and MODIS (e.g., Kovacs et al. 2004).

Analysis of land-cover change from 1990 to 2001 shows some continuation, yet also distinct divergence in LCLUC-environmental relationships in the first decade of the post-Soviet era when compared with the Soviet era. This finding held across all three sites (Tomsk, Krasnoyarsk, and Irkutsk). Specifically, these patterns included the following: (1) a decrease in logging activity, (2) a decline in the proportion of land under active cultivation at a slightly greater rate than during the Soviet period, (3) a continued decrease in the proportion of mature conifer, and (4) the prevalence of forest succession (continued regrowth) dominated by deciduous forests (Bergen et al. 2008). In the Irkutsk site, increased fire was also found (Peterson et al. 2009).

Independent data on logging trends were compiled from Russian forest statistics for the provinces of Tomsk, Krasnoyarsk, and Irkutsk. These data indicated consistently high rates of logging in the broader central Siberia study region in the 1960s–1980s; these rates peaked in 1989 and dropped sharply in 1990–1991 (Bergen et al. 2008; Fig. 7.9). These data corroborated the decrease in logging activity documented through remotely sensed analysis of the central Siberian sites. In terms of agriculture, Oldfield (2006) reported that actively used arable land declined in the Urals and Siberia by 8 and 11 %, respectively, between 1990 and 1998. Using SPOT VEGETATION 1 km land-cover data during the early post-Soviet transition (~1990 and ~2000) Achard et al. (2006) characterized types of rapid early post-Soviet era forest-cover change over boreal Eurasia and validated these using Landsat

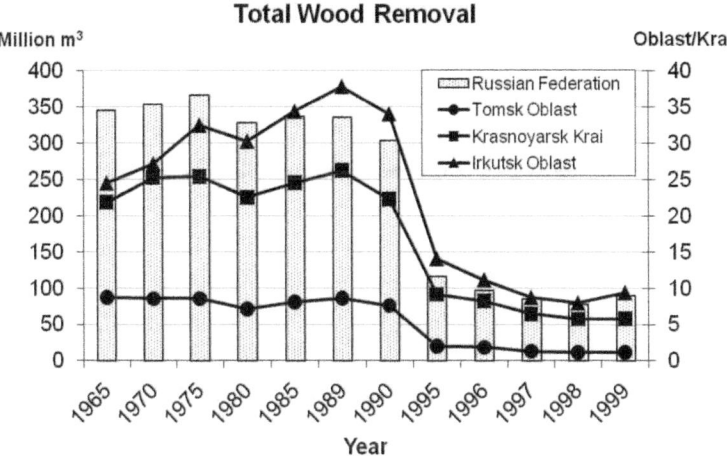

Fig. 7.9 Russian official statistics showing total wood removal trends from 1965 to 2000 (series incomplete) (All Union Scientific Research Institute of Economics 1991; Goskomstat Russia, annual; modified from Bergen et al. 2008, with permission from ASPRS)

and field data. In several Landsat and field-based validation sites in Krasnoyarsk Krai in central Siberia, the analysis found dominant change types to be logging in intact forests and regrowth of deciduous species in early successional stages. Furthermore, the highest rates of logging and the largest clear-cuts in these sites occurred in Scots pine forests; lower logging rates (and smaller clear-cuts) were observed in other conifer types. In their sites near Lake Baikal, the following were observed using Landsat imagery: moderate intensity clear-cut or selective logging producing small to medium change rates; moderate levels of change in species composition, with an increase in deciduous species (including natural regrowth); and increased fire frequency (Achard et al. 2006). Overall, the types of changes documented at Landsat validation sites by Achard et al. (2006) during the early post-Soviet transition were largely similar to those observed in the central Siberian study (Bergen et al. 2008).

Forest change trends may also shed light on carbon sequestration in the region. Land-cover change data for the central Siberian case study sites were combined with carbon values estimated from the Far East model (Yan and Shugart 2005) for each forest and land-cover category in the sites and pooled for change categories (Zhao et al. 2009). Results (Table 7.3) suggest several trends. First, from 1990 to 2000, the Tomsk and Krasnoyarsk sites functioned as a carbon sink. The Irkutsk site, which was considerably more impacted by fires during the same time period (see also Achard et al. 2006 above), was a small source of carbon. Because the uncertainty was greater in the land-cover change data for the Irkutsk site (due to poorer image clarity resulting from clouds and haze), this site also contains greater uncertainty in the carbon estimate. Even without this uncertainty, however, it is unlikely that this site could have been a carbon sink or as much of a carbon sink as the other two sites

Table 7.3 Land-cover change (percent of study site area) and area-weighted above-ground bio-mass change (Mg C ha^{-1}) for the three central Siberian case study sites

LCLUC	Tomsk Area (%)	Tomsk Area-weighted Δbiomass	Krasnoyarsk Area (%)	Krasnoyarsk Area-weighted Δbiomass	Irkutsk Area (%)	Irkutsk Area-weighted Δbiomass
Forest disturbance	4.82	−3.45	11.37	−4.98	3.98	−3.59
Forest regrowth	15.25	+6.46	20.91	+11.35	9.00	+3.11
Total		**+3.01**		**+6.37**		**−0.48**
Constant	75.35		60.75		74.26	
Unknown	4.52	−0.34	6.97	+0.34	12.76	−1.87

Modified from Zhao et al. (2009), with permission from Springer
The total is the sum of Δbiomass between post-Soviet years 1990 and 2000 weighted by proportions of area occupied by forest disturbance (change in land-cover class resulting in biomass loss) and forest regrowth (change in land-cover class resulting in biomass gain). Forest disturbance included logged, burned, insect damaged, and developed lands. Forest regrowth included regeneration following disturbance and forest succession

given the observed greater fire occurrence during the time period. Potapov et al. (2011), in a study of the northern boreal biome extending from European Russia to the RFE, found that East Siberia (the location of the Krasnoyarsk and Irkutsk sites) exhibited the greatest amount of high-intensity forest loss due to expansive wildfires in 2000–2005. The Krasnoyarsk site, which was heavily logged during the Soviet era, had the highest rate of carbon accrual due largely to continued forest growth and succession. Greater carbon accrual would have occurred if this site had not been impacted by serious insect infestation (Siberian silkmoth; *Dendrolimus superans sibiricus* Tschetw; Kharuk et al. 2004, 2007).

The results from Zhao et al. (2009) appear to agree with hypotheses that natural variability between years may cause local areas to switch between sink and source, even against the backdrop of the boreal forest as a moderate carbon sink (Harden et al. 2000; Goodale et al. 2002). Other recent work (Hayes et al. 2011) based on modeling using coarse (0.5°) spatial resolution, however, suggests that northern Eurasian boreal landscapes were a carbon source between 1997 and 2006 with LCLUC and fire accounting for the largest sources. Results from Zhao et al. (2009) for limited case study sites at high spatial resolution (60 m) confirmed that the occurrence of fairly significant fires pushes the carbon balance toward source. The latter study also quantified the influence of spatial resolution (ranging from 60 m to 1 km) in remotely-sensed data on carbon estimates, demonstrating that the amount of carbon released was generally overestimated in the case of large fires mapped at coarser spatial resolutions.

In terms of landscape pattern, new logging regulations instituted in 1994 established a limit of 50 ha for clear-cutting and placed new restrictions on the adjacency of clear-cut sites (Zaslavskaja 1994). In the 2001 Landsat scenes for all three central Siberian case study sites, the resulting checkerboard pattern of small neat rectangular areas that touch at most only at their corners is clearly evident (shown in Fig. 7.4 for Tomsk).

Landsat-derived data and logistic regressions methods were used to develop further relationships between disturbance and natural and human variables (Zhao and Bergen 2006) for the early post-Soviet transition era. Results showed that while logging was more prevalent at lower elevations during the Soviet era, in the post-Soviet era this relationship was no longer significant in the pooled data for the three sites (Table 7.2). Post-Soviet data showed a continuation of the significant relationship between logging and closeness to roads first seen in the ~1990 data, and a weakened or nonexistent relationship between logging and closeness to rivers. This finding is possibly explained by the decreased importance of (and regulations discouraging) water transportation of felled logs in the post-Soviet era due to enormous timber losses in river transport (Shvidenko and Nilsson 1994). The relationship between logging and distance to urban areas was not significant in the 1975 data, yet showed a significant positive association in 1990 and 2000, indicating that logging occurred away from human settlements. New logging in 1990 and 2000 showed a strong negative correlation with the deciduous forest type, and also with the mixed forest type in 2000. There was a significant positive correlation between logging and conifer forests in the 1990 pooled data (2000 was not significant). In the Irkutsk site, however, there was a significant positive correlation between logging and conifer forests in both ~1989 and 2001.

The post-Soviet era central Siberian site results are somewhat more complex with respect to the relationship between fires and closeness to roads and human settlements. Pooled data show no significant relationships (~2000) or a positive correlation of fire with increasing distance from roads (~1990). Other early post-Soviet era studies in the broader region have also investigated relationships between fires and human infrastructure (Korovin 1996; Crevoisier et al. 2007; Mollicone et al. 2006; Achard et al. 2008). Kovacs et al. (2004) used MODIS and found a positive spatial correlation of fire with closeness to roads in central Siberia in 2001–2003, though considered over greater spatial scales than the central Siberia Landsat case sites. Crevoisier et al. (2007) also found that fire occurrence had a positive association with road density until road density exceeded a threshold, above which fire had a much lower probability of occurring, which they attributed to greater fire suppression enabled by infrastructure. This finding agrees with pooled data for the central Siberia study sites in 2000 where fire occurred at somewhat greater distances from settlements, which in turn have denser transportation networks. Mollicone et al. (2006) and Achard et al. (2008) took a different approach to understanding the potential contribution of human activities and infrastructure to fire events. Their studies divided forests into most intact (i.e., interior primary forests least likely to be subject to human influence), intact, and non-intact. They used these categories to test the null hypothesis that fire frequency would be equal in all forest types if fires were largely natural in origin. Their results confirm that human influence is likely responsible for 87 % of fires in the nonintact forests and between 72 % (2002) and 78 % (2003) of the total area burned in both nonintact and intact forests.

Markov methods were again employed in order to create future projections of early post-Soviet era land-cover change trends for the central Siberian case study sites (Zhao and Bergen 2006; Peterson et al. 2009). Again, using the Irkutsk site as an example based on early post-Soviet era rates, 2013 proportions of logging showed a continued decrease of 10 % from proportions observed in 2001. Proportions of agricultural lands continued to decrease by about 13 % with respect to proportions observed in 2001, and conifer proportions decreased by ~5 %, a rate less than predicted by Soviet-era trends. These 2013 projections do not take into account continuing change in socioeconomic drivers that may have altered actual rates subsequent to 2001. An analysis of the difference between these projected and actual 2013 land-cover changes would shed light on emerging era LCLUC characteristics.

7.3.2.2 Synthesis Case Study: Continuing Effects of Industrial Mining on Sub-Arctic Forest Landscapes in the Post-Soviet Era

As discussed in Sect. 7.2.2.2, Norilsk Nickel was established as one of the country's leading industrial works during the Soviet era. It also became associated with significant ecosystem damage. At the beginning of the post-Soviet era (roughly corresponding to the dates of the last forest inventory maps; see Fig. 7.7), forest damage had been detected at a distance of up to 200 km, and forest mortality at a distance of 120 km. The sustained operation of Norilsk Nickel has continued to leave its mark on the surrounding landscapes up until the present time. The lowest annual output of SO_2 coincided with the dissolution of the Soviet Union in 1991 at ~1.6 million tons (Fig. 7.6). A saturation trend after about 1993 seen in the total SO_2 data is due to a relatively constant level of SO_2 output during this time; however, rates have been high since the late 1970s when stands that experienced lethal SO_2 impact were killed. Mortality of trees weakened by sub-lethal doses was also observed, leading to the conclusion of ongoing damage over time.

To examine the continuing influence of pollution on forests in the post-Soviet era, additional and new types of remote sensing data were analyzed. The AVHRR and MODIS sensors plus ancillary data were used for mapping forest damage patterns in 1995 and 2001. Despite its coarser spatial resolution, AVHRR data has proved useful in mapping boreal forest vigor (Kharuk et al. 2004) and damage from earlier eras at the Norilsk site (Kharuk et al. 1996a, b). The forest inventory map was used to guide a supervised maximum likelihood image classification of the summer 1995 AVHRR imagery. AVHRR-derived map is presented on Fig. 7.10. Agreement of generated and reference maps was estimated by the Kappa statistic ($\kappa = 0.75$; Rosenfield and Fitzpatric-Lins 1986). Results showed that AVHRR was able to discern four damage categories on the early post-Soviet era landscape (Fig. 7.10). Separately, the AVHRR-derived normalized difference vegetation index (NDVI) showed minimal greenness values within the zone of maximal pollution load, gradually increasing with a distance from Norilsk (Zubareva et al. 2003).

More recent post-Soviet era observations for the Norilsk study site were assessed based on two composite MODIS images (MOD09 Surface Reflectance bands 1–7) from 2001: a summer scene (June 4–July 11, 2001) and a winter scene (October 16–23, 2001). The winter scene was used to detect evergreen (spruce) stands. The MODIS data for the zone of Norilsk industry impact were classified and results overlaid

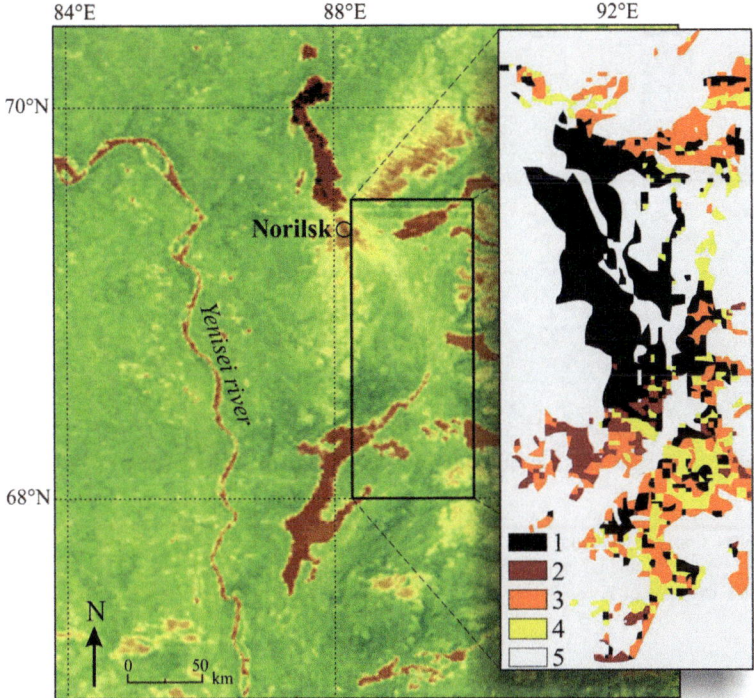

Fig. 7.10 The inset is an AVHRR-derived map of the majority of the most pollution-affected territory around the Norilsk Nickel site, and shows four forest vigor categories: *1* dead, *2* heavily damaged, *3* moderately damaged, and *4* slightly damaged and undamaged; *5* is non-forest dominated. Overall Kappa = 0.75. Background: A sketch-map of Norilsk site (Sukachev Institute of Forest)

on topographic data (Fig. 7.11). The MODIS classification showed that the majority of forest damage was observed along the Ribnaya River valley in 2001. This valley is in the direction of prevailing winds, which results in maximal impact of a smoke plume on the vegetation (Fig. 7.5). A mountain ridge (average elevation heights of 800–900 m) limits the continuous expansion of pollution to the south; however, pockets of degradation are found here and other data demonstrate that lichens beyond the ridge have shown signs of damage at a distance of 200–250 km (Kharuk 1992). These patterns are relatively similar to those mapped through a separate study based on SPOT vegetation data for 1998 and 2003 (Bartalev in Derome and Lukina 2011, Fig. 11.5). The latter study derived quantitative levels of forest degradation (0–100 %). An analysis of 1998 SPOT data revealed the greatest forest and land degradation in the vicinity of the city of Norilsk; at the same time, degradation extended down the Ribnaya valley (as in the MODIS-derived classification). In the 2003 analysis, the former patterns were preserved, and additional degradation appeared in the plateau area south of Norilsk. Estimates of total area damaged from Norilsk Nickel pollution range from 400,000 ha (Kozlov and Zvereva 2007) to two million ha (Derome and Lukina 2011).

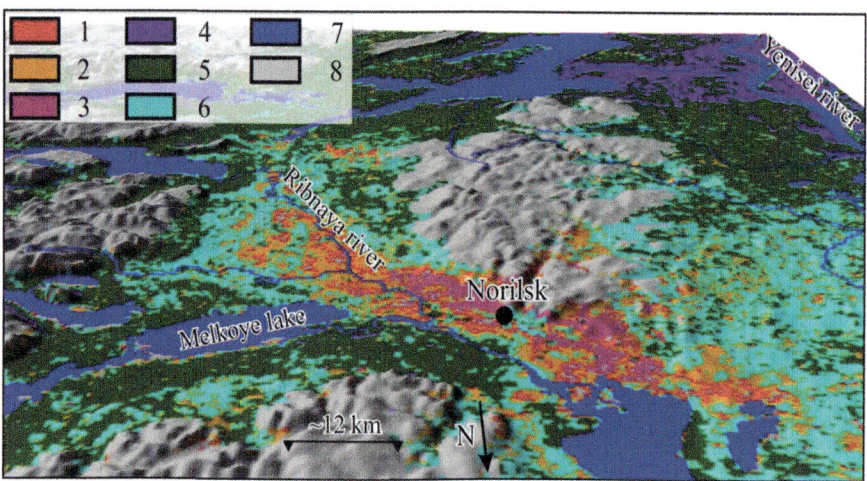

Fig. 7.11 Classified map of the Norilsk impact zone based on July and October 2001 MODIS composites: *1* severely damaged area, *2* moderately damaged area, *3* industrial area and technogenic desert, *4* spruce dominated stands, *5* larch dominated stands, *6* tundra, *7* water bodies, *8* mountains (Sukachev Institute of Forest)

7.4 Emerging Era Human Dimensions

7.4.1 Introduction

As is evident in the preceding discussions, there have been considerable changes in human demographics, institutions, and natural resources management in Siberia through both the Soviet and early post-Soviet transition eras. Furthermore, these have influenced the Siberian environment, to the extent where landscape alterations are readily observed in remotely sensed imagery with moderate spatial resolution. The ultimate political and socioeconomic drivers of these factors continue to evolve in Siberia and its broader contexts, the Russian Federation and northern Eurasia. While future environmental changes cannot be fully predicted, there appear to be emerging patterns related to the forest and energy sectors that are affecting the landscape and that are likely to be detected by remote sensing. These are (1) a resumption of higher timber harvests, (2) a geographic shift in oil and gas drilling and transport toward East Siberia (and the RFE), and (3) the relatively new and growing phenomenon of transnationalism. Clearly, transnationalism represents a paradigm that differs almost completely from that of the Soviet era and is increasingly predictive of the future as Siberia moves past the early post-Soviet transformation. In this final section we first detail the most recent changes in sociopolitical institutions that are likely at some scale to drive new environmental changes in the Siberian forestry and oil and gas sectors. This part is followed by synthesis case studies in Siberia, one on forest-cover change along the Russian-Chinese border, and the other on influence of oil and gas exploration on East Siberian landscapes.

7.4.2 Forestry and Emerging Environmental Change in Siberia

7.4.2.1 Recent Changes in Forest Sector Institutions and Resource Management

Recently, Siberia's forestry sector has been experiencing profound and dynamic change due to the restructuring of governance arising from the introduction of the new Forest Code of the Russian Federation in December 2006. This code was intended to clarify the 1997 Forest Code, which was opaque in delineating the rights and responsibilities of individuals and of the federal and regional governments, resulting in confusion in the forestry sector over who had jurisdiction over a forest area or enforced the forestry laws. Moreover, the 1997 Forest Code did not set reliable and enforceable standards, particularly in regards to penalties for illegal logging (Crowley 2005). Both codes stipulate the federal government as owner of the forests; however, the original intent of the new code was to privatize forests through the transitional stage of forest leases (Moiseev 2011, quoted in Hitztaler 2011). Overall, Russia's latest forestry legislation is one piece of a wider campaign to decentralize the government and shift its responsibilities to the private sector.

The main distinguishing features of the current Forest Code include decentralization, separation of administrative and management functions in the Forest Service, a shift from a permit to a declarative system of forest use, and increased responsibilities for leasers in terms of forest management (forest regeneration, forest fire prevention) (Hitztaler 2011). To begin, decentralization has been fully realized under this code. The *Rosleskhoz* (federal Forest Service) today has retained little authority in the administration, control, and supervision of the forests. Rather, this power has been transferred to regional administrations (provinces) that now must formulate their own policies on forest management. This devolution of power has resulted in a highly fragmented forestry structure throughout Russia.

Most provinces have divided administration of the new code into three categories: management, inspection, and commercial exploitation, all under the oversight of a regional administration (Lesniewska et al. 2008). Formerly, the *leskhozy* (Forest Service branches) undertook all these lines of work; however, beginning on January 1, 2008 the liquidation of 3,500 *leskhozy* across the country took place. Their functions were redistributed among different organizations, for instance, administrative work was assigned to organizations called *lesnichestva* that formed on the base of the *leskhozy* (these differ from the *lesnichestva* in the old system). Arrangements were made to contract out forest regeneration and fire prevention work in nonleased forests to other organizations that compete for this work through auctions. In several regions, these contracts go almost exclusively to newly formed state unitary enterprises (*gosudarstvennye unitarnye predpriyatiya*). (This type of enterprise is unique to Russia and some other post-Soviet states; one defining characteristic is that it claims no ownership rights to the assets used in its operations.) Many former *leskhoz* employees have found new jobs through these enterprises.

Another marked difference in the Forest Code is the switch from a system of government-issued logging permits (*lesnye bilety*) to a declarative one in which timber companies and other forest users state annually where and how much timber or other resources they will log or use in a year. Moreover, while the Forest Code gives substantial leeway to leasers, it also places heavy management obligations on them, including forest regeneration and sanitary cutting; leasers are also legally bound to take fire prevention measures and to fight fires should they break out on the land under their lease. The sheer volume of work that must be done at the leaser's expense raises the question of whether they will actually undertake this work.

The transition to decentralized forest management, and to new rules on forest classification, planning, and inventory, has not been seamless. Rather, it has created confusion among all forest stakeholders, including foresters, regional authorities, and local people (Lesniewska et al. 2008). A prominent problem has been the inability to recognize what is legal, leading to a convoluted situation where it has been virtually impossible to follow forestry laws. This problem stems in part from the differing forest management systems that have emerged in each province. Moreover, the Forest Code's promotion of large-scale private investment overshadows conservation measures and the needs of local communities. The new Forest Code also appears to have intensified instead of remedied some of the issues it set out to address. In just one year after implementation, illegal logging increased by 30 % (Lesniewska et al. 2008). Overall, illegal logging has stifled the collection of taxes that could boost economic growth, amounting to losses estimated at one to three billion dollars (USD) annually (Crowley 2005). These losses further perpetuate the cycle of illegality, depriving the government of the tax revenue needed to enforce existing legislation on sustainable logging practices, or to invest in domestic wood-processing facilities.

In addition to a rise in illegal logging, the dearth of effective and coordinated management resulting from the new code is thought to have perpetuated severe forest fires that ravaged European Russia in the record-breaking heat of summer 2010. The fires escalated in large part due to dramatic cutbacks in forest surveillance, which has left more than a billion hectares of forests unprotected. Only the forest land under lease, which is approximately one-third of the total in Russia (Strakov 2010, quoted in Hitztaler 2011), is monitored. The actual amount of forests under protection, however, is likely to be lower, assuming that companies are often not in full compliance with their management responsibilities. To address this serious shortcoming, then-president Medvedev mandated new amendments to the code in the aftermath of the fires. These were intended to ameliorate forestry administration and fire-fighting procedures, and to restore forest protection. The greatest postfire development was the move of the *Rosleskhoz* out of the Ministry of Agriculture and the subsequent reestablishment of it as a separate division directly under the jurisdiction of the federal government. This decision increased the visibility and power of the *Rosleskhoz*. Yet, the question remains whether this relocation will improve the current critical situation (Kuznetsov 2010, quoted in Hitztaler 2011).

7.4.2.2 Emerging Implications of Transnationalism for Siberian Forests

Russia's emergence into the global economy following the dissolution of the Soviet Union has made it susceptible to international forces that have deep implications – both positive and problematic – for its forestry sector and for forested landscapes. These forces are particularly apparent in Siberia where a common border with China extends for over 3,500 km from the Primorsky Krai and Amur Oblast in the southern part of the RFE to Zabaykalsky Krai in East Siberia. Numerous crossings have appeared along this border, facilitating an influx of raw timber into China by rail and truck. China's flourishing economy has created a spike in its demand for timber, a phenomenon that beginning in 2001 has diverted the flow of Russian exports away from Japan, which was the primary destination in the 1990s (Lankin 2005).

Besides a thriving economy in China, several other factors have contributed to its rising timber demand. These include (1) a long-standing disparity in timber resources (Siberia and China are approximately 75 and 20 % forested, respectively); (2) the onset of trade liberalization in Russia in the mid-1990s; and (3) China's National Logging Ban issued in 1998 after the country experienced the worst flooding in half a century exacerbated by forest degradation. The National Forest Conservation Program (NFCP) in China established protected areas to help enforce this new legislation. Subsequent to the ban, China became the second largest importer of logs in the world (Crowley 2005).

China's increased consumption of Russian timber is clearly reflected in statistics. Based on official statistics of legally exported timber alone, in 1996 Russia exported <5 million cubic meters of industrial round wood, sawn wood, and wood pulp to China; however, by 2002 this figure had risen to almost 30 million cubic meters (Mayer et al. 2005). This trend is likely to continue as China increasingly depends on Russian exports to fill its timber deficit, which is predicted to expand to 200 million cubic meters annually by 2025, according to the Center for International Trade in Forest Products (Newell et al. 2000).

It has been difficult, however, to determine accurately the actual amount of timber flowing into China since exports from Russia, especially those along the numerous rail and road entry points to China, are extensively unreported (Newell et al. 2000). Recent data reveal that illegal logging and trade in Siberia and the RFE may account for up to 70 % of forestry operations (Lesniewska et al. 2008). Other figures put illegal timber harvests at 40–50 % of total harvests (Crowley 2005). Illegal logging encompasses the following activities: logging without a license; logging in protected areas; logging outside of boundaries indicated on the forest lease agreement; misclassification of species or grade; smuggling timber; false documentation; bribing officials; and obtaining illegal permits (Newell et al. 2000; Crowley 2005; Lebedev 2005). One highly perfected illegal practice occurs under the auspices of legal sanitary logging, which is intended to improve forest health through the removal of fire-damaged or diseased timber from the forests.

China's swelling timber consumption, and the further decentralization of the Russian forestry sector both stand as clear threats to intact forested landscapes in

Siberia (and the RFE), especially as they have spurred the rise of illegal logging in these regions. At the same time these international and national forces have also intensified more systematic and ostensibly legal forest exploitation. Currently, the Russian federal government intends to open up previously inaccessible areas through a one billion ruble investment in new transportation infrastructure in Eastern Russia (Siberia and the RFE). This investment would ease the costs of transporting timber from remote areas to markets, thereby resulting in a projected 90 % increase in timber harvests from 2005 to 2030 (Northway et al. 2009). The majority of these harvests would take place in northern regions where forest regeneration is protracted and where there is a high release of carbon from permafrost areas during logging.

Not all effects of transnationalism have had negative ecological impacts: Russia's forestry sector has attracted the attention of the international environmental community. Organizations (such as USAID, US Forest Service, the World Bank, and the World Wildlife Fund) have been working with Russian partners to increase forest productivity and improve management of forest resources and landscapes (Henry and Douhovnikoff 2008). According to Henry (2010), "the transnational nature of the timber market has correspondingly led to transnational efforts to encourage sustainable forestry." These efforts include forest certification through the Forest Stewardship Council (FSC), which involves a third party inspection to verify that timber enterprises adhere to a set of ecological, economic, and social criteria, including the preservation of ecologically valuable forests. This system gives enterprises a competitive edge in the international market as official certified wood may be of greater value. Thus far, forest certification campaigns have attained much higher success rates in the Russian Northwest (Karelia and Komi) than they have in Siberia where there have been mixed outcomes (Henry 2010).

7.4.3 Energy Resources and Emerging Environmental Change in Siberia

7.4.3.1 Changes in Energy Sector Politics and Economics

In August 2003 Russia instituted a new energy strategy centered on its influential oil and gas complex, which was to serve as the foundation of economic development, and as a means of formulating internal and external policy (Hashim 2010). Unlike Boris Yeltsin's regime, under President Vladimir Putin, the state reverted back to a "central planning mentality" in its energy policy characterized by renewed control of the oil and gas sectors for political purposes (Milov et al. 2007, quoted in Hashim 2010). The state's heightened control over the energy sector, along with rising oil prices and an improved tax collection system, has granted it increased leverage in developing policies and undertaking developments. The state has also begun to wield control over European countries that depend on its oil and gas, thereby helping Russia reemerge on the world stage.

Besides its political influence, the energy sector's economic importance in Russia cannot be overstated. This sector fully provides for the nation's needs; and it accounts for more than two-thirds of Russia's export revenue and more than 15 % of GDP (2008 World Bank Report quoted in Henry 2010). Consequently, Russia's economic advancement pivots almost entirely on the production and selling of its energy resources (Bykovsky 2002). Russia is second, after the United States, in the number of oil wells drilled and total cumulative oil withdrawn from its oil fields (Dienes 2004). In terms of world crude oil production, Russia ranks just behind Saudi Arabia (11.7 % versus 11.8 %, respectively), and is responsible for 14.4 % of global oil exports (Hashim 2010). Overall, Russia ranks sixth in global crude oil reserves. Today, the European Union (EU) depends on Russia to fill 30 % of its oil and 50 % of its gas needs (Hashim 2010); seven East European countries fill 90 % of their crude oil demand with Russian supplies.

In contrast to oil production, gas development in Russia is relatively new, especially in northwest Siberia, which was targeted for gas extraction in the 1980s (Espiritu 1999). Because oil is easier to extract and process than gas, it was initially given priority. Nonetheless, Russia's natural gas sector has now risen considerably in national and international significance. Russia has approximately 27 % of the world's known natural gas reserves (Ahrend and Tompson 2005). By the end of 2003, these known reserves measured approximately 47 trillion cubic meters, making them the largest in the world. This prodigious amount enables Russia to provide 26.7 % of the world's gas exports; thus, gas has become its second most important export commodity (Hashim 2010; Ahrend and Tompson 2005). Currently, six EU countries fully rely on Russian imports for their natural gas needs (Baran 2007, quoted in Hashim 2010). This growing dependence on Russia's natural gas comes as the North Sea gas fields are being exhausted. Within Russia, gas constitutes around 50 % of the country's primary energy supply.

Despite the global financial and economic crisis that began in 2008, the prognosis for the continued growth of Russia's energy sector remains optimistic. The Russian government has not curtailed its development programs slated for East Siberia and the RFE (discussed below), citing the vast deposits of oil and gas in these regions. Moreover, it has ensured profitability at current oil prices by removing the export duty on oil extracted from East Siberian fields. These steps have created an environment conducive to investment and to rising production. In other regions, such as the Khanty-Mansi Autonomous District of West Siberia, the prominent Russian oil company Surgutneftegaz posted a net profit in the first half of 2009 exceeding that of the same period in 2008 by 6.3 %. Russian crude oil output is expected to rise at an average rate of 0.3 % per year through 2020 based on studies of production, consumption, and exports (Reed 2009).

7.4.3.2   Changing Oil/Gas Geographies and Transnationalism

The historic large oil wells of the West Siberian basin are now considered to be up to 75 % tapped, leaving significantly smaller deposits in much more difficult geologic strata and more remote geographic terrain (Dienes 2004). Thus, the strategy for Siberian oil production in the coming decades is expected to focus on the largely untapped reservoirs in East Siberia and the RFE, and on greater expansion into the Yamal peninsula in West Siberia. The oil fields of East Siberia are expected to provide at least half of all new oil in Siberia starting in 2010 (Dienes 2004). This development is also associated with the aim put forth by President Putin in 2006 to broaden the Asian portion of Russia's energy exports from 3 to 30 % within 15 years (Hashim 2010).

The East Siberian oil fields span parts of Krasnoyarsk Krai, Irkutsk Oblast, and the Sakha Republic (Yakutia), and hold approximately 337 million tons primarily in three locations, with an additional 55 million tons significantly further east. Currently, the fields that lie north of Irkutsk are slated for prospecting as part of a joint agreement between Russia and Japan to develop oil production in East Siberia. This plan, however, is fraught by the region's distant location from either the infrastructure in the west or the Pacific ports in the east, epitomizing the long-standing Siberian problem of isolation and long stretches to transportation and markets. Dienes (2004) noted that: "remoteness from the domestic centers of consumption and existing export outlets, combined with the intervening location of the much more developed West Siberian Basin, rule out the westward shipment of East Siberian oil." The construction of the East Siberia-Pacific Ocean (ESPO) oil pipeline, which originated in an agreement signed between Russia and China in 2003, is a recent key development in the Siberian oil and gas sector that offers a feasible solution to this dilemma.

The now partially completed pipeline complex will provide a direct connection for Russian oil from both West and East Siberia to Japanese and Chinese markets. The first and main pipeline commences from Tayshet (approximately 680 km northwest of Irkutsk and the point of divergence of the Trans-Siberian and Baikal-Amur Mainline Railways), runs north into Yakutia to the vicinity of the city of Lensk, then eastward to Aldan and eventually turns south to the town of Skovorodino in the northeastern part of the Amur Oblast. Currently, oil must be transported from the latter city to the port city of Nakhodka on the Pacific Ocean (near Vladivostok). The second route is a branch pipeline from Skovorodino through Mohe on the Amur River (here the Amur River forms the Russian-China border) to Daqing, China, the site of China's largest oil field and a pre-existing industrial complex (Fig. 7.1). These pipelines were completed in 2009–2010. Remaining to be completed are pipelines from Skovorodino to the vicinities of Khabarovsk and Nakhodka-Vladivostok. The projected date of final completion is 2014.

The intensification of oil and gas production in the Yamal Peninsula is now underway to exploit the largest share of gas reserves in northern West Siberia. In August 2009, Gazprom Neft, the oil branch of the giant Russian conglomerate Gazprom, signed a comprehensive agreement with the Yamal-Nenets Autonomous District to prospect for both oil and gas fields over the period 2009–2011. In this same month, then-prime minister Putin opened drilling in the Vankor oil field, commencing a significant first step toward the broad development of hydrocarbon fields in this region. Oil production at this field is expected to increase from 18,000 to 30,000 tons per day. Besides this impressive output, Vankor is strategically located at the crossroads of pipelines running to the east and northwest.

7.4.3.3   Environmental Implications of Oil and Gas Development

Several practices of the energy industry in Russia are likely to have landscape-level environmental consequences, which can be categorized into two groups. The first result from the unintended leaks and spills that spread into the tundra, swamps, and rivers. Significant oil and gas pipeline spills are estimated to occur in Russia each day (Pelley 2001, quoted in Henry 2010), presenting a major ecological threat. The second stem from "business as usual" procedures in energy prospecting and production, including the use of heavy machinery and explosives (Balzar 2006), and gas flaring methods (Henry 2010). These lead to forest fragmentation and accelerate the occurrence of forest exploitation and wildfires.

Besides extractive activities, the transportation of oil and gas leaves a deep ecological footprint, exemplified by the ESPO pipeline construction. Initially, this oil pipeline was proposed to skirt the west coast of Lake Baikal, a World Heritage Site. Protests of environmental groups resulted in the rerouting of pipeline away from this highly sensitive area to the north of Baikal through the Sakha Republic and closer to the oilfields. There is also concern about the possibility of earthquakes (explaining the preference for above-ground, flexible pipelines rather than underground). Today, the 2,757 km long main ESPO pipeline already includes 32 pumping stations, 13 tank farms, and has necessitated the building of a large power complex at Olyokminsk in the Sakha Republic. These continued developments, including the plan to lay a parallel Yakutia–Khabarovsk–Vladivostok gas pipeline alongside the oil pipeline (scheduled to begin in 2016), will undoubtedly lead to population redistribution and its associated impacts on the environment.

To date, negative side-effects of oil production in Siberia have been most widely observed in the Khanty-Mansi Autonomous District (Balzar 2006). Here, the oil giants Surgutneftegaz and the former Yukos have been responsible for the most intensive extractive activity. The Russian government has done little to mitigate the environmental and social impact of energy production, especially in relation to indigenous peoples in the region (Espiritu 1999). This same story is also being played out in other energy-rich regions, for instance, in the Yamal peninsula where oil and gas extraction threaten the traditional livelihood of the Yamalo-Nenets indigenous reindeer herders. Seasonal migratory herds are especially vulnerable to vegetation destruction and to the main gas pipelines that act as barriers to movement. These consequences extend over a vast territory (estimated at 2500 square km) that encompasses oil and gas fields, and the area along the main gas pipelines (Maynard et al. 2011; Walker et al. 2011; Forbes 1999). Damage to the landscape is expected to spread with the opening of the Vankor oil field; however, part of the development agreement between Gazprom and the Yamal-Nenets Autonomous District stipulates socioeconomic development in the district, and the establishment of programs to promote natural resource conservation.

7.4.4 Emerging Era: Synthesis Case Studies of Environmental Change

7.4.4.1 Synthesis Case Study: Land Change in the Amur Site on the Siberian-Chinese Border

The border between Siberia and China extends for almost 2,000 miles, a significant portion of which is formed by the Amur River. This border divides an ecologically similar landscape into two different management regimes where different trajectories of forest change are evident (Fu et al. 2008). The Amur case study site is located in the southern taiga of East Siberia and Northern China. The site spans the Amur River between the Amur Oblast (Russia) and Daxinganling Prefecture (China). The elevation ranges between 200 and 1,400 m above sea level (ASL) with a mean elevation of 600 m ASL. The climate is characterized by long cold winters and short hot summers. Needle-leaf forests are dominated by larch (*Larix gmelinii*) and pine (*Pinus sylvestris* L.), while the deciduous broad-leaf forests are dominated by oak (*Quercus mongolica*), aspen (*Populus davidiana*), and birch (*Betula* spp.). This area is one of the prime timber production sites in China; timber harvesting also occurs in the Siberian portion. Moreover, a significant portion of the landscape on the Chinese side is used for crop production and rangelands.

Landsat MSS (WRS-1 p132 r 23) and TM and ETM+ (WRS-2 p122 r23) images present the primary source of data for land cover mapping and change detection over the time period between 1974 and 2006. According to the classified Landsat ETM+ image from 2002 (Fig. 7.12), closed mixed tree stands dominate the land cover (~30 %). Tree-dominated land cover accounts for ~70 % of the total area with the remaining area covered by shrubs (21 %), herbaceous (7 %), bare and sparse (1 %) vegetation, and water (1 %). Small human settlements are distributed along major rivers with bare/built category accounting for 0.2 % of the total scene. Three Landsat TM/ETM+ scenes for the study site were selected to use in land-cover change detection: (a) June 15, 1987; (b) May 15, 2002; (c) July 5, 2006.

Timber harvesting represents the major type of anthropogenic disturbance on both sides of the border while wildfires and insect infestation are the cause of natural disturbances. The catastrophic fire of 1987 by far exceeds the impact from subsequent wildfires (Fig. 7.13). It is widely believed that the impact from the 1987 fire was greater in China; however, spatial analysis of the Amur study site comprised of the entire Landsat scene reveals a greater proportion of fire damage on the Russian side (Table 7.4). While the forests on the Russian side recovered from 1987 fire naturally and experienced significant new fires, especially in 2003 (Fig. 7.13), the Chinese government implemented a 10-year program to suppress forest fire and plant trees in the burned area. Subsequently, the Chinese government implemented a new forest policy, the Natural Forest Conservation Program (NFCP) (Zhang et al. 2000; Zhao and Shao 2002). The fire control appears to have been very effective (Fig. 7.13), yet the rate of forest disturbance on the Chinese side was still extremely high during 2002–2006: in just 4 years over 50 % of forest cover that remained undisturbed by

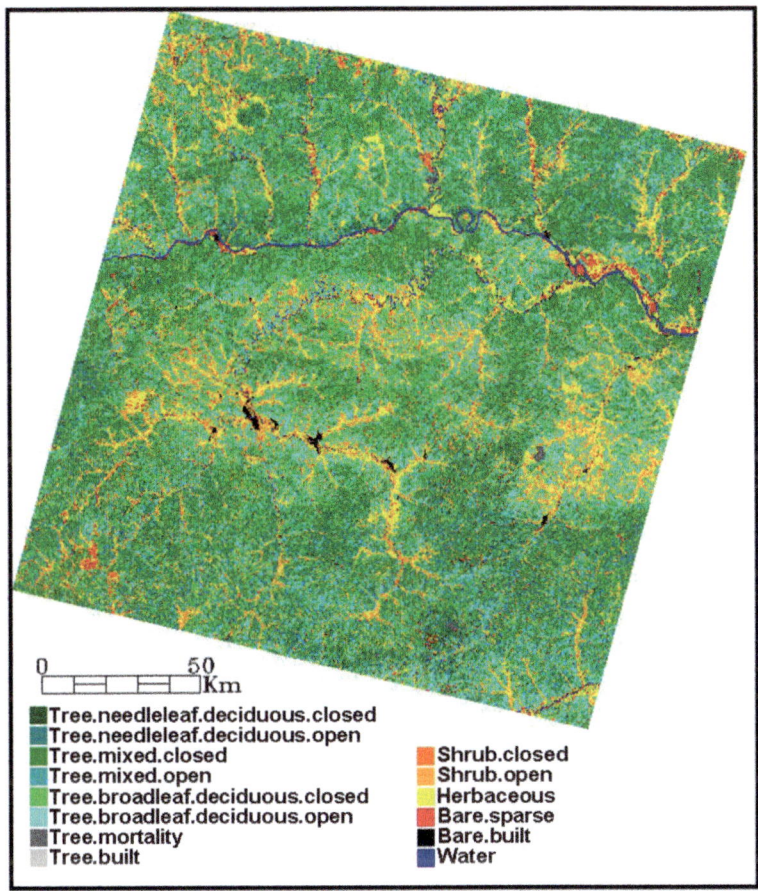

Fig. 7.12 Land cover at the Amur site in 2002 (Loboda et al. 2008)

1987 fire and subsequent logging was removed. Nearly all of this detected forest disturbance was due to timber harvest since the impact of fire during this time was minimal on the Chinese side (Fig. 7.13). At this rate of removal virtually all mature forests on the Chinese side of the study area are likely gone. As the area can no longer serve as a significant source of timber the demand had to shift elsewhere, most likely to Siberia. On the Siberian side of the study area, however, the fraction of forests left undisturbed in the past 20 years was even smaller due to continued wildfires (16 % vs. 22 % on the Chinese side, Table 7.4). The regenerating forests in the Amur site will be unsuitable for timber harvest for many decades.

The analysis of the 1987–2002 disturbed areas in comparison with land cover mapped from 2002 image (Figs. 7.12 and 7.13) shows that nearly 56 % of previously disturbed areas were in a "shrub-dominated" stage of the community regrowth and did not return to tree-dominated communities. Of those that retuned to tree-dominated communities, most were represented by broadleaf species; only 11 % had significant

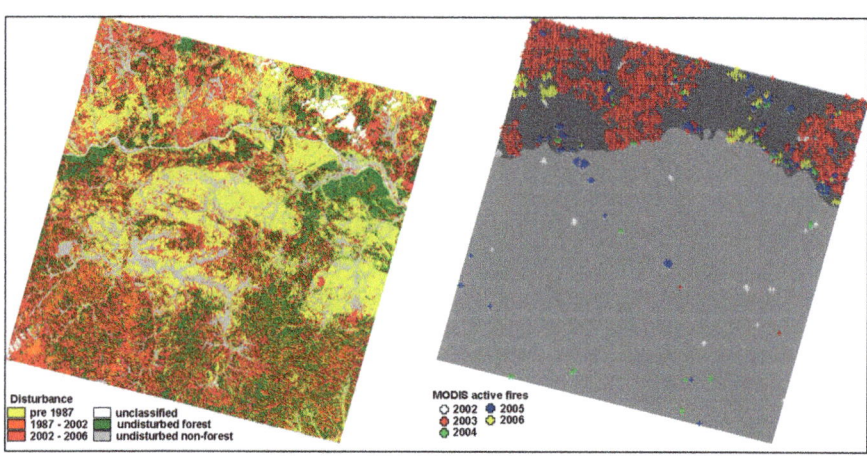

Fig. 7.13 Forest disturbance at the Amur site. *Left*: Disturbance pre-1987 (*yellow* includes 1987 fire), 1987–2002 (*orange*), 2002–2006 (*red*); *Right*: MODIS active fire detections June 2002–June 2006 (*colors* indicate different years) (Loboda et al. 2008)

Table 7.4 Forest disturbance on the Russian and on the Chinese side of Amur study area (Loboda et al. 2008)

Disturbance	China (mil ha)	Russia (mil ha)	China (%)	Russia (%)
Pre-1987	6.0	2.8	26	34
1987–2002	3.3	0.1	14	1
2002–2006	5.3	2.7	23	33
Nonforest	3.3	1.3	14	16
Undisturbed forest	5.2	1.3	22	16
Total	23.1	8.2	100	100

presence of conifer species on the Chinese side and only 8 % on the Siberian side, which is indicative of reforestation efforts in China and predominantly natural regeneration in Siberia.

Because of ongoing forest regeneration throughout the site, there was a net gain of approximately 10 % forest cover between 2002 and 2006. A large proportion of the new forested area represents continued regrowth of the forests disturbed in 1987 and in prior years. Results from the Amur case study indicate how focusing on net change in forest cover without accounting for disturbance and disturbance type can mask the ongoing loss of mature forests and dwindling timber supply. For example, analysis of forest loss in northeastern China between 1990 and 2000 based on Landsat and MODIS data showed only minor forest loss (about 0.2 % per year) in certain areas (Krankina et al. 2004). Achard et al. (2006) showed an overall trend of slight forest increase attributed to reforestation in specific zones on the Chinese side of the border. Unlike these encouraging results, the analysis at the Amur study site indicates rapid loss of mature conifer forests along the Amur River as a result of aggressive logging in China that far exceeds sustainable rates, and of widespread wildfires in Russia. If this finding represents a wider pattern, then long-term loss of timber supply from this area is inevitable, as are the economic, environmental, and social consequences of this loss.

7.4.4.2 Synthesis Case Study: Impact of Oil and Gas Reconnaissance on Taiga Landscapes

Remotely sensed data were used to analyze the impact of exploration and development associated with the Yurubchen-Tohomo oil and gas fields on larch-dominant vegetative communities in Evenkiya, Siberia (Kharuk et al. 2003). While this oil and gas field in East Siberia was first explored in an earlier decade (first explored for oil in the 1970s), recent analyses provide a useful synthesis of the environmental consequences that may become more widespread as oil and gas development becomes intensified in East Siberia in the coming decades. The particular location studied is also likely to be a key region for greater exploitation due to the emerging geographic shift of oil/gas extraction from West Siberia into East Siberian fields. Oil and gas field reconnaissance and exploration has the potential to disturb taiga forests directly through strip-cuts, oil rigs, pipelines, spills, and soil erosion, as well as indirectly through human-caused wildfires Ecologists have primarily concentrated on the direct effects (e.g., Scott 1994). Analysis of remotely sensed imagery combined with ground data has facilitated an evaluation of the direct impact caused by the initial phases of reconnaissance and exploitation, as well as an estimation of human-induced wildfires resulting indirectly from these activities.

The Yurubchen-Tohomo oil and gas fields in Evenkiya, Siberia (Fig. 7.1) are located within Lena-Tunguska oil-gas field. Their projected deposit size is comparable with the exploited western Siberian fields. Although gas is present, to date exploration in the region has focused primarily on oil. This large, remote, and sparsely populated area is located within the bounds of the mid-Siberian plateau. Its physiography consists of elevated sloping plains and low hills of 500–600 m ASL elevation dissected by dense stream networks. The regional climate is severe continental, with long, cold winters (mid-October–mid-April) and hot, short summers (June–August). The area lies within the middle taiga forest zone with widespread bogs along the river valleys. The soils are loamy, clay sod-podzol, rubbly, and frozen-taiga types, with discontinuous permafrost located along the river valleys and north-facing slopes, reaching 40–70 m deep (Zuy and Volod'ko 1991). Taiga vegetation in the area includes larch-dominated (*Larix sibirica*) forests, dark-coniferous forest (*Pinus sibirica, Picea obovata*, and *Abies sibirica*), and mixed forests with presence of birch (*Betula pendula*) and aspen (*Populus tremula*). Scots pine (*Pinus silvestris*) grows sparsely on southerly slopes. Willows (*Salix* spp.) and bush-form alder (*Alnus* spp.) form the forest undergrowth.

A combination of data from the Russian Resurs KFA-1000 satellite camera (June 1984), Landsat ETM+ (October 3, 1999), AVHRR (July 1995), MODIS (July and October 2001), and a ground survey were processed and analyzed. The region of

Evenkiya was divided into two strata: impacted area and background area (~4 and ~26 million ha, respectively). The ground survey covered 28 test plots (each 19.6 m in diameter). Ground data collected included: site location, land-use history (e.g., burn or clear-cut), and vegetative (including regeneration) composition and structure. The KFA-1000 false-color scene with 5 m spatial resolution was used to supplement the ground data in analyses of the Landsat and MODIS/AVHRR moderate and coarse spatial resolution data. MODIS and AVHRR spectral data were used for detection of large fire scars, defined as burns of 200 ha or more. Other studies in Siberia have shown that ~90 % of total burned area and ~10 % of the total number of fire scars is typically attributable to such large-sized fire scars (Kharuk et al. 2008). Burns were detected based on the temperature characteristics of the ground vegetation cover and NDVI (Normalized Difference Vegetation Index). The burned area, based on detected fire scars, was then normalized (i.e., burned area within the impacted and background zones were divided by the area of those zones) to assess anthropogenic impact.

Analysis of the moderate and fine spatial resolution remote sensing data showed that oilfield reconnaissance activities included several characteristic features. Strip forest cutting used to allow access for geological testing was detected on the Landsat ETM+ image in an area of active geological reconnaissance as a linear network with transects spaced at 150 and 600 m. Follow-on oil well drilling is based on the results of this geological reconnaissance. In another part of the Landsat ETM+ image, an active oil well drilling zone features included oil rigs and some industrial infrastructure (storehouses, repair shops, garages, etc.). The oil drilling stations appeared as ~200 m × 200 m (or 200 m × 400 m) rectangular clearings within the forest. The ground survey showed that these stations belonged to two main categories: mineralized oil rig (a term used for active or recently deactivated oil rigs) and vegetated oilrig (abandoned sites covered with grass, shrub, or tree regeneration). Remote sensing time-series data allowed for the tracking of changes during reconnaissance and exploitation of the oil field. A comparative data analysis based on the KFA-1000 (1984) and Landsat ETM+ scenes (1999) revealed a more than fourfold increase in the number of oil rigs between 1984 and 1999, indicating intensification of oil well drilling within the study area (Fig. 7.14). Overall, estimates of the direct impact (oil rigs, cleared area, other industrial infrastructure) of the reconnaissance and oil field exploitation were not high, reaching about 1.5 % by area for the reconnaissance area, and ~1.0 % for the area of exploitation.

In contrast, considerably higher damage appeared to be caused by wildfires. Generally, wildfires are considered an essential element of forest ecosystem dynamics, as evident in this study site where fires (predominantly ground fires) have formed the forest mosaic. Stands (patches) are either even-aged (regenerating after stand-replacing fires), or are mixed stands consisting of mature larch which survived fire, accompanied by dense understory. Larch is highly fire resistant due to its thick bark; however, ground fires are a significant danger for larch forests since they can damage roots that are concentrated in a thin soil layer above the permafrost. An analysis of the time-series using AVHRR and MODIS data revealed a higher frequency of wildfires within the impacted part of the study area (i.e., the zone of reconnaissance

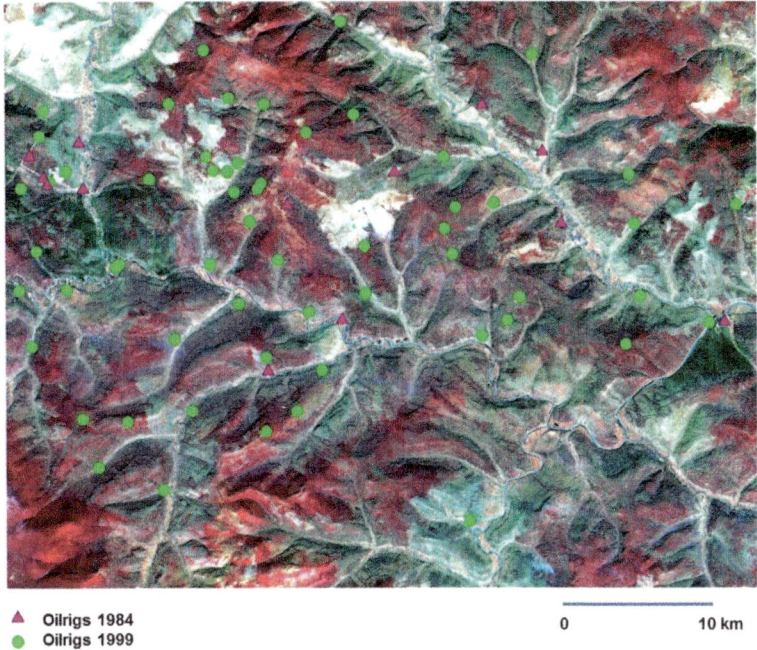

▲ **Oilrigs 1984**   0   10 km
● **Oilrigs 1999**

Fig. 7.14 Change in oilrigs presence in the Evenkiya central Siberia oil and gas LCLUC study site. Oilrigs in 1984 and 1999 are shown overlaid on a Landsat ETM+1999 false color infrared composite (Modified from Kharuk et al. 2003, with permission from SPIE)

and exploratory oil exploitation) compared with the background area (Fig. 7.15). The relative (normalized) area of burns in the impacted area was about two times higher than the otherwise similar background territory. Furthermore this has increased over time, with fires within the impacted area 1.8 times more frequent in 2001 than in 1995.

Overall, the remote sensing-based analysis showed disturbance patterns that are typically associated with oil and gas reconnaissance and exploitation. The results demonstrated that at the initial phases of oil field exploitation the greatest environmental change was not caused by direct human impact (i.e., infrastructure); instead it occurred indirectly through increased wildfire frequency at about twice the rate in comparison with background areas. Similar increases in wildfire frequency have been reported for gold mining areas along the Yenisey Ridge (Kharuk et al. 2010). These results are also consistent with observed positive relationships between fire frequency and accessibility by road or other form of infrastructure (Kovacs et al. 2004). A clear understanding of these relationships is important as human-induced fires in combination with extreme weather conditions, similar to those in European Russia in 2010, could potentially lead to serious or even catastrophic consequences for the taiga forests and oil extraction infrastructure, especially for forested areas along pipelines and within areas of exploration.

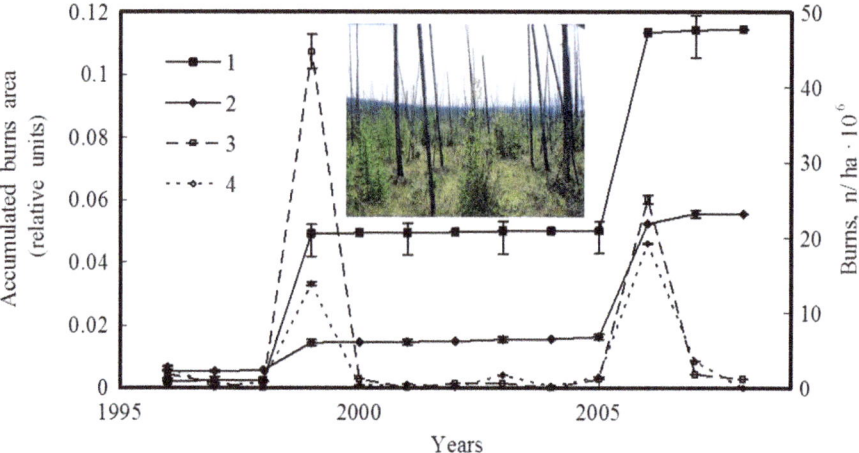

Fig. 7.15 The dynamics of cumulative burned areas (*1*, *2*) and fire events (*3*, *4*) within the impacted zone (*1*, *3*) and background (*2*, *4*) in the Evenkiya central Siberia oil and gas LCLUC study site. The data are per one million ha. *Inset*: A view on a regenerating burn (Sukachev Institute of Forest)

7.5 Summary

Throughout the history of Siberia the drive by its peoples to exploit the region's natural resources has propelled resource extraction, development of associated industry, and settlement expansion. In the preindustrial era, the economic activities of trapping (furs), farming, and mining were instrumental in encouraging expansion from western Russia into West Siberia and then East Siberia. As the industrial era began in the nineteenth century, mining, and then the development of energy resources, required infrastructure in the form of railways and roads, which led to the rise of early industrial cities and complexes. These hubs of activity dominated Siberian expansion in the nineteenth and into the early twentieth century.

The Soviet era began in the early decades of the twentieth century and continued until 1991. During this era, the sociopolitical twin goals of firmly strengthening the state's domain to the Pacific Ocean and supplying the rapidly growing industrial regions of European Russia dominated human dimensions of environmental change in Siberia. Specifically, these changes involved (1) large influxes of population through migration; (2) the growth of urban areas vastly out of proportion to the remaining sparsely-inhabited expanses of Siberia; (3) the conversion of steppe and southern taiga regions to industrial agriculture; (4) the expansion of mining (coal, bauxite) activities into oil, gas, and other minerals and metals sectors; (5) the harnessing of hydropower to supply industrialization; (6) the extension of railways; and finally (7) the beginning of industrial forestry in Siberia. The relatively short timeframe in which these goals were instituted is remarkable. Russia's rise to modernization required no more than 50 years: by the beginning of the 1950s, Russia's progress was on track with that of the rest of Europe.

Satellite remote sensing has been able to provide data documenting some of the most important influences of Soviet-era human-driven environmental change on the Siberian landscape. Two Soviet-era synthesis case studies were presented in this chapter, one focusing on change in a predominantly forested, yet multi-use landscape in central Siberia, and the other on environmental impacts of mining in northern Siberia. In central Siberia, Soviet-era land-use and land-cover change was quantified through Landsat MSS/TM-based change detection between 1974 and 1990. Predominant changes included high amounts of logging (equivalent to those reported in official statistics), high amounts of logging in mature conifer, growing proportions of deciduous forest, and the beginnings of agriculture abandonment. Older logged areas (indicated detected by the presence of both cuts and regenerating cuts at the ~1975 date) were statistically correlated with closeness to rivers, a major mode of timber transportation during the Soviet era.

The Norilsk case study site has been the location of the greatest pollution in the northern Eurasian boreal regions, starting in the Soviet era. Ongoing analysis of pollution effects using multiple types of remote sensing data combined with field data has been undertaken to understand the impact of Norilsk Nickel on the surrounding environment. Results show that mining and industrial complexes as an epicenter of a transformed landscape are clearly evident from space, as are the effects of their emitted industrial pollutants (dominantly SO_2) on the surrounding land cover. Use of multiple remote sensing data over time showed signs of degradation and mortality in the larch forests surrounding Norilsk that began in the 1940s and peaked in the 1980s.

The post-Soviet era began abruptly in 1991 following the dissolution of the Soviet Union. At this time production from large industrial enterprises (especially those related to forestry) were brought nearly to a halt. Many collective agriculture sites dissolved, replaced by smaller, privatized farms. Industrial development did not cease, but limited funds meant that there was little support for maintenance of aged infrastructure. Deep economic hardship led to population loss in Siberia and necessitated the transformation of people's livelihoods and their relationship to natural resources. Reorganization of natural resource institutions and forest management was undertaken with the intent to encourage privatization, yet unwittingly also fostered the emergence of local oligarchies and illegal logging operations.

In the central Siberian study sites, land-cover and land-use change and biomass change analyses were extended into the post-Soviet period 1990–2001. New trends observed via Landsat TM/ETM+ included depressed rates of logging. At the same time, the continued increase in regenerating deciduous forests was observed in all three study sites. These forests were the product of large scale logging (or logging and fire) that had occurred prior to 1990, especially in the late Soviet era at the peak of timber production in the region. These trends were corroborated by official Russian statistics for the study regions. Remote sensing-based results also showed that collective agriculture decreased between 1990 and 2000, although the trend in the region had actually begun before 1990. When change analyses were combined with biomass quantities, the overall trends indicated that the study sites were moderate carbon sinks, yet the occurrence of high fire years could tip the balance from sink to source. Continued investigations at the Norilsk site and analysis of AVHRR and MODIS for post-Soviet landscapes show a continued high level of industrial pollution affecting the larch-dominated forests in the region (Kharuk et al. 1996a, b; Zubareva et al. 2003).

As Russia (and Siberia) regrouped after the early post-Soviet transformation era, emerging developments in the forest and energy sectors have begun to come to the fore. Although evidence of illegal logging operations surfaced in the early post-Soviet years, this phenomenon is now acknowledged to be a significant and growing issue, even despite difficulties in documenting and observing it directly through remote sensing. Neither the Forest Code of 1997 nor the new Forest Code of 2006 with their liberalization and decentralization aims has been able to limit this activity. The geographic location of forest exploitation is also shifting to areas that can easily supply and transport logs to the Asian market. In a rising era of transnationalism, Siberia faces challenges. There are, however, several factors in place that could promote sustainable development in the forestry sector. Foremost is the vast resource base of the Siberian forests, presenting opportunities for biodiversity conservation and carbon sequestration. Other factors include a relatively well-developed forest management infrastructure, highly qualified scientists, and a network of forest enterprises (Danilin and Crow 2008). Today practices with the potential to promote sustainability, such as the gathering of non-timber forest products (e.g., mushrooms and berries), hunting, fishing, and tourism, are gaining greater visibility in Siberia (and the RFE) (Hitztaler 2010). Together, these activities could play a key role in creating a broader base of forest use.

While the energy sector has long been a feature of both the Siberian economy and its landscapes, this sector is growing in emphasis and spreading into new geographic areas of Siberia. Historically, production has been centered on the large oilfields of West Siberia, and is now shifting toward largely undeveloped regions of East Siberia. New pipelines traversing East Siberia and the RFE have been constructed, and a direct link to China's largest oil processing center through an offshoot pipeline has just been completed. These pipelines, plus new gas developments in the far north of both West and East Siberia, will likely spur new landscape changes.

The above trends in the forestry and energy sectors are evidenced in two synthesis case studies, one comparing the differing trajectories of land-cover change on either side of the Russian-Chinese border, and the other characterizing disturbance types and the increase of human influences (especially fire) that have accompanied new oil and gas developments in East Siberia. In the Amur case study site, changes in forest policy in China have led to a dichotomy between Siberian and Chinese landscapes that border the Amur River. Whereas replanting and the almost complete removal of mature conifer are evident on the Chinese side, fire predominates on the Siberian side. These patterns have occurred in spite of legislation intended to protect forests and restrict logging. Furthermore, frequently cited increases in total forest area tend to mask the ongoing loss of mature conifer forest both in China and Siberia. The second synthesis study provides data on both the direct (e.g., from oil rigs, clearing, roads, etc) and indirect (human-induced wildfires) effects of oil and gas reconnaissance activities. Most significantly, the results reveal an approximately twofold increase in fires in reconnaissance and exploitation areas compared to nonexplored, yet ecologically similar sites nearby. The results underscore the potential for serious environmental consequences to taiga forests resulting from a combination of high-risk anthropogenic activity and fire-conducive climate conditions.

As stated in the Introduction, the focus of this chapter is on human dimensions and on developing an understanding of *both* socio-economics and environmental change and their relationships in Siberia over important eras. Scientific understanding has been greatly aided by spaceborne remote sensing and the development of its spatial, spectral, and temporal records. In particular, moderate resolution sensors such as Landsat provide landscape-level information that can be linked to the scale of local human drivers and activities. These data, appropriate for case study sites, are complemented by coarser spatial or temporal, or both, resolution satellite data that provide synoptic observations of land cover over large regions. In the coming decades, remote sensing-based analyses will be integral in providing data leading to increased knowledge of emerging trends in human dimensions of environmental change and LCLUC across the vast geographic realm of Siberia. Satellite observation and monitoring capabilities are now greatly expanded over those of previous eras. Growing networks and programs that could contribute to this are discussed in Chap. 2.

References

Achard F, Mollicone D, Stibig HJ, Aksenov D, Laestadius L, Li ZY et al (2006) Areas of rapid forest-cover change in boreal Eurasia. For Ecol Manage 237:322–334

Achard F, Eva HD, Mollicone D, Beuchle R (2008) The effect of climate anomalies and human ignition factor on wildfires in Russian boreal forests. Philos Trans R Soc Lond B Biol Sci 363:2331–2339

Afontsev S, Kessler G, Markevich A, Tyazhelnikova V, Valetov T (2008) The urban household in Russia and the Soviet Union, 1900–2000: patterns of family formation in a turbulent century. Hist Family 13:178–194

Ahrend R, Tompson W (2005) Unnatural monopoly: the endless wait for gas sector reform in Russia. Europe-Asia Stud 57:801–821

All Union Scientific Research Institute of Economics (1991) Unpublished database. Academgorodok, Novosibirsk

Anderson DG (2002) Entitlements, identity and time: addressing aboriginal rights and nature protection in Siberia's new resource colonies. In: Kasten E (ed) People and the land: pathways to reform in post-Soviet Siberia. Dietrich Reimer Verlag, Berlin, pp 99–123

Balzar MM (2006) The tension between might and rights: Siberians and energy developers in post-socialist binds. Europe-Asia Stud 58:567–588

Barr BM, Braden KE (1988) The disappearing Russian forest: a dilemma in Soviet resource management. Rowman & Littlefield, London, 252 p

Bergen KM, Conard SG, Houghton RA, Kasischke ES, Kharuk VI, Krankina ON et al (2003) NASA and Russian scientists observe land-cover and land-use change and carbon in Russian forests. J For 101:34–41

Bergen KM, Zhao T, Kharuk V, Blam Y, Brown DG, Peterson LK et al (2008) Changing regimes: forested land cover dynamics in central Siberia 1974 to 2001. Photogramm Eng Remote Sens 74:787–798

Brown DG, Walker R, Manson S, Seto K (2004) Modeling land-use and land-cover change. In: Gutman G, Janetos AC, Justice CO et al (eds) Land change science: observing, monitoring and understanding trajectories of change on the Earth's surface. Kluwer Academic Publishers, Dordrecht, pp 395–409

Bykovsky VA (2002) Problemy investitsii v Rossiiskuyu ekonomiku i neftegazovuyu otrasl' severa zapadnoi Sibiri. Neftyanoe Khozyaistvo 11:17–21

Chernina NV (1996) Economic transition and social exclusion in Russia. International Labour Organization, Geneva, 103 p

Coale AJ, Anderson BA, Harm E (1979) Human fertility in Russia since the nineteenth century. Princeton University Press, Princeton, 285 p

Crevoisier C, Shevliakova E, Gloor M, Wirth C, Pacala S (2007) Drivers of fire in the boreal forests: data constrained design of a prognostic model of burned area for use in dynamic global vegetation models. J Geophys Res Atmos 112(D24):D24112. doi:10.1029/2006JD008372

Crowley RM (2005) Stepping on to a moving train: the collision of illegal logging, forestry policy, and emerging free trade in the Russian Far East. Pacific Rim Law Policy J 14:425–453

Danilin IM, Crow TR (2008) The great Siberian forest: challenges and opportunities of scale. In: Lafortezza R, Chen J (eds) Patterns and processes in forest landscapes. Springer, Dordrecht, pp 47–66

Davis S (2003) The Russian Far East: the last frontier? Routledge, New York, 155 p

Derome J, Lukina N (2011) Environmental pollution and land-cover/land-use change in arctic areas. In: Gutman G, Reissell A (eds) Eurasian arctic land use and land cover in a changing climate. Springer, Dordrecht, pp 269–289

Dienes L (2004) Observations on the problematic potential of Russian oil and the complexities of Siberia. Eurasian Geogr Econ 45:319–345

Eikeland S, Riabova L (2002) Transition in a cold climate: management regimes and rural marginalisation in northwest Russia. Soc Ruralis 42:250–266

Eikeland S, Eythorsson E, Ivanova L (2004) From management to mediation: local forestry management and the forestry crisis in post-socialist Russia. Environ Manage Health 33:285–293

Espiritu AA (1999) The impact of industrialization and resource development on indigenous peoples of Northwest Siberia: the Khanty, Mansi and Iamalo-Nenets. Dissertation, University of Alberta, Edmonton, 223 p

Europa Publications (2010) Eastern Europe, Russia and Central Asia. Europa Publications, London, 680 p

Forbes BC (1999) Land use and climate change on the Yamal Peninsula of north-west Siberia: some ecological and socio-economic implications. Polar Res 18:367–373

Fu AM, Sun GQ, Guo ZF, Wang DZ (2008) Forest cover classification from MODIS images in northeastern Asia. International workshop on earth observation and remote sensing applications, Beijing, Peoples Republic of China, pp 178–189

Gerber TP (2006) Regional economic performance and net migration rates in Russia, 1993–2002. Int Migr Rev 40:661–697

Goodale CL, Apps MJ, Birdsey RA, Field CB, Heath LS, Houghton RA et al (2002) Forest carbon sinks in the Northern Hemisphere. Ecol Appl 12:891–899

Goskomstat Russia (1965–1989) The USSR in figures; [annual] statistical handbook. Statistika Publishers, Moscow

Granasen J, Nilsson S, Zackrisson U (1997) Russian forest sector – human resources. International Institute for Applied System Analyses (IIASA), Laxenburg, 89 p

Groisman PY, Clark EA, Kattsov VM, Lettenmaier DP, Sokolik IN, Aizen VB et al (2009) The Northern Eurasia Earth Science Partnership: an example of science applied to societal needs. Bull Am Meteorol Soc 90:671–688

Gutman G (2007) Contribution of the NASA land-cover/land-use change program to the Northern Eurasia Earth Science Partnership Initiative: an overview. Global Planet Change 56:235–247

Gutman G, Reissell A (eds) (2011) Eurasian arctic land cover and land use in a changing climate. Springer, Dordrecht, 306 p

Gutman G, Janetos AC, Justice CO, Moran EF, Mustard JF, Rindfuss RR et al (2004) Land change science: observing, monitoring and understanding trajectories of change on the earth's surface. Kluwer Academic Publishers, Dordrecht/London, 459 p

Harden JW, Trumbore SE, Stocks BJ, Hirsch A, Gower ST, O'Neill KP et al (2000) The role of fire in the boreal carbon budget. Global Change Biol 6:174–184

Hashim SM (2010) Power-loss of power-transition? Assessing the limits of using the energy sector in reviving Russia's geopolitical stature. Communist Post-Communist Stud 43:263–274

Hayes DJ, McGuire AD, Kicklighter DW, Burnside TJ, Melillo JM (2011) The effects of land cover and land use change on the contemporary carbon balance of the arctic and boreal terrestrial ecosystems of northern Eurasia. In: Gutman G, Reissell A (eds) Eurasian arctic land cover and land use in a changing climate. Springer, Dordrecht, pp 109–136

Heleniak T (2001) Demographic change in the Russian Far East. In: Bradshaw MJ (ed) The Russian Far East and Pacific Asia: unfulfilled potential. Curzon, Richmond, pp 127–153

Henry LA (2010) Between transnationalism and state power: the development of Russia's post-Soviet environmental movement. Environ Polit 19:756–781

Henry LA, Douhovnikoff V (2008) Environmental issues in Russia. Ann Rev Environ Resour 33:437–460

Hill F, Gaddy C (2003) The Siberian curse: how communist planners left Russia out in the cold. The Brookings Institution, Washington, DC, 326 p

Hitztaler S (2010) An ethnography of landscape: exploring the dynamics among people, forests, and resource use in post-Soviet central Kamchatka. Dissertation, University of Michigan, Ann Arbor, 352 p

Hitztaler S (2011) Policy transformed: an examination of Russia's latest forest code and its effects on the forestry sector. IREX Scholar Policy Brief, Washington, DC, 6 p

Hoffmann DL (2000) Mothers in the motherland: Stalinist pronatalism in its Pan-European context. J Soc Hist 34:35–54

Houghton RA (1995) Land-use change and the carbon-cycle. Global Change Biol 1:275–287

Houghton RA, Joos F, Asner GP (2004) The effects of land use and management on the global carbon cycle. In: Gutman G, Janetos AC, Justice CO et al (eds) Land change science: observing, monitoring and understanding trajectories of change on the Earth's surface. Kluwer Academic Publishers, Dordrecht, pp 237–256

Janetos AC (2004) Research directions in land-cover and land-use change. In: Gutman G, Janetos AC, Justice CO et al (eds) Land change science: observing, monitoring and understanding trajectories of change on the earth's surface. Kluwer Academic Publishers, Dordrecht, pp 449–457

Kasten E (2002) Introduction. In: Kasten E (ed) People and the land: pathways to reform in post-Soviet Siberia. Dietrich Reimer Verlag, Berlin, pp 1–4

Kharuk, VI (1992) Monitoring of forest stands and soils of Norilsk region. Annual report, Krasnoyarsk, 112 p (in Russian)

Kharuk S (2000) Air pollution impact on subarctic forests at Norilsk, Siberia. In: Innes JL, Oleksyn J (eds) Forest dynamics in heavily polluted regions. CAB International, Wallingford, pp 77–86

Kharuk VI, Vinterberger K, Tsibulsky GM (1996a) Satellite data analysis of pollution-induced subtundra forest decline. Earth Obs Remote Sens 13:631–640

Kharuk VI, Winterberger K, Tsibulskii GM, Yakhimovich AP, Moroz SN (1996b) Technogenic disturbance of pretundra forests in Noril'sk valley. Russ J Ecol 27:406–410

Kharuk VI, Ranson KJ, Im ST (2003) Landsat-7 for evaluation of oilfield exploitation impacts on the south Evenkiya larch dominant communities. In: Larar AM, Tong Q, Suzuki M (eds) Proceedings of SPIE, multispectral and hyperspectral remote sensing instruments and applications conference, Hangzhou, CN, October 2002, pp 272–278

Kharuk VI, Ranson KJ, Kozuhovskaya AG, Kondakov YP, Pestunov IA (2004) NOAA/AVHRR satellite detection of Siberian silkmoth outbreaks in eastern Siberia. Int J Remote Sens 25:5543–5555

Kharuk VI, Ranson KJ, Fedotova EV (2007) Spatial pattern of Siberian silkmoth outbreak and taiga mortality. Scand J For Res 22:531–536

Kharuk VI, Ranson KJ, Dvinskaya ML (2008) Wildfires dynamic in the larch dominance zone. Geophys Res Lett 35:1–6

Kharuk V, Ranson KJ, Im ST, Vdovin AS (2010) Spatial distribution and temporal dynamics of high-elevation forest stands in southern Siberia. Global Ecol Biogeogr. doi:10.1111/j.1466-8238.2010.00555.x

Korovin GN (1995) Problems of forest management in Russia. Water Air Soil Pollut 82:13–23

Korovin GN (1996) Analysis of the distribution of forest fires in Russia. In: Goldammer JG, Furyaev VV (eds) Fire in ecosystems of boreal Eurasia. Kluwer Academic Publishers, Boston, pp 112–128

Kovacs K, Ranson KJ, Sun G, Kharuk VI (2004) The relationship of the Terra MODIS fire product and anthropogenic features in the Central Siberian landscape. Earth Interact 8:1–25

Kozlov M, Zvereva E (2007) Industrial barrens: extreme habitats created by non-ferrous metallurgy. Rev Environ Sci Biotechnol 6:233–259

Krankina ON, Dixon RK (1992) Forest management in Russia – challenges and opportunities in the era of Perestroika. J For 90:29–34

Krankina ON, Sun G, Shugart HH, Kasischke E, Kharuk VI, Bergen KM et al (2004) Northern Eurasia: remote sensing of boreal forest in selected regions. In: Gutman G, Janetos AC, Justice CO et al (eds) Land change science: observing, monitoring, and understanding trajectories of change on the earth's surface. Kluwer Academic Publishers, Dordrecht, pp 123–138

Kukuev YA, Krankina ON, Harmon ME (1997). The forest inventory system in Russia: a wealth of data for western researchers. J For 95:15–20

Kumo K (2007) Inter-regional population migration in Russia: using an origin-to-destination matrix. Post-Communist Econ 19:131–152

Lankin A (2005) Forest product exports from the Russian Far East and Eastern Siberia to China: status and trends. Forest Trends, Washington, DC, 58 p

Lebedev A (2005) Siberian and Russian Far East timber for China. Forest Trends, Washington, DC, 40 p

Lesniewska F, Laletin A, Lebedev A, Harris K (2008) Transition in the taiga: the Russian Forest Code 2006 and its implementation process. Taiga Rescue Network and FERN, Brussels, 38 p

Loboda T, Sun G, Zhang Z (2008) NELDA test site report: Amur site. University of Maryland Department of Geography, College Park, 22 p

Malmlöf T (1998) The institutional framework of the Russian forest sector: a historical background. International Institute for Applied System Analyses (IIASA), Laxenburg, 84 p

Marsh GP (1885) The earth as modified by human action. C. Scribner's Sons, New York, 629 p

Mathieson RS (1979) Japan's role in Soviet economic growth: transfer of technology since 1965. Praeger Publishers, New York, 277 p

Mayer AL, Kauppi PE, Angelstam PK, Zhang Y, Tikka PM (2005) Importing timber, exporting ecological impact. Science 308:359–360

Maynard N, Oskal A, Turi J, Mathiesen S, Eira I, Yurchak B et al (2011) Impacts of arctic climate and land use changes on reindeer pastoralism: indigenous knowledge and remote sensing. In: Gutman G, Reissell A (eds) Eurasian arctic land cover and land use in a changing climate. Springer, Dordrecht, pp 177–205

Metzo K (2003) "It didn't used to be this way": households, resources, and economic transformation in Tunka Valley, Buriatia, Russian Federation. Dissertation, Indiana University, Bloomington, 238 p

Mollicone D, Eva HD, Achard F (2006) Ecology – human role in Russian wild fires. Nature 440:436–437

Naumov IV, Collins DN (2006) The history of Siberia. Routledge, London/New York, 242 p

Newell J, Lebedev A, Gordon D (2000) Plundering Russia's Far Eastern taiga: illegal logging corruption and trade. Friends of the Earth-Japan/Pacific Environment & Resources Center, Tokyo, 47 p

Nilsson S, Blauberg K, Samarskaia E, Kharuk VI (1998) Pollution stress of Siberian forests. In: Linkov I, Wilson R (eds) Air pollution in the Ural Mountains. Kluwer Academic Publishers, Dordrecht, pp 31–54

Northway S, Bull GQ, Shvidenko A, Bailey L (2009) Recent developments in forest products trade between Russia & China: potential production, processing, consumption and trade scenarios. Forest Trends, Washington DC, 15 p

Obersteiner M (1995) Status and structure of the forest industry in Siberia. International Institute for Applied System Analyses (IIASA), Laxenburg, 38 p

Oldfield JD (2006) Russian nature: exploring the environmental consequences of societal change. Ashgate, Burlington, 158 p

Pallot J, Moran D (2000) Surviving in the margins in post-soviet Russia: forestry villages in Northern Perm Oblast. Post-Soviet Geogr Econ 41:341–364

Perevedentsev V (1999) The demographic situation in post-soviet Russia. In: Demko GJ, Ioffe G, Zayonchkovskaya Z (eds) Population under duress. Westview Press, Boulder, pp 17–38

Peterson LK, Bergen KM, Brown DG, Vashchuk L, Blam Y (2009) Forested land-cover patterns and trends over changing forest management areas in the Siberian Baikal region. For Ecol Manage 257:911–922

Potapov PV, Hansen MC, Stehman SV (2011) High-latitude forest cover loss in northern Eurasia, 2000–2005. In: Gutman G, Reissell A (eds) Eurasian arctic land cover and land use in a changing climate. Springer, Dordrecht, pp 37–51

Precoda N (1970) Opposing statements on Soviet forestry. J For 68:129

Ranson KJ, Kovacs K, Sun G, Kharuk VI (2003) Disturbance recognition in the boreal forest using radar and Landsat-7. Can J Remote Sens 29:271–285

Reed A (2009) Coming from Russia: more crude, lighter and sweeter. Oil Gas J 107:20–22

Rindfuss RR, Walsh SJ, Turner BL, Moran EF, Entwisle B (2004) Linking pixels and people. In: Gutman G, Janetos AC, Justice CO et al (eds) Land change science: observing, monitoring and understanding trajectories of change on the earth's surface. Kluwer Academic Publishers, Dordrecht, pp 379–394

Rosenfield GH, Fitzpatric-Lins K (1986) A coefficient of agreement as a measure of thematic classification accuracy. Photogram Eng Remote Sens 52:223–227

Russian Federal State Statistics Service (2005) The 2002 all-Russian population census, 14 vols. East View Publications, Minneapolis

Russian Federal State Statistics Service (2010) Russia in figures [annual]. Federal State Statistics Service, Moscow, 558 p

Goskomstat Russia (1996) 1989 USSR population census. East View Publications, Minneapolis, CD-ROM

Ryabushkin T (1978) Social policy and demography in the Soviet Union. Popul Dev Rev 4:715–720

Savchenko VA (1998) Ekologicheskie problemy Taimyra (Ecological problems of Taimyr). SIP RIA, Moscow, 97 p

Scott A (1994) Oil exploitation and ecological problems in Siberia. Institute for Applied System Analysis (IIASA), Laxenburg, 64 p

Shvidenko A, Nilsson S (1994) What do we know about the Siberian forests? Ambio 23:396–404

Shvidenko A, Schepaschenko D, McCallum I, Nilsson S (2007) Russian forests and forestry CD-ROM. International Institute for Applied Systems Analysis (IIASA) and the Russian Academy of Science (RAS). Online database. Last accessed 15 Feb 2011

Stolbovoi V, McCallum I (2002) Land resources of Russia. International Institute for Applied Systems Analysis (IIASA) and the Russian Academy of Science (RAS). Online database. Last accessed 11 Feb 2011

Thomas WL (1956) Man's role in changing the face of the Earth. University of Chicago Press, Chicago, 1193 p

Turner BL (1990) The Earth as transformed by human action. Cambridge University, Cambridge, 713 p

Turner BL, Moran E, Rindfuss R (2004) Integrated land-change science and its relevance to the human sciences. In: Gutman G, Janetos AC, Justice CO et al (eds) Land change science: observing, monitoring and understanding trajectories of change on the Earth's surface. Kluwer Academic Publishers, Dordrecht, pp 431–447

Vitousek PM (1994) Beyond global warming – ecology and global change. Ecology 75:1861–1876

Vlasova T, Klein D (1992) Lichens, a unique forage resource threatened by air pollution. Rangifer 12:21–27

Voinova VD, Zakharova OD, Rybakovsky LL (1993) Sovremennyi Rossiisky sever i ego naselenie. In: Rybakovsky LL (ed) Sotsial' no – Demografischeskoe Razvitie Rossiiskogo Severa. Rossiiskaya Akademiya Nauk, Institut Sotsial' no – Ekonomicheskikh Problem Narodonaseleniya, Moscow, pp 7–25

Walker D, Forbes B, Leibman M, Epstein H (2011) Cumulative effects of rapid land-cover and land-use changes on the Yamal Peninsula, Russia. In: Gutman G, Reissell A (eds) Eurasian arctic land cover and land use in a changing climate. Springer, Dordrecht, pp 207–236

Walter H, Breckle SW (2002) Walter's vegetation of the Earth: the ecological systems of the geo-biosphere. Springer, Berlin/New York, 527 p

Williams C (1996) Economic reform and political change in Russia, 1991–1996. In: Williams C, Chuprov V, Staroverov V (eds) Russian society in transition. Dartmouth Publishing Co, Aldershot, pp 9–36

Williams RA, Kinard JC (2003) A strategy for economic development of the forestry sector in Tomsk, Russia. J For 101:36–41

Yan X, Shugart HH (2005) FAREAST: a forest gap model to simulate dynamics and patterns of eastern Eurasian forests. J Biogeogr 32:1641–1658

Yaroshenko AY, Potapov PV, Turubanova SA (2001) The last intact forest landscapes of northern European Russia. Greenpeace - Global Forest Watch Russia, Moscow, 77 p

Zaslavskaja LA (1994) Forest codes and laws in the republics of the Russian Federation. Leg Econ (3–4):105–109

Zayonchkovskaya Z (1999) Recent migration trends in Russia. In: Demko GJ, Ioffe G, Zayonchkovskaya Z (eds) Population under duress. Westview Press, Boulder, pp 107–136

Zhang P, Shao G, Zhao G, LeMaster DC, Parker GR, Dunning JB et al (2000) China's forest policy for the 21st century. Science 288:2135–2136

Zhao TT, Bergen KM (2006) Unpublished models. University of Michigan, Ann Arbor

Zhao G, Shao GF (2002) Logging restrictions in China – a turning point for forest sustainability. J For 100:34–37

Zhao TT, Bergen KM, Brown DG, Shugart HH (2009) Scale dependence in quantification of land-cover and biomass change over Siberian boreal forest landscapes. Landsc Ecol 24:1299–1313

Zubareva ON, Skripal'shchikova LN, Greshilova NV, Kharuk VI (2003) Zoning of landscapes exposed to technogenic emissions from the Norilsk Mining and Smelting Works. Russ J Ecol 34:375–380

Zuy AN, Volod'ko BV (1991) Frost-and-geothermal analysis of oil-and-gas deposits in Krasnoyarsk Kray. Institute of Permafrost SIB RAS, Yakutsk, 84 p

Chapter 8
Aspects of Atmospheric Pollution in Siberia

Alexander A. Baklanov, Vladimir V. Penenko, Alexander G. Mahura,
Anna A. Vinogradova, Nikolai F. Elansky, Elena A. Tsvetova,
Olga Yu. Rigina, Leonid O. Maksimenkov, Roman B. Nuterman,
Fedor A. Pogarskii, and Ashraf Zakey

Abstract This chapter considers specific atmospheric pollution problems in Siberia, the current state of studies and strategic activities, and peculiarities of Siberian environmental protection problems, risk assessment, and tendencies in atmospheric pollution in Siberia, including health-affecting pollutants, greenhouse gases, aerosols, etc. The chapter does not presume to cover all the aspects of atmospheric pollution in Siberia. Its main focus is a short general overview of the existing problems of airborne pollution in Siberia and methodological aspects of air pollution impact assessments followed by several examples of such studies for Siberia. In particular, the following issues are described: (1) sources and characteristics of air pollution in Siberia, (2) air quality and atmospheric composition characterization, (3) assessment of airborne pollution in Siberia from air and space, (4) methodology and models for air pollution assessment on different scales, and (5) case studies of long-range atmospheric transport of heavy metals from industries of the Ural and Norilsk regions.

In this chapter, we are considering specific atmospheric pollution problems in Siberia including the current state and projections, peculiarities of Siberian environmental protection problems, risk assessment, and tendencies in atmospheric pollution,

A.A. Baklanov (✉) • A. G. Mahura • R.B. Nuterman • A. Zakey
Danish Meteorological Institute, Copenhagen, Kingdom of Denmark
e-mail: alb@dmi.dk

V.V. Penenko • E.A. Tsvetova
Institute of Computational Mathematics and Mathematical Geophysics of Siberian Branch
of Russian Academy of Sciences Novosibirsk, Russia

A.A. Vinogradova • N.F. Elansky • L.O. Maksimenkov • F.A. Pogarskii
A.M. Obukhov Institute of Atmospheric Physics, Russian Academy of Sciences
Moscow, Russia

O.Yu. Rigina
Institute of the Northern Ecology Problems (INEP), Kola Scientific Center
of Russian Academy of Sciences, Apatity, Russia

P.Ya. Groisman and G. Gutman (eds.), *Regional Environmental Changes in Siberia*
and Their Global Consequences, Springer Environmental Science and Engineering,
DOI 10.1007/978-94-007-4569-8_8, © Springer Science+Business Media Dordrecht 2013

including health-affecting pollutants, greenhouse gases (GHGs), aerosols, etc. We do not pretend to cover all the aspects of atmospheric pollution in Siberia. The main focus is on a short general overview of the airborne pollution problem in Siberia, methodological aspects of air pollution impact assessments, and examples of such studies for Siberia.

There are the book in preparation entitled *Man-Induced Environmental Risks in Siberia* (Springer) and already published FP6 Enviro-RISKS project Scientific Report (Baklanov et al. 2008a) which also assessed atmospheric pollution issues in Siberia focusing on the risk of anthropogenic impact on the Siberian environment and population. Another relevant overview publication we would like to recommend is Gutman and Reissell (2011), and especially the book chapters by Derome and Lukina (2011) and by Sokolik et al. (2011).

8.1 Sources and Characteristics of Air Pollution in Siberia

8.1.1 Introduction

In general, the atmospheric environment over Siberia is relatively clear compared with other surrounding regions of Asia and Eastern Europe (see, e.g., an example of sulfur air pollution in Fig. 8.4). However, air pollution from industrial centers in Siberia pose observable environmental threats. Siberian ecosystems have begun to show stress from the accumulation of pollution depositions that come from cities and industrial plants. Urban air quality in several Siberian cities (e.g., Norilsk, Barnaul, Novokuznetsk) is one of the worst among the Russian and European cities. To sketch a scope of environmental problems of Siberian cities, it is necessary to underline an essential dependence of air quality from climatic conditions typical for Siberia. A stable atmospheric stratification and temperature inversions are predominant weather conditions for more than half a year. This contributes to accumulation of pollutants in the low layers of the atmosphere, namely where ecosystems function and people live. In addition to the severe climatic conditions, man-made impacts on the environment in industrial areas and large cities have strengthened more and more. The impacts manifest themselves in pollution of environment, change of the land use, hydrology, and hydrodynamic regimes of the atmosphere. Ultimately, these impacts feed back to population affecting its health and well-being.

8.1.2 Sources of Atmospheric Pollution

The main atmospheric pollution sources in Siberia can be distinguished on anthropogenic (industrial, transport, combustion, etc.) and natural (biogenic emissions, wildfires, dust storms, sea aerosols, volcano eruptions, pollen, etc.) emission sources.

These sources are characterized by different chemical composition, temporal dynamics, which include local, regional, and remote sources, accidental releases, etc. The main pollutants (gases and aerosols) can be classified as health harmful, climate effecting, and ecosystem damaging.

Among the natural emission sources, the biomass, soils, and peatlands of Siberia contain one of the largest pools of terrestrial carbon (cf., Chaps. 1 and 6). In Siberia under current climate change scenarios, some of the largest temperature increases have occurred and are expected to occur (Chap. 3); the stored carbon has the potential to be released with associated changes in fire regimes.

There are also natural sources of gas compounds like methane (CH_4), sulfur dioxide (SO_2), nitrogen oxides (NO_x), and ammonia (NH_3) in the region. These gases are locked (e.g., inside frozen soil) and can be released to the atmosphere gradually (e.g., during the permafrost thaw with warming) or abruptly (e.g., during wildfires). The frequency, duration, and severity of forest fires in the boreal forest zone, which is a major source of SO_2 and sulfur aerosols in Siberia, appear to have increased in the late twentieth century (Shvidenko and Goldhammer 2001; McGuire et al. 2004; Derome and Lukina 2011), and there are no reasons to assume that this increase will not continue in the near future. Ammonia and NO_x emissions, derived from diffuse sources such as agriculture and transport, are of more local concern in Siberia.

The main acidifying compounds in atmospheric deposition are SO_2, sulfate (SO_4), NO_x, and NH_3. Sulfur dioxide emissions are mainly associated with point sources such as power plants, the pulp and paper industry, non-ferrous metal smelters, and oil and gas processing. The latter involves the emission of exhaust gases containing CO_2, NO_x, SO_x, and volatile organic compounds (VOCs).

The Siberian environment has also sources of atmospheric particulates and their gaseous precursors such as SO_2, dimethylsulfide (DMS), and VOCs. Natural aerosols include sulfates, which are formed through gas-to-particle conversion of SO_2 derived from oxidation of DMS, and sea-salt aerosols, both originating from ice-free ocean and sea-water (Sokolik et al. 2011). Organic aerosols formed from biogenic emission of VOCs are also common. Local sources also include volcanoes (outside Siberia) and wildfires, either natural or human-related. Open biomass fires produce several important types of aerosols, including organic carbon (OC) and black carbon (BC).

Anthropogenic sources. Warner-Merl (1998) analyzed a pollution database for Siberia (covering the 1992–1993 period) provided by the Russian government. A pollution profile was revealed with acute spots of emissions exposure and large areas of less affected environment. This analysis was based upon data collected at the time of sharp decline of Russian industry and all (or most) of the estimates provided below should be considered as low boundaries of actual industrial pollution. Among the areas at risk for significant damage due to air pollution, East Siberia has the highest and the Russian Far East has the lowest levels of emissions. Observations show that atmospheric emissions tend to be the highest around the major industrial centers, with their highest levels in the south-central part of Siberia (Irkutsk, Krasnoyarsk, and Novosibirsk regions). Two areas are distinguished as Siberia's

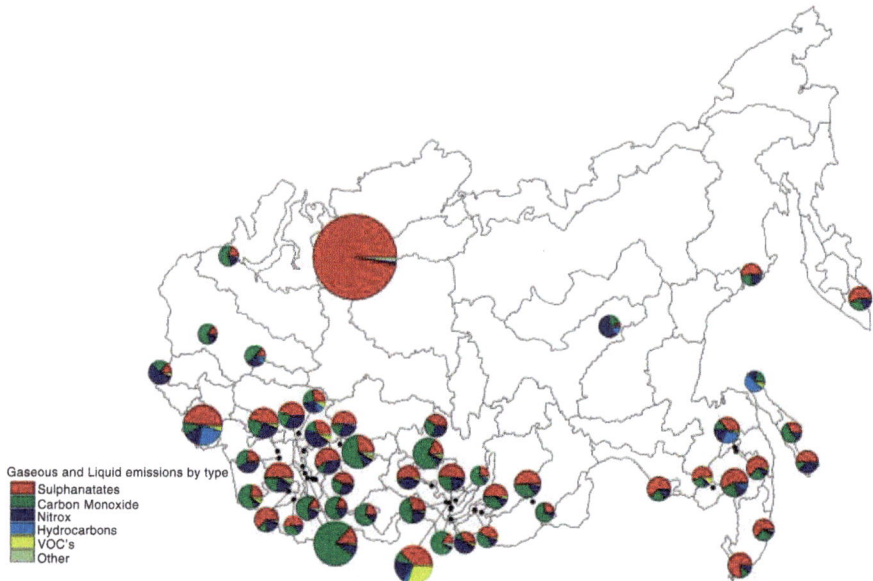

Fig. 8.1 Emissions by volume, location, and type in Siberia (Source: IIASA, Sustainable Boreal Forest Resources, from Warner-Merl 1998)

most polluted areas: Tyumen oblast (West Siberia) and Krasnoyarsk Krai (East Siberia). In these areas, anthropogenic activities have contributed over two times the amount of pollution in any other area in Siberia, amounting between 1992 and 1993 to about 6 million tons. The next most polluted areas are Irkutsk and Kemerovo oblasts located in East and West Siberia, respectively. Figure 8.1 shows that, considering only pollutant by sheer volume, SO_2 appears to be one of the most serious threats to the Siberian environment, and in particular around Norilsk. In West Siberia, SO_2, CO_2, and NO_x contribute more equitably to environmental degradation.

Due to economic crisis of the 1990s, the ecological situation in Russia, and in particular, in Siberia, has temporarily improved. During 1992–2001, the emissions of pollutants decreased by 32 %. Namely, the emissions of SO_2, NO and NO_2, CO and CO_2, and particulate matter (PM) decreased by 35.5, 38, 41, and 46 %, respectively (National estimate... 2002). However, recently, the level of atmospheric pollution began to increase again. This tendency is especially pronounced in cities and industrial regions where the enterprises function more intensively and number of cars continues to increase rather quickly (cf., Sect. 8.1.4). In the south of the Eastern Siberia, significant atmospheric transport of pollutants including organics from China is observed (National estimate... 2002). From the Taimyr Peninsula, Urals and Southern Siberian region, atmospheric pollutants are transported by air flows to neighboring territories. During the 2000s, total anthropogenic emissions into the atmosphere from stationary sources have increased by more than 10 %, from traffic – by 30 %, and toxic wastes production – by 35 %. Simultaneously, air composition in rural locations of Siberia and in the Arctic is changing quite fast due to long-range transport from Russian industrial centers and neighboring countries (CIS, EU, China, Japan, and others). In 2006, the general emission of GHG in Siberia (without agriculture and forestry) is estimated as 2.19 billion tons of carbon dioxide. This represents 107 % increase of the emissions in 2000 and 66 % increase of the emissions in 1990. The main producer of GHG emissions is the energy sector (82 % in 2006). Coal used in power production in Siberia is dominating, and its proportion is much higher than on average for Russia (Gromov 2008). Fuel consumption (in % of 1,000 ton of carbon equivalent) for 2000 in Siberian regions was the following: Krasnoyarsk region: coal – 98 %, heavy oil – 2 %, and gas – 0 %; Irkutsk region: 99, 1, and 0 %. These numbers should be compared to similar for all Russian power plants: 28, 9, and 63 %, respectively (Fig. 8.2).

8.1.3 Air Quality and Atmospheric Composition Characterization

There are several localized industrial regions in Siberia (e.g., Ural, Norilsk, Kuzbass, and a few more smaller regions) which produce intense atmospheric emissions being of importance for the Siberia and on a global/hemispheric scale (Rahn and Lowenthal 1984). Due to prevalence of westlies, the main large industrial emission sources for Siberia are the Norilsk and Ural regions with their mineral resources, metal mining industries, and metal manufacturing plants. These sources emit into the atmosphere NO_x, O_3, sulfur, lead, black carbon, non-ferrous metals, and other pollutants. Anthropogenic pollution is transported by air masses (air emissions) and river waters (industrial, agricultural, and residential waste waters). As a result, environment, wild life, and people are affected by anthropogenic pollution even far from its sources. Real monitoring of the environmental state throughout the large territory of Siberia is still limited due to economic and technical problems. Thus, modeling of atmospheric transport and estimation of anthropogenic pollution is a useful approach for partial solution of this task.

In situ observations of atmospheric pollution in Siberia, conducted as a part of the Russian State Air Monitoring Programme, are carried out by the Russian Hydrometeorological Service (Roshydromet) jointly with the Rospotrebnadzor and other agencies, with participation of the Federal Government and local government entities of the Russian Federation. Several research institutions of the Siberian Branch of the Russian Academy of Sciences (RAS) are also involved in this monitoring. The Russian national monitoring network includes 234 Roshydromet stations with regular observations in 35 main Russian cities, including 12 Siberian cities. Additionally, under supervision of Roshydromet and with a support of different research institutes, several air quality monitoring networks have stations in Siberia including a few established networks related to acid deposition monitoring: (1) national precipitation chemistry (PC) network (nationwide, since 1960s); (2) WMO-GAW (nationwide, since 1980s), (3) IBMoN (nationwide, since 1980s), and (4) EANET (South-Eastern Siberia and Far East, since 1998).

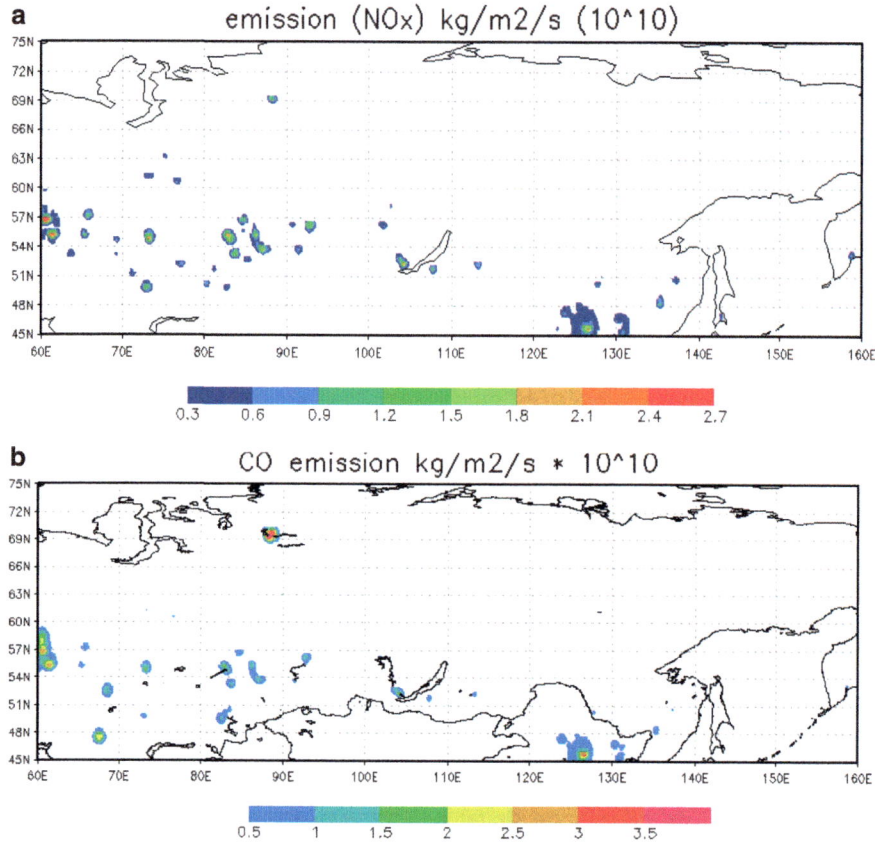

Fig. 8.2 Anthropogenic emissions of main pollutants in Siberia based on different global emission inventories: (**a**) NOx, (**b**) CO, (**c**) total particular matter (*TPM*), (**d**) SO_2 in Siberia based on different global emission inventories

The main disadvantages of the existing national system of atmospheric monitoring in Siberia are deterioration/outdated conditions of measuring instruments, insufficiently developed system to upkeep and control the measurements quality, almost total lack of continuous observations (including automatic ones), insufficient equipment for data processing and transmission, its incompatibility with the existing international observation networks, and the loss of skilled researchers during the past two decades (Elansky 2004). Reconstruction of the Russian system of atmospheric monitoring is to be performed with prospects that it will be a part of the Integral Global Observation System and, in particular, of the Global Atmospheric Watch system (WMO 2007).

The A.M. Obukhov Institute of Atmospheric Physics of RAS together with the Max-Planck Institute for Chemistry (MPIC, Germany) with numerous local and international collaborators initiated a comprehensive research and modern mobile monitoring program called the TRans-continental Observations. Into the Chemistry of the Atmosphere (TROICA; Elansky 2009b). During 1995–2009, among 13 TROICA

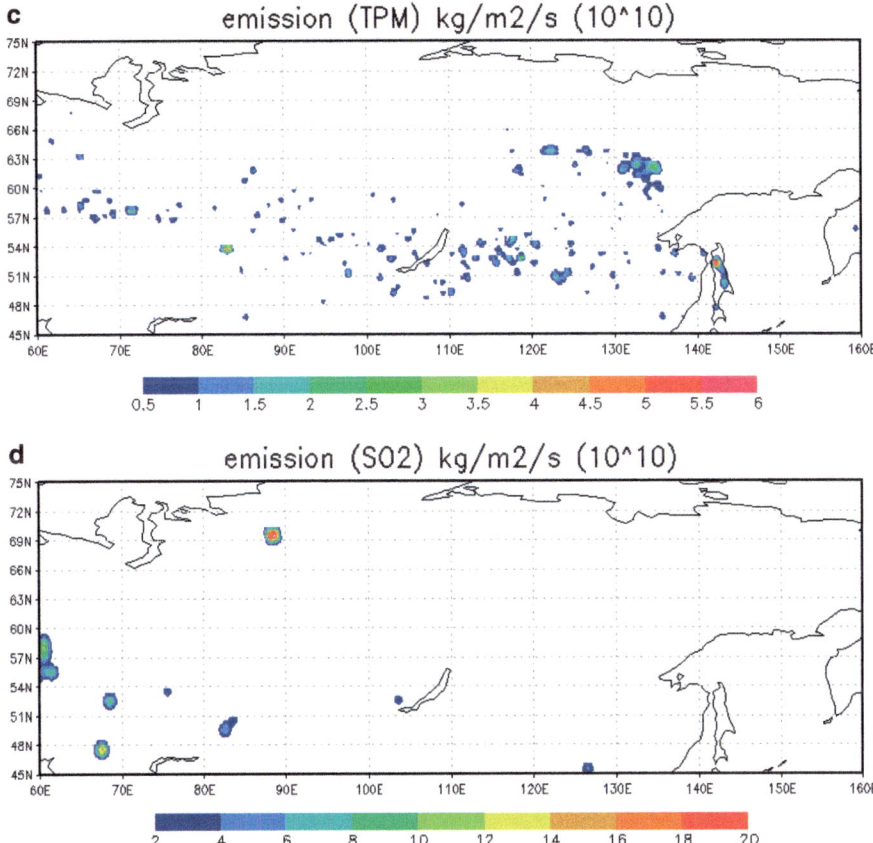

Fig. 8.2 (continued)

roundtrips, 11 trips took place across Siberia from Moscow to Vladivostok. The measurements included gases: O_3, NO, NO_2, CO_2, SO_2, NH_3, THC, VOC (20 compounds), remote sensing of O_3 and NO_2 total content and vertical profiles, isotope composition of CO, CO_2, CH_4, chemical, elementary and morphological composition of aerosols (2 nm to 10 mcm). Aerosols were studied for the single scattering albedo, mass concentration, and black carbon (BC). Meteorological measurements package that accompanied special TROICA observations includes atmospheric pressure, temperature, humidity, wind speed, and direction, and lately the air temperature vertical profiles up to 600 m (Fig. 8.3).

Smelting of sulfide ores. Smelting of nonferrous metals from sulfide ores is the largest source of locally generated acidifying emissions (SO_2 and NO_x) in the Arctic and Siberia, as well as a significant generator of heavy metal contamination. It is a major source of environmental damage as well. The largest smelter (over 10^9 kg, which is over 1 % of the estimated global anthropogenic sulfur emission) is located at the Krasnoyarsk Krai in North-Central Siberia. Figure 8.4 shows the distribution of SO_2 sources in the Northern Hemisphere and the modeled distribution of deposition.

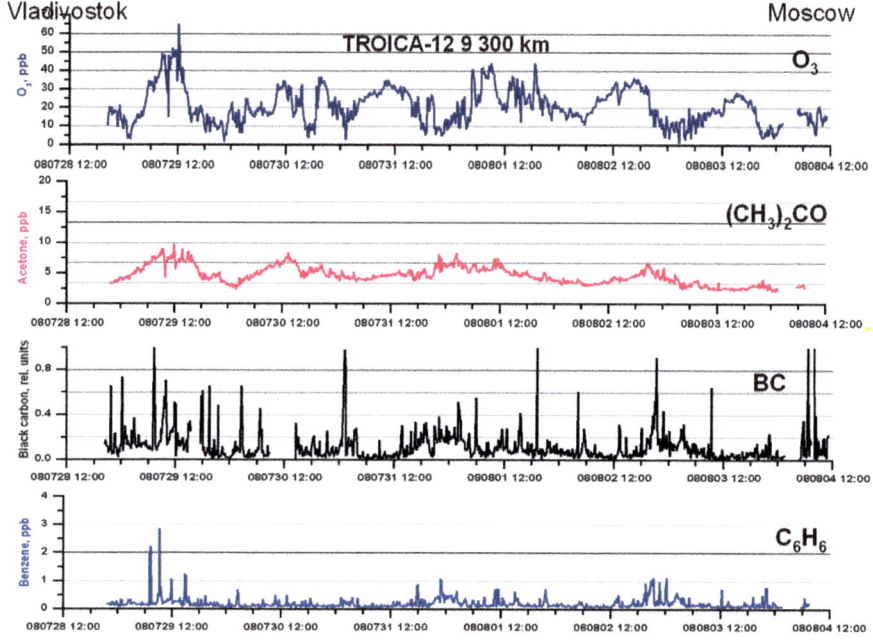

Fig. 8.3 Species concentrations (10 min averages) measured during Vladivostok – Moscow trip TROICA-12 (From Elansky 2009b)

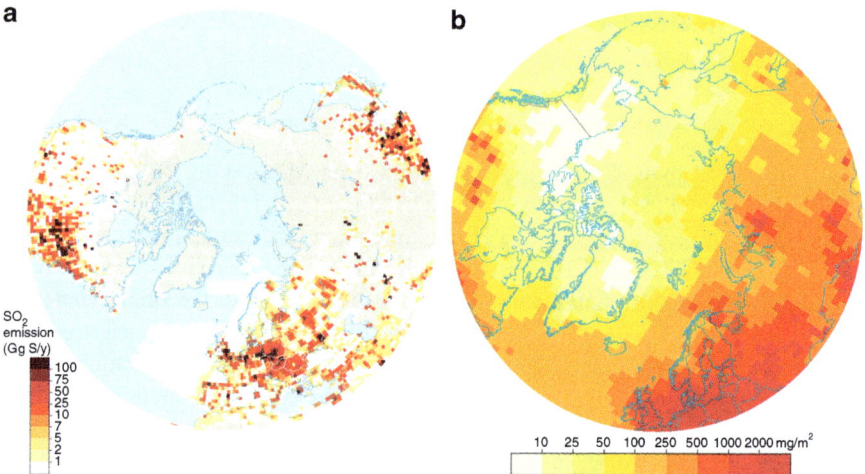

Fig. 8.4 (**a**) Sulfur dioxide emissions in 1985 and (**b**) modeled total (wet and dry) sulfur deposition for 1988 (Data from Benkovitz et al. 1996 and the Norwegian Meteorological Institute, reproduced from Kämäri et al. 1998)

Table 8.1 The approximate annual average Pb, Ni, and Cu atmospheric emissions from the industrial regions of Norilsk and Ural (Sverdlovsk and Chelyabinsk oblasts)

Source region	Decade	Annual emissions [ton year^{-1}]		
		Pb	Ni	Cu
Norilsk	1981–1990	100	3,000	3,500
	1990–1999	40	1,400	2,000
	1999–2008	26	600	700
Ural	1981–1990	2,000	90	4,500
	1990–1999	1,200	30	2,000
	1999–2008	500	150	1,000

This figure shows that in most parts of the Arctic and Siberia, the deposited sulfur has been generated far to the south, in the industrialized regions (mainly as a result of fossil fuel burning). However, there are local hotspots of sulfur deposition in Siberia and the largest of these arise from the nonferrous metal smelting activities in the Norilsk region.

Heavy metals contribution. Industries of the Ural and Norilsk regions are powerful emitters of nickel, copper, and lead into the atmosphere. These heavy metals (HM) are the markers of different man-made processes, and the most dangerous ecotoxicants (Persistent… 2004). In addition, due to their long residence time in the atmosphere, these metals are tracers of long-range atmospheric pollution transport (Rahn and Lowenthal 1984; Vinogradova 2000; Pacyna et al. 1985). The approximate annual source emissions during the last three decades are summarized in Table 8.1 (generalized data published by Roshydromet of Russia from 1990 (1989 Annual 1990) to 2009. The ratios of three chosen HM emitted from these two sources allow analyzing three different combinations: lead originated almost entirely from the Ural sources, and nickel, on the contrary, from Norilsk, whereas copper is emitted by these two sources in approximately equal amounts. Important to note, that emissions from both sources have been decreasing during almost all of the three recent decades. It was caused by both Russian industry reduction in the 1990s, and by improvement of air emission refinement methods.

On an annual scale, the variations in atmospheric circulation causes the same absolute values of environmental pollution change as the effects due to emission reduction in these decades (Vinogradova et al. 2008a, b). Under HM emission average decreasing by 38 % in the 2000s compared with 1990s (Table 8.1), the pollution decrease from the Ural and Norilsk sources (averaged for three metals) could vary by 10–50 % at different sites in Siberia.

8.1.4   Urban Air Quality

According to Russian urban emission statistical analysis for 1998–2007 (GGO 2009), the highest atmospheric emissions of particulate matter (PM) are observed in Siberian and Ural cities. For Barnaul, Krasnoyarsk, Novokuznetsk, Omsk, and

Chelyabinsk, they have exceeded on annual scale 30,000 tons. For gaseous compounds, the maximum emissions included the following: sulfur dioxide is emitted at a 30 tons year^{-1} rate in Novokuznetsk, Novosibirsk, Omsk, and Ufa, and nitrogen dioxide is emitted at more than 35,000 tons year^{-1} rate in Krasnoyarsk and Omsk. The density of PM emissions per capita is maximal in Novokuznetsk, and per unit area in Krasnoyarsk and Novokuznetsk. The highest densities of sulfur dioxide emissions per capita and per unit area are in Novokuznetsk and Omsk. One of the highest densities of nitrogen dioxide emissions per capita is observed in Novokuznetsk.

Effects of the high urban emissions on air pollution levels in the cities are increased by a specific feature of the Siberian environment. Meso-climates in the cities promote pollution accumulation. Urban biocenosis develops there in the extreme and evolutionary nonproviding conditions. Therefore, it consists from peculiar and poorly studied ecosystem types. Here, the natural and industrial (energy, transport) complexes are closely connected. There are severe contradictions between the growing chemistry components in all branches of the industry and the low level of general chemistry competence even at the level of decision-making of high responsibility. For example, the harm of unfinished technologies without the final stages of the waste utilization was often not realized until now. That is why a potential risk of man-made catastrophes remains high. The latter may provoke ecological disasters by atmospheric emission of heat, humidity, and toxic pollutants.

For most of the above-mentioned Siberian cities, annual concentrations of several chemical compounds can exceed the national permissible threshold values. The integrated air pollution index (API) calculated for five major pollutants (CO, NO_2, NO, O_3, and formaldehyde), yielded in 2007 a high or very high degree level of air pollution in more than 71 % of the cities; and during the 1998–2007 period it increased by 26.5 % (GGO 2009). An example of temporal dynamics of the annual mean concentration of PM in main Siberian cities (GGO 2009) is demonstrated in Fig. 8.5. As seen, the average concentrations of PM10 remained almost unchanged for most of the cities with some tendency to increase in Irkutsk during the last 6 years. During the 1990–2003 period, there has been a clear increase (as much as 2–3 times) in NO_x concentrations in Russian urban cities settlements (Nikel, Norilsk, and Salekhard), clearly reflecting a sharp increase in the number of private vehicles in Russia in general, and in Siberia, in particular.

Contamination of the surface air layer to a large extent depends on meteorological conditions. On average, the air pollution potential in Siberian cities is higher than in European cities of Russia due to a lower dispersion potential (Bezuglaya 1999). In certain periods, when meteorological conditions trigger an accumulation of harmful substances in the surface layer, the pollution concentrations may drastically increase – leading to high pollution/smog episodes. Both summer and winter episodes with high concentrations occur often in most of industrial Siberian cities.

It should be noted that cities are not isolated systems. They may distribute as much pollution over surroundings as receive it from the outside. Thus, the problem of mutual risk assessment for the cities and their surroundings should not be underestimated in the specific Siberian conditions. The influence of the transboundary atmospheric transport between Russia, China, Kazakhstan, and Mongolia should

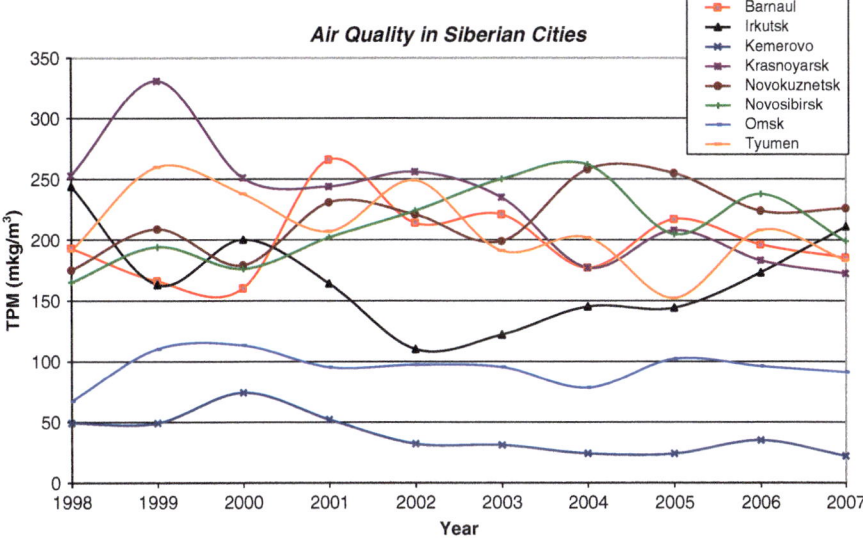

Fig. 8.5 Annual average concentrations of PM (μg m⁻³) in main Siberian cities for the 1998–2007 period (According to GGO 2009)

not be ignored as well. All these problems are connected with the Siberian ecological safety and quality of life. Moreover, there are indications (cf., Jaffe et al. 2004) that the long-range transport of Siberian biomass burning emissions could produce pollution episodes and impact on surface ozone as far as in western North America.

Close links between environment quality and human health, working capacity, and life expectancy are beyond any doubt. A general worsening of the ecological situation in Siberian cities causes increases in accumulation of the toxic products in human organs and tissues. This affects the functioning of the human organism as a whole and leads to metabolic imbalances.

There are unique ecological and environmental conditions in Siberian cities responsible for inherent complications of atmospheric pollution problems. For instance, Novosibirsk, Krasnoyarsk, Irkutsk, Ulan-Ude, and Ust'-Kamenogorsk are in the impact zones of the meso-climates provided by the interplay of the urban heat island and the nearby water objects (lakes and reservoirs). This increases the vulnerability of the air quality to pollution and demands additional studies/analysis while planning there the future economic activity. Furthermore in West Siberia, many cities are surrounded by vast wetland (bog) territories of the gas-oil provinces where methane is naturally emitted into the atmosphere. Because methane is a greenhouse gas, West Siberia is the focus of attention when considering the problem of climate change. Emissions resulting from mining and processing hydrocarbons contribute to the increase in the GHG concentrations. It is known that a series of toxic secondary products, such as formaldehyde, formic acid, etc., is generated when methane transforms. It is not yet quite clear as to what extent these products can be dangerous for Siberian environment and population.

8.1.5 Background Levels of Tropospheric Ozone

Great interest to the problem of climate change has induced the development of monitoring of the atmospheric composition and, in particular, of ozone and ozone-active compounds in remote areas of Siberia being a priority (Belan 2010). In situ ozone observations were carried out at the Russian background stations: Mondy (51.2°N, 100.9°E, elevation: 2030 m.a.s.l.), Tomsk (56.5°N, 84.9°E), Danki (54.5°N, 37.5°E), Kislovodsk High Mountain Station (KHMS) (43.7°N, 42.7°E, elevation: 2070 m.a.s.l.), Lovozero (68.0°N, 35.0°E), and Shepelevo (59.6°N, 29.1°E). The Obukhov Institute of Atmospheric Physics of the Russian Academy of Sciences (IAP RAS) equipped a new background station Zotino (60.7°N, 89.4°E) in the Krasnoyarsk Krai in Central Siberia. Regular measurements of the O_3, NO, and NO_2 concentrations at the lower level (4 m) of the 300-m tower in Zotino started in 2007 in the framework of the ZOTTO international project (Vivchar et al. 2009; Vivchar et al. 2010). All the background stations are equipped with the same sets of network instrumentation calibrated regularly against the ENV 03-41M No 1298 mobile standard (at Mondy station, the Japan national standard) (Elansky 2009a).

The longest series (since 1989) of surface ozone concentration (SOC) data was obtained at the KHMS. The yearly mean SOC values decreased abruptly over the period from 1989 to 1996; then, up to 2006, the decrease decelerated; in the last years, the SOC practically did not change (Tarasova et al. 2009a, b; Elansky et al. 2010). The linear SOC trend is equal to -1.53 ± 0.14 ppb/year during the 1989–1996 period and to -0.44 ± 0.06 ppb/year during the entire observation period 1989–2010. At the Alpine high-mountain stations, no significant trends were observed for the last decade. The opposite SOC trends, i.e., a SOC decrease at the KHMS and a SOC increase at the Alpine stations, were observed only during the period up to 1999, when industrial production and industrial and transport emissions of ozone precursors in the Former Soviet Union decreased abruptly.

The long-term SOC series (from late 1996 till 2010) is available for the Mondy station (Timofeeva et al. 2008). As compared to the KHMS, this station is located at nearly the same altitude. For these stations, the SOC values averaged over the period 1996–2010 are almost equal, 44 ppb (Mondy) and 42 ppb (KHMS). No significant trend was observed at these stations over this period; the amplitudes of the seasonal and daily variations were almost the same. The differences between the observations at these two stations are in the presence and the absence of the August secondary seasonal maximum at the KHMS and at the Mondy station, respectively. Furthermore, the SOC daily minima were observed at 12–13 and at 10–11 local time at the KHMS and at the Mondy station, respectively. These differences manifest the specificities in the uphill-downhill, regional, and large-scale atmospheric circulations and, to a smaller extent, the anthropogenic effect (Elansky et al. 2010). The surface ozone concentrations measured under background conditions over the Siberian plain differ significantly from that measured at the high-mountain stations. The annual mean SOC observed at the Zotino station is 26.2 ppb, while the annual mean SOC level measured in the TROICA experiments under background conditions (within the latitudinal belt 48°N–58°N between Moscow and Vladivostok) is about 27 ppb (Pankratova et al. 2011). The daily-variation amplitudes and seasonal variations obtained at the plain stations are much greater than those obtained at the high-mountain stations. The main causes of the SOC sink in Siberia are intensive dry ozone deposition on the surface covered with vegetation under conditions of highly stable nighttime atmospheric boundary layer and mild daytime vertical air mixing over the zone of boreal forests. In Siberia, such conditions are quite frequent. Therefore, the mean SOC level observable over these regions is lower in comparison with high-mountain stations, as well as with island and coastal stations where inversions are rather rare.

The contribution of the processes of ozone transport and ozone sink on the ground into the ozone variability in Siberia were estimated using the observational data obtained at the Zotino station (Skorokhod and Verkhovets 2006) and in the TROICA train and YAK-AEROSIB flight experiments (2006–2008) (Belan et al. 2010). To specify the advective flows of minor atmospheric components, the potential emission sensitivity was calculated by using the FLEXPART Lagrangian dispersion model. It is shown that the Siberian region is characterized by an unusually high vertical ozone gradient, namely, the summer ozone concentrations measured aboard the aircraft at altitudes over and below 3 km are 67 and 32 ppb, respectively, while the SOC is 18–27 ppb. The contact of air with forest-covered surfaces under conditions of slight air mixing leads to ozone depletion. The ozone sink in Siberia is influenced not only by the dry ozone deposition but also by VOCs. On the whole, the ozone sinks in Siberia prevail over the sources connected with the advective transport of ozone and ozone precursors from southern and western regions (China, Europe, etc.). Thus, Siberia, due to its large area, is capable of contributing significantly to the ozone global balance (Elansky 2009b; Pankratova et al. 2011) (Fig. 8.6).

The results of an analysis of observational data obtained in the TROICA experiments are published in (Elansky 2009b; Elansky et al. 2010; Pankratova et al. 2011; Turnbull et al. 2009; Tarasova et al. 2007; Timkovsky et al. 2010; Vartiainen et al. 2007). The detailed structure of the spatial distribution of about 20 VOCs playing the most important role in the surface-ozone chemistry over continental Russia was first obtained by using a proton mass-spectrometer. In particular, high concentrations of terpenes and isoprene were recorded in the zones of boreal forests (the Siberian taiga) and of broad-leaved forests in the Primorsky Krai (between Khabarovsk and Vladivostok), respectively (Timkovsky et al. 2010). The peculiar features in spatial distribution of minor atmospheric species over Northern Eurasia are revealed. Along the latitudinal belt 48°N–58°N, SOC increases eastward in all seasons except winter (Elansky et al. 2009b; Pankratova et al. 2011). The mean gradient is 0.47 ± 0.02 ppb per $10°$ of longitude. The SOC daily-behavior amplitude also increases eastward. These data are influenced by the eastward decrease in NO_x emissions, long-term intensive near-surface temperature inversions in the mountain areas of eastern Siberia, frequent forest fires, transport of pollutants from the northeastern region of China, high isoprene concentrations in the Primorsky Krai and the occurrence of the intensive subtropical high frontal zone with characteristic stratospheric-tropospheric intrusions over the eastern regions of the continent.

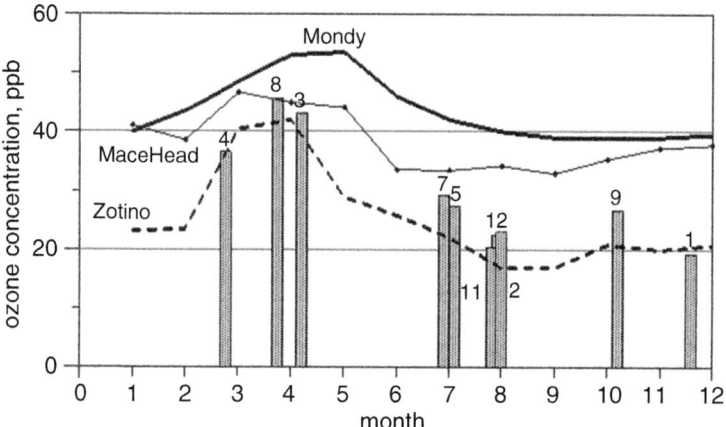

Fig. 8.6 Averaged O_3 concentrations over Moscow-Vladivostok route for TROICA experiments (1–12 – the number of experiment) compared with monthly mean O_3 concentration values at Mondy, Zotino, Mace Head (Ireland)

In Siberia, moderate concentrations of ozone precursors, NO_x, CO, and VOCs of biogenic and anthropogenic origins as well as a rather low level of UV radiation do not favor active chemical formation and destruction of ozone. The seasonal and daily SOC variations depend largely on the vertical air mixing and ozone sink on the ground. Their character and amplitudes obtained for Siberian background areas (along the Moscow-Vladivostok route, background conditions covered from 35 to 52 % of the total observation time) differ from those obtained at the Mace-Head station (Ireland) just because of high stability of the atmospheric boundary layer over the central area of the continent.

The SOC distribution over the Northern Eurasia may be somewhat influenced by the significant methane emissions from the industrial and natural sources. A comparison of the observational data on the methane isotopic composition ($\delta^{13}C$) with the results of numerical simulation applying the values of emissions of different isotopic compositions showed that CH_4 sources located in the Western Siberia are underestimated (Tarasova et al. 2007). The inventory of the CH_4 sources over the Russian area should be revised toward an increase in the emissions from the "light" sources (swamps, cud-chewing animals, waste recycling, etc.) and toward a decrease in the emissions from the "heavy" sources (extraction and processing of oil, gas, and coal, biomass burning, etc.).

8.2   Assessment of Airborne Pollution in Siberia from Air and Space

Monitoring and assessments of air pollution and its environment consequences from air and space have a strong and continuously increasing potential, especially for such large territories like Siberia.

8.2.1 Observations of Aerosols

Satellite observations of aerosols yield valuable information on vertically integrated columns of aerosols although retrievals have been in particular difficult over continents and it is difficult to quantify the anthropogenic fraction. Up to last decade, estimates of the direct aerosol radiative forcing were available only from models (Haywood and Boucher 2000). Currently, the studies based on observations have become available (Bellouin et al. 2005; Chung et al. 2005; Kaufman et al. 2005). Global mean direct aerosol radiative forcing with this approach has been estimated between −0.8 and −0.35 W m^{-2} (Bellouin et al. 2005; Chung et al. 2005), which shows that uncertainties are still significant.

A range of aircraft campaigns (e.g., TARFOX, ACE-1, INDOEX, ACE-Asia, SAFARI, SAFARI-2000, SCAR-B, SHADE, MINOS) have provided a large amount of more detailed than from satellites data on aerosol distributions and chemical and physical properties. However, only a few aircraft aerosol observations have been made in Siberia. Knowledge of the vertical distribution of aerosols is crucial for studying the climate effect of aerosols (especially of absorbing aerosols). Therefore, to provide vertical profiles of aerosol distributions and properties and to quantify aerosol climate effect over Siberia, a joint assessment is urgently needed of aircraft and LIDAR instruments (both at the surface and satellite borne, e.g., CALIPSO).

One of the important tasks is the preparation for Siberia of an integrated dataset from the observational satellite records with different and complementary information content. Several such studies were realized in Europe and America and this is possible in Siberia too. For example, TERRA- and AQUA-MODIS provide twice daily global coverage with 1×1 km^2 and 10×10 km^2 of Aerosol Optical Depth (AOD) and coarse/fine fraction since 2000/2003 from multispectral dark field retrievals. TERRA-MISR provides a weekly global coverage with 16-km resolution of AOD and aerosol type (selected from a set of predefined aerosol types) since 2000 based on multi-angular observations. Utilizing the synergy of the radiometer AATSR and the spectrometer SCIAMACHY aboard ENVISAT, biweekly AOD and aerosol speciation (from a set of predefined aerosol compositions) can be derived since 2002 from different pixel sizes on 60×30 km pixels. Additionally, AOD and aerosol type information can be provided by MERIS measurements which will complement the suite of ENVISAT data.

One of the main problems related to climate radiative forcing is associated with a low level of understanding of indirect effect of tropospheric aerosols. This problem is also complicated by a poor predictability of nucleation and new particle formation events. It is known that nucleation is widely observed over forested regions (Mäkelä et al. 1997; Dal Maso et al. 2008). Siberia occupies the vast forested areas of the Northern Eurasia, but only a few research data on ultrafine aerosol particles and their size distribution are available for the Siberian region (Bashurova et al. 1992; Koutsenogii and Jaenicke 1994; Koutsenogii 1997). The most detailed data on new particle formation and growth in the troposphere over Siberian forests were reported by Dal Maso et al. (2008).

In 2010, the V.E. Zuev Institute of Atmospheric Optics of SB RAS and Tomsk State University began continuous measurements of atmospheric aerosols in a wide range of sizes in order to fill up this gap in data (Arshinov and Belan 2011). An automated diffusion battery (Ankilov et al. 2002a, b) coupled with a TSI (TSI Inc., USA) Model 3781Water-based Condensation Particle Counter measures the size distribution of aerosol nanoparticles in the size range of 3–200 nm as well as the total number concentration of particles. First year-long monitoring conducted in the Tomsk area of West Siberia showed that new particle formation events in Siberia were more often observed during spring (from March till May) and autumn (secondary frequency peak in September). The strongest nucleation bursts occurred in April. The highest formation and growth rates of ultrafine particles measured in Siberia reached values of 2.3 cm^{-3} s^{-1} and ≈ 25 nm h^{-1} (in April) and 1.1 cm^{-3} s^{-1} and ≈ 9 nm h^{-1} (in September), respectively (Arshinov and Belan 2011).

8.2.2 Satellite Remote Sensing of Wildfire Emissions and Pollution

Satellite remote sensing methods proved to be very useful for Siberia in estimation and monitoring of wildfire emissions and aerosol concentrations. Emissions of air pollutants from fires can be estimated using measurements performed by MODIS (Moderate Resolution Imaging Spectroradiometer) instruments on board of NASA Aqua and Terra satellites. Specifically, the Fire Radiative Power data can be used (e.g., Sofiev et al. 2011), which are publicly available from the Land Processes Distributed Active Archive Center (LP DAAC) through the Earth Observing System Clearing House (ECHO, URL: http://www.echo.nasa.gov/) system as Level 2 (MOD14/ MYD14) products.

The global fire emission inventory has been compiled using the Fire Assimilation System (FAS) of FMI (Sofiev et al. 2009). It uses two remotely sensed wild-land fire characteristics: 4-μm Brightness Temperature Anomaly (TA) and Fire Radiative Power (FRP) – for the needs of emission estimation from wild-land fires. Two treatments of the TA and FRP data comprise the methodology that is applied for evaluating the emission fluxes from the MODIS level-2 fire products. The FAS does not contain a complicated analysis of vegetation state, fuel load, burning efficiency and related factors, which are uncertain but inevitably involved in approaches based on burnt-area scars or similar products. The core of the current methodology is based on the fire intensity expressed either via TA or FRP values, which are converted into emission of atmospheric pollutants via empirical emission factors derived from the analysis of several fire episodes. The satellite systems are successfully used in different countries to control and prevent natural fires. The detection of fire sites is usually done by IR-radiometers that are part of on-board complexes of a range of satellite systems. The algorithms of detection are based on the registration of the radiating temperature in the spectral range 3.5–3.7 μm and difference of radiating temperatures in this channel and in the spectral channel ~11.0 μm (Bondur 2010).

In Siberia there is a plenty of forested territories that are difficult to access. Therefore, the space information for early detection and estimation of consequences of natural fires is here especially important. The emergency space monitoring is carried out for the purposes of early detection, prognosis of development dynamics, and estimation of consequences of natural fires, as well as for the purposes of urgent creation and transmission to clients of different information about these natural disasters. This monitoring is conducted by several Institutions, private firms, and Agencies: The Federal Agency of Forestry of Russia employs an informational system based on remote sensing, which was created by a consortium of several institutions (SE "Avialesohrana," Space Research Institute, Scientific research center "Planeta" and a few other organizations). Ministry of Handling the Emergency Situations of Russia also uses satellite data for fire control. The engineering and technology center SCANEX is carrying out space monitoring of fires. A system of emergency space monitoring of natural fires has been successfully used in "Aerokosmos" Scientific Center for Aerospace monitoring (www.aerocosmos.info). The perspectives of development of a unified system for urgent monitoring of natural fires in Russia, and in particular in Siberia, are related to the wide use of space information from different existing and perspective space tools. These tools include radars, microwave-radiometers, etc., as well as methods and technologies for the monitoring of fires from space and processing these and other data in order to obtain a bigger number of informational products. At present, an important task is related to increasing their fidelity and efficiency for the purpose of timely fire detection and prevention, and mitigation of the damage (Bondur 2010). Examples of such studies for Moscow and Siberia wildfires can be found in Bondur (2010), Konovalov et al. (2011), Sofiev et al. (2011), etc. Figure 8.7 shows an example of detections of wildfires in Siberia (in Irkutsk and Krasnoyarsk Regions), Russia, in June 2011 by the Russian space monitoring system.

8.2.3 Remote Sensing of Air Pollution Impacts in Siberia

In the context of airborne pollution environmental impacts and consequences studies, remotely sensed data can be analyzed in terms of land cover (especially, but not exclusively, the distribution and health of different vegetation types), water color, and the presence of smoke, dust, or other particulate material in the atmosphere. This has implications for the types of industrial impact that can be monitored, and whether this monitoring is direct or indirect, through their consequences (Rees and Rigina 2003).

Sulfur deposition may result in acidification of soil and water. In its turn, acidification of soil can cause loss of soil fertility, loss of plant nutrients and desiccation of plant tissues, leading in extreme cases to the complete destruction of forest and loss of lichens (Rees and Rigina 2003). These effects can be measured as far as 60 km from Norilsk industrial centers. Emission of sulfur from metal smelters is accompanied by heavy metals, which exhibit a range of toxic effects. Higher organisms are affected directly by toxicity effects and indirectly through loss of habitat and food supplies. The area affected by emissions from Norilsk has been estimated

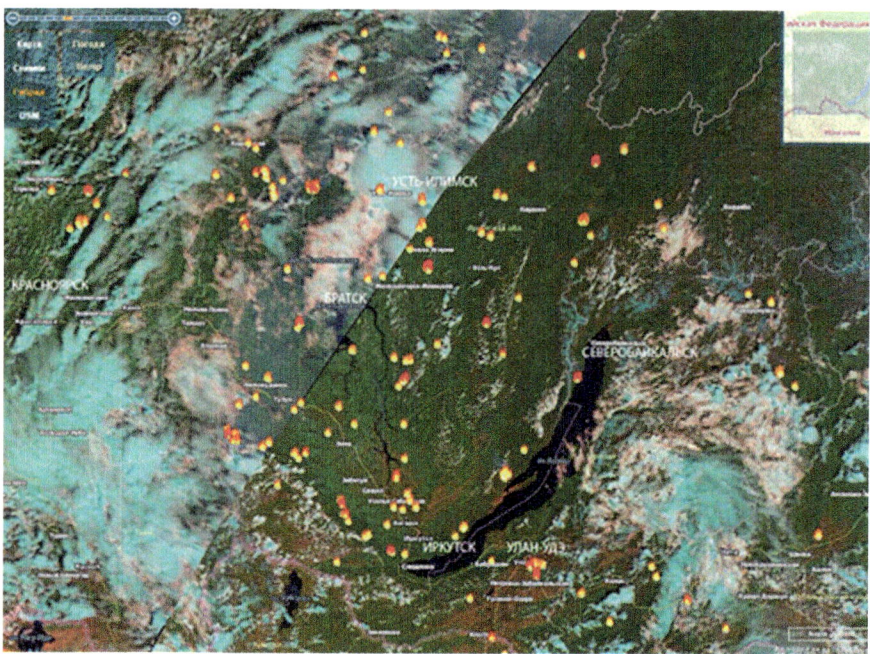

Fig. 8.7 Detection of wildfires in Siberia (Irkutsk and Krasnoyarsk regions), Russia, June 2011, by the Russian space monitoring system (*red spots*) (Source: http://ria.ru/trend/forest_fire_siberia_24042011)

between 6,800 km^2 (Mel'nikov et al. 1996) and 20,000 km^2 (Kharuk 1998; Derome and Lukina 2011). The main environmental impact of sulfide ore smelting that is detectable in remotely sensed imagery is damage to the surrounding vegetation. The airborne SO2 and heavy metal particles cannot be detected directly, but only through their impact. For the boreal forest, the early stages of damage are displayed as chlorosis and necrosis of leaves and needles and premature defoliation. These consequences can be monitored and assessed by means of remote sensing since they alter the spectral reflectance (Colwell 1983; Häme 1991; Ekstrand 1993; Litinksy 1996). In most cases, spectral signatures.

The principal environmental impact of sulfide ore smelting that is detectable in remotely sensed imagery is damage to the surrounding vegetation. The pollutants themselves (airborne SO$_2$ and heavy metal particles) cannot be detected this way, so this is an indirect approach to the problem. It is, nevertheless, a well-developed and powerful methodology. For the boreal forest, the early stages of damage are displayed as chlorosis and necrosis of leaves and needles and premature defoliation. These consequences can be monitored and assessed by means of remote sensing with the aid of ground truth data since they alter the spectral reflectance (Colwell 1983; Ekstrand 1993; Häme 1991; Litinksy 1996). In most cases, spectral signatures are identified for damaged and healthy forest on the basis of ground truth data, and used to extrapolate from the areas studied in fieldwork to much larger areas.

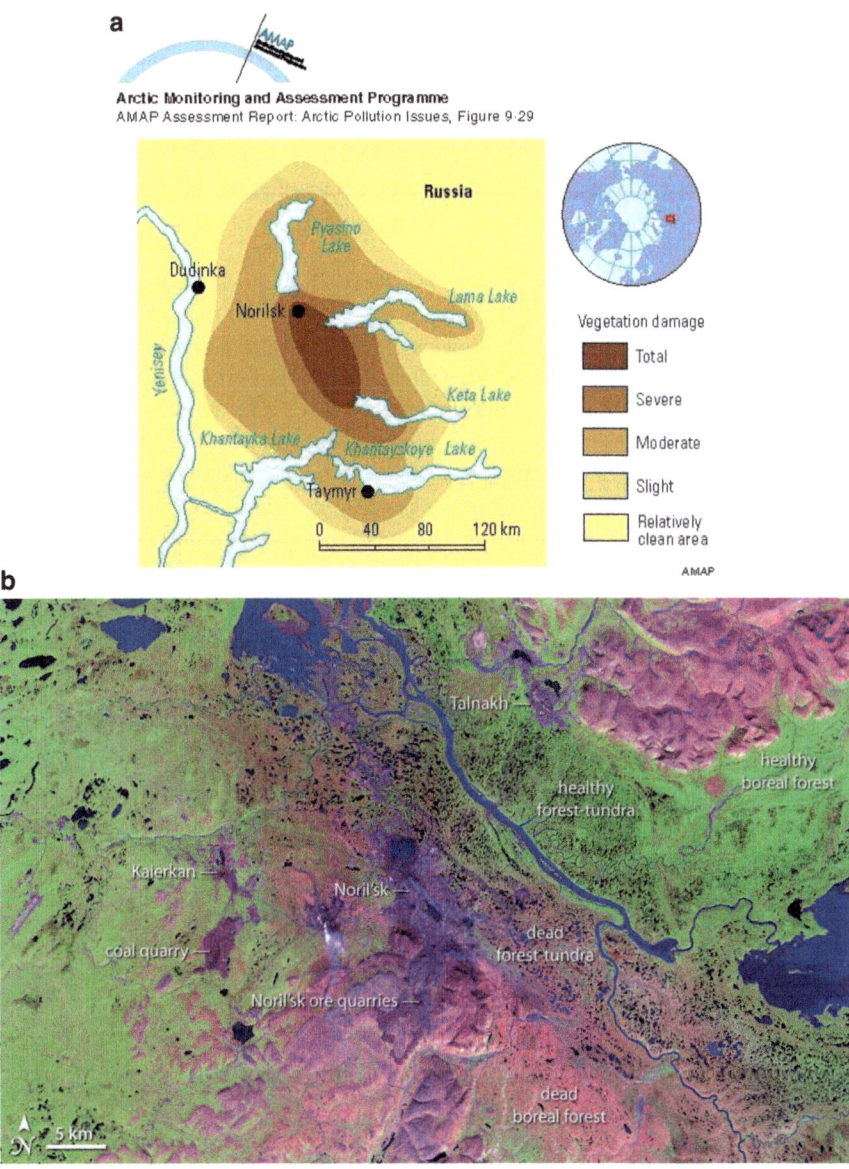

Fig. 8.8 (**a**) Vegetation damage zones around Norilsk (Reproduced from Kämäri et al. 1998 and originally from Vlasova et al. 1990). The colors, from *darkest* to *lightest*, denote total, severe, moderate, and slight levels of damage, with the *lightest color* representing relatively clean areas (**b**) Classified Landsat TM image showing the area indicated at *left* (Modified from Toutoubalina and Rees 1999). *Black*: unvegetated areas; *gray*: sparsely vegetated; *red*: heavily damaged tundra vegetation; *orange*: moderately damaged tundra vegetation; *yellow*: relatively healthy tundra vegetation; *pale green*: moderately damaged forest; *green*: heavily damaged forest; *dark green*: dead forest; *light purple*: moderately damaged dog; *dark purple*: heavily damaged bog; *blue*: water; *white*: snow, cloud, smoke, ice, no data (Source: http://visibleearth.nasa.gov/view_rec.php?id=17532)

The spectral signatures are normally species-dependent, although this is less important at the higher damage levels. Typically, five or six zones of pollution impact, from "anthropogenic barren" to undisturbed areas, can be distinguished around a smelter. Kravtsova (1999) gives a detailed review of the methods used and results obtained, and an example of such a classification is shown in Fig. 8.8a. However, the effect of industrial emissions can be masked by a wide variety of other factors, and, in general, it can be estimated by remote sensing only if the forest has been severely damaged or killed.

This false-color image in Fig. 8.8b shows data collected by Landsat 7's Enhanced Thematic Mapper Plus (ETM+) instrument on August 9, 2001 (http://visibleearth. nasa.gov/view_rec.php?id=17532). Shades of pink and purple indicate bare ground: rock formations, cities, quarries, and places where pollution has damaged the vegetation. Brilliant greens show predominately healthy mixed tundra-boreal forest. Water appears in shades of blue. The Norilskaya River cuts through the scene, and the forest-tundra terrain on either side is riddled with small "pothole" lakes. Blue-white plumes of smoke (just left of the image center) drift southeastward from smokestacks of Norilsk. The deep and pale pinks downwind of the city, as well as the deep purple in the hillsides immediately outside Norilsk, are moderately to severely damaged ecosystems. To the northeast of the River, the ecosystems appear to be healthier. A large coal quarry, one of the sources of electricity in Norilsk, is seen as a purple patch south of the city of Kaierkan.

Similar remote sensing studies of forest damage have been performed for the nonferrous metal smelters at Sudbury, Canada by Pitblado and Amiro (1982) and at Norilsk, Russia, by Toutoubalina and Rees (1999, 2001). Pitblado and Amiro (1982) also found that vegetation indices can be successfully used in discriminating different damage classes, but Toutoubalina and Rees (1999) found these indices to be less effective in the more heterogeneous land cover conditions of Norilsk.

8.3 Methodology and Models for Air Pollution Assessment on Different Scales

Estimates of environmental risks stipulated by hydrodynamics processes in the atmosphere, pollutants' emission as well as by changing the Earth's surface characteristics can be made using numerical models combined with observations. This section describes methodology and these models used for environmental forecasting in Siberia and some results of their implementation.

8.3.1 Concepts of Environmental Modeling and Forecasting

The main quantitative characteristics of the environment state are some distributed 4-D space-time functions describing the behavior of the atmosphere as a dynamical system and the concentrations of the multicomponent gas substances and aerosols.

To calculate them, prognostic informative modeling complexes have been designed. The databases, in which the measurement results are collected, are the essential part of the model complex.

The events with abrupt and significant changes of parameters, which characterize the evolution of air pollution, can be considered as catastrophes. Most often, the catastrophic ecological situations do not have clear manifestations, and at each moment the input of the pollutants' emission influencing the air quality is minor. But their aggregated impact can "suddenly" lead to the catastrophic consequences due to accumulative and synergic effects. A representative example of such development is methane emission from the peatbogs and wetlands in the Northern Hemisphere and, in particular, in Siberia. The problem is that besides the properties of methane as a greenhouse gas, its transformation in the atmosphere leads to high levels of reactive species like formaldehyde, carbon oxide, and many other highly toxic products of the secondary pollution, the concentration of which may exceed the acceptable safe limits. Probability of such development increases with increases of the surface air temperature and solar radiation in the summer months. Other examples of catastrophes' stimulation are deliberate or unintended consequences of the destruction of ecologically dangerous objects. That is why a modeling approach was developed to efficiently estimate the tendencies in the development of chemical and aerosol situations and to reveal the prerequisites for the formation of ecologically hazardous situations (Penenko 1981, 2010; Penenko and Tsvetova 2009a, b; Penenko et al. 2002; Baklanov 2000, 2007). Within this approach, the mathematical models of the atmospheric processes serve as interconnections between the state of the atmosphere variables and external parameters. These connections are necessary for estimation of goal functionals of generalized characteristics defined as functionals on the spaces of the state variables and input parameters of the models. They can describe the air quality, social and economic criteria for environment, admissible restrictions for loads, etc. Data assimilation technique is applied to improve the information supply of the models (Penenko 2009; Penenko and Tsvetova 2009b). If the goal functionals are linear with respect to the state variables, it is possible to use inverse modeling assessing the sensitivity of the goal functionals to external forcing. Thereafter, regions are categorized with respect to the degree of the risk and vulnerability to detrimental impact of air pollution and relative contributions of different sources into the integral air quality functional are evaluated. Hence, the modeling output deliver projections of the regions (their boundaries and major sources) that can send the amount of pollutants of the given level to each receptor-region.

8.3.2 Scenario Approach

Formation of hydrodynamic background for calculation of prognostic scenarios is usually accomplished by numerical models of atmospheric circulation with the use of available actual data. To take into account the influence of the global processes at the Siberian regional level, a special class of models with leading phase spaces is used (Penenko and Tsvetova 2008). This allows differentiation of the scales of

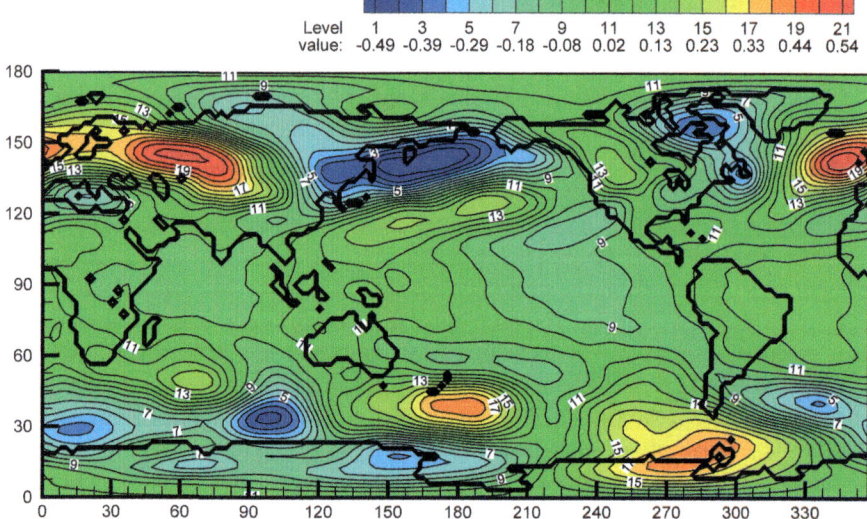

Fig. 8.9 One of the 62 fragments of the leading OBV-1 for 500-hPa geopotential height corresponding to 00:00 UTC on April 15

processes between the climatic component and "weather noises." The calculations show that the leading factor for long-term forecasting is so-called "long-term memory" of the climatic system. The input data for these calculations are derived from the NCEP/NCAR reanalysis (Kalnay et al. 1996) for 1950–2005. The reanalysis information is converted into two types of databases: principal components (PC) and orthogonal base vectors (OBV). Principal components characterize the interannual variability of an initial sample (reanalysis data) with respect to an OBV-system. The system consists of 4D (space-time) orthogonal base vectors. Penenko and Tsvetova (2008, 2009b), used $2 \times N$ time fragments for each month with a time step of 12 h in each vector, where N is the number of days of the current month. The spatial resolution was taken in accordance with the reanalysis data: horizontal resolution of $2.5° \times 2.5°$ in spherical co-ordinates and standard p-levels in the vertical direction. In Figs. 8.9 and 8.10, the fragments of leading OBVs for the 500-hPa geopotential height and horizontal velocity vectors are shown at 00:00 UTC on 15 April. These informative bases (actually, the prevailing patterns) have been used for the construction of various hydrodynamic scenarios for environmental applications in Siberia. Moreover, a general analysis of PC and OBV base subspaces gives a possibility to detect the regions of the climatically stipulated increase of ecological risks.

8.3.3 Inverse Numerical Modeling Approach to Risk Forecasting

Here a scenario of long-term risk forecasting in the East Siberian region (47.5–60°N; 95–115°E; see Fig. 8.11) is considered as an example. This zone includes a specially

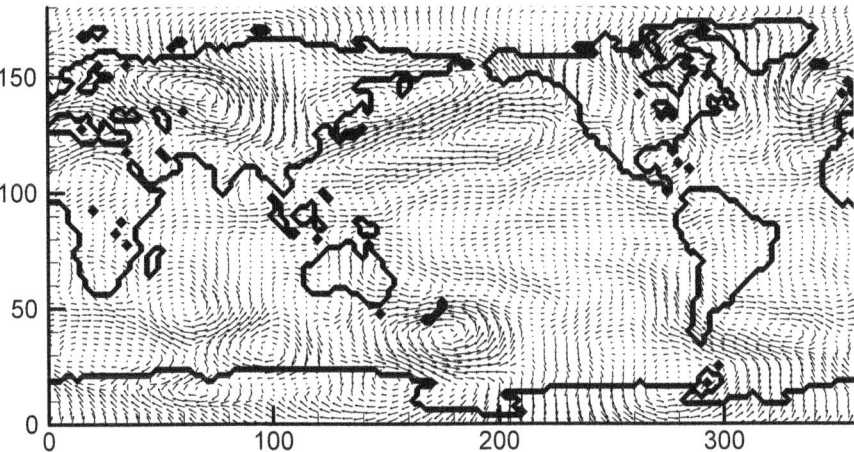

Fig. 8.10 The same as in Fig. 8.9, but for the horizontal velocity vector

protected territory – Lake Baikal, which is the world-level heritage. East Siberia is rich in natural resources and has a well-developed industrial complex with a high man-caused load on the environment. In addition, this region is located in the area of the Altai–Sayan cyclogenesis (Chung et al. 1976; Kasahara 1980; Chen et al. 1991) and high seismic activity. The Altai–Sayan cyclogenesis area is classified as one of the most active continental zones in the Northern Hemisphere, its high intensity part being located in the region of 40–50°N and 95–125°E from Altai and Sayan on the west to Great Xinganling on the east. In Figs. 8.9 and 8.10, this zone is within the longitudinal band of 130–140° on the ordinate axis.

The problem consists in assessment of the prognostic goal functional, which is chosen to be linear in a four-dimensional domain of the receptor area. The latter is a three-dimensional area of the atmosphere above the Baikal Lake extended in the vertical direction to 10 hPa. The value of this functional is the total amount of pollutants located in the selected region in the indicated time interval. Solving this problem by traditional methods of forward modeling is difficult. The pollutants' field is formed in the atmosphere from different sources that can be located almost anywhere on the Hemisphere, while the quantitative information about the pollutant emissions is unavailable. Air masses carrying the pollutants pass through the region and the receptor area from one boundary to another within a time interval of the order of 24 h. This requires prescribing the transboundary fluxes of meteorological elements and pollutants from the adjacent territories.

The problem is to estimate the amount of pollutants passing through the domain above the lake on a monthly time scale and estimate the pollutants that become deposited on the underlying surface. The methods of inverse modeling and methods of organizing prognostic scenarios (Penenko and Tsvetova 2007; Penenko 2009) allow obtaining results that are acceptable for practical applications without specifying information about emissions. The calculations are performed with a set of mathematical

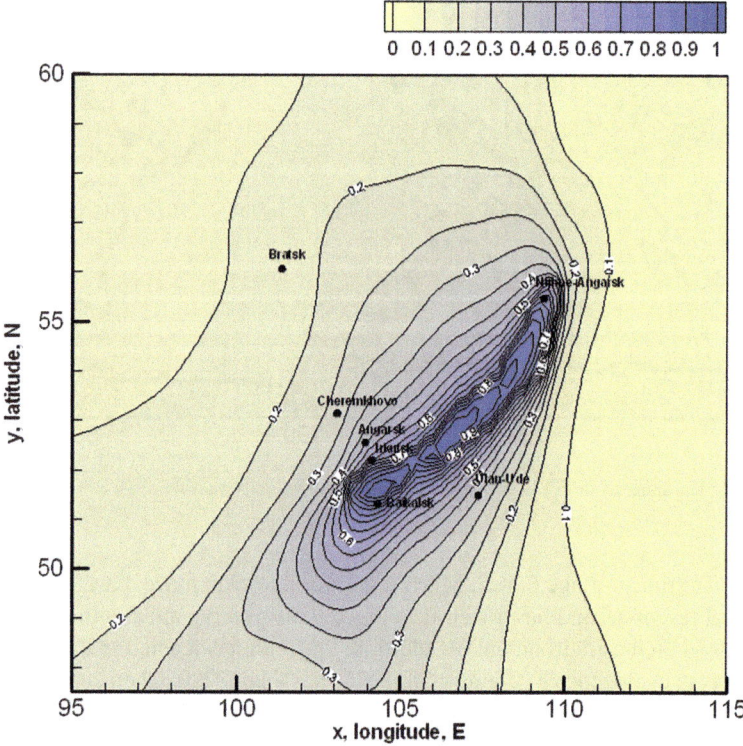

Fig. 8.11 Total (during October) estimates of the relative contribution of pollutant emission from acting and potentially possible sources in the region (47.5–60°N, 95–115°E) into the atmospheric quality functional over the Lake Baikal water area

models whose general structure was described in Penenko and Tsvetova (1999, 2007). The hydrodynamic part of the scenario is formed under assumption that the regional atmosphere is a constituent of the global long-term (climatic) circulation of the atmosphere. For this case, the analysis is organized as in Penenko and Tsvetova (2008),[1] and the reanalysis data for the period of October 1–31 are assimilated in the regional model for reconstruction of the hydrodynamic background for the scenario.

Figure 8.11 shows the calculated SF of the total (during a month) relative contribution of emission of pollutants from acting and potentially possible sources located on the Earth's surface in the East Siberian region into the atmosphere quality functional in the receptor area (i.e., the Lake area). The fragment in the figure corresponds to

---

[1] The dynamics of four-dimensional fields of meteo-elements is calculated with a 30-min time step in the domain (47.5°N–60°N; 95°E–115°E) in spherical co-ordinates on a latitude–longitude grid with a step of 15′ in each co-ordinate direction. The vertical resolution of the grid domain is 19 levels in hybrid p–σ coordinates from the Earth's surface to a level corresponding to a pressure of 10 hPa. The vertical coordinate follows the relief of the underlying surface.

the model level, which coincides with the upper boundary of the surface layer. This fragment is a 2D cross-section of the 3D function obtained by temporal integration of the 4D SF of the goal functional with respect to variations of source function in the region. The values on the isolines are obtained by normalization of the integrated SF to its maximum in the entire three-dimensional atmospheric domain of the region. The higher the SF values are, the greater will be the risk of contamination of the receptor area from the sources located in the areas-carriers of these values.

To estimate the risk of contamination of the receptor area owing to transboundary transfers, it is necessary to take into account the information contained in the SFs on the boundaries of the domain.

An analysis of data plotted in the figure and results calculated for other time intervals confirm the previously made conclusion that almost all cities and industrial areas of the East Siberian region are in the zone of a high environmental risk for Lake Baikal. The city of Baikalsk should be specially noted because almost all its contaminating gases and aerosols are injected into the atmosphere above the lake. Generalized information about the degree of environmental risks at the long-time intervals obtained by inverse modeling methods is quite convenient and, therefore, can be used for management decisions, in particular, during design of new objects in the lake region and developing there the strategy for environment protection and risk control.

8.3.4 Environmental Risk Assessment of Long-Term Air Pollution Using Dispersion Modeling

The risk assessment strategy can be realized using published results from real events, and by physical and theoretical modeling. In particular, Baklanov et al. (2008b) and Mahura et al. (2006a, b) suggested a methodology for multidisciplinary probabilistic environmental risk and vulnerability assessments. It considers several stages with respect to long-term modeling for atmospheric pollutants from potential risk sites. Analysis and selection of potential risk sites (of nuclear, chemical, or biological nature) representing a potential danger for the environment and population of the considered region is a starting point in such a methodology. Sources like chemical and metallurgical enterprises, smelters, processing plants, dumping sites, nuclear power plants and weapons testing sites, nuclear submarine bases, etc., are identified in the Arctic and Siberia regions. A grouping of sources of similar nature and located geographically close to each other can be done (without consideration of accidents' probabilities) because the individual atmospheric transport patterns will be relatively similar.

Thereafter, trajectory and dispersion models can be used to simulate long-term atmospheric transport, dispersion, and deposition of, for example, radionuclides, chemical species, and heavy metals from selected risk sites on local, regional, and global scales. One of such models is the Danish Emergency Response Model for Atmosphere (DERMA; Sørensen et al. 2007; Baklanov et al. 2008b) that is a numerical 3D atmospheric model of Lagrangian type. This model was employed to perform simulations of air concentration, time integrated air concentration, dry and wet deposition patterns resulting from continuous emissions, and/or accidental releases in many parts of the world including Siberia. The geographical locations of the Siberian chemical and metallurgical enterprises situated near the industrial cities were selected as representative sources of such emissions. To perform simulations, different 3D meteorological gridded datasets can be employed such as the High Resolution Limited Area Model (HIRLAM), the European Center for Medium-Range Weather Forecasts (ECMWF), or NCAR/NCEP models outputs of different resolution and scales. To minimize computational resources, the long-term simulations can be done for a climatologically typical year and a year with a significant deviation of atmospheric circulation pattern (e.g., with respect to the North Atlantic Oscillation Index). Different types of emissions can be considered in such simulations – short vs. long-term or episodic/accidental vs. continuous, hypothetical scenarios vs. real events, nuclear vs. chemical, etc. For each daily release, the followed transport through the atmosphere and deposition on the underlying surface due to dry and wet removal processes were estimated on an interval ranging from a few hours up to 2 weeks depending on a task or an application. The trajectory modeling results can be analyzed and represented in a form of different indicators. Among them are:

- Airflow probability fields, typical transport time from the source to receptor;
- Maximum reaching distance (represents the possibility of event when, at least, one trajectory originating over the source arrives the receptor);
- Maximum possible impact zone (underlines possibility of the highest impact from the source to receptor);
- Fast transport probability fields (shows scale of potential impact due to atmospheric transport during the first 12 and 24 h after the release at the site with respect to the area where such impact can be the highest).

The dispersion modeling results are more valuable since in addition to the trajectory-related indicators, these show also detailed information about potential levels of air and surface contamination. The results of the probabilistic analysis of dispersion modeling results for the risk sites can be presented as a set of indicators of the risk sites' impacts on the geographical regions of interest such as the time integrated air concentration (TIAC) at ground level, dry deposition (DD), and wet deposition (WD) patterns. The indicators can be constructed using the distribution of the total sum of daily continuous emissions from the sites. This type of field (summary field) shows the most probable geographical distribution at the surface if the releases occur during a long-term period. Another approach is based on calculating the average value from the summary field. This type of field (average field) shows the climatologically averaged spatial distribution during any given average day of the assessed period. Finally, the indicators can be further integrated into GIS environment and used as an essential input for evaluation of doses, impacts, risks, vulnerability, short- and long-term consequences for population and environment from potential sources of continuous/accidental emissions. Studies by Mahura and Baklanov (2003), Mahura et al. (2005), and Lauritzen et al. (2007) showed that a combination of different types of analysis and using results of probabilistic long-term trajectory and dispersion modeling from the selected risk sites can provide a more detailed level, quality, and accuracy in evaluation and ranking of potential impact on both geographical localized area (city, site, etc.) and region (country, county, etc.).

Summary of evaluation of dispersion modeling results for the Cu-Ni smelters (located on the Kola Peninsula and Krasnoyarsk region) of the Russian Arctic is given below for the Norilsk Nickel enterprise. Modeling allows estimation of concentration and deposition fields and their follow-up evaluation for any potential sources both individual (Fig. 8.12a) and combined (Fig. 8.13). In general, the TIAC and DD (Fig. 8.13a, b) have a similar structure compared with WD (Fig. 8.13c). Around the site both have distributions which are closer to elliptical than circular, and the shape of these distributions reflects the presence of dominating atmospheric transport pattern throughout the year. The WD field is less smooth and often has a cellular structure, because it reflects the irregularity of the rainfall pattern. Moreover, these fields have different monthly spatial structures. Detailed analyses of simulated concentration and deposition fields for each site allow evaluating spatial and temporal variability of these resulting patterns on different scales.

The geographical extension and levels of potential impact due to atmospheric transport and deposition of sulfates on the population and environment resulting from continuous emissions from smelters (of the Pechenganickel, Severonickel, and Norilsk Nickel enterprises) are shown in Fig. 8.13. For example, for the "mild scenario" of SO_2 emissions (approximately 32,000 tons annually), for the Norilsk Nickel smelters, the annual average daily dry deposition value is 10.6 ton. The highest average DD (17.8 ton) is in May and the lowest (4.5 ton) in August. For wet deposition it is 28.3 tons, and a strong month-to-month variability is observed compared with dry deposition. The contribution of WD into the total deposition is comparable to DD during May–August, but is almost eight times higher in February. The highest average WD (57.2 ton) is in February and the lowest (7 ton) in August. Considering the monthly amount deposited into the annual total, the mean monthly average is 45 % ranging from 13.2 to 70.6 % in August and February, respectively. Larger amounts are deposited from the Norilsk Nickel emissions compared with the Kola Peninsula sources (cf., Fig. 8.12b).

In a separate modeling experiment, large populated urban areas – Ekaterinburg (1.29 million), Novosibirsk (1.43), Chelyabinsk (1.08), Omsk (1.14), Tumen (0.538), and Tomsk (0.487) – in the Ural and West Siberia (in total more than 30 million) were selected, and levels of the time integrated concentration and deposition were estimated taking into account the amount of annual emissions and its intensity. Analysis showed that the influence of the Kola Peninsula sources (enterprises Severonickel and Pechenganickel) is lower by almost two orders of magnitude compared to the Norilsk Nickel. Therefore, the latter plant is the main source for both geographical regions studied. It was found that on an annual time scale among all cities considered, in Tomsk the summary time integrated air concentration (34 µg·h m^{-3}) and the dry and wet depositions (182 and 954 µg m^{-2}, respectively) are the highest. For other assessed cities, the annual concentrations and depositions estimated are significantly lower being the lowest (below 5) for Chelyabinsk and Ekaterinburg.

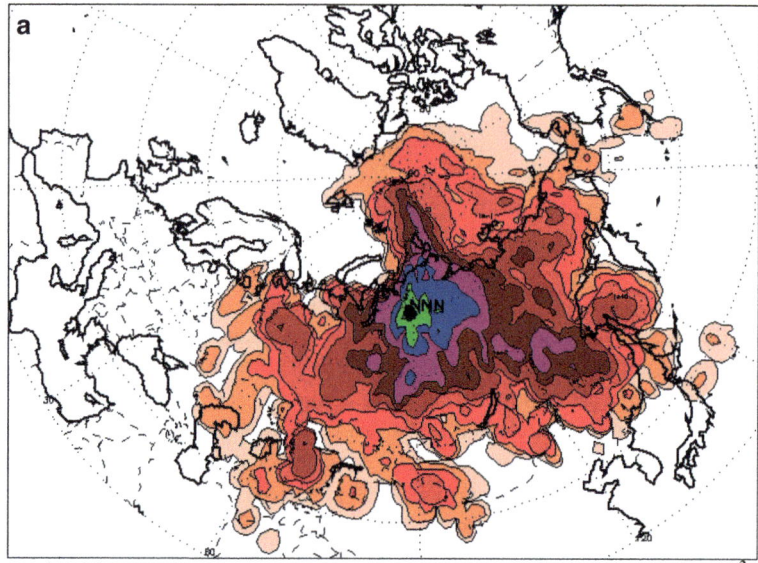

Norilsk Nickel (NNN) : June : Summary Wet Deposition (WD) of Sulfates [µg/m²]

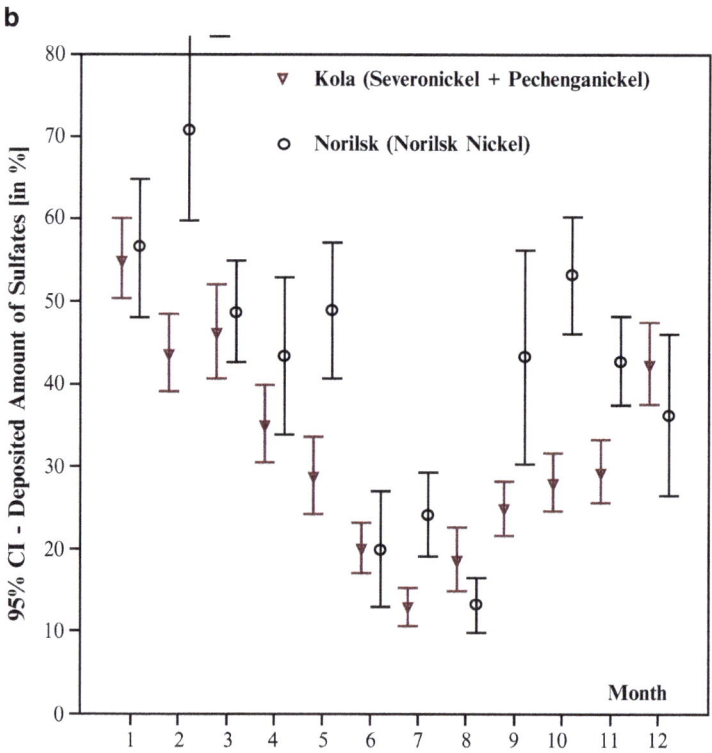

Fig. 8.12 (**a**) Summary wet deposition patterns resulted from the Norilsk Nickel emissions of sulfates during month of June and (**b**) monthly variability of deposited amount of sulfates (in % from daily released) at 95 % confidence interval during atmospheric transport and deposition from the Cu-Ni smelters of the Russian Arctic

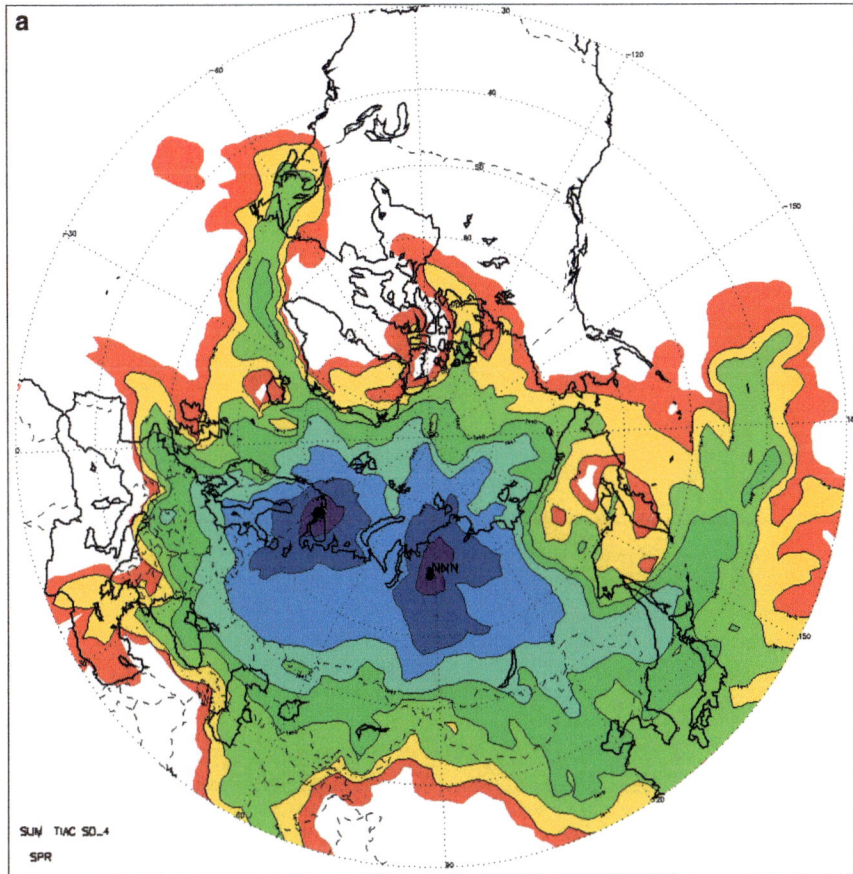

Spring : Summary Time Integrated Air Concentration (TIAC) of Sulfates [μg/m³]

Fig. 8.13 Results of the long-term dispersion modeling: spring summary (**a**) time integrated air concentration, (**b**) dry deposition, and (**c**) wet deposition patterns resulting from the continuous emissions of sulfates from the Pechenganickel, Severonickel, and Norilsk Nickel enterprises isolines are drawn by order of magnitude and stating at 1e-4 (red color)

8.3.5   Long-Range Atmospheric Transport of Heavy Metals from Industries of Ural and Norilsk

To study long-range atmospheric transport of heavy metals (HM) – Ni, Cu, Pb – from the Ural and Norilsk industries to Siberian environment, the decadal (1980s, 1990s, and 2000s; cf., Table 8.1) averaged HM concentrations in the air, averaged fluxes onto the surface, as well as seasonal and long-term variations were estimated. Two key points of approach used (Vinogradova 2000; Vinogradova et al. 2008a) are: (1) an assumption of small size of sources compared with transport distances and (2) analyses of multiyear arrays of air mass transport trajectories. Atmospheric

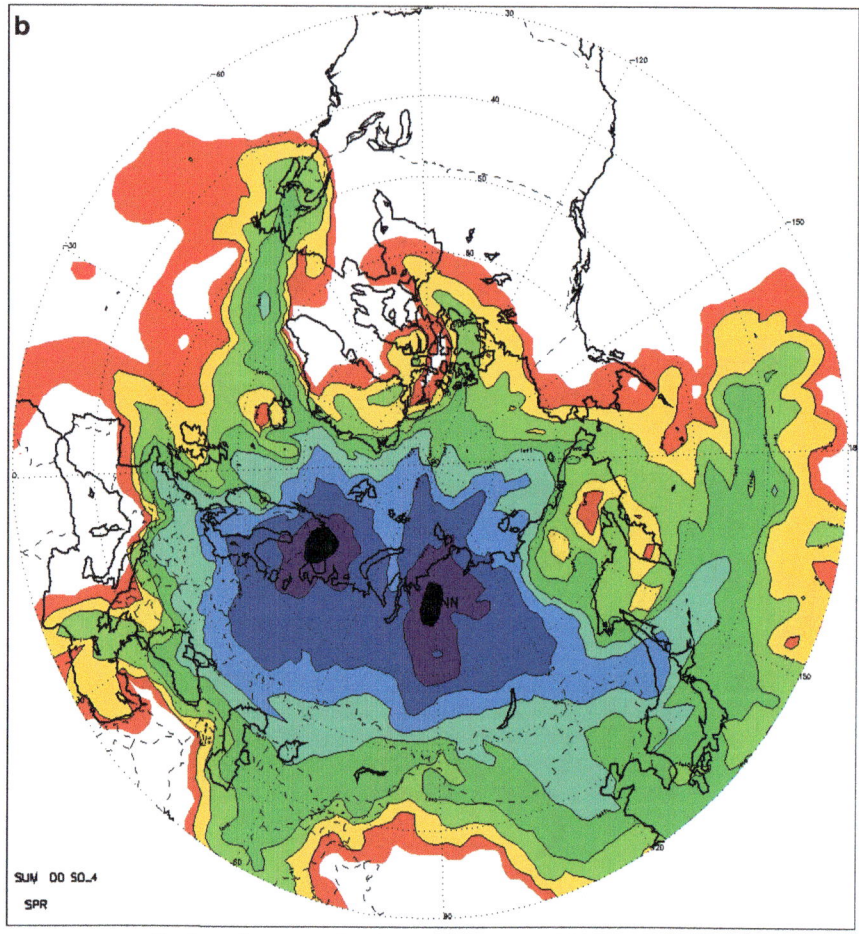

Spring : Summary Dry Deposition (DD) of Sulfates [μg/m²]

Fig. 8.13 (continued)

inputs of anthropogenic HM from both the Ural and Norilsk sources to the Lake Baikal, Kara and Laptev Seas, and to basins of the large Siberian rivers (Ob, Yenisei, and Lena) were estimated and compared with annual fluxes moving by river waters. For simplicity, the vast territory of Siberia was divided into eight zones (Fig. 8.14) with different climatic and soil regimes. The deposition and precipitation parameters for these zones are summarized in Table 8.2. The changes in precipitation pattern over Northern Eurasia during the last 15 years (Kitaev et al. 2004) did not show large trends that could lead to dramatic changes in our estimates.

The forward trajectories with a duration of 5 days were computed for selected locations: 53°N, 58°E (Ural 1) and 57°N, 61°E (Ural 2) as for a line source, and 69°N, 88°E (Norilsk) as for a point source. The Hybrid Single Particle Lagrangian

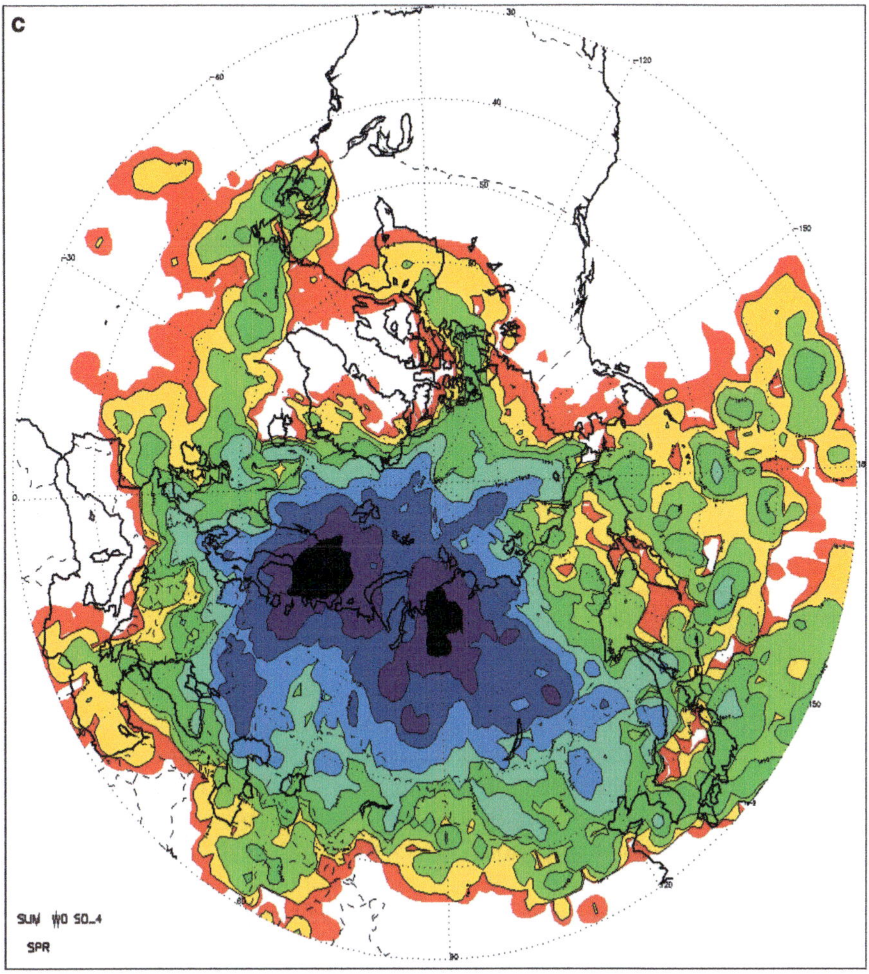

Spring : Summary Wet Deposition (WD) of Sulfates [μg/m²]

Fig. 8.13 (continued)

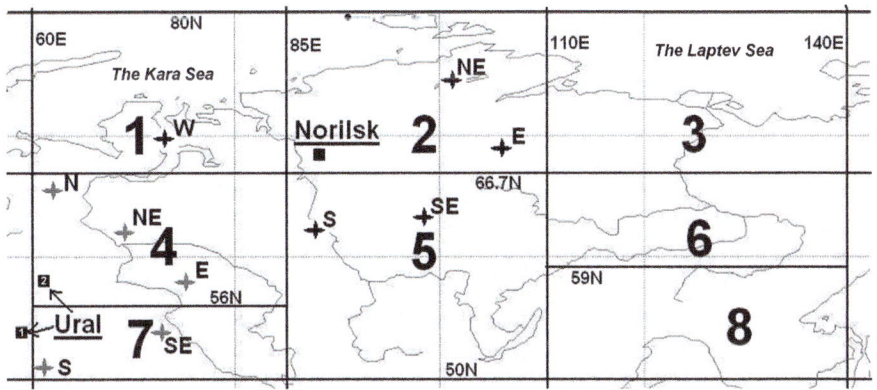

Fig. 8.14 Climatic zoning of Siberian territory (– positions of sources; *black* and *gray stars* – selected points for estimates near the Norilsk and Ural sources, respectively)

Table 8.2 Monthly variability of averaged deposition characteristics of heavy metals for selected climatic zones (see Fig. 8.14)

Zone number	1	2	3	4	5	6	7	8
Month	*Average monthly precipitation amount, mm/month*							
Jan	15	15	15	20	15	10	10	5
Apr	20	15	20	25	20	20	15	10
Jul	70	50	50	80	80	80	60	100
Oct	60	30	30	60	50	50	30	20
Month	*Average HM deposition velocity onto the surface K, cm/s*							
Jan	0.10	0.09	0.09	0.14	0.15	0.13	0.14	0.11
Apr	0.16	0.12	0.12	0.26	0.39	0.38	0.34	0.27
Jul	0.96	0.73	0.73	1.74	1.74	1.10	1.32	1.34
Oct	0.36	0.17	0.18	0.84	0.72	0.70	0.70	0.37
Month	*Average share of wet deposition, %*							
Jan	18	7	7	16	19	15	28	9
Apr	24	15	19	38	59	61	51	44
Jul	84	79	79	89	89	84	88	87
Oct	64	34	33	82	81	83	83	62

Integrated Trajectory Model (HYSPLIT-4; http://www.arl.noaa.gov/ready; Draxler and Rolph 2003) and NCEP/NCAR Reanalysis Data Files were used for simulations. Calculations were done for 925 and 850 hPa levels for all trajectories started at 00 UTC (with calculating interval of 1 h). In total, trajectories were computed for every day in January, April, July, and October (assuming that these months are the most representative for corresponding seasons) during the last three decades. The decadal averaged spatial distributions of trajectories and pollution characteristics from each source for selected months were calculated in each cell of spatial grid ($1° \times 1°$) and mapped. Assuming equal duration of seasons, annual means and totals were calculated, and long-term variations were assessed in seasonal and annual values by comparing results for selected decades.

Atmospheric transport of heavy metals from the Ural and Norilsk sources was evaluated using approach by Vinogradova (2000), Galperin et al. (1995), and Rovinskii et al. (1994). For simplicity it has been assumed that about 20 % of emissions have deposited onto the surface near (within a range of 30–50 km) the sources, and 80 % have been involved into the long-range atmospheric transport. Pollutants are assumed to be vertically uniformly distributed inside the mixing layer H of about 1 km height. In cold seasons, this height is determined by surface air temperature inversion. The longer the time pollution moves along a calculated trajectory, the lesser amount will remain in the air. It is assumed also that the remaining amount is nearly proportional to $\exp(-Kt/H)$, where K is the deposition velocity, and t is the real time of air transport from the source.

The deposition velocities (sum of dry and wet components: $K = K_D + K_W$) onto the surface are assumed to be the same for three metals because they are transported for long distances mainly as submicron aerosol particles (Galperin et al. 1995), and they have seasonal and spatial variations in different climatic zones (Table 8.2).

The HM air concentrations were compared with measurements in different Siberian sites. Near Norilsk, during 1990s the HM variability was very high (Belan et al. 2007): for Ni – from 8 to 50, for Cu – from 5 to 500, and for Pb – from 1 to 10 ng m^{-3}. Therefore, their mean estimated values (28, 40, and 1.2 ng m^{-3} for Ni, Cu, and Pb, respectively) in the last decade are not surprising and justify the applicability of modeling.

Air concentrations and fluxes onto the surface at selected points (placed at distances of 800–1,300 km around two sources, see Fig. 8.14) are shown in Fig. 8.15. For each point the seasonal variation in concentrations and depositions is a result of long-range air mass transport from the sources and regional conditions influencing deposition velocities. The summer minimal (winter maximal) concentrations are connected with high (low) efficiency of deposition processes (i.e., precipitation influences the deposition velocities; cf., Table 8.2). Higher fluxes onto the surface are often observed in spring and autumn as a result of higher precipitation (mainly in a liquid form) in combination with still (or already) high concentrations of heavy metals in the air. The territories near Ural are characterized by very similar climatic conditions with almost similar deposition velocities (as seen in Table 8.2). Thus, the HM fluxes onto the surface are approximately proportional to concentrations at different points and decades (Fig. 8.15a, b). But near Norilsk the situation is different (Fig. 8.15 c, d). There is much more variability in climatic characteristics and deposition velocities, in particular, in latitudinal direction (Table 8.2).

Figure 8.16 illustrates spatial distribution of anthropogenic loading (HM fluxes onto the surface) from Ural and Norilsk sources on the Siberian environment. Evidently, fluxes are the highest near the sources providing maximal anthropogenic loading, and at larger distances the influence is weaker. The Pb spatial distribution (Fig. 8.16a) is typical for transport by air masses from the Ural region. On the contrary, the Ni distribution (Fig. 8.16b) shows domination of atmospheric transport from the Norilsk region. The Cu distribution is more complex (Fig. 8.16c). Approximately equal inputs can be identified in West Siberia. The area of Lake Baikal is more polluted from the Norilsk than from the Ural sources.

Long-term variations in anthropogenic loading through the atmosphere to the surface environment are determined by variations of the following main factors: (1) sources' emissions, (2) atmospheric circulation processes, and (3) atmospheric precipitation that affect deposition velocities of pollutants. In general, atmospheric circulation over the Northern Eurasia has significant spatial and temporal variability, and hence, pollution may also have significant month-to-month variations (as it can be seen in Fig. 8.17 for different months).

The HM input from the atmosphere onto the surface of the Kara and Laptev Seas and onto the Ob, Yenisei, and Lena River Basins were computed using the data on annual flow and water composition (Gordeev 2002) and compared with HM fluxes moving with rivers' waters into the Arctic Ocean. The fluxes incoming to the Kara and Laptev Seas from both sources through the atmosphere are comparable with those brought by the river inflow (Fig. 8.17). Although HM concentrations in the air, water, and soil within the Arctic region do not exceed maximum permissible concentrations (MPC), the detected amounts of Pb and Cd in muscles, livers, and

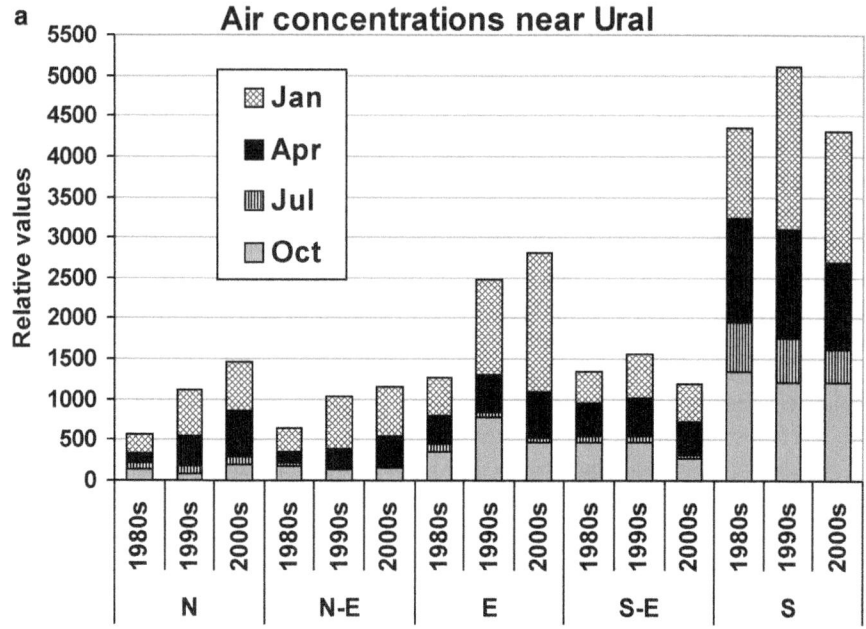

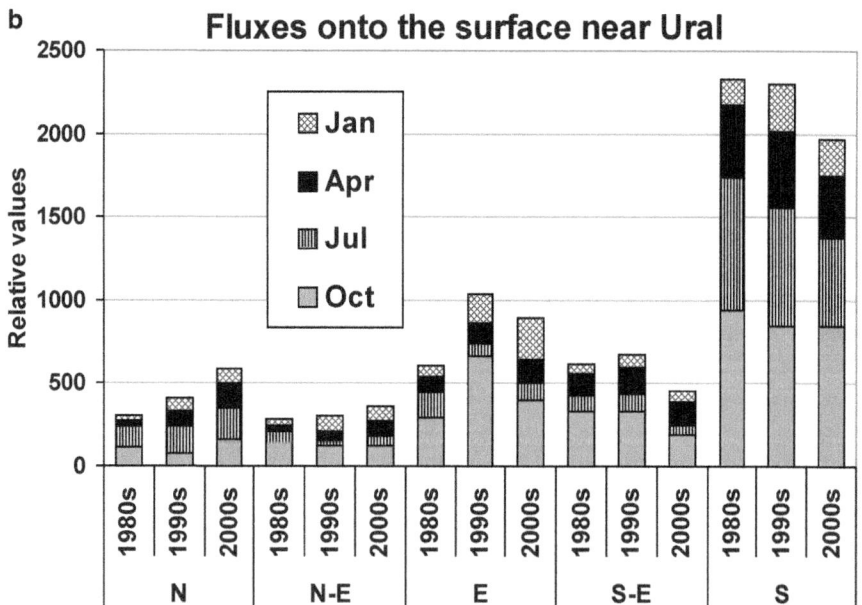

Fig. 8.15 Long-term (decadal – 10-year) variations of averaged relative values of heavy metals (**a**, **c**) air concentrations and (**b**, **d**) fluxes onto the surface for selected months at points (*W*, *NE*, *E*, *SE*, *S*) surrounding the Ural and Norilsk sources

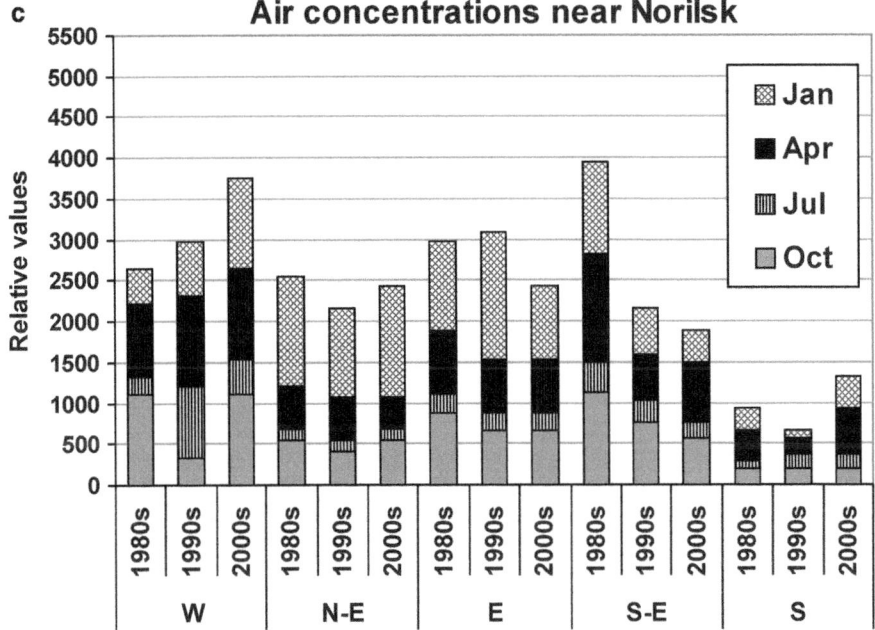

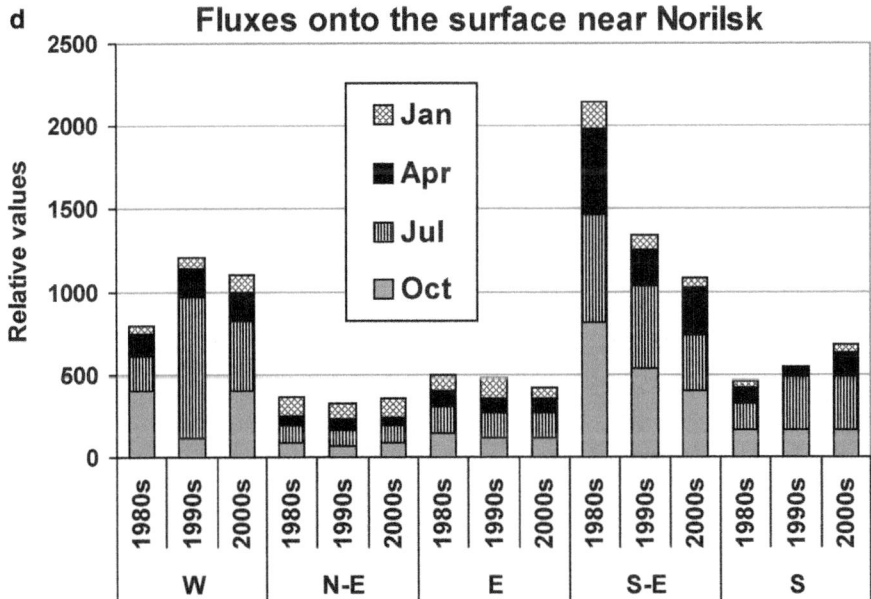

Fig. 8.15 (continued)

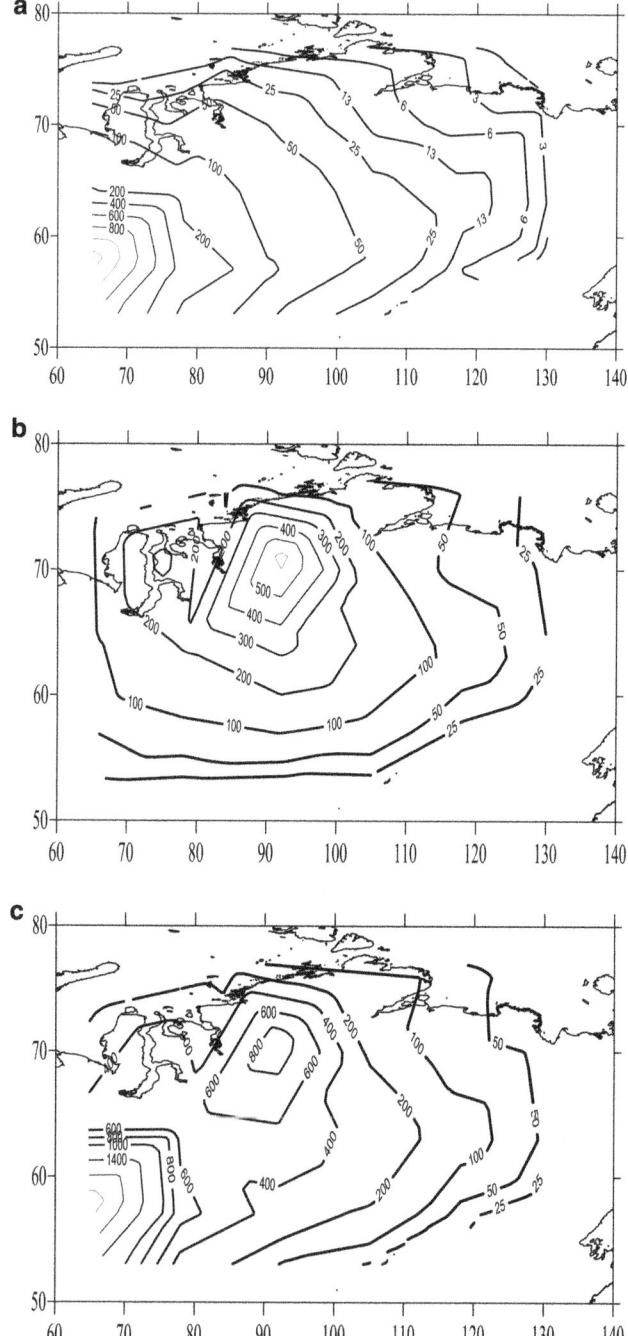

Fig. 8.16 Spatial distribution of decadal (2000s) averaged fluxes [g/km^2/year] of (**a**) Pb, (**b**) Ni, and (**c**) Cu from the Ural and Norilsk sources due to atmospheric transport and deposition onto the surface

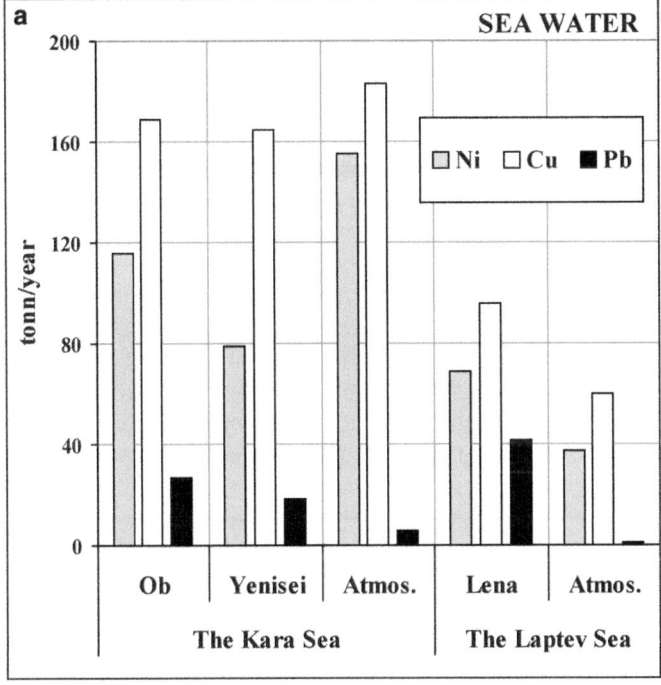

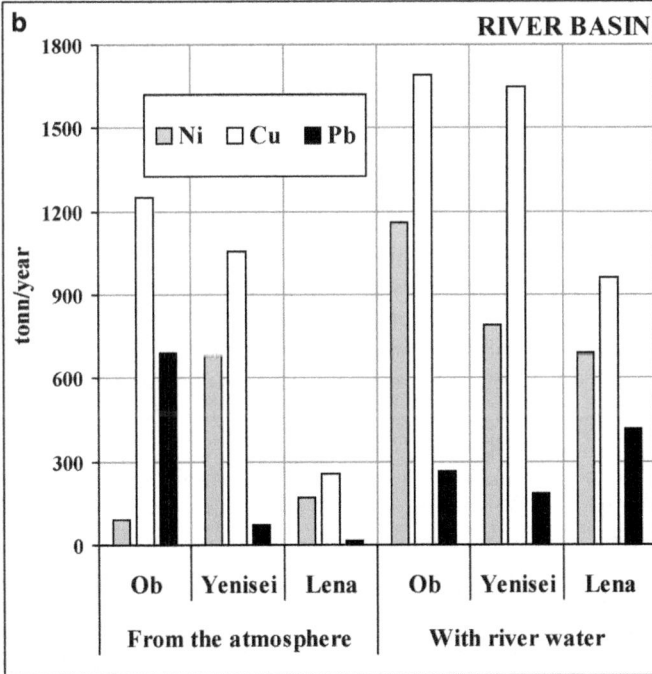

Fig. 8.17 Heavy metals (Ni, Cu, Pb) fluxes in 1990s coming from the Ural and Norilsk sources through the atmosphere (**a**) to central waters of the Kara and Laptev Seas vs. fluxes inflowing by Ob, Yenisei, and Lena rivers' waters after coastal zone, and (**b**) onto river basins vs. fluxes bringing by rivers' waters to the Arctic Ocean

kidneys of Arctic reindeers, predatory animals, and birds have exceeded MPC (Review 2007; Persistent 2004), and hence, the native population has to consume contaminated food. As for long-term variations in atmospheric transport from these two sources onto the Siberian river basins, a small increase of Pb fluxes to the Lena and Yenisei rivers has been observed in the 2000s compared to the 1990s (although source emissions have decreased during the last three decades). This is a good example of the influence of atmospheric circulation variations on resulting patterns of nonequal distribution in the atmosphere and on the surface. It shows that the effects of atmospheric circulation can be even larger than the effects of decreasing emissions.

At present, the HM migration and their transformation within the surface ecosystems still need to be further evaluated in order to more accurately estimate their impact. There are many relevant properties of HM (chemical form, mobility, solubility, toxicity, etc.) as well as habitat characteristics (temperature, water hardness, soil type, etc.) which determine complex behavior and migration ways of heavy metals in the Earth system as well as their impact on the population health.

8.4 Conclusions

Despite the atmospheric environment over Siberia being relatively clear in comparison with neighboring regions of Asia and Eastern Europe, air pollution from industrial centers in Siberia poses observable environmental threats. Siberian ecosystems have begun to show stress from the accumulation of pollution depositions that come from cities and industrial plants. Urban air quality in several Siberian cities (e.g., Norilsk, Barnaul, and Novokuznetsk) is considered among the worst among Russian and European cities.

The main disadvantages of the existing national system of atmospheric monitoring in Siberia are deterioration of measuring instruments, insufficient development of the maintenance capabilities that would allow to upkeep and control the quality of measurements, lack of continuous observations for a large fraction of its territory, insufficient equipment with means of data processing and transmission, and its incompatibility with the existing international observation networks. At present, reconstruction of the Russian System of Atmospheric Monitoring (RSAM) is on the way. It includes both the modernization of research equipment and a broader use of remote sensing methods with prospects that in the future RSAM will be a part of the Integrated Global Earth Observing System.

Existing technology and methods provide the possibility to receive estimates of risk/vulnerability for different objects and regions of Siberia. Various direct and inverse modeling tools have been applied here to diagnosing, designing, monitoring, and forecasting of air pollution. Special approaches have been designed for sparsely populated Siberian and Arctic areas with restricted amount of observational and emission data. In particular, regional models coupled with the global atmospheric models using the orthogonal decomposition method allows correct transfer (downscaling) of the global processes information at the regional level for further use in the environmental quality control analyses.

Analyses of the atmospheric transport and deposition of sulfur and heavy metals (Ni, Cu, and Pb) from the industries of Norilsk and Ural regions over the territory of Siberia provided in this Chapter can be used for assessment of anthropogenic influence of these two source regions on the environment of Siberia as well as beyond the Siberian boundaries. Analysis of spatial, seasonal, and long-term variations in heavy metals concentrations in near-the-surface air and their fluxes onto the surface shows that maximum air concentrations are in the cold part of a year, whereas the maximum fluxes onto the surface occur during the warm season. Air pollution level is not the single factor that determines the ecological state of environment, but it also depends on deposition processes. It is possible that the cleaner air will not guarantee the safer surface environment.

Presented maps of spatial distributions of heavy metals fluxes onto the surface may be used for estimation of the anthropogenic loading of Siberian environment from two selected sources (Ural and Norilsk). In general, the pollution from these sources has been decreasing during the last three decades, but the spatial variations are sizeable. Heavy metals (HM) fluxes incoming per year from these two anthropogenic sources through the atmosphere to the Kara and the Laptev Seas are quite comparable with fluxes delivered there by rivers across the coastal zone. Moreover, for three largest Siberian rivers (Ob, Yenisei, and Lena), such atmospheric fluxes of HM into the river basins are also comparable with the HM fluxes brought by these rivers to the Arctic Ocean. This implies that atmospheric transport of heavy metals should be taken into account in analyses of HM pollution for both, the marine and surface ecosystems.

The surface ozone concentrations distribution over Northern Eurasia may be somewhat influenced by the significant methane emission from the industrial and natural sources. A comparison of the observational data on the methane isotopic composition with the results of numerical simulation (applying the values of emissions of different isotopic compositions) showed that the CH_4 sources of West Siberia are underestimated. The inventory of the CH_4 sources over Russia should be revised. Specifically, an increase in the emissions from "light" sources (swamps, cud-chewing animals, waste recycling, etc.) and a decrease in the emissions from "heavy" sources (extraction and processing of oil, gas, and coal, biomass burning, etc.) are suggested.

It is suggested t"that the developed and tested methodology for risk assessment using dispersion modelling results can be successfully applied for other areas of concern such as chemical, biological, and natural hazards, for assessments of long-term impacts from existing emission/release sources of different kinds of pollutants as well as for environmental problems of wider spectra. In particular, for the sites of chemical and nuclear risk worldwide, the long-term and short-term modeling of atmospheric transport and deposition patterns can be simulated and evaluated on different temporal and spatial scales.

References

1989 annual state of air pollution and pollutant emissions into the atmosphere of towns and industrial centers of the Soviet Union, vol "Pollutant Emissions". Leningrad, 1990, 486 pp (in Russian)

Ankilov A et al (2002a) Particle size dependent response of aerosol counters. Atmos Res 62:209–237

Ankilov A et al (2002b) Intercomparison of number concentration measurements by various aerosol particle counters. Atmos Res 62:177–207

Arshinov MYu, Belan BD (2011) New particle formation events in the Siberian boreal zone. Geophys Res Abstr 13, EGU2011–5582–2, EGU General Assembly 2011

Baklanov A (2000) Modeling of the atmospheric radionuclide transport: local to regional scale. Numer Math Math Model, RAS Inst Num Math, Moscow 2:244–266

Baklanov A (2007) Environmental risk and assessment modeling: scientific needs and expected advancements. In: Ebel A, Davitashvili T (eds) Air, water and soil quality modeling for risk and impact assessment, Security through science, series C – environmental security. Springer/Elsevier Publishers, Dordrecht, pp 29–44

Baklanov AA, Co-Authors (2008a) In: Baklanov A, Gordov E (eds) Enviro-RISKS: man-induced environmental risks: monitoring, management and remediation of man-made changes in Siberia. DMI Scientific Report 08–05 in 4 volumes: vol 1 Atmoispheric pollution and risk; vol 2 Climate and global change and risks; vol 3 Terrestrial ecosystems and hydrology; vol 4 Information systems, integration and synthesis. ISBN: 978–87–7478–571–2 (to be published as a Springer book)

Baklanov A, Sørensen JH, Mahura A (2008b) Methodology for probabilistic atmospheric studies using long-term dispersion modelling. Environ Model Assess 13:541–552. doi:10.1007/s10666-007-9124-4

Bashurova VS, Dreiling V, Hodger TV, Jaenicke R, Koutsenogii KP, Koutsenogii PK, Kraemer M, Makarov VI, Obolkin VA, Potjomkin VL, Pusep AY (1992) Measurements of atmospheric condensation nuclei size distributions in Siberia. J Aerosol Sci 23:191–199

Belan BD (2010) Ozone in the troposphere. Institute of Atmospheric Optics of Siberian Branch of the Russian Academy of Sciences, Tomsk, 487 pp

Belan BD, Zadde GO, Ivlev GA et al (2007) Complex assessment of the conditions of the air basin over Norilsk industrial region. Part 5. Impurities in the atmospheric boundary layer. The correspondence of air composition to hygienic norms. Recommendations. Atmos Ocean Opt 20(6):119–129 (Engl. Transl)

Belan BD, Tolmachev GN, Fofonov ÀV (2010) Vertical ozone distribution in troposphere above south regions of West Siberia. Atmos Ocean Opt 23:777–783

Bellouin N, Boucher O, Haywood J, Reddy MS (2005) Global estimate of aerosol direct radiative forcing from satellite measurements. Nature 438:1138–1141

Benkovitz CM, Scholtz MT, Pacyna J, Tawasou L, Dignon J, Voldner EC, Spiro PA, Logan JA, Graedel TE (1996) Global gridded inventories of anthropogenic emissions of sulfur and nitrogen. J Geophys Res 101:29239–29252

Bezuglaya EYu (ed) (1999) Air quality in largest cities of Russia for 10 years (1988–1997). Hydrometeoizdat, St-Petersburg, 146 p (in Russian)

Bondur VG (2010) Importance of aerospace remote sensing approach to natural fires monitoring in Russia. Vestnik Otd. Nauk o Zemle RAS, Moscow, 2, NZ11001. doi:10.2205/2010NZ000062

Chen S-J, Kuo Y-H, Zhang P-Z, Bai Q-F (1991) Synoptic climatology of ciclogenesis over East Asia, 1958–1987. Mon Wea Rev 119:1407–1418

Chung Y-S, Hage KD, Reinelt ER (1976) On lee cyclogenesis and airflow in the Canadian Rocky mountains and the East Asian mountains. Mon Wea Rev 104:879–891

Chung CE, Ramanathan V, Kim D, Podgorny IA (2005) Global anthropogenic aerosol direct forcing derived from satellite and ground-based observations. J Geophys Res 110:D24207. doi:10.1029/2005JD006356

Colwell R (ed) (1983) Manual of remote sensing, vol 1. American Society of Photogrammetry, Falls Church

Dal Maso M et al (2008) Aerosol particle formation events at two Siberian stations inside the boreal forest. Boreal Environ Res 13(2):81–92

Derome J, Lukina N (2011) Interaction between environmental pollution and land-cover/land-use change in Arctic areas. In: Gutman G, Reissell A (eds) Eurasian arctic land cover and land use in a changing climate, VI. Springer, Amsterdam, pp 269–290, 306 pp

Draxler RR, Rolph GD (2003) HYSPLIT (HYbrid Single-Particle Lagrangian Integrated Trajectory) Model, NOAA ARL READY Website: http://www.arl.noaa.gov/ready/hysplit4.html

Ekstrand S (1993) Assessment of forest damage using Landsat TM, elevation models and digital forest maps. PhD thesis, Royal Institute of Technology, Stockholm

Elansky NF (2004) Atmospheric monitoring: Russian contribution. Sci Russ 6:20–26

Elansky NF (2009a) Russian studies of the atmospheric ozone in 2003–2006. Izv Atmos Ocean Phys 45:218–231

Elansky NF (ed) (2009b) Atmospheric composition observations over Northern Eurasia using the mobile laboratory TROICA experiments. The International Science and Technology Center, Moscow, 74 p. ISBN 978-5-904610-03-6

Elansky NF, Belikov IB, Golitsyn GS, Grisenko AM, Lavrova OV, Pankratova NV, Safronov AN, Skorokhod AI, Shumckii RA (2010) Observations of the atmosphere composition in the Moscow megapolis from a mobile laboratory. Doklady Earth Sci 432:649–655

Galperin MV, Sofiev M, Gusev A et al (1995) Approaches to simulation of heavy metals transboundary pollution of the European atmosphere. Report 7/95. EMEP/MSC East, Moscow, 85 pp (in Russian)

GGO (2009) Air quality in large cities of Russia during ten years 1998–2007. Analytical review. Voeikov Main Geophysical Observatory, RosHydroMet, St. Petersburg, 133 pp. ISBN 978-5-94856-583-5

Gordeev V (2002) Heavy metals in the Russian Arctic Rivers and some of their estuaries: concentrations and fluxes. In: Pacyna JM (ed) Proceedings of the AMAP workshop on sources, emissions and discharges, Kjeller, Norway. NILU OR 3/2002, pp 79–100

Gromov S (2008) The impact of Russian coal-burning power plants in Siberia on atmospheric environment and their possible role in the long range transport of sulphur. Institute of Global Climate and Ecology, Roshydromet and RAS. Presented at the conference on trans-boundary air pollution in Northeast Asia, 17 Dec 2008

Gutman G, Reissell A (eds) (2011) Eurasian arctic land cover and land use in a changing climate, VI. Springer, Amsterdam, 306 pp

Häme T (1991) Spectral interpretation of changes in forest using satellite scanner images. Acta For Fenn 222:111

Haywood J, Boucher O (2000) Estimates of the direct and indirect radiative forcing due to tropospheric aerosols: a review. Rev Geophys 38:513–543

Jaffe D, Bertschi I, Jaegle L, Novelli P, Reid JS, Tanimoto H, Vingarzan R, Westphal DL (2004) Long-range transport of Siberian biomass burning emissions and impact on surface ozone in western North America. Geophys Res Lett 31:L16106. doi:10.1029/2004GL020093

Kalnay E, Kanamitsu M, Kistler R et al (1996) The NCEP/NCAR 40-year reanalysis project. Bull Am Meteor Soc 77:437–471

Kämäri J et al (1998) Acidifying pollutants, Arctic haze, and acidification in the Arctic. In: Wilson SJ, Murray JL, Huntington HP (eds) AMAP assessment report: Arctic pollution issues. Arctic Monitoring and Assessment Programme, Oslo, Norway, pp 621–659

Kasahara A (1980) Influence of orography on the atmospheric general circulation. GARP Publ Ser No 23, pp 7–52

Kaufman YJ, Boucher O, Tanre D, Chin M, Remer LA, Takemura T (2005) Aerosol anthropogenic component estimated from satellite data. Geophys Res Lett 32, Art. No. L17804. doi:10.1029/2005GL023125

Kharuk VI (1998) O razrabotke GIS tekhnogennikh vozdeistviy na lesa Sibiri. Sibirskii Ekologicheskiy Zhurnal 5:25–30 (in Russian)

Kitaev LM, Radionov VF, Forland E et al (2004) Duration of snow cover in Northern Eurasia in conditions of present climate change. Russ Meteorol Hydrol 11:65–72

Konovalov IB, Beekmann M, Kuznetsova IN, Yurova A, Zvyagintsev AM (2011) Atmospheric impacts of the 2010 Russian wildfires: integrating modelling and measurements of the extreme air pollution episode in the Moscow megacity region. Atmos Chem Phys Discuss 11: 12141–12205

Koutsenogii PK (1997) Aerosol measurements in Siberia. Atmos Res 44:167–173

Koutsenogii P, Jaenicke R (1994) Number concentration and size distribution of atmospheric aerosol in Siberia. J Aeros Sci 25:377–383

Kravtsova VI (1999) Izucheniye promyshlennogo vozdeistviya na severnuja rastitelnost' po kosmicheskim snimkam: trudnosti i nereshennye problemy [Investigation of industrial damage on northern vegetation using satellite images: difficulties and unresolved problems]. Issledovanie Zemli iz Kosmosa 1:112–121 (in Russian)

Lauritzen B, Baklanov A, Mahura A, Mikkelsen T, Sørensen JH (2007) Probabilistic risk assessment for long-range atmospheric transport of radionuclides. J Environ Radioact 96:110–115

Litinksy P (1996) Assessment of boreal pine forests decline around ore/dressing mill using satellite images. Russian Academy of Sciences, Petrozavodsk (in Russian)

Mahura A, Baklanov A (2003) Assessment of possible airborne impact from nuclear risk sites. Part II: probabilistic analysis of atmospheric transport patterns in Euro-Arctic region. Atmos Chem Phys Discuss 3:5319–5356

Mahura AG, Baklanov A, Sørensen JH, Parker FL, Novikov V, Brown K, Compton KL (2005) Assessment of potential atmospheric transport and deposition patterns due to Russian Pacific fleet operations. Environ Monit Assess 101:261–287

Mahura A, Baklanov A, Sørensen JH (2006a) Long-term dispersion modelling. Part II: assessment of atmospheric transport and deposition patterns from nuclear risk sites in Euro-Arctic region. J Comput Technol 10:112–134

Mahura A, Baklanov A, Sørensen JH (2006b) Influence of long-range and long-term continuous and accidental anthropogenic emissions from Eurasian sources on Greenland environment. In: Proceedings of the international conference "The Greenlandic environment: pollution and solutions", 21–23 Feb 2006, Sisimiut in Greenland, pp 38–44

Mäkela JM, Aalto P, Jokinen V, Pohja T, Nissinen A, Palmroth S, Markkannen T, Seitsonen K, Lihavainen H, Kulmala M (1997) Observations of ultrafine aerosol particle formation and growth in boreal forest. Geophys Res Lett 24:1219–1222

McGuire A, Apps M, Chapin FS III, Dagraville R, Flannigan M, Kasischke E, Kicklighter D, Kimball J, Kurz W, McRae D, McDonald K, Melilli J, Myneni R, Stocks B, Verbyla D, Zhuang Q (2004) Land cover disturbances and feedbacks to the climate system in Canada and Alaska. In: Gutman G, Janetos A, Justice C, Moran E, Mustard J, Rindfuss R, Skole D, Turner B II, Cochrane M (eds) Land change science. Observing, monitoring and understanding trajectories of change on the earth surface. Kluwer Academic Publishers, Dordrecht, pp 139–163

Mel'nikov YO, Rzhanitsyn PV, Yakovlev AO (1996) Geologo-ekologicheskoye kartirovaniye masshtaba 1:1000000 Noril'skaya rayona list R-45-V.G. RAO Norilsky Nikel, Noril'sk

National estimate for the development of the Russian Federation on the way to gradual development (2002) Russian Ministry of Economy Development, Moscow, 46 pp

Pacyna JM, Ottar B, Tomza U, Maenhaut W (1985) Long-range transport of trace elements to Ny-Alesund, Spitsbergen. Atmos Environ 19(6):857–864

Pankratova NV, Elansky NF, Belikov IB, Lavrova OV, Skorokhod AI, Shumcky RA (2011) Ozone and nitrogen oxides in the surface air over Northern Eurasia from the observations in TROICA experiments. Izv Atmos Ocean Phys 17(3):343–358

Penenko VV (1981) Methods of numerical modelling of atmospheric processes. Gidrometeoizdat, Leningrad, 352 pp (in Russian)

Penenko VV (2009) Variation methods of data assimilation and inverse problems for studying the atmosphere, ocean, and environment. Numer Anal Appl 2(4):341–351

Penenko VV (2010) On a concept of environmental forecasting. Atmos Ocean Opt 23(6):432–438

Penenko VV, Tsvetova EA (1999) Mathematical models for the study of interactions in the system Lake Baikal–atmosphere of the region. J Appl Mech Tech Phys 40(2):308–316

Penenko V, Tsvetova E (2007) Mathematic models of environmental forecasting. J Appl Mech Tech Phys 48(3):428–436

Penenko V, Tsvetova E (2008) Orthogonal decomposition methods for inclusion of climatic data into environmental studies. Ecol Model 217:279–291

Penenko V, Tsvetova E (2009a) Discrete-analytical methods for the implementation of variational principles in environmental applications. J Comput Appl Math 226:319–330

Penenko VV, Tsvetova EA (2009b) Optimal forecasting of natural processes with uncertainty assessment. J Appl Mech Tech Phys 50(2):300–308

Penenko V, Baklanov A, Tsvetova E (2002) Methods of sensitivity theory and inverse modelling for estimation of source term. Future Gener Comput Syst 18:661–671

Persistent toxic substances, Food security and indigenous peoples of the Russian North (2004) Final report. AMAP, Oslo, Norway, 192 pp

Pitblado JR, Amiro BD (1982) Landsat mapping of the industrially disturbed vegetation communities of Sudbury, Canada. Can J Remote Sens 8:16–29

Rahn KA, Lowenthal DH (1984) Elemental tracers of distant regional pollution aerosols. Science 223:132–139

Rees WG, Rigina O (2003) Methodologies for remote sensing of the environmental impacts of industrial activity in the Arctic and Sub-arctic. In: Rasmussen RO, Koroleva NE (eds) Social and environmental impacts in the North. Kluwer, Dordrecht, pp 67–88

Review of environmental state and pollution in Russian Federation in 2006 (2007) Hydrometeoizdat, St.-Petersburg, 162 pp (in Russian)

Rovinskii FYa, Gromov SA, Burtseva LV et al (1994) Heavy metals: long-range transport in the atmosphere and their deposition. Russ Meteorol Hydrol 1994(10):5–14

Shvidenko A, Goldhammer J (2001) Fire situation in Russia. Int Fire News 23:49–65

Skorokhod A, Verkhovets S (2006) Study of reactive atmospheric constituents and of ecosystem parameters in the area of Zotino tall tower (Central Siberia). In: Proceedings of international workshop ISTC "Baikal-2006", 15–19 Aug 2006, Irkutsk, Russia. Publishing house "V-Spectr", Tomsk, pp 79–83

Sofiev M, Vankevich R, Lotjonen M, Prank M, Petukhov V, Ermakova T, Koskinen J, Kukkonen J (2009) An operational system for the assimilation of satellite information on wild-land fires for the needs of air quality modelling and forecasting. Atmos Chem Phys 9:6833–6847. http://www.atmos-chem-phys.net/9/6833/2009/acp-9–6833–2009.html

Sofiev M, Prank M, Baklanov A (eds) (2011) Influence of regional scale emissions on megacity air quality. Deliverable D5.5, MEGAPOLI scientific report 11–12, MEGAPOLI-38-REP-2011–06, 28p, ISBN: 978–87–92731–16–6. http://megapoli.dmi.dk/publ/MEGAPOLI_sr11–12.pdf

Sokolik IN, Curry J, Radionov V (2011) Interactions of Arctic aerosols with land-cover and land-use changes in Northern Eurasia and their role in the Arctic climate system. In: Gutman G, Reissell A (eds) Eurasian Arctic land cover and land use in a changing climate, VI. Springer, Amsterdam, pp 237–268, 306 pp

Sørensen JH, Baklanov A, Hoe S (2007) The Danish emergency response model of the atmosphere. J Environ Radioact 96:122–129

Tarasova OA, Brenninkmeijer CAM, Assonov SS, Elansky NF, Röckmann T, Sofiev MA (2007) Atmospheric CO along the trans-Siberian railroad and river Ob: source identification using isotope analysis. J Atmos Chem 57:135–152

Tarasova OA, Senik IA, Sosonkin MG, Cui J, Staehelin J, Prevot ASH (2009a) Surface ozone at the Caucasian site Kislovodsk High Mountain Station and the Swiss Alpine site Jungfraujoch: data analysis and trends (1990–2006). Atmos Chem Phys 9(12):4157–4175

Tarasova OA, Houweling S, Elansky N, Brenninkmeijer CAM (2009b) Application of stable isotope analysis for improved understanding of the methane budget: comparison of TROICA measurements with TM3 model simulations. J Atmos Chem 63:49–71

Timkovsky I, Elansky NF, Skorokhod AI, Shumskiy RA (2010) Studying of biogenic volatile organic compounds in the atmosphere over Russia. IZV Atmos Ocean Phys 46:319–327

Timofeeva AA, Latysheva IV, Potemkin VL (2008) Dynamics of lightning activity and its influence on ozone variations in Baikal region. Proc IRGTU 2(34):24–27

Toutoubalina OV, Rees WG (1999) Remote sensing of industrial impact on Arctic vegetation around Noril'sk, Northern Siberia: preliminary results. Int J Remote Sens 20:2979–2990

Toutubalina OV, Rees WG (2001) Vegetation degradation in a permafrost region as seen from space: Noril'sk, 1961–1999. Cold Reg Sci Technol 32:191–203

Turnbull JC, Miller JB, Lehman SJ, Hurst D, Tans PP, Southon J, Montzka S, Elkins J, Mondeel DJ, Romashkin PA, Elansky N, Skorokhod A (2009) Spatial distribution of $^{14}CO_2$ across Eurasia: measurements from the TROICA-8 expedition. Atmos Chem Phys 9:175–187

Vartiainen E, Kulmala M, Ehn M, Hirsikko A, Junninen H, Petäjä T, Sogacheva L, Kuokka S, Hillamo R, Skorokhod A, Belikov I, Elansky N, Kerminen V-M (2007) Ion and particle number concentrations and size distributions along the Trans-Siberian railroad. Boreal Environ Res 12:375–396

Vinogradova AA (2000) Anthropogenic pollutants in the Russian Arctic atmosphere: sources and sinks in spring and summer. Atmos Environ 34(29–30):5151–5160

Vinogradova AA, Maksimenkov LO, Pogarskii FA (2008a) Atmospheric transport of anthropogenic heavy metals from the Kola Peninsula to the surfaces of the White and Barents seas. Izv Atmos Ocean Phys 44(6):753–762 (Engl. Transl)

Vinogradova AA, Maksimenkov LO, Pogarsky FA (2008b) The influence of Norilsk and Ural industry on the environment of different Siberian regions. Atmos Ocean Opt 21(6):415–420 (Engl. Transl)

Vivchar AV, Moiseenko KB, Shumskii RA, Skorokhod AI (2009) Identifying anthropogenic sources of nitrogen oxide emissions from calculations of lagrangian trajectories and the observational data from a tall tower in Siberia during the spring–summer period of 2007. Izv Atmos Ocean Phys 45:325–336

Vivchar AV, Moiseenko KB, Pankratova NV (2010) Estimates of carbon monoxide emissions from wildfires in Northern Eurasia for Air quality assessment and climate modeling. Izv Atmos Ocean Phys 46:281–293

Vlasova TM, Kovalev NI, Filipchuk AN (1990) Effects of point source atmospheric pollution on boreal forest vegetation of Northern Siberia. In: Weller G, Wilson CL, Severin BAB (eds) International conference on the role of the polar regions in global change. University of Alaska-Fairbanks, Fairbanks, pp 423–428

Warner-Merl NK (1998) Air pollution in Siberia: a volume and risk-weighted analysis of a Siberian pollution database. IIASA Interim Report IR-98–059/October

WMO (2007) WMO global atmosphere watch strategic plan: 2008–2015. GAW report 172, 104 pp

Chapter 9
Summary and Outstanding Scientific Challenges for Research of Environmental Changes in Siberia

Garik Gutman, Pavel Ya. Groisman, Evgeny P. Gordov, Alexander I. Shiklomanov, Nikolay I. Shiklomanov, Anatoly Z. Shvidenko, Kathleen M. Bergen, and Alexander A. Baklanov

Abstract This chapter summarizes the volume content focusing on land change in Siberia. The volume is compilation of results of the most recent international studies of Earth's system interactions including biogeochemical and water cycles, natural ecosystems changes, and human impacts on environment. Outstanding scientific challenges are outlined as they were discussed in each chapter.

G. Gutman (✉)
The NASA Land-Cover/Land-Use Change Program, NASA Headquarters, Washington, DC, USA
e-mail: ggutman@nasa.gov

P.Ya. Groisman
NOAA National Climatic Data Center, Asheville, NC, USA

State Hydrological Institute, St. Petersburg, Russian Federation

E.P. Gordov
Siberian Center for Environmental Research and Training and Institute of Monitoring of Climatic and Ecological Systems SB RAS, Tomsk, Russian Federation

A.I. Shiklomanov
Water Systems Analysis Group, Complex Systems Research Center, Institute for the Study of Earth, Ocean, and Space, University of New Hampshire, Morse Hall, Durham, NH, USA

N.I. Shiklomanov
Department of Geography, George Washington University, Washington, DC, USA

A.Z. Shvidenko
Forestry Program, International Institute for Applied Systems Analysis, Laxenburg, Austria

Forestry Institute, Siberian Branch, Russian Academy of Science, Krasnoyarsk, Russian Federation

K.M. Bergen
School of Natural Resources and Environment, University of Michigan, Ann Arbor, MI, USA

A.A. Baklanov
Danish Meteorological Institute, Copenhagen, Denmark

P.Ya. Groisman and G. Gutman (eds.), *Regional Environmental Changes in Siberia and Their Global Consequences*, Springer Environmental Science and Engineering, DOI 10.1007/978-94-007-4569-8_9, © Springer Science+Business Media Dordrecht 2013

9.1 Introduction

In Chap. 1 (Groisman et al. 2012a), we started with a brief description of general bioclimatic features of Siberia that make it a special and important area of the globe. In summary, this region is characterized by continental climate and is expected to experience potentially rapid changes of its land cover from taiga to steppe or even to semidesert under projected climate changes. The region has had a unique storage of carbon in the frozen ground and across the expansive taiga zone that could change (i.e., be burned, or simply be released from thawing permafrost). These changes would create a positive feedback to the global Earth system.

Given the importance of the region and its risks associated with global change, there has been an effort to jointly develop monitoring and information networks. A description of the tools (information systems and their content) developed by Russian, US, EU, and Japanese institutions representing information-computational infrastructure required to study the Siberian region was presented in Chap. 2. (Gordov et al. 2012) Although these tools were directed at the needs of researchers working on Siberia, they can be easily adapted to needs of different regions in the extratropics.

Past climates of the region were briefly discussed in Chap. 3 (Groisman et al. 2012b). In high latitudes, the warmer climates of the past were mostly accompanied by higher humidity, and in the south of Siberia, they could be both wet and dry in different past warm periods. During the past century, the climate of Siberia became much warmer. There is an indication that the cold season precipitation north of 55°N increased unlike rainfall over most of Siberia. A general increase in water demand for evapotranspiration led to drier summer conditions and more frequent extreme dry events. On the other hand, projections of the future climate for the end of the twentieth century indicate further temperature increases, with greater increase in the cold season (by up to +8 °C in high latitudes) and less in the warm season (by up to +5 °C in southern Siberia). Precipitation is also projected to increase but only in the cold season. Changes in hydrology regime (as described in Chap. 4; Shiklomanov et al. 2012) may lead to a significant shift in terrestrial ecosystems accompanied by changes in the regional carbon cycle (Chap. 6; Shvidenko et al. 2012) and changes in the permafrost distribution and stability with potential for detrimental impact on the regional infrastructure (Chaps. 5 and 7; Shiklomanov and Streletskiy 2012 and Bergen et al. 2012, respectively). The summer warming during the period of instrumental observations (past 130 years) occurred only in the past two decades (however, in autumn, winter, and spring, a significant warming in 1930–1960s had been documented, particularly in the Arctic), and the vegetation and permafrost (and therefore, positive climatic feedbacks associated with them) had not yet fully reacted to this warming. Projections suggest that frequencies of various extreme events will further increase.

We further discussed changes in different components of the hydrological regime in Siberia based on long-term observational records and modeled future river runoff projections (Chap. 4). Although the dramatic changes over the last few decades in Siberia have not been uniform, some general patterns have emerged and were briefly analyzed in this volume. The increase of annual river discharge to the Arctic Ocean is very consistent with the rise of regional air temperature. The discharge increase is strongly associated with significant increases in low flow, often (but not exclusively) in regions underlain by permafrost. Warmer Siberian air temperatures have also led to thinning river ice, shortening of ice cover duration, deepening permafrost active layer thickness, and earlier snowmelt and spring floods.

Up-to-date analysis of future projections reveals that annual runoff will significantly increase in basins of large rivers flowing to the Arctic Ocean but will not change in the drier areas of southern Siberia. The most significant runoff increases will be observed during winter and spring periods. A significant shift to earlier snowmelt and earlier spring flood across the entire region is expected. For more reliable regional-scale projections of future hydrology, application of higher resolution regional climate models along with land-surface models accounting for changes in permafrost and vegetation is required. The above physical mechanisms are only partially understood. There is also a severe lack of regional experimental investigation of interactions between ground and surface water, especially in the permafrost zone. Such studies would help in better understanding changes in intra-annual runoff and would improve hydrological simulations in numerous land-surface models.

In Chap. 5, we demonstrated that a significant reduction in bearing capacity of foundations due to climatically driven permafrost warming is expected. During the early 1990s, increases in permafrost temperature and ground salinity and the resulting decrease in the soil's ability to support foundations resulted in serious deformation of many structures. We emphasized that the full evaluation of infrastructure stability at local scales requires a comprehensive engendering assessment of every structure. The broad geographic assessments described in this volume can serve as a basis for developing regional adaptation, mitigation, and risk management options, including changes to construction codes designed to mitigate the adverse impacts that climate change may have on infrastructure and socioeconomic life.

Further, we focused on how Siberian terrestrial ecosystems might respond to ongoing and anticipated climate and environmental changes (Chap. 6). For this, we analyzed expected ecosystem change in vitality and productivity, acceleration of natural disturbance, migration and alteration of the land, and the character and intensity of anthropogenic pressure. Major drivers of increasing ecosystems productivity are elevated atmospheric CO_2 concentration and increased nitrogen deposition in combination with longer and warmer growing seasons. The integral compound effect of these processes is not clear. There is much empirical and modeling evidence of changes in ecosystems in boreal Asia, including productivity, both increasing and decreasing, particularly at northern and southern tree lines and altitudinal transition zones.

The overall effect is that terrestrial ecosystems of Siberia can still be considered a net carbon sink of about 0.3–0.4 Pg C annually. However, this conclusion seems likely rather in a short-period than long-period run. Increase of CO_2 and CH_4 emissions from permafrost and wetlands; aridization of vast continental areas and following alteration of fire regimes; impoverishment of forests, particularly on the southern ecotone of the taiga zone; and further unregulated industrial pressure on terrestrial ecosystems may dramatically change the future carbon budget (cf., Chap. 6).

Some empirical and modeling results show that ongoing climate and environmental change, together with increasing natural disturbance, could have a clearly negative impact on vegetation and could change some ecosystems from C sinks to C sources. This appears to be particularly the case in disturbed forests and ecosystems on permafrost. These changes also serve as prerequisites to feedbacks to weather and climate, which could accelerate future disturbance regimes in Siberia. As noted in Chap. 6, future trajectories of Siberian ecosystems will strongly depend upon a number of large-scale social and economic decisions including transition to integrated land management, implementation of an ecologically friendly paradigm of industrial development of Siberian territories, as well as the overall governance of renewable natural resources and transition to adaptive forest management.

We listed the risks to terrestrial ecosystems in Siberia (steppe, forest, tundra, and agricultural lands) that may occur due to climate change and anthropogenic pressure (see Chap. 6). But we also noted that many ecological processes and tendencies are poorly understood. There are outstanding problems that require urgent investigation. Setting thresholds of acceptable (nondestructive) impacts on ecosystems taking into account nonlinear and multivariant responses of ecosystems to a long-term accumulation of stress is important. Also needed is system (holistic) analysis of a complicated dynamic system to include assessment of ecosystem impacts, responses, and feedbacks in ecosystems that evolved ecologically, socially, economically, and distinctly. Development of integrated observing systems is urgently needed. This high uncertainty in climatic and terrestrial systems projections levies requirements on future strategies of coevolution of human and nature in high latitudes.

Chapter 7 addressed the anthropogenic drivers that affect landscapes of Siberia. They include natural resource extraction, development of associated industry, and settlement expansion. In the preindustrial era, the economic activities of trapping (furs), farming, and mining were instrumental in encouraging expansion from western Russia into West Siberia and then East Siberia. As the industrial era began in the nineteenth century, mining, and then the development of energy resources, required infrastructure in the form of railways and roads, which led to the rise of early industrial cities and complexes and devastating transformation of surrounding landscapes. These hubs of activity dominated Siberian expansion in the nineteenth and into the early twentieth century.

During the Soviet era, there were significant and rapid demographic changes, which had a large impact on environment. These included large influxes of population through organized migration; the growth of anthropogenic impact vastly out of proportion to the remaining virtually uninhabited expanses of Siberia; the conversion of steppe and southern taiga regions to industrial agriculture; the expansion of mining (coal, bauxite) activities into oil, gas, and other minerals and metals sectors; the harnessing of hydropower to supply industrialization; the extension of railways; and the beginning (and eventual peak) of the industrial forest sector in Siberia.

Deep economic hardship immediately after 1991 led to population loss in Siberia and necessitated the transformation of people's livelihoods and their relationship to natural resources. Reorganization of natural resource institutions and forest management was undertaken with the intent to encourage privatization, yet unwittingly fostered the emergence of local oligarchies and illegal logging operations. Additionally, during the more recent post-Soviet era, the geographic location of natural resources exploitation has been shifting to areas that can easily supply and transport logs and oil/gas to the growing Asian market. In a rising era of transnationalism, Siberia faces challenges. While official statistics from both Russia and China record a significant increase in the export of Russian logs to China, there remains a need to better understand, quantify, and model the impact of this activity on patterns and processes of Siberian landscapes, especially those in East Siberia.

There are several factors in place that could promote sustainable development in the forestry sector. Foremost is the vast resource base of the Siberian forests, presenting opportunities for biodiversity conservation and carbon sequestration along with sustainable forest harvest. Today, additional practices with the potential to promote sustainability, such as the gathering of nontimber forest products, hunting, fishing, and tourism, are gaining greater visibility in Siberia. Together, these activities could play a key role in creating a broader base of sustainable forest use. Greater investigation and quantification of their success and influence would be useful.

Despite the importance of forests, the energy (oil/gas) sector is today the most important economic base in Russia. During the Soviet era, Siberian oil/gas extraction occurred mainly in West Siberia. More recently, focus is shifting to large oil/gas fields in East Siberia and the construction of pipelines and infrastructure to connect directly to China.

Results of pilot remote sensing-based case studies in East Siberia reveal an approximately twofold increase in fires in oil/gas reconnaissance and exploitation areas compared to nonexplored, yet ecologically similar sites nearby. The results underscore the potential for serious environmental consequences to taiga forests resulting from a combination of high-risk anthropogenic activity related to energy extraction and fire-conducive climate conditions. Additional research is needed on potential impacts of new oil/gas fields, transportation, and other infrastructure on land cover and land use in the East Siberian region.

During the Soviet era, some of the world's largest industrial complexes were built in Siberia. Remote sensing time series data have been used in analyzing pollution effects of Norilsk Nickel on the surrounding northern Siberia environment. Results have shown signs of degradation and mortality in the larch forests surrounding Norilsk, beginning in the 1940s and peaking in the 1980s. Continued investigations at the Norilsk site and analysis of AVHRR and MODIS for post-Soviet landscapes show a continued high level of industrial pollution affecting the larch-dominated forests in the region (e.g., Chaps. 7 and 8).

Atmospheric pollution and its whole ecosystem effects in Siberia are further discussed in more detail in Chap. 8 (Baklanov et al. 2012). Siberian ecosystems have begun to show stress from the accumulation of pollution depositions that come from cities and industrial plants. Urban air quality in several Siberian cities (e.g., Norilsk, Barnaul, Novokuznetsk) is considered among the worst in Russian and European cities.

A combination of direct and inverse modeling has been applied to diagnosing, designing, monitoring, and forecasting of air pollution in Siberia. Regional models coupled with the global one by means of orthogonal decomposition methods allow one to correctly introduce data about the global processes onto the regional level where environmental quality control strategies are typically implemented.

Atmospheric transport and deposition of sulfur and heavy metals (Ni, Cu, Pb) from the industries of Norilsk and Ural regions over the territory of Siberia were analyzed. Using meteorological input from the NCEP/NCAR reanalysis data in the HYSPLIT model, 5-day air mass transport trajectories from the sources were calculated for the 1981–2008 period. Analysis of spatial, seasonal, and long-term variations in selected heavy metal concentrations in near-surface air and precipitation, as well as fluxes onto the surface, showed that on an annual basis for three large Siberian rivers – Ob, Yenisei, and Lena – the heavy metal fluxes from selected source regions through the atmosphere onto the territory of river basins correspond well to fluxes carried away by these rivers to the Arctic Ocean.

To determine the ecological state of the environment in Siberia, air pollution level should be considered in combination with deposition processes. That is, cleaner air is not a guarantee of the safer environment at the surface.

In general, the pollution of the Siberian environment from major anthropogenic sources (like the Ural industrial region, Irkutsk, and Norilsk) has been decreasing during the last two decades, but the spatial variations are sizeable. Heavy metal fluxes incoming per year from these sources through the atmosphere to the central waters of the Kara and the Laptev seas are quite comparable with fluxes inflowing there by rivers along the coastal zone. Moreover, for three largest Siberian rivers (Ob, Yenisei, and Lena), such atmospheric fluxes of HM into river basins are also comparable with fluxes brought by rivers to the Arctic Ocean. Thus, atmospheric transport of heavy metals should be taken into account in analysis of modeled vs. measured data both for the atmosphere and for the surface ecosystems.

A key question remains as to the overall role that anthropogenic aerosols and the associated feedbacks play in modulating the greenhouse warming. This is an urgent question because Siberia and northern latitude region, in general, is expected to experience temperature changes 40 % greater than the global mean while being large enough to feedback to regional and global climate systems. Improved understanding of the impacts of aerosols on regional and global climate requires further research, addressing the dynamics of aerosol sources, variability in transport to high latitudes due to changes in general circulation, as well as changes in other climatic characteristics that might be involved in aerosol-induced feedbacks. Our understanding of the relevant physical processes has been hampered by a lack of concurrent measurements of aerosols, clouds, radiation, snow, and sea ice processes. Isolated measurements of the individual components of this system cannot be used to elucidate the relevant physical processes or to validate model simulations.

Scientific understanding of the Siberian environmental changes discussed in the chapters of this book has been greatly aided by spaceborne remote sensing and the development of its spatial, spectral, and temporal records. We have accumulated sufficient information on the environment of Siberia from space to study long-term processes and to conduct meaningful analyses of the interannual variability in land-cover changes. For example, the coarse spatial (1–4 km) but high temporal (daily) resolution NOAA-AVHRR observational data record has now been accumulated over 30 years. From these observations, long-term datasets have been constructed on a global as well as regional basis. Less frequent but much higher spatial resolution (30 m) sensor data from Landsat represent the longest data record (over the last 40 years) for land-cover change studies, providing landscape-level information that can be linked to the scale of local human drivers and activities.

The contributions in this volume demonstrate the utility of satellite data for monitoring changes and trends over time, including the associations among vegetation productivity, surface temperature, biogeochemical and water cycles, and air pollution under changing climate conditions and their relation to human population dynamics in Siberia. Satellite remote sensing has been able to provide data documenting some of the most important influences of Soviet-era human-driven environmental change on the Siberian landscape. Two case studies were presented in Chap. 7: one focused on change in a predominantly forested, yet multiuse landscape in central Siberia, and the other on environmental impacts of mining in northern Siberia. In central Siberia, Soviet-era land-use and land-cover change was quantified through Landsat MSS/TM-based change detection between 1974 and 1990. During the recent years, a number of instruments onboard of ENVISAT, GOSAT, and other satellites present new possibilities for measurements of important biophysical parameters of terrestrial ecosystems.

An expanded program of satellite and in situ observations is needed to provide the data to better understand the processes and to quantify the changes. Satellite imagery can help overcome the sparseness of in situ observations, of particular importance over the vast and often remote areas of Siberia. The need to secure consistent long-term satellite data records to detect and monitor surface changes in remote regions of Siberia is partially fulfilled by the recent launch of the NASA National Polar-orbiting Operational Environmental Satellite System Preparatory Project (Suomi-NPP) with VIIRS onboard to extend the AVHRR-MODIS sensor data record and the anticipated launch of the next Landsat in January 2013 which will continue the 40-year record of the Landsat program. An increasing number of Earth observation satellite systems are also being put into service by the international community. International cooperation needs to be enhanced to secure the necessary observations and continue the exchange of data, in order to fortify a better understanding of this important and rapidly changing region. Therefore, the concept of international constellations will play an important role in creating future datasets with less spatial and temporal gaps than what has been possible to date. However, issues of coordination of data acquisition (coverage), data interoperability, and cross calibration and access become critical under this new paradigm. Additionally, designing appropriate systems for data integration and processing is of increasing importance (see Chap. 2).

Research on environmental and land-change issues requires a stronger role of social scientists familiar with the region. Development of strategies for adaptation to, and mitigation of, the negative impacts of global change and recommendations to the federal government and regional authorities should be undertaken by joint effort of physical and social scientists.

References

Baklanov AL et al (2012) Chapter 8. Aspects of atmospheric pollution in Siberia. In: Regional environmental changes in Siberia and their global consequences, Springer environmental science and engineering. Springer, Dordrecht, pp 303–346

Bergen KM, Hitztaler SK, Kharuk V, Krankina ON, Loboda T, Zhao T, Shugart HH, Sun G (2012) Chapter 7. Human dimensions of environmental change in Siberia. In: Regional environmental changes in Siberia and their global consequences, Springer environmental science and engineering. Springer, Dordrecht, pp 250–302

Gordov EP et al (2012) Chapter 2. Development of information-computational infrastructure for environmental research in Siberia as a baseline component of the Northern Eurasia earth science partnership initiative (NEESPI) studies. In: Regional environmental changes in Siberia and their global consequences, Springer environmental science and engineering. Springer, Dordrecht, pp 19–55

Groisman PYa, Gutman G, Shvideenko AZ, Bergen K, Baklanov AL, Stackhouse P Jr (2012a) Chapter 1. Introduction: regional features of Siberia. In: Regional environmental changes in Siberia and their global consequences, Springer environmental science and engineering. Springer, Dordrecht, pp 1–17

Groisman PYa, Blyakharchuk TA, Chernokulsky AV et al (2012b) Chapter 3. Climate changes in Siberia. In: Regional environmental changes in Siberia and their global consequences, Springer environmental science and engineering. Springer, Dordrecht, pp 56–109

Shiklomanov NI, Streletskiy DA (2012) Chapter 5. Effect of climate change on Siberian infrastructure. In: Regional environmental changes in Siberia and their global consequences, Springer environmental science and engineering. Springer, Dordrecht, pp 155–170

Shiklomanov AI, Lammers RB, Lettenmaier D, Yu Polischuk, Savichev O, Smith LC (2012) Chapter 4. Hydrological changes: historical analysis, contemporary status and future projections. In: Regional environmental changes in Siberia and their global consequences, Springer environmental science and engineering. Springer, Dordrecht, pp 111–154

Shvidenko AZ et al (2012) Chapter 6. Terrestrial ecosystems and their change. In: Regional environmental changes in Siberia and their global consequences, Springer environmental science and engineering. Springer, Dordrecht, pp 171–249

Index

P.Ya. Groisman and G. Gutman (eds.), *Regional Environmental Changes in Siberia
and Their Global Consequences*, Springer Environmental Science and Engineering,
DOI 10.1007/978-94-007-4569-8, © Springer Science+Business Media Dordrecht 2013

Printed by Printforce, the Netherlands